Developments in Petroleum Science, 15A

fundamentals of
well-log interpretation

1. the acquisition of logging data

Developments in Petroleum Science, 15A

fundamentals of well–log interpretation

1. the acquisition of logging data

O. SERRA

Former Chef du Service Diagraphies Differées à la Direction Exploration de la SNEA (P), Pau, France

and

Geological Interpretation Development Manager, Schlumberger Technical Services Inc., Singapore

(Translated from the French by *Peter Westaway* and *Haydn Abbott*)

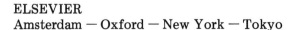

ELSEVIER
Amsterdam — Oxford — New York — Tokyo   1984

ELF AQUITAINE
Pau

ELSEVIER SCIENCE PUBLISHERS B.V.
Molenwerf 1
P.O. Box 211, 1000 AE Amsterdam, The Netherlands

Distributors for the United States and Canada:

ELSEVIER SCIENCE PUBLISHING COMPANY INC.
52, Vanderbilt Avenue
New York, NY 10017, U.S.A.

First edition 1984
Second impression 1985

Library of Congress Cataloging in Publication Data

Serra, Oberto.
    Fundamentals of well-log interpretation.

    (Developments in petroleum science ; 15A-
    Translation of: Diagraphies différées.
    Bibliography: p.
    Includes index.
    Contents: v. 1. The acquisition of logging data.
    1. Oil well logging. I. Title. II. Series.
TN871.35.S4713 1984   622'.18282   83-20571
ISBN 0-444-42132-7 (U.S. : v. 1)

ISBN 0-444-42132-7 (Vol. 15A)
ISBN 0-444-41625-0 (Series)

© Elsevier Science Publishers B.V., 1984

Printed in The Netherlands

PREFACE

The relentless search for elusive hydrocarbon reserves demands that geologists and reservoir engineers bring into play more and more expertise, inventiveness and ingenuity. To obtain new data from the subsurface requires the continual refinement of equipment and techniques.

This book describes the various well-logging equipment at the disposal of geologists and reservoir engineers today. It follows two volumes on carbonates, also published by Elf Aquitaine.

One can never over-emphasize the importance to the geological analysis of basins, and sedimentology in general, of the information which drilling a borehole makes available to us. But this data would be incomplete, even useless, if not complemented by certain new techniques—well-logging in particular—which represent a tremendous source of information both about hydrocarbons and the fundamental geology of the rocks.

It required considerable enthusiasm and a determination to succeed on the part of the author to bring the present work to its culmination, while at the same time performing the daily duties of Manager of the Log Analysis Section of the Exploration Dept. at Elf Aquitaine.

Such qualities, indeed, earned Oberto Serra the first Marcel Roubault award on March 21st, 1974, in just recognition of "work concerned with methods and techniques, or with ideas and concepts, which have led to important progress in the exploration for, and development of, natural energy resources, the discovery of new reservoirs, or the accomplishment of major works". The judges recognized in particular an invaluable liaison between the spirit of the naturalist, constantly tempered by reality, and the rigorous training of a physicist and informatician—bringing to the field of geological analysis all the facets of modern technology.

Since this award, O. Serra has, by virtue of his talents as an instructor and author, acquired a very large audience both in professional circles and in the universities studying geological sciences.

This book is the result of a fruitful collaboration involving several companies. It testifies to the desire of Elf Aquitaine to further an active policy of research and training, disseminating knowledge of modern technical developments amongst its field personnel.

It is hoped, that, in publishing this work, and continuing the series, we may contribute to a better understanding of geological sciences and their methods, fundamental to our search for energy and mineral resources.

We also hope to strengthen the bonds between the "thinkers" and the "doers", those involved in research, and those on the operational side. Progress depends as much on the deliberations of the one, as on the techniques and skill of the other, and a close collaboration between the two is essential. Exploration is no longer an exclusive domain of guarded secrets; information gathered in the course of major projects must, wherever feasible, be made available to all interested, particularly those most likely to put it to good use.

Finally, let us hope that, with this publication, we might make a modest contribution to Marcel Roubault's proud concept: "Geology at the service of Man".

ALAIN PERRODON, Paris, March 1978

FOREWORD TO THE FRENCH EDITION

The development of hydrocarbon reservoirs follows complex natural laws dependent on a number of factors.

The geologist's goal is to understand the processes of hydrocarbon accumulation. For, simply, the better these are understood, the better are the chances of discovering new hydrocarbon reserves.

The fields of geological study are several: sedimentology, structural geology, geochemistry, fluid geology, geophysics. Techniques are becoming continually more sophisticated to keep pace with the demands of modern hydrocarbon exploration.

Well-logging plays a particularly important role in geophysics:

–well-logs provide an objective, continuous record of a number of properties of the rocks which have been drilled through;

–they are the link between geophysical measurements on the surface, and subsurface geology;

–they provide numerical data, introducing the possibility of fairly rigorous quantification in the description of sedimentological processes.

It is no longer realistic to consider the geological description of a reservoir without incorporating log data. Its omission would effectively exclude most of the information potentially available from drill-holes, which itself represents a significant fraction of the total evidence on which the description can be based.

This book was conceived and written by a geologist. It is hoped it will provide geologists, (and, indeed, all engineers involved in hydrocarbon exploration and reservoir development) with a good understanding of well-logging techniques, and an appreciation of the wealth of information available from log measurements, and their relevance in reservoir description.

The present work is the first of a two-volume series on well-logging. It deals with the acquisition of log data (tool principles, logging techniques), and describes how the measurements are influenced by the many aspects of the geology of the rocks.

The second volume, currently under preparation, will cover in detail log interpretation and applications.

ACKNOWLEDGEMENTS

I wish to express my thanks to my colleagues C. Gras, L. Sulpice and C. Augier of Elf Aquitaine, for their continual encouragement and constructive criticism of the first volume during its initial preparation in French for publication; to G. Hervé for his help in preparing the texts and figures for publication; to P. Pain for his humorous illustrations; and to all anonymous draft-men and typists who contributed to this book.

I also wish to express my gratitude to Schlumberger for their permission to reproduce figures and texts from their various publications.

My thanks are also extended to Dresser-Atlas, SPE of AIME, and SPWLA from whose publications I have borrowed numerous figures.

Finally, I thank B. Vivet, Ph. Souhaite, J. Piger, J. Gartner and L. Dupal of Schlumberger Technical Services for their advice and criticism of the original French text; A. Perrodon for his constant support and encouragement; D. Bugnicourt for his advice on numerous points; H. Oertli for his invaluable assistance in editing and correcting the French text; and Elf Aquitaine for their permission to publish this book.

OBERTO SERRA, Paris, March 1978

FOREWORD TO THE ENGLISH EDITION

Following the suggestion of several people, I decided to translate my French book on Well Logging into English. At the same time I took the opportunity to improve and update the content by revising the original texts, correcting some errors which had eluded me, and adding sections on the most recent techniques. The critisms of reviewers of the original French text have been taken into account.

It considered that this translation could best be performed by my English colleagues within Schlumberger. I wish to thank Peter Westaway and Haydn Abbott for undertaking this formidable task. Their contribution has been invaluable, both in translating the French text and in bringing the present volume up to date.

I should also like to mention the typists responsible for preparing the final draft for editing.

Once again, I am indebted to H. Oertli for his help in the publication of this work, and to Elf-Aquitaine for their financial contribution without which this book could not have been published, and to Schlumberger for their permission to reproduce some of their material.

OBERTO SERRA, Singapore, May 1982

CONTENTS

1. REVIEW OF BASIC CONCEPTS

1.1. THE DEFINITION OF A "WELL-LOG"

When we speak of a log in the oil industry we mean "a recording against depth of any of the characteristics of the rock formations traversed by a measuring apparatus in the well-bore."

The logs we shall be discussing in this book, sometimes referred to as "wireline logs" or "well-logs", are obtained by means of measuring equipment (logging tools) lowered on cable (wireline) into the well. Measurements are transmitted up the cable (which contains one or several conductors) to a surface laboratory or computer unit. The recording of this information on film or paper constitutes the well-log. Log data may also be recorded on magnetic tape. A large number of different logs may be run, each recording a different property of the rocks penetrated by the well.

Wireline logging is performed after an interruption (or the termination) of drilling activity, and is thus distinguished from "drilling-logs" (of such things as drilling-rate, mud-loss, torque, etc.) and "mud-logs" (drilling mud salinity, pH, mud-weight, etc) obtained during drilling operations.

1.2. THE IMPORTANCE OF WELL-LOGS

Geology is the study of the rocks making up the Earth's crust. The field of geology that is of most importance to the oil industry is sedimentology, for it is in certain sedimentary environments that hydrocarbons are formed. It entails a precise and detailed study of the composition, texture and structure of the rocks, the colour of the constituents, and identification of any traces of animal and plant organisms. This enables the geologist: (a) to identify the physical, chemical and biological conditions prevalent at the time of deposition; and (b) to describe the transformations that the sedimentary series has undergone since deposition. He must also consider the organisation of the different strata into series, and their possible deformation by faulting, folding, and so on.

The geologist depends on rock samples for this basic information. On the surface, these are cut from rock outcrops. Their point of origin is, obviously, precisely known, and in principle a sample of any desired size can be taken, or repeated.

Sampling from the subsurface is rather more problematic. Rock samples are obtained as cores or cuttings.

Cores obtained while drilling (using a core-barrel), by virtue of their size and continuous nature, permit a thorough geological analysis over a chosen interval. Unfortunately, for economical and technical reasons, this form of coring is not common practice, and is restricted to certain drilling conditions and types of formation.

"Sidewall-cores", extracted with a core-gun, sample-taker or core-cutter from the wall of the hole after drilling, present fewer practical difficulties. They are smaller samples, and, being taken at discrete depths, they do not provide continuous information. However, they frequently replace drill-coring, and are invaluable in zones of lost-circulation.

Cuttings (the fragments of rock flushed to surface during drilling) are the principle source of subsurface sampling. Unfortunately, reconstruction of a lithological sequence in terms of thickness and composition, from cuttings that have undergone mixing, leaching, and general contamination, during their transportation by the drilling-mud to the surface, cannot always be performed with confidence. Where mud circulation is lost, analysis of whole sections of formation is precluded by the total absence of cuttings. In addition, the smallness of this kind of rock sample does not allow all the desired tests to be performed.

As a result of these limitations, it is quite possible that the subsurface geologist may find himself with insufficient good quality, representative samples, or with none at all. Consequently, he is unable to answer with any confidence the questions fundamental to oil exploration:
(a) Has a potential reservoir structure been located?
(b) If so, is it hydrocarbon-bearing?
(c) Can we infer the presence of a nearby reservoir?

An alternative, and very effective, approach to this problem is to take in situ measurements, by running well-logs. In this way, parameters related to porosity, lithology, hydrocarbons, and other rock properties of interest to the geologist, can be obtained.

The first well-log, a measurement of electrical resistivity, devised by Marcel and Conrad Schlumberger, was run in September 1927 in Pechelbronn (France). They called this, with great foresight, "electrical coring". (The following text will demonstrate the significance of this prophetic reference to coring.) Since then scientific and technological advances have led to the development of a vast range of highly sophisticated measuring techniques and equipment, supported by powerful interpretation procedures.

Well-log measurements have firmly established applications in the evaluation of the porosities and saturations of reservoir rocks, and for depth correlations.

More recently, however, there has been an increasing appreciation of the value of log data as a source of more general geological information. Geologists have realised, in fact, that well-logs can be to the subsurface rock what the eyes and geological instruments are to the surface outcrop. Through logging we measure a number of physical parameters related to both the geological and petrophysical properties of the strata that have been penetrated; properties which are conventionally studied in the laboratory from rock-samples. In addition, logs tell us about the fluids in the pores of the reservoir rocks.

The rather special kind of picture provided by logs is sometimes incomplete or distorted, but always permanent, continuous and objective. It is an objective translation of a state of things that one cannot change the statement (on paper!) of a scientific fact.

Log data constitute, therefore, a "signature" of the rock; the physical characteristics they represent are the consequences of physical, chemical and biological (particularly geographical and climatic...) conditions prevalent during deposition; and its evolution during the course of geological history.

Log interpretation should be aimed towards the same objectives as those of conventional laboratory core-analyses. Obviously, this is only feasible if there exist well-defined relationships between what is measured by logs, and rock parameters of interest to the geologist and reservoir engineer.

The descriptions of the various logging techniques contained in this book will show that such relationships do indeed exist, and that we may assume: (a) a significant change in any geological characteristic will generally manifest itself through at least one physical parameter which can be detected by one or more logs; and (b) any change in log response indicates a change in at least one geological parameter.

1.3. DETERMINATION OF ROCK COMPOSITION

This is the geologist's first task. Interpretation of the well-logs will reveal both the mineralogy and proportions of the solid constituents of the rock (i.e. grains, matrix * and cement), and the nature and

proportions (porosity, saturations) of the interstitial fluids.

Log analysts distinguish only two categories of solid component in a rock-"matrix" ** and "shale". This classification is based on the sharply contrasting effects they have, not only on the logs themselves, but on the petrophysical properties of reservoir rocks (permeability, saturation, etc.). Shale is in certain cases treated in terms of two constituents, "clay" and "silt". We will discuss this log-analyst terminology in more detail.

1.3.1. Matrix

For the log analyst, matrix encompasses all the solid constituents of the rock (grains, matrix *, cement), excluding shale. A simple matrix lithology consists of single mineral (calcite or quartz, for example). A complex lithology contains a mixture of minerals: for instance, a cement of a different nature from the grains (such as a quartz sand with calcitic cement).

A clean formation is one containing no appreciable amount of clay or shale.

(Thus we may speak of a simple shaly sand lithology, or a clean complex lithology, and so on).

Table 1-1 summarizes the log characteristics (radioactivity, resistivity, hydrogen index, bulk density, acoustic wave velocity...) of some of the principle minerals found in sedimentary rocks.

1.3.2. Shale, silt and clay

A *shale* is a fine-grained, indurated sedimentary rock formed by the consolidation of clay or silt. It is characterized by a finely stratified structure (laminae 0.1–0.4 mm thick) and/or fissility approximately parallel to the bedding. It normally contains at least 50% silt with, typically, 35% clay or fine mica and 15% chemical or authigenic minerals.

A *silt* is a rock fragment or detrital particle having a diameter in the range of 1/256 mm to 1/16 mm. It has commonly a high content of clay minerals associated with quartz, feldspar and heavy minerals such as mica, zircon, apatite, tourmaline, etc...

A *clay* is an extremely fine-grained natural sediment or soft rock consisting of particles smaller than 1/256 mm diameter. It contains clay minerals (hydrous silicates, essentially of aluminium, and sometimes of magnesium and iron) and minor quantities of finely divided quartz, decomposed feldspars, carbonates, iron oxides and other impurities such as

---

* For a sedimentologist, matrix is "The smaller or finer-grained, continuous material enclosing, or filling the interstices between the larger grains or particles of a sediment or sedimentary rock; the natural material in which a sedimentary particle is embedded" (Glossary of Geology, A.G.I., 1977).

** For a log analyst, matrix is "all the solid framework of rock which surrounds pore volume", excluding clay or shale (glossary of terms and expressions used in well logging, S.P.W.L.A., 1975).

SEDIMENTARY ROCKS

ATL MARKETING November 1982

SILICATES

NAME	FORMULA	M Mol. Wt.	Z	ρ g/cm³	ρFDC g/cm³	ρma g/cm³	φ SNP (LIME) p.u.	φ CNL (LIME) p.u.	φa p.u.	Δt μs/ft	Δts μs/ft	Δtma μs/ft	M	N	Σ c.u.	Pe barn/elec.	U barn/cm³	Uma barn/cm³	εr	tp (nsec/m)	GR API	Th ppm	K % wt	Ur ppm	R Ωm²/m	CEC meq/100g
Quartz	SiO_2	60.09	11.78	2.65	2.64	2.65	-1.	-2.	0	55.5	88	55.5	0.81	0.64	4.26	1.81	4.79	4.79	4.65	7.2	-	2	0	0.7	10^4–10^{12}	
β Cristobalite	SiO_2	60.09	11.78	2.19	2.15	2.15	-2.	-3.	18					0.90	3.52	1.81	3.89	4.74			-					
Opal 3.5%H₂O	$SiO_2(H_2O)_{.1209}$	62.26	11.68	2.16	2.13	2.18	4.	2.	22	58		55.5	1.16	0.87	5.03	1.75	3.72	4.77			-					
Opal 6.33%H₂O	$SiO_2(H_2O)_{.2253}$	64.15	11.60	2.10	2.07	2.18	7.	6.	26					0.88	6.12	1.70	3.52	4.75			-					
Opal 8.97%H₂O	$SiO_2(H_2O)_{.3286}$	66.00	11.52	2.04	2.01	2.18	11.	11.	30					0.88	7.05	1.66	3.34	4.77			-					
Grenat	$Fe_3Al_2(SiO_4)_3$	497.76	19.51	4.32	4.31		3.	7.						0.28	44.91	11.09	47.80				-					
Hornblende	$Ca_2NaMg_5Fe_4AlSi_8O_{22}(O,OH)_2$	900.13	16.44	3.2	3.2	3.13	4.	8.	2	43.8	81.5	41.5	0.66	0.42	18.12	5.99	19.17	19.5	7.5	9.1	-	5–50		1–30		
Tourmaline	$NaMg_3Al_3Si_6O_2(OH)_4$	558.78	12.35	3.03	3.02	3.15	16	22	5					0.39	7449.82	2.14	6.46	6.8			-					
Zircon	$ZrSiO_4$	183.31	32.43	4.67	4.5		-1.	-3.						0.30	6.92	69.10	311.				-	50 to 4,000		300 to 3,000		

CARBONATES

NAME	FORMULA	M Mol. Wt.	Z	ρ g/cm³	ρFDC g/cm³	ρma g/cm³	φ SNP (LIME) p.u.	φ CNL (LIME) p.u.	φa p.u.	Δt μs/ft	Δts μs/ft	Δtma μs/ft	M	N	Σ c.u.	Pe barn/elec.	U barn/cm³	Uma barn/cm³	εr	tp (nsec/m)	GR API	Th ppm	K % wt	Ur ppm	R Ωm²/m	CEC meq/100g
Calcite	$CaCO_3$	100.09	15.71	2.71	2.71	2.71	0	-1.	0	47–49	88.4	46.5	0.83	0.59	7.08	5.08	13.77	13.77	7.5	9.1	-		0–7 (aver 0.3) >0		10^7–10^{12}	
Dolomite	$CaCO_3 \cdot MgCO_3$	184.41	13.74	2.87	2.88	2.88	2.	1.	1	43.5	72	41.5	0.78	0.50	4.70	3.14	9.00	9.00	6.8	8.7	-					
Ankerite	$Ca(Mg,Fe)(CO_3)_2$	240.25	18.59	2.9	2.86	2.86	0.	1.	1					0.54	22.18	9.32	26.65	26.6			-					
Siderite	$FeCO_3$	115.86	21.09	3.94	3.89		5	12		47				0.30	52.31	14.69	57.14		6.8–7.5	8.8–9.1	-					

OXIDES

NAME	FORMULA	M Mol. Wt.	Z	ρ g/cm³	ρFDC g/cm³	ρma g/cm³	φ SNP (LIME) p.u.	φ CNL (LIME) p.u.	φa p.u.	Δt μs/ft	Δts μs/ft	Δtma μs/ft	M	N	Σ c.u.	Pe barn/elec.	U barn/cm³	Uma barn/cm³	εr	tp (nsec/m)	GR API	Th ppm	K % wt	Ur ppm	R Ωm²/m	CEC meq/100g
Hematite	Fe_2O_3	159.69	23.44	5.27	5.18	5.18	4	11	2	42.9	79.3	40.0	0.35	0.21	101.37	21.48	111.27	18.8			-					
Magnetite	Fe_3O_4	231.54	23.67	5.18	5.08	5.08	3.	9			73	55.5	0.28	0.22	103.08	22.24	112.98	19.27			-	3–20		1–30		
Goethite	$FeO(OH)$	88.86	22.66	4.37	4.34	4.34	50*	60+							85.37	19.02	82.55	18.68			-					
Limonite	$FeO(OH)(H_2O)_{.205}$	125.79	20.39	3.5	3.59		50*	60+	2	56.9	102.6		0.51	0.30	71.12	13.00	46.67	17.8	9.9–10.9	10.5–11.0	-					

PHOSPHATES

NAME	FORMULA	M Mol. Wt.	Z	ρ g/cm³	ρFDC g/cm³	ρma g/cm³	φ SNP (LIME) p.u.	φ CNL (LIME) p.u.	φa p.u.	Δt μs/ft	Δts μs/ft	Δtma μs/ft	M	N	Σ c.u.	Pe barn/elec.	U barn/cm³	Uma barn/cm³	εr	tp (nsec/m)	GR API	Th ppm	K % wt	Ur ppm	R Ωm²/m	CEC meq/100g
Hydroxyapatite	$Ca_5(PO_4)_3OH$	502.33	16.30	3.15	3.17	2.96	5	8	2	42				0.42	9.60	5.81	18.4	18.8			-	P R E S E N T				
Chlorapatite	$Ca_5(PO_4)_3Cl$	520.78	16.50	3.18	3.18	3.39	-1.	-1	0	42				0.46	130.21	6.06	19.27	19.27			-					
Fluorapatite	$Ca_5(PO_4)_3F$	504.32	16.31	3.20	3.21	2.90	-1.	-2	0	42				0.46	8.48	5.82	18.68	18.68			-					
Carbonatapatite	$[Ca_5(PO_4)_{2.5}(CO_3)_{0.5}\cdot \tfrac12 H_2O]$	1048.67	16.20	3.11	3.13	2.88	5	8	2					0.43	9.09	5.58	17.47	17.8			-					

FELDSPARS – Alkali

NAME	FORMULA	M Mol. Wt.	Z	ρ g/cm³	ρFDC g/cm³	ρma g/cm³	φ SNP (LIME) p.u.	φ CNL (LIME) p.u.	φa p.u.	Δt μs/ft	Δts μs/ft	Δtma μs/ft	M	N	Σ c.u.	Pe barn/elec.	U barn/cm³	Uma barn/cm³	εr	tp (nsec/m)	GR API	Th ppm	K % wt	Ur ppm	R Ωm²/m	CEC meq/100g
Orthoclase	$KAlSi_3O_8$	278.34	13.39	2.55	2.52	2.52	-2	-3.	4	69		71	0.79	0.68	15.51	2.86	7.21	7.5	4.4–6.0	7–8.2	~220	8	10.5			
Anorthoclase	$KAlSi_3O_8$	278.34	13.39	2.62	2.59	2.59	-2.	-2	3.8					0.64	15.91	2.86	7.41	7.7			~220	10	10			
Microcline	$KAlSi_3O_8$	278.34	13.39	2.56	2.53	2.53	-2.	-3	4					0.67	15.58	2.86	7.24	7.5			~220	12	16			

FELDSPARS – Plagioclases

NAME	FORMULA	M Mol. Wt.	Z	ρ g/cm³	ρFDC g/cm³	ρma g/cm³	φ SNP (LIME) p.u.	φ CNL (LIME) p.u.	φa p.u.	Δt μs/ft	Δts μs/ft	Δtma μs/ft	M	N	Σ c.u.	Pe barn/elec.	U barn/cm³	Uma barn/cm³	εr	tp (nsec/m)	GR API	Th ppm	K % wt	Ur ppm	R Ωm²/m	CEC meq/100g
Albite	$NaAlSi_3O_8$	262.23	11.55	2.62	2.59	2.59	-1	-2	3.8	49	85	51	0.88	0.64	7.47	1.68	4.35	4.5			-					
Anorthite	$CaAl_2Si_2O_8$	278.22	13.73	2.76	2.74	2.74	-1	-2	2.1	45		48	0.83	0.59	7.24	3.13	8.58	8.8			-					

CLAYS

NAME	FORMULA	M Mol. Wt.	Z	ρ g/cm³	ρFDC g/cm³	ρma g/cm³	φ SNP (LIME) p.u.	φ CNL (LIME) p.u.	φa p.u.	Δt μs/ft	Δts μs/ft	Δtma μs/ft	M	N	Σ c.u.	Pe barn/elec.	U barn/cm³	Uma barn/cm³	εr	tp (nsec/m)	GR API	Th ppm	K % wt	Ur ppm	R Ωm²/m	CEC meq/100g
Kaolinite H***	$Al_4Si_4O_{10}(OH)_8$	-	11.83	2.42	2.41	2.96	34.	37.	28	49–60				0.45	14.12	1.83	4.44	6.17		DRY.8	80–130	6–19	0–0.5	4.4–7	3–25	
Chlorite H	$(Mg,Fe)_5Al(Si_3Al)O_{10}(OH)_8$	-	16.67	2.77	2.76	3.39	37.	52.	26					0.27	24.87	6.30	17.38	23.49	5.8		180–250	0–8	0–0.3	17.4–362	10–40	
Illite H	$K_{1-1.5}Al_4(Si_{7-6.5}Al_{1-1.5})O_{20}(OH)_4$	-	14.10	2.53	2.52	2.90	20.	30.	21					0.46	17.58	3.45	8.73	11.05			250–300	10–25	3.51–8.31	8.7–12.4	10–40	
Montmorillonite H	$(Ca,Na)_7(Al,Mg,Fe)_4(Si,Al)_8O_{20}(OH)_4(H_2O)_n$	-	13.19	2.12	2.12	2.88	10.	44.	41					0.50	14.12	2.04	4.40	7.48			150–200	14–24	0–1.5	4.3–7.7	80–150	

MICAS

NAME	FORMULA	M Mol. Wt.	Z	ρ g/cm³	ρFDC g/cm³	ρma g/cm³	φ SNP (LIME) p.u.	φ CNL (LIME) p.u.	φa p.u.	Δt μs/ft	Δts μs/ft	Δtma μs/ft	M	N	Σ c.u.	Pe barn/elec.	U barn/cm³	Uma barn/cm³	εr	tp (nsec/m)	GR API	Th ppm	K % wt	Ur ppm	R Ωm²/m	CEC meq/100g
Muscovite	$KAl_2(Si_3Al)O_{10}(OH)_2$	398.32	12.76	2.83	2.82	2.97	12.	20–25	8	49–60	149	39.5	0.77	0.44	18.85	2.40	6.74	7.33	6.2–7.9	8.3–9.4	~270	20–25	7.9–9.8	2–8		
Glauconite** H	$K_2(Mg,Fe)_2Al_6(Si_4O_{10})_3(OH)_{12}$	-	16.71	2.56	2.54	3.04	23.	38.	25		224.1	40		0.40	24.79	6.37	16.24	21.65			~220	2–4	3.2–5.8			
Biotite**	$K_2(Mg,Fe)_6(Si_6Al_2)O_{20}(OH)_4$	-	16.65	3.01	2.99	3.1	11.	21.	5.5	50.8				0.40	29.83	5.27	18.75	19.8	4.8–6.0		200–350	5–50	6.2–10	1–40		

LOGGING TOOL CHARACTERISTIC READINGS OF MAIN MINERALS FOUND IN SEDIMENTARY ROCKS

NAME	FORMULA	M Mol. Wt	Z	ρ g/cm³	ρFDC g/cm³	ρma g/cm³	Ø SNP (LIME) p.u.	Ø CNL (LIME) p.u.	Øa p.u.	Δt μs/ft	Δts μs/ft	Δtma μs/ft	M $\frac{119\,\Delta t\times10^{3}}{\rho_b-1.07}$	N $\frac{1-Ø CNL}{\rho_b-1}$	Σ c.u.	Pe barn/elec.	U barn/cm³	Uma barn/cm³	εr	tp (nsec/m)	GR API	Th ppm	K % wt	Ur ppm	R Ωm²/m	CEC meq/100g
EVAPORITES																										
Halite	NaCl	58.44	15.33	2.17	2.04	2.04	-2	-3.	23.5	67	120	70	1.17	0.99			4.65	9.48	124	5.6-6.3	7.9-8.4	-				10^4-10^{14}	
Anhydrite	CaSO₄	136.14	15.68	2.96	2.98	2.98	-1	-2.	0	50		52	0.70	0.51	12.45	5.05	14.93	14.93	6.3	8.4	-				10^4-10^{10}	
Gypsum	CaSO₄(H₂O)₂	172.17	14.68	2.32	2.35	3.59	50⁺	60⁺	48.5	52			1.02	0.38	18.5	3.99	9.37	18.38	4.1	6.8	-				10^3	
Trona	Na₂CO₃NaHCO₃H₂O	208.02	9.08	2.12	2.08	2.70	24	35.	36	65		38.2	1.15	0.60	15.92	0.71	1.48	2.31			-					
Tachydrite	CaCl₂(MgCl₂)₂(H₂O)₁₂	517.60	14.53	1.68	1.66		50⁺	60⁺		92			1.47		406.02	3.84	6.37	6.32			-					
Kieserite	MgSO₄H₂O	138.38	11.82	2.57	2.59	3.15	38	43.	25					0.36	13.96	1.83	4.74				-					
Epsomite	MgSO₄(H₂O)₇	246.48	10.39	1.68	1.71		50⁺	60⁺							21.48	1.15	1.97				-					
Bischofite	MgCl₂(H₂O)₆	203.31	13.02	1.55	1.54		50⁺	60⁺		100			1.65		323.44	2.59	3.99				-					
Sylvite	KCl	74.55	18.13	1.98	1.86	1.86	-2	-3.	30	74		76	1.34	1.20	564.57	8.51	15.83	22.6	4.6-4.8	7.2-7.3	>500		53		10^{14}-10^{15}	
Carnallite	KClMgCl₂(H₂O)₆	277.86	14.79	1.61	1.57		41	60⁺		83			1.87		368.99	4.09	6.42				~220		14			
Langbeinite	K₂SO₄(MgSO₄)₂	414.99	14.23	2.83	2.82	2.8	-1	-2.	0	52		54	0.75	0.56	24.19	3.56	10.04	10.04			~290		18.5			
Polyhalite	K₂SO₄MgSO₄(CaSO₄)₂(H₂O)₂	602.94	15.01	2.78	2.79	3.0	14	25		57.5		40.5	0.78	0.42	23.70	4.32	12.05	13.54			~200		13			
Kainite	MgSO₄KCl(H₂O)₃	248.97	14.17	2.13	2.12		40	60⁺	11						195.14	3.50	7.42				~245		15.5			
SECONDARY MINERALS																										
Barite	BaSO₄	233.40	47.20	4.48	4.09		-1.	-2						0.33	6.77	266.82	1091				-					
Celestite	SrSO₄	183.68	30.47	3.97	3.79		-1.	-1						0.36	7.96	55.19	209				-					
Sulphur	S	32.06	16.00	2.07	2.02	2.07	-2.	-3	22.0	122		122		1.01	20.22	5.43	10.97	14.1			-				10^9-10^{16}	
Pyrite	FeS₂	119.97	21.96	5.0	4.99		-2.	-3		39.2	62.1		0.66	0.26	90.10	1697	84.68				-				10^{-1}-10^{-4}	
Marcassite	FeS₂	119.97		4.89	4.87		-2.	-3						0.27	88.12	1697	82.64				-					
Pyrrhotite	Fe₇S₈	647.41	21.16	4.6	4.53		-2.	-3.					0.38	0.29	94.18	20.55	93.09				-					
Sphalerite	ZnS	97.44	27.04	4.0	3.85		-2.	-3.						0.36	25.34	35.93	15.13		7.8-8.1	9.3-9.5	-					
Chalcopyrite	CuFeS₂	183.51	24.91	4.2	4.07		-2.	-3.						0.34	102.13	26.72	108.75				-					
Galena	PbS	239.26	78.05	7.5	6.36		-3.	-3.					1.52	0.19	13.36	1631.37	10424				-					
COAL																										
Anthracite	CH.358N.009O.022	-	6.0	1.51	1.47		37	38		105			1.78	0.89	8.65	0.16	0.23				-					
Bituminous	CH.793N.015O.078	-	6.09	1.27	1.24		50⁺	60⁺		120			2.87	0.36	14.30	0.17	0.21				-					
Lignite	CH.849N.015O.211	-	6.41	1.23	1.19		47	52.		160			1.52	2.53	12.79	0.20	0.24				-					
FLUIDS																										
Pure Water	H₂O@80°F	18.02	7.52	1.	1.		50⁺ (100)	60⁺ (100)		189				2.22		0.36	0.36		80	30	-					
Salt Water	330.000 ppm NaCl@80°F	-	11.47	1.22	1.19		50⁺	60⁺		185					157.6	1.64	1.95		56	25	-				$3\cdot10^{-2}$	
Oil	(CH₂)n 30°API	-	5.54	0.88	0.88		50⁺	60⁺		235			3.83		25.13	0.12	0.11		2.2	4.9	-				10^6-10^9	
Methane Gaz	CH₄@200°F, 7.000 psi	16.04	5.21	0.25	0.15		-3.	0.							12.49	0.1	0.015		1.0	3.3	-					
TUBING CASING																										
Iron	Casing (@0.5% C)		25.96	7.74	7.62		-1.	0.		57.8		57	0.20	0.15	211.66	31.02	233.27				-				$15\cdot10^{-8}$	
Aluminium	Al	26.98	13.	2.69	2.54	2.54	-2	-3.	4	60.9		64	0.67		13.81	2.57	6.53	6.80			-					

POTASSIUM SALTS {Sylvite … Kainite}

* Sonde 3"⁵/₈ , dₕ=8", Wₘᵤd = 10lb/gal
** = average
*** = H:humid

organic matter. Clays form pasty, plastic impermeable masses.

The crystal lattices of clay minerals take the form of platelets. The stacking of these platelets gives the minerals a flaky nature.

There are several types of clay minerals, classified according to the thickness of the platelets or the spacing of the crystal lattice (Fig. 1-1).

Table 1-2 is a summary of the mineralogical and chemical properties, and log characteristics of the various minerals. Clays have a high water hydrogen content, which has an effect on log response. The hydrogen is present: (a) in the hydroxyl ions in the clay mineral molecule; (b) in a layer of adsorbed (or "bound") water on the clay particle surface; and (c) in water contained in the spaces between the clay platelets. The quantity of this "free-water" varies with the arrangement and compaction of the platelets.

The clays found in sedimentary series will rarely show the theoretical log characteristics appearing in Table 1-2. They usually include several mineral types, and may be mixed with silts and carbonates in varying proportions. Their pore-space will depend on the arrangement of the particles and on the degree of compaction of the rock. These pores generally contain water, but it is quite possible to find solid or liquid hydrocarbons or gases in them. So the log response to a clay depends very much on its composition, porosity and hydrocarbon content.

Log response to a rock containing clay minerals will vary according to the clay fraction and its characteristics.

1.3.3. Fluids

The arrangement of the grains usually leaves spaces (pores and channels) which are filled with fluids: water, air, gas, oil, tar, etc. (Fig. 1-2). Just how much fluid is contained in a rock depends on the space, or *porosity*, available.

With the exception of water, these pore-fluids have one important property in common with the large majority of matrix minerals—they are poor electrical conductors. Water, on the other hand, conducts electricity by virtue of dissolved salts.

The electrical properties of a rock are therefore strongly influenced by the water it contains. The quantity of water in the rock is a function of the porosity, and the extent to which that porosity is filled with water (as opposed to hydrocarbons). This explains why the *resistivity* of a formation is such an important log measurement. From the resistivity we can determine the percentage of water in the rock (provided we know the resistivity of the water itself). If we also know the porosity, we may deduce the percentage of hydrocarbons present (the hydrocarbon saturation).

1.3.3.1. Porosity

Definition: Porosity is the fraction of the total volume of a rock that is not occupied by the solid constituents.

There are several kinds of porosity:

(a) Total porosity, ϕ_t, consists of all the void spaces (pores, channels, fissures, vugs) between the

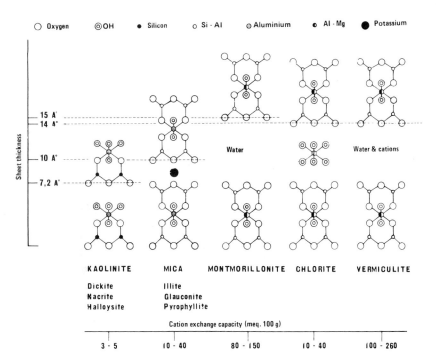

Fig. 1-1. Schematic representation of the structures of the main plately minerals (from Brindley, 1951).

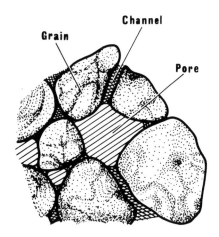

Fig. 1-2. Interstitial spaces in a clastic rock.

solid components:

$$\phi_t = \frac{V_t - V_s}{V_t} = \frac{V_p}{V_t} \qquad (1\text{-}1)$$

where:

V_p = volume of all the empty spaces (generally occupied by oil, gas or water);

V_s = volume of the solid materials;

V_t = total volume of the rock.

We distinguish two components in the total porosity:

$$\phi_t = \phi_1 + \phi_2 \qquad (1\text{-}2)$$

ϕ_1 is the primary porosity, which is intergranular or intercrystalline. It depends on the shape, size and arrangement of the solids, and is the type of porosity encountered in clastic rocks. ϕ_2 is the secondary porosity, made up of vugs caused by dissolution of the matrix, and fissures or cracks caused by mechanical forces. It is a common feature of rocks of chemical or organic (biochemical) origin.

(b) Interconnected porosity, $\phi_{connect}$, is made up only of those spaces which are in communication. This may be considerably less than the total porosity *. (Consider pumice-stone for instance, where ϕ_t is of the order of 50%, but $\phi_{connect}$ is zero because each pore-space is isolated from the others: there are no interconnecting channels.)

(c) Potential porosity, ϕ_{pot}, is that part of the interconnected porosity in which the diameter of the connecting channels is large enough to permit fluid to flow (greater than 50 μm for oil, 5 μm for gas).

* Pores are considered to be unconnected when electrical current and fluids cannot flow between them.

TABLE 1-2

Classification of clay minerals

GROUP	Main minerals	Structure	Unit layer Å	Chemical composition	Origin
Kaolinites	Kaolinite Dickite Nacrite Halloysite	1 silica, 1 gibbsite 1 silica, 1 gibbsite	7.2 10.1	$Al_4Si_4O_{10}(OH)_8$ $Al_4Si_4O_{10}(OH)_8 , \eta H_2O$	Feldspars, micas (low pH) or hydrothermal
Illites	Illite Sericite, damourite Pyrophyllite Glauconite	2 silica, 1 gibbsite	10	$K_{1-1.5}Al_4(Si_{7-6.5}Al_{1-1.5})O_{20}(OH)_4$ $Si_4Al_2O_{10}(OH)_2$ $K_2(Mg,Fe)_2Al_6(Si_4O_{10})_3(OH)_{12}$	Feldspars, micas (high pH) Diagenetic
Smectites	Montmorillonite Bentonite Beidellite Nontronite Saponite	2 silica, 1 gibbsite	9.7 - 17.2	$(Ca,Na)_7(Al,Mg,Fe)_4(Si,Al)_8O_{20}(OH)_4$ $(H_2O)_n$	Volcanic ashes
Chlorites	Chlorite	2 silica, 1 gibbsite, 1 brucite	14	$(Mg,Fe,Al)_6(Si,Al)_4O_{10}(OH)_8$	
Vermiculites	Vermiculite	2 silica, 1 brucite	14.4 (variable)		Biotite, hornblende
Polygorskites	Attapulgite	Chain			
Mixed - layers or interstratified	Illite - montmorillonite Vermiculite - chlorite Mica - chlorite Montmorillonite - chlorite Saponite - chlorite Montmorillonite - kaolinite / corrensite				

*From Edmundson, 1979

ϕ_{pot} may in some cases be considerably smaller than $\phi_{connect}$. Clays or shales, for instance, have a very high connected porosity (40–50% when compacted, and as much as 90% for newly deposited muds). However, owing to their very small pores and channels, molecular attraction prevents fluid circulation.

(d) Effective porosity, ϕ_e, is a term used specifically in log analysis. It is the porosity that is accessible to free fluids, and excludes, therefore, non-connected porosity and the volume occupied by the clay-bound water or clay-hydration water (adsorbed water, hydration water of the exchange cations) surrounding the clay particles.

N.B. Porosity is a dimensionless quantity, being by definition a fraction or ratio. It is expressed either as a percentage (e.g. 30%), as a decimal (e.g. 0.30) or in porosity units (e.g. 30 p.u.).

The geological and sedimentological factors influencing the porosity of a rock will be discussed later on in this book.

1.3.3.2. *Resistivity and conductivity*

The *resistivity* (R) of a substance is the measure of its opposition to the passage of electrical current. It is expressed in units of ohm m^2/m. A cube of material of sides measuring 1 metre, with a resistivity $R = 1$ ohm m^2/m, would have a resistance of 1 ohm between opposite faces.

The *electrical conductivity* (C) is the measure of the material's ability to conduct electricity. It is the inverse of the resistivity, and is usually expressed in units of millimhos/m (mmho/m) or mS/m (milli Siemens per metre)

$$C \text{ (mmho/m)} = 1000/R \text{ (ohm } m^2/m) \qquad (1\text{-}3)$$

There are two types of conductivity:

(a) Electronic conductivity is a property of solids such as graphite, metals (copper, silver, etc.), haematite, metal sulphides (pyrite, galena) etc.

(b) Electrolytic conductivity is a property of, for instance, water containing dissolved salts.
Dry rocks, with the exception of those mentioned above, have extremely high resistivities.

The conductive properties of sedimentary rocks are of electrolytic origin—the presence of water (or mixtures of water and hydrocarbons) in the porespace. The water phase must of course be continuous in order to contribute to the conductivity.

The resistivity of a rock depends on:

(a) The resistivity of the water in the pores. This

C.E.C (meq./100g)	Log characteristics *										K (% weight)
	ρ (g/cm³)	ρ_{FDC} (g/cm³)	Pe	Z	U	Σ (c.u.)	$(I_H)_{SNP}$ (p.u.)	$(I_H)_{CNL}$ (p.u.)	$GR_K 10^{15}$ (at/g)	
3-25	2.42 2-2.2	2.41	1.83	11.83	4.41-7	14.12	34	37	7.79	0-0.6
10-40	2.53 2.8-2.9 2.51-2.65	2.52 2.49-2.63	3.45 5.32-7.04	14.1 15.91-17.2	8.69-12.4	17.58 22.37-26.7	20 21-26	30 37-42	86.68 90.51-113.22	3.51-8.31(av. 5.2) 3.2-5.8 (av. 4.5)
80-150	2.12	2.12	2.04	12.19	4.32-7.71	14.12	40	44	3.89	0-4.9 (av. 1.6)
10-40	2.77	2.76	6.3	16.67	17.38-36.2	24.87	37	52	1.06	0-0.35
100-260										
20-36	2.29-2.30									
	2.37-2.62	2.37-2.62	2.23-3.96	12.50-14.66	5.28-18.51	15.22-18.80	13-35	19-44	25.91-105.07	

will vary with the nature and concentration of its dissolved salts.

(b) The quantity of water present; that is, the porosity and the saturation.

(c) Lithology, i.e. the nature and percentage of clays present, and traces of conductive minerals.

(d) The texture of the rock; i.e. distribution of pores, clays and conductive minerals.

(e) The temperature.

Resistivity may be anisotropic by virtue of stratification (layering) in the rock, caused, for instance, by deposition of elongated or flat particles, oriented in the direction of a prevailing current. This creates preferential paths for current flow (and fluid movement), and electrical conductivity is not the same in all directions.

We define horizontal resistivity (R_H) in the direction of layering, and vertical resistivity (R_V) perpendicular to this.

The *anisotropy* coefficient, λ, is:

$$\lambda = (R_V/R_H)^{1/2} \qquad (1\text{-}4)$$

This can vary between 1.0 and 2.5, with R_V generally larger than R_H. It is R_H that we measure with the laterolog and other resistivity tools (induction), whereas the classical electrical survey reads somewhere between R_H and R_V.

The mean resistivity of an anisotropic formation is:

$$R = (R_H R_V)^{1/2} \qquad (1\text{-}5)$$

The anisotropy of a single uniform layer is microscopic anisotropy. When we talk of the overall characteristic of a sequence of thin resistive layers, this is macroscopic anisotropy. In such a series, the current tends to flow more easily along the layers than across them, and this anisotropy will affect any tool reading resistivity over a volume of formation containing these fine layers.

Microscopic anisotropy occurs in clays, and mudcakes. In the second case, the resistivity measured through the mudcake perpendicular to the wall of the hole is higher than that parallel to the axis. This has an effect on the focussed micro-resistivity tools (MLL, PML) which must be taken into account in their interpretation. A mud-cake of anisotropy, λ, and thickness, h_{mc}, is electrically equivalent to an isotropic mud-cake having a resistivity equal to the mean, $R_H R_V$ with thickness, λh_{mc}.

Summarizing, what we call the true resistivity (R_t) of a formation is a resistivity dependent on the fluid content and the nature and configuration of the solid matrix.

1.3.3.2.1. *The relationship between resistivity and salinity*

We have mentioned that the resistivity of an elec-

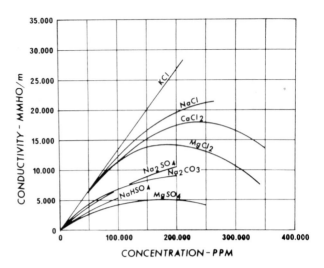

Fig. 1-3. The dependence of resistivity on dissolved salt concentration.

trolyte depends on the concentration and type of dissolved salts. Referring to Fig. 1-3, notice that the resistivity decreases as concentration increases, up to a certain maximum beyond which undissolved, and therefore non-conducting, salts impede the passage of current-carrying ions.

The salinity is a measure of the concentration of dissolved salts. It can be expressed in several ways: (a) parts per million (ppm or $\mu g/g$ of solution); (b) g/litre of solvent; and (c) g/litre of solution. The chart of Fig. 1-4 permits easy conversion between these units. Sodium chloride (NaCl) is the most common salt contained in formation waters and drilling muds. It is customary to express the concentrations of other dissolved salts in terms of equivalent NaCl for evaluation of the resistivity of a solution. Figure 1-5 shows the multipliers (Dunlap coefficients) for conversion of some ionic concentrations to their NaCl equivalents. For a solution of mixed salts simply take

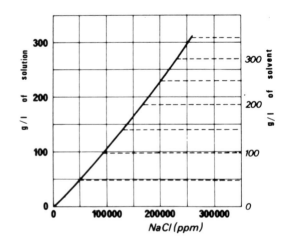

Fig. 1-4. The equivalence between the various units of concentration.

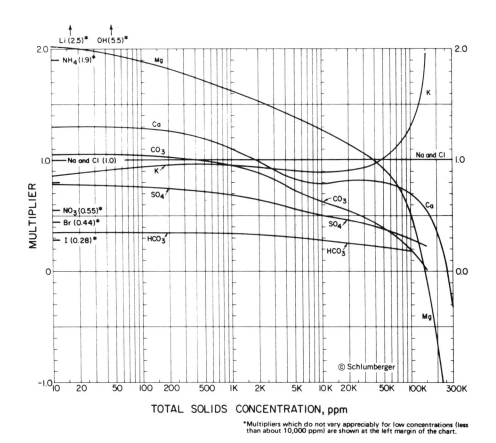

Fig. 1-5. The multipliers to convert ionic concentration to equivalent NaCl concentration (courtesy of Schlumberger).

the sum of the equivalent concentrations of each type of ion present.

1.3.3.2.2. *Relationship between resistivity and temperature*

The resistivity of a solution decreases as the temperature increases. The chart of Fig. 1-6 can be used to convert the resistivity at a given temperature to that at any other temperature.

Arps' formula approximates this relationship by:

$$R_{wT°F} = R_{75°F}\left(\frac{75°F + 6.77}{wT°F + 6.77}\right) \qquad (1\text{-}6)$$

where R_{wT} is the solution resistivity at formation temperature $T(°F)$. 75°F is a commonly used reference temperature. More generally:

$$R_{wT_2} = R_{wT_1}\left[\frac{T_1 + 6.77}{T_2 + 6.77}\right] \text{ in } (°F) \qquad (1\text{-}7a)$$

$$R_{wT_2} = R_{wT_1}\left[\frac{T_1 + 21.5}{T_2 + 21.5}\right] \text{ in } (°C) \qquad (1\text{-}7b)$$

1.3.3.3. *The resistivity of clays*

With the exception of pyrite, haematite, graphite and a few others, the dry minerals have infinite resistivity.

Certain minerals do exist that appear to be solid conductors. Clay minerals are an example. According to Waxman and Smits (1967), a clayey sediment behaves like a clean formation of the same porosity, tortuosity and fluid saturation, except that the water appears to be more conductive than expected from its bulk salinity (Fig. 1-7a).

Clays are sheet-like particles, very thin (a few Ångstroms), but have a large surface area depending on the clay mineral type (Fig. 1-7b). There is a deficiency of positive electrical charge within the clay sheet. This creates a strong negative electrical field perpendicular to the surface of the clay sheet which attracts positive ions (Na^+, K^+, Ca^{2+} ...) and repels the negative ions (Cl^-,...) present in the water (Clavier et al., 1977). The amounts of these compensating ions constitute the Cation Exchange Capacity which is commonly referred to as the CEC (meq/g dry rock) or Q_v (meq per cm³ total pore volume). CEC is related to the specific area of clay and so depends on the clay mineral type. It has its lowest value in kaolinite and its highest values in montmorillonite and vermiculite (Table 1-2).

The excess of conductivity observed within clays is due to these additional cations held loosely captive in a diffuse layer surrounding the clay particles (Fig. 1-8).

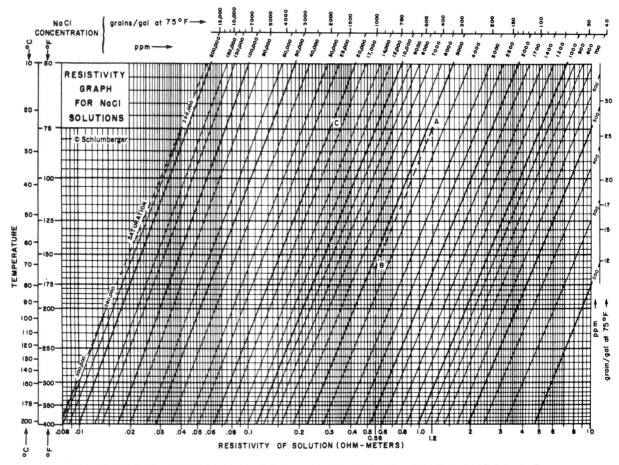

Example: R_m is 1.2 at 75°F (point A on chart). Follow trend of slanting lines (constant salinities) to find R_m at other temperatures; for example, at Formation Temperature (FT) = 160°F (point B) read R_m = 0.56. The conversion shown in this chart is approximated by the Arps formula: $R_{FT} = R_{75°} \times (75° + 7)/(FT\ (in\ °F) + 7)$.

Fig. 1-6. The relationship between resistivity, salinity and temperature (courtesy of Schlumberger).

Thus the conductivity of a clayey sediment is the sum of two terms: (a) One associated with free water or the water-filled porosity (indeed this type of sediment has porosity as high as 80% at the time of deposition. Subsequently, as they become compacted, some of the free water is expelled. However, this porosity is never reduced to zero in sedimentary rocks which have not yet reached the metamorphic phase).

(b) The other associated with the CEC. This can be expressed in another way: a clayey sediment has a conductivity which depends on its porosity on one hand, and on the effective conductivity, C_{we}, of the water it contains on the other. So we can write:

$$C_{cl} = (\phi_{cl})^m C_{we} \tag{1-8}$$

Following Clavier et al. (1977), for C_{we} we have:

$$C_{we} = (f\phi)_{fw} C_w + (f\phi)_{cw} C_{cw} \tag{1-9}$$

where:

$(f\phi)_{fw}$ = fraction of total pore volume occupied by "far water":

$$(f\phi)_{fw} = 1 - (f\phi)_{cw} = 1 - \alpha v_q Q_v; \tag{1-10}$$

$(f\phi)_{cw}$ = fraction of total pore volume (ϕ_{cl}) occupied by "clay water";

C_w = (formation) water conductivity;

C_{cw} = conductivity of clay water which is a universal parameter for sodium clays, which depends only on temperature:

$$C_{cw} = \frac{\beta}{\alpha v_q} = \frac{\beta Q_v}{(f\phi) C_w} \tag{1-11}$$

With the new values for $(f\phi)_{fw}$ and C_{cw} we can write:

$$C_{we} = (1 - \alpha v_q Q_v) C_w + \beta Q_v \tag{1-12}$$

where:

α = expansion factor for diffuse layer;

v_q = $\nu x_H = A_v x_H / Q_v$ (cm^3/meq);

Q_v = concentration of clay counter-ions per unit pore volume (meq/cm^3);

A_v = clay surface area per unit of pore volume (m^2/cm^3);

x_H = distance of "outer Helmholtz plane" from clay surface (Fig. 1-9);

β = equivalent conductivity of sodium counter-ions, dual-water model (mho/m) (meq/cm^3);

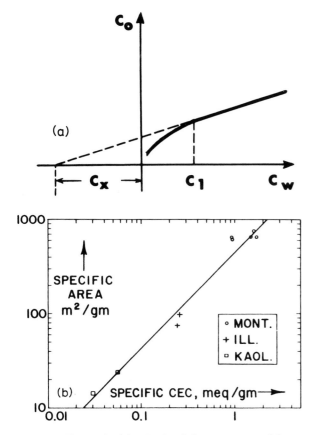

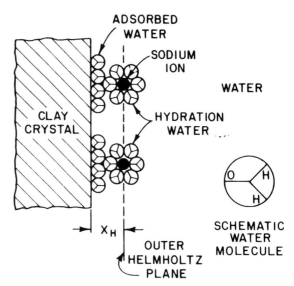

Fig. 1-9. Schematic view of outer Helmholtz plane (from Clavier et al., 1977).

Fig. 1-7. a. The conductivity C_o of a shaly water-saturated formation as a function of the conductivity C_w of the water. b. Relation between specific CEC and specific area for API standard clays (from Patchett, 1975).

ν = specific clay-area coefficient (m^2/meq).

Clays are widely encountered not only in discrete layers but in mixtures with other rocks such as sands, limestones.

When dealing with formations containing clays, therefore, we can no longer consider the solid matrix to be non-conductive. This has to be taken into account in calculations based on resistivity measurements (formation factor, porosity, saturation...). The influence of the clay will depend on the percentage present, its physical properties and the manner in which it is distributed in the formation.

1.4. ROCK TEXTURE AND STRUCTURE *

The shape and size of the rock grains, their degree of sorting, the manner in which they are cemented, and the relative importance of the cement itself, have three important consequences. They determine the porosity; the size of the pores and connecting channels influences the permeability, and hence the saturation; and the distribution of the porosity decides the tortuosity.

The internal structure of layers (homogeneity, heterogeneity, lamination, continuous gradation), the configuration of individual sedimentary structures, the thickness of the nature of their interfacing, their arrangement in sequences **, and the boundaries and trends of these sequences provide valuable information about the depositional environment. This is why the study of the many different characteristics of sedimentary formations is of such interest to the geologist and reservoir engineer.

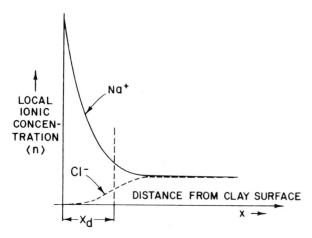

Fig. 1-8. Schematic of diffuse-layer ionic concentrations (Gouy model) (from Clavier et al., 1977).

---

* When describing the texture of a rock, we look at the grains and cement, and their intrinsic properties.
The structure refers to the 3-dimensional arrangement of these constituents, and their anisotropy. (ref: "Méthodes modernes de géologie de terrain", Technip, Paris, 1974).
** A. Lombard defined a lithological sequence as "at least two geological strata forming a natural series, in which they occur no significant discontinuities other than the interfaces between the strata themselves".

12

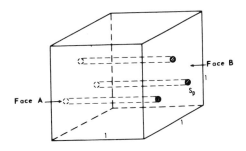

Fig. 1-10. Theoretical case—a cube of unit volume containing parallel cylindrical capillaries of cross-sectional area S_p.

Well-log interpretation will provide much of this information. Before embarking on a detailed study of how this is achieved, we must understand the basic relationships between the log measurements and the physical parameters.

1.4.1. The relationship between porosity and resistivity: the formation factor

In clean, porous aquifers, the formation resistivity R_0 is proportional to that of the interstitial water, R_w:

$$R_0 = F_R R_w \qquad (1\text{-}13a)$$

F_R is the formation resistivity factor, a function of the rock texture. It is the ratio of the resistivity of the formation to that of the water with which it is 100%

saturated:

$$F_R = R_0/R_w \qquad (1\text{-}13b)$$

If the interstitial spaces were made up solely of parallel cylindrical channels (Fig. 1-10), R_0 would be inversely proportional to the porosity. The resistance between faces A and B of the elementary cube of Fig. 1-10 is, by definition:

$$R_0 = R_w \frac{l}{S_p} \qquad (1\text{-}14)$$

S_p being the cross-sectional area of each of the channels. Now S_p is equal to ϕ_t:

$$\phi_t = \frac{V_p}{V_t} = (\text{see section } 1.3.3.1.) = \frac{1 \times S_p}{1 \times 1 \times 1}$$
$$= S_p \ (l = 1) \qquad (1\text{-}15)$$

So: $R = R_w/\phi_t$, which means that $F_R = 1/\phi_t$.

In reality, the electrical current is constrained to follow complex meandering paths, whose lengths increase with their tortuosity, and whose sectional areas (and hence resistances) vary erratically between the pores and fine interconnecting capillaries. The nature of these paths depends on the rock texture.

A large number of measurements on rock samples has shown that the formation factor of a shale-free rock is related reasonably consistently to porosity by:

$$F_R = a/\phi^m \qquad (1\text{-}16)$$

a is a coefficient between 0.6 and 2.0 depending on

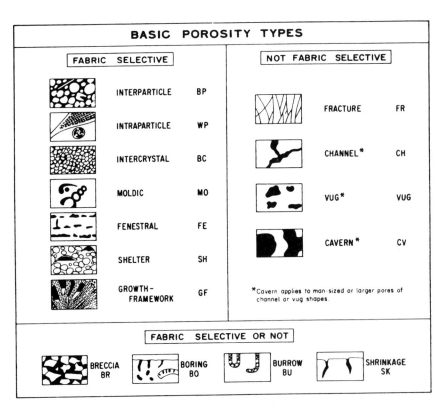

Fig. 1-11. The principle types of porosity (from Choquette and Pray, 1970).

TABLE 1-3

Classification of porosity according to its evolution (Choquette and Pray, 1970).

MODIFYING TERMS				
GENETIC MODIFIERS		**SIZE* MODIFIERS**		
PROCESS	DIRECTION OR STAGE	CLASSES		mm†
SOLUTION — s	ENLARGED — x	MEGAPORE mg	large — lmg / small — smg	256 / 32 / 4
CEMENTATION — c	REDUCED — r	MESOPORE ms	large — lms / small — sms	1/2
INTERNAL SEDIMENT — i	FILLED — f	MICROPORE mc		1/16
TIME OF FORMATION				
PRIMARY — P		Use size prefixes with basic porosity types:		
pre-depositional — Pp		mesovug — msVUG		
depositional — Pd		small mesomold — smsMO		
SECONDARY — S		microinterparticle — mcBP		
eogenetic — Se		*For regular-shaped pores smaller than cavern size.		
mesogenetic — Sm		†Measures refer to average pore diameter of a single pore or the range in size of a pore assemblage. For tubular pores use average cross-section. For platy pores use width and note shape.		
telogenetic — St				
Genetic modifiers are combined as follows:		**ABUNDANCE MODIFIERS**		
PROCESS + DIRECTION + TIME		percent porosity — (15%)		
		or		
EXAMPLES: solution-enlarged — sx		ratio of porosity types — (1:2)		
cement-reduced primary — crP		or		
sediment-filled eogenetic — ifSe		ratio and percent — (1:2) (15%)		

lithology. m is the cementation or tortuosity factor. It varies between 1 and 3 according to the type of sediment, the pore-shape, the type of porosity and its distribution, and to some extent the degree of compaction.

Choquette and Pray (1970) have proposed a means of classifying the porosity of carbonates based on several factors, such as particle shape, size, distribution, origin, evolution, as shown in Fig. 1-11 and Table 1-3. It can be seen from this classification that m may vary over a considerable range.

The idealized case in Fig. 1-10 is closely approximated by the "fenestral" category, where $m \simeq 1$. Fracture porosity, although usually very small, presents short-circuits to current flow, and once again, $m \simeq 1$, or $F_R \simeq 1/\phi$.

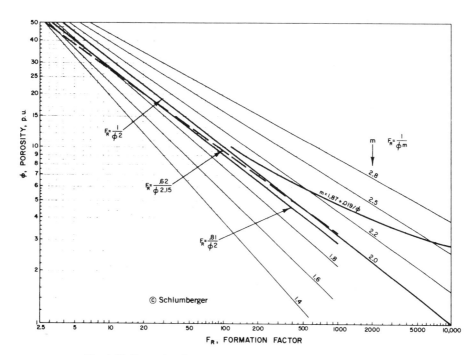

Fig. 1-12. Formation factor and porosity (courtesy of Schlumberger).

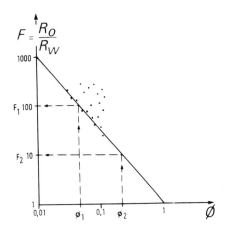

Fig. 1-13. Plot of F $(= R_0/R_w)$ against porosity for the determination of a and m (using core or neutron-density porosities).

Sandstones and detrital quartz formations almost always have intergranular porosity. The Humble formula: $F_R = 0.62/\phi^{2.15}$ is often used.

In well-consolidated formations, $F_R = 1/\phi^m$ is preferable. For non-fissured carbonates of low porosity, the Shell formula is applied: $F_R = 1/\phi^m$ with $m = 1.87 + 0.019/\phi$.

The chart in Fig. 1-12 shows the dependence of F_R on ϕ for various values of a and m.

Precise evaluation of m and a is obtained from the classical plot, on log-log scales, of R_0/R_w against porosity, taken from a clean, water-bearing zone in the reservoir (Fig. 1-13). The porosity is measured from cores, or from well-logs (a neutron-density combination, for instance).

The straight line passing through the most "southwesterly" of the points (and therefore having the least chance of containing any hydrocarbon) represents the "water-line" for the formation, and we can derive m from its slope, and a from its intercept on the ϕ axis at: $\phi = 1.0$. Since:

$$R_0/R_w = F_r = a/\phi m \qquad (1\text{-}17)$$

then

$$m = (\log F_1 - \log F_2)/(\log \phi_2 - \log \phi_1) \qquad (1\text{-}18)$$

where $[\phi_1, F_1]$ and $[\phi_2, F_2]$ are any two points on the

water-line, as shown in the Fig. 1-13. Now $a = F_R \phi^m$, therefore, a is given by the value of F_R when $\phi = 1.0$ (in Fig. 1-13, $a = 1.0$). Alternatively:

$$a = F_1 \phi_1^m = F_2 \phi_2^m$$

So it can be seen that measurements of R_0 (deep resistivity tools) or R_{xo} (micro-resistivity tools) can be used to derive porosity once a and m are known (in addition, of course, to R_w and R_{mf}). Note, however, that resistivity is dependent on connected porosity only.

1.4.2. The relationship between saturation and resistivity: Archie's formula

Quite often (fortunately!) porous rocks contain quantities of liquid or gaseous hydrocarbons. Being infinitely resistive, these hydrocarbons are electrically equivalent to the solid, clean matrix. So, for the same porosity, a formation will have a higher resistivity if it is hydrocarbon-bearing than if it is water-bearing (Fig. 1-14).

The *saturation* of a fluid in a formation is the ratio of the volume occupied by the fluid to the total pore volume, i.e. it is the fraction of the porosity occupied by that particular fluid. If the fluid is formation water, then its saturation S_w is:

$$S_w = V_w/V_p \qquad (1\text{-}19)$$

If there are no other fluids present, then $V_w = V_p$, and $S_w = 1.0$. If some hydrocarbon (V_{hy}) is present, then $V_{hy} = V_p - V_w$, and now:

$$S_w = \frac{V_p - V_{hy}}{V_p} = \frac{V_w}{V_p} \qquad (1\text{-}20)$$

N.B. Saturation is a dimensionless quantity, being a ratio. It is expressed either as a decimal ($0 < S_w < 1.0$) or as a percentage ($0\% < S_w < 100\%$): $S_w(\%) = 100\, V_w/V_p$

A large number of laboratory measurements has shown that S_w is related to resistivity by an equation

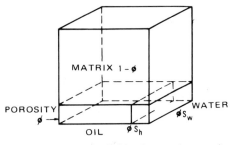

Fig. 1-14. The volumes occupied by the constituents of a rock of unit volume.

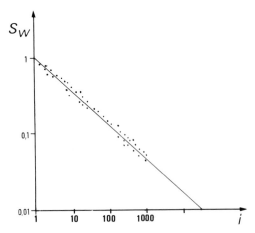

Fig. 1-15. Plot of S_w against R_t/R_0 to determine n.

of the form:

$$S_w^n = R_0/R_t = 1/I \qquad (1\text{-}21)$$

$-R_0$, as we have already seen, is the true resistivity of the formation of porosity ϕ_e totally saturated with water of resistivity R_w:

$-R_t$ is the true resistivity of the same formation, containing water at saturation S_w, and hydrocarbons (saturation $1 - S_w$).

$-n$ is the saturation exponent. It is determined empirically, and lies between 1.2 and 2.2.

N.B. The ratio R_t/R_0 is often referred to as the resistivity index, I. It is equal to 1.0 when $S_w = 1.0$, and is larger than 1 when hydrocarbons are present.

Figure 1-15 demonstrates the method for determining n from laboratory measurements. S_w is plotted against R_t/R_0 on logarithmic scales. Rewriting the saturation equation in logarithmic form:

$$n \log S_w = \log(1/I) = -\log I \qquad (1\text{-}22)$$

so that:

$$\log S_w = -(1/n) \log I$$

which means that the slope of the line in Fig. 1-15 has a gradient $-1/n$, passing through $R_t/R_0 = 1$ when $S_w = 1$. The slope may be written as:

$$-1/n = \log S_w / \log I$$

from which:

$$n = -\log I / \log S_w \qquad (1\text{-}23)$$

Generally, $n = 2$ may be taken as a first approximation.

We can replace R_0 by $F_R R_w$ (section 1.4.1.), giving:

$$R_t = F_R R_w / S_w^n \qquad (1\text{-}24)$$

This is Archie's equation for clean formations.

Hydrocarbon saturation never attains 100%. Even after migration of hydrocarbons into the reservoir rock, a small fraction of water always remains, retained by capillary forces (see section 1.4.5. for a fuller discussion).

This is referred to as the irreducible water saturation, $(S_w)_{irr}$. Its magnitude will depend on the type of porosity, the pore-size, the diameter of the interconnecting channels and the nature of the rock grains, some solids tending to retain water more effectively than others (Fig. 1-16).

Just as incoming hydrocarbon cannot possibly displace all the formation water, so incoming water will not necessarily displace all the hydrocarbons should water imbibition occur. A *residual hydrocarbon saturation* S_{rh} will remain. This, too, is discussed in 1.4.5.

1.4.3. The effect of shaliness on the resistivity

The presence of conductive shales in a formation influences the resistivity measurement. Consider two formations; one is clean, one is shaly, but otherwise they have the same effective porosities, and contain identical saturations of the same water in the pores. They will not exhibit the same resistivities. The resistivity of the shaly formation will depend on the shale-type, the percentage present, and its manner of distribution in the rock.

1.4.4. The effect of shale distribution

Log-analysts customarily distinguish three modes of shale distribution, as shown in Fig. 1-17. Each mode has a different effect on the resistivity, spontaneous potential and acoustic velocity, and influences reservoir permeability and saturation in a different way.

1.4.4.1. *Laminar shales*

These are thin beds or streaks of shale deposited between layers of reservoir rock (sand, limestone,

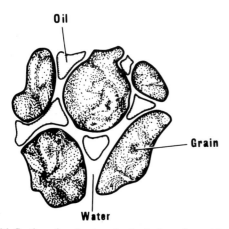

Fig. 1-16. Section of rock saturation by hydrocarbons. This particular rock is water-wet, the hydrocarbon is the non-wetting phase.

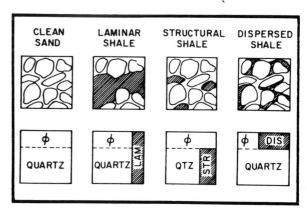

Fig. 1-17. The different modes of clay distribution; volumetric representations.

16

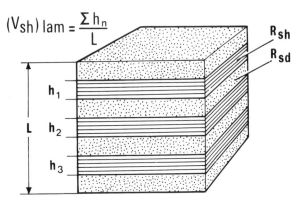

$$(V_{sh})\,lam = \frac{\Sigma h_n}{L}$$

Fig. 1-18. Distribution of resistivities in an idealized series of laminations.

etc), as in Fig. 1-18. Such shales do not alter the effective porosity, saturation or permeability of each intermediate reservoir layer, provided they do not form lateral permeability barriers. They do, of course, impede vertical permeability between porous beds. Laminar shales can be considered to have the same properties as neighbouring thick shale beds, since they have been subjected to the same conditions of evolution.

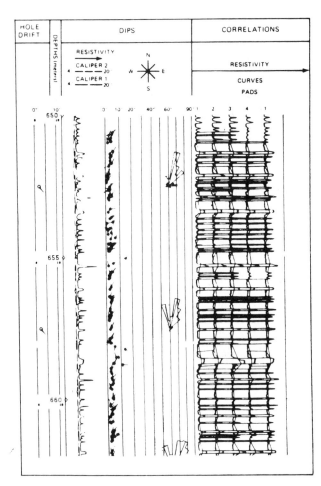

Fig. 1-19. Laminated shale-sand series as shown by the dipmeter.

Electrically, laminar shales produce a system of conductive circuits in parallel with the more or less conductive porous beds.

Using this model of parallel circuits we can write the resistivity R_t of a formation containing laminar shales as:

$$\frac{1}{R_t} = \frac{V_{shl}}{R_{shl}} + \frac{1 - V_{shl}}{R_{sd}} \qquad (1\text{-}25)$$

where R_{shl} is the resistivity of the shale, and V_{shl}, the shale fraction, is defined in Fig. 1-18. R_{sd}, the resistivity of the clean sand, is simply:

$$R_{sd} = F_{sd} R_w / S_w^2 \qquad (1\text{-}26)$$

F_{sd} is the formation factor of the clean sand. Therefore:

$$\frac{1}{R_t} = \frac{V_{shl}}{R_{shl}} + \frac{(1 - V_{shl}) S_w^2}{F_{sd} R_w} \qquad (1\text{-}27)$$

from which:

$$S_w = \left[\left(\frac{1}{R_t} - \frac{V_{shl}}{R_{shl}} \right) \frac{F_{sd} R_w}{1 - V_{shl}} \right]^{1/2} \qquad (1\text{-}28)$$

Comment: Laminar shales can be identified from micro-resistivity logs such as the microlog or, more particularly, the dipmeter, where they produce an "arrow plot" of fairly regular dips of constant azimuth (Fig. 1-19).

1.4.4.2. *Dispersed clays*

Clays of this category adhere to the rock grains, either coating them or partially filling the pore-spaces (Fig. 1-20). They exhibit different properties from laminar shales, being subjected to different constraints.

Permeability is considerably reduced, firstly because the space available for fluid movement in the pores and channels is now restricted; secondly because the wettability of the clay with respect to water is generally higher than that of quartz. The consequences are an increase in water saturation and a reduction in fluid moveability.

Electrically, a dispersed-clayey formation acts like an assembly of conductors consisting of the pore-fluid and the dispersed clays.

De Witte (1950) proposed approximating this clay-water mixture by a single electrolyte of resistivity, R_z, the harmonic mean of the constituent resistivities R_{shd} and R_w:

$$\frac{1}{R_z} = \frac{(q/S_z)}{R_{shd}} + \frac{(S_z - q)/S_z}{R_w} = \frac{1}{S_z} \left(\frac{q}{R_{shd}} + \frac{S_z - q}{R_w} \right)$$
$$(1\text{-}29)$$

where:
q = fraction of the porosity ϕ_z occupied by the dispersed clay;

a – "DISCRETE PARTICLE" KAOLINITE

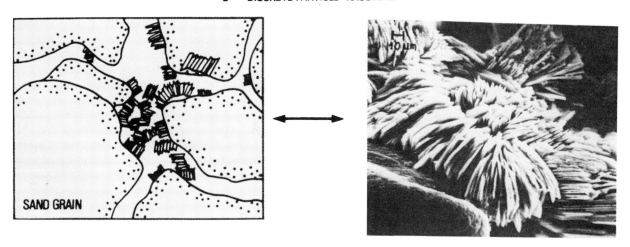

b – "PORE LINING" CHLORITE

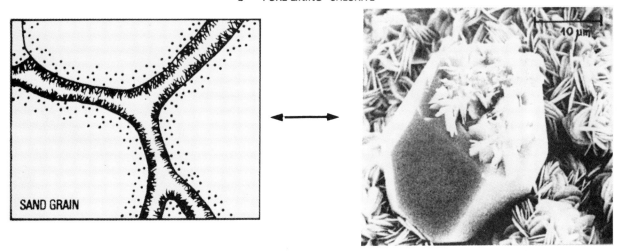

c – "PORE BRIDGING" ILLITE

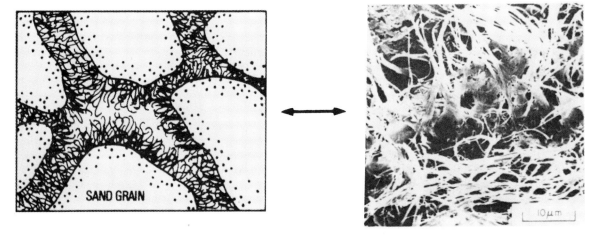

Fig. 1-20. Three general types of dispersed clay in a sandstone reservoir (from Neasham, 1977).

S_z = fraction of the porosity ϕ_z occupied by the clay-water mixture, so that:

$$S_z = S_w(1 - q) + q; \qquad (1-30)$$

q/S_z = proportion of dispersed clay in the mixture;
$(S_z - q)/S_z$ = proportion of water in the mixture;
R_{shd} = resistivity of the dispersed clay;

ϕ_z = porosity of the sand matrix. It consists of all the space occupied by fluids and dispersed clay;

F_z = formation factor corresponding to ϕ_z ($= a/\phi_z^m$). Archie's equation gives:

$$R_t = F_z R_z / S_z^2 \qquad (1\text{-}31)$$

Therefore:

$$\frac{1}{R_t} = \frac{S_z^2}{F_z} \frac{1}{S_z}\left(\frac{q}{R_{shd}} + \frac{S_z - q}{R_w}\right) \qquad (1\text{-}32)$$

replacing R_z by its value. Combining eqs. 1-30 and 1-32:

$$S_w = \frac{\left[\dfrac{aR_w}{\phi_z^2 R_t} + \left(\dfrac{q(R_{shd} - R_w)}{2 R_{shd}}\right)^2 - \dfrac{q(R_{shd} + R_w)}{2 R_{shd}}\right]^{1/2}}{1 - q} \qquad (1\text{-}33)$$

The resistivity R_{shd} of dispersed clays is difficult to determine. It is generally taken as $0.4 R_{shl}$ (laminar). If it is much larger than R_w, its exact value is unimportant, since eq. 1-33 simplifies to:

$$S_w = \frac{\left(\dfrac{aR_w}{\phi_z^2 R_t} + \dfrac{q^2}{4} - \dfrac{q}{2}\right)^{1/2}}{1 - q} \qquad (1\text{-}34)$$

ϕ_z is measured by the sonic porosity tool. q is derived from:

$$q = (\phi_S - \phi_D)/\phi_S \qquad (1\text{-}35)$$

The sonic (ϕ_S) "sees" only intergranular porosity regardless of whether it contains fluid or clay, while the density (ϕ_D) sees effective porosity (assuming the wet clay density to be the same as that of the clean sand grains).

1.4.4.3. *Structural shales*

Structural shales are grains or nodules of shale forming part of the solid matrix along with the quartz or other grains. They are considered to have many characteristics in common with laminar shales since they have been subjected to similar constraints. However, their effects on permeability and resistivity resemble more closely those of dispersed clays, although they are somewhat weaker for the same shale fraction.

Laminar and structural shales are essentially of depositional origin, while dispersed clays evolve from alteration, in situ, of other minerals (feldspars, for example), or precipitation. All three shale types may be encountered in the same shaly formation.

Comment: Note that in this discussion we are using the word "shale" rather than "clay". Referring again to section 1.3.2, you will see that a shale contains

other minerals along with clay, and has been compacted. Where a laminar shale contains silts, for instance, the term "shale fraction", V_{sh}, used in the preceding equations, refers to the sum of the clay mineral and silt fractions:

$$V_{shale} = V_{silt} + V_{clay} \qquad (1\text{-}36)$$

1.4.5. **Permeability**

A permeable rock must have connected porosity. The *permeability* of a rock is a measure of the ease with which fluid of a certain viscosity can flow through it, under a pressure gradient.

The absolute permeability k describes the flow of a homogeneous fluid, having no chemical interaction with the rock through which it is flowing. Darcy's law describes this flow as:

$$Q = k\frac{1}{\mu}\frac{S}{h}(P_1 - P_2) \qquad (1\text{-}37)$$

where:

Q = flow rate in cm^3/s;

μ = fluid viscosity in centipoise;

S = surface area in cm^2 across which flow occurs;

h = thickness in cm of material through which flow occurs, in the direction of flow;

P_1, P_2 = pressures, in atmospheres, at the upflow and downflow faces of the material, respectively;

k = absolute permeability in darcies.

N.B. The millidarcy (mD) is the commonly used unit of permeability. 1 darcy = 1000 millidarcies.

1.4.5.1. *The relationship between permeability and porosity*

There is no general mathematical relationship expressing permeability in terms of porosity, that can be applied to all cases. In clastic formations, trends such as those of Figs. 1-21 and 1-22 are often observed. However, it is possible to have a very high porosity without any permeability at all, as in the case of pumice-stone (where there is no interconnected porosity) and clays and shales (where the pores and channels are so fine that surface tension forces are strong enough to prevent fluid movement).

The controlling factor, therefore, is not the porosity itself, but the size of the pores and connecting channels, if indeed the latter exist. The geometry of the pore-space, in effect, determines a whole series of physical relationships between the solids and fluids of the rock. This is an aspect of the rock texture.

1.4.5.1.1. *Contact phenomena between water and rock—capillarity*

The pores are often connected by fine channels of

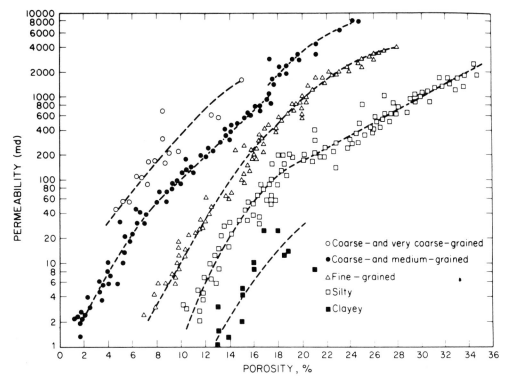

Fig. 1-21. The relationship between permeability and porosity (from Chilingar, 1964).

diameter less than a micron. These act as capillary tubes; the pore fluids are subjected to capillary forces.

At a liquid—solid interface, the liquid may be attracted or repelled, depending on the surface tension, which is a property of the liquid in contact with that particular solid. A liquid that is attracted is said to wet the solid surface.

For example, when a glass capillary tube is dipped in water (Fig. 1-23), the water rises in the tube. The surface tension forces for the water-glass contact are attractive and the water is drawn upwards (it wets the glass). The height of the resulting water column represents an equilibrium between the weight of the water in the column, and the capillary pressure. Laplace's equation states that:

$$P_c = 2\,T\cos\theta/r \tag{1-38}$$

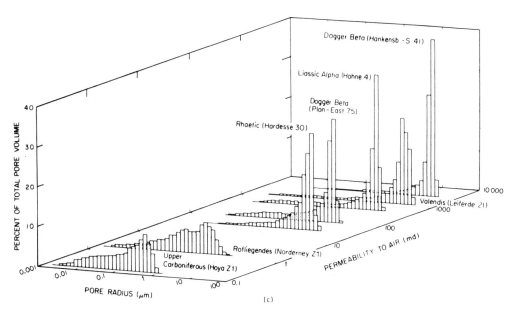

Fig. 1-22. Relationship between porosity, permeability and poreradius (from Gaida).

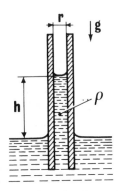

Fig. 1-23. Capillary rise of water in a tube.

where:

P_c = capillary pressure (dynes/cm^2);

T = surface tension (in dynes/cm) of water in contact with glass, (with a water–air interface in this case);

θ = contact angle of the meniscus with the capillary tube (degrees);

r = radius of the tube (cm).

In equilibrium:

$$P_c = \frac{2\,T\cos\theta}{r} = h\rho g \qquad (1\text{-}39)$$

where:

h = height of the water column (cm);

ρ = density of water (g/cm^3);

g = acceleration due to gravity (cm/s^2).

from which:

$$h = \frac{2\,T\cos\theta}{r\rho g} \quad \text{(Jurin's law)} \qquad (1\text{-}40)$$

We can see at once that the finer the capillary tube, the higher the water will rise (because of the larger capillary pressure; Fig. 1-24). For water: if $r = 1$ mm, then $h = 1.5$ cm; and if $r = 1\ \mu$m, then $h = 15$ m, with $T = 73$ dynes/cm at 20°C and $\theta = 0°$ for a capillary that is wettable by water.

Guillemot (1964) has provided reference values of capillary pressures as a function of sand grain size:

3,000 dynes/cm^2 for a coarse sand

12,000 dynes/cm^2 for a medium sand

30,000 dynes/cm^2 for a fine sand

60,000 dynes/cm^2 for a very fine sand

1.4.5.1.2. *Interfacial tensions*

Rocks are, in most cases, water-wet, which means that water occupies the angular corners of the pores and exists as a film covering the solid particles. Oil-wet rocks do exist but are rare, and oil is usually not in direct contact with the rock.

When considering the height to which water will rise in a zone impregnated with oil, we must modify eq. 1-40 (which was written for the case of a water–air

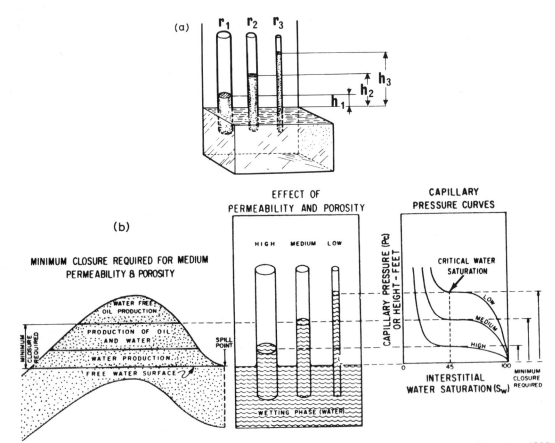

Fig. 1-24. a. Effect of tube radius on capillary rise. b. Effect of tube radius on capillary pressure curves (from Arps, 1964).

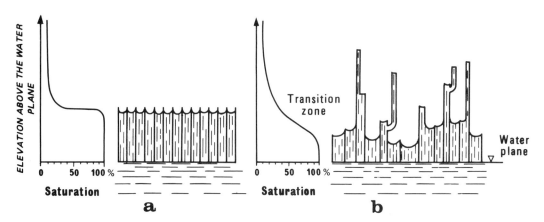

Fig. 1-25. The effect of sorting on the height of the transition zone. a. Good sorting, capillaries have the same radius. b. Poor sorting, radii of the capillaries are mixed.

interface), to include the interfacial tension of the two immiscible liquids, and the density of the oil as well as the water.

To a good approximation, the interfacial tension is equal to the difference between the surface tension of each liquid in contact with air.

$$T_{1-2} \simeq T_1 - T_2 \tag{1-41}$$

Equation 1-40 now becomes:

$$h = \frac{2(T_1 - T_2) \cos \theta}{r(\rho_1 - \rho_2)g} \tag{1-42}$$

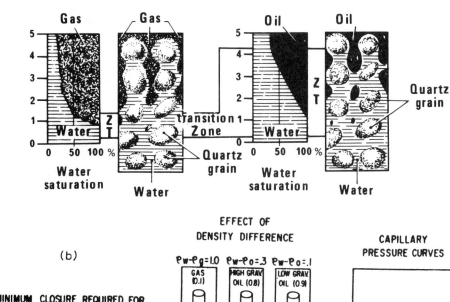

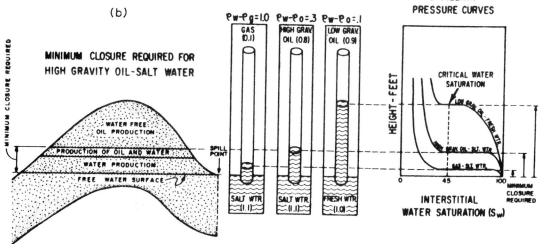

Fig. 1-26. a. The distribution of water and gas or oil in a transition zone. For a given set of conditions, the smaller the difference between the densities of the fluids in contact, the higher the transition zone. b. Influence of density difference on the transition zone and capillary pressure curves (from Arps, 1964).

The subscript 1 refers to the water, 2 to the oil.

The water will rise to a greater height the smaller the difference between the densities of the two liquids, and the finer the capillaries. The surface tension (T_1) of water is two to three times that of most oils (T_2).

In a clastic reservoir, the size of the capillaries depends on the grain-size. When the sorting of the grains is very regular, the diameters of the channels are practically identical, and the oil–water contact will be quite sharply defined, as in Fig. 1-25a. On the other hand, if the sorting is poor, there will be a wide range of capillary diameters and, since the height to which the water will rise depends on capillary size, there will be an ill-defined transition zone rather than a sharp contact (Fig. 1-25b). This transition will be longer the poorer the sorting, and the denser the oil (Fig. 1-26).

Quoting from Perrodon (1966) "During the migration of hydrocarbons from their source rock, and in the subsequent operations to produce these hydrocarbons, their displacement seems to be governed largely by the characteristics of the oil–water interface; in particular the interfacial tensions at the contact of the two fluids. These are generally of the order of 30 dynes/cm, but vary with the nature of the fluids.

The larger these interfacial tensions are the greater will be the difficulty in displacing the fluids....

... Levorsen (1956) summarises some of the factors which affect these oil–water interfacial tensions:

(1) Temperature: as it increases, so the tension decreases.

(2) Pressure: similarly, as it increases, the tension decreases.

(3) Gas dissolved in the oil or water: above the bubble-point, the tension is lower the more gas there is in the solution; below the bubble-point, the tension is increased by the presence of more gas in solution.

(4) Viscosity: as the difference between the viscosities of the oil and water becomes smaller, the interfacial tension is reduced.

(5) Tensio-active agents: their presence in the water or oil decreases the tension. It is important (for an efficient water-flood recovery project, for instance) that the water displays the oil effectively. It should be remembered that the wetting phase is in contact with the rock in the pores and fine interstices; the non-wetting phase occupies the interior (middle) of the pore-spaces".

1.4.5.1.3. *Effective and relative permeabilities*

"In the majority of sediments, initially impregnated with water, oil can only penetrate the water-filled pore-space under a driving force superior to the capillary pressure at the oil–water interface (Fig. 1-27). In other words, in formations possessing very fine capillaries, where capillary forces are high, a very high driving pressure would be required to cause the oil to displace the water. Under ordinary conditions, such formations would be impermeable to oil. Thus the concept of permeability is a relative one; the same rock being permeable to water, impermeable to oil, at a certain pressure, but permeable to both fluids if one of them is submitted to a force greater than the capillary forces acting" (Perrodon, 1966).

Darcy's law (eq. 1-37) in fact assumes a single fluid. Now, a reservoir can quite well contain two or even three fluids (water, oil and gas). In such cases, we must consider diphasic flow and relative permeability; the flows of the individual fluids interfere and their effective permeabilities are less than absolute permeability k defined in Darcy's equation.

The effective permeability describes the passage of a fluid through a rock, in the presence of other pore fluids. It depends not only on the rock itself, but on the percentages of fluids present in the pores; that is, their saturations.

The relative permeabilities (k_{rw}, k_{ro}) are simply the ratios of the effective permeabilities (k_w, k_o) to the absolute (singe-fluid) permeability, k. They vary between 0 and 1, and can also be expressed as percentages:

$$k_{ro} = k_o/k \text{ for oil}$$

$$k_{rw} = k_w/k \text{ for water}$$

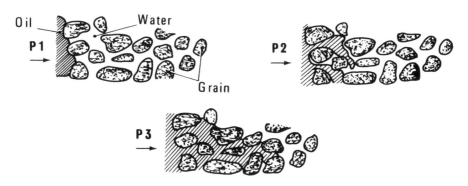

Fig. 1-27. The advance of oil through the pores of a water-bearing rock as pressure, P, is increased.

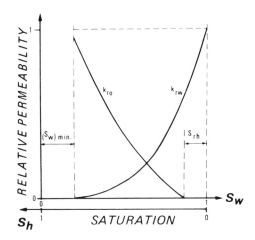

Fig. 1-28. Relative permeability as a function of saturation.

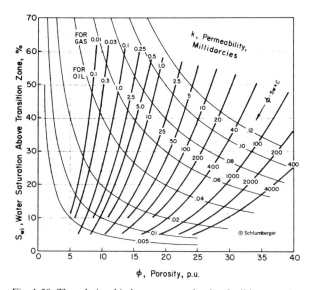

Fig. 1-30. The relationship between porosity, irreducible saturation and permeability.

Figure 1-28 shows a typical variation of relative permeability with saturation for an oil–water system, in a water-wet rock.

As the water saturation increases, the relative permeability to oil, (k_{ro}), decreases, while for the water, (k_{rw}), it increases. Water, therefore, flows with greater ease through the rock and would be produced in increasing quantities at surface. When the oil saturation has decreases to residual, S_{rh}, (discussed in section 1.4.2.) its permeability is zero, and only water will flow.

The water saturation above which a commercially unacceptable quantity of water (or water-cut) is produced, is called the limiting saturation, $(S_w)_{lim}$.

The converse applies in an oil imbibition situation. As the oil saturation increases, so does its relative permeability, k_{ro}. k_{rw} decreases, and it reaches zero at the irreducible water saturation $(S_w)_{irr}$; no further displacement of water is now possible.

1.4.5.2. Relationship between permeability and saturation

Capillary action is responsible for the retention of water in the sharp angles and capillaries between the grains. This irreducible water cannot be displaced by the forces acting on the fluids in the large pores. The irreducible water saturation $(S_w)_{irr}$ tends to be larger in low-permeability rocks, where the capillaries are fine (Fig. 1-29).

Using log data for porosity and irreducible water saturation measured above the water–oil transition zone, the chart of Fig. 1-30 has been derived for the estimation of permeability in a clastic reservoir.

Wyllie and Rose (1950) proposed this empirical formula to calculate absolute permeability:

$$k = C\phi^3/(S_w)_{irr}^2 \qquad (1\text{-}43)$$

with C a constant depending on the hydrocarbon density. ($C = 250$ for a medium density oil, $C = 79$ for a dry gas.)

Timur (1968) proposed a fairly similar relationship:

$$k = 0.136\, \phi^{4,4}/(S_w)_{irr}^2 \qquad (1\text{-}44)$$

A general equation can be written:

$$\log k = \alpha \log \phi + \beta \log(S_w)_{irr} + \gamma \qquad (1\text{-}45)$$

In a given well or field, the coefficients α, β, γ may be determined experimentally from cores and log measurements.

1.4.6. Thickness and internal structure of strata

(a) The thickness of a bed can be clearly discerned from well-logs, provided it is greater than the spacing

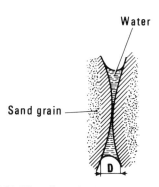

Fig. 1-29. Water is retained in narrow capillaries.

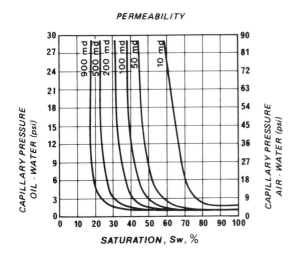

Fig. 1-31. The relationship between water saturation, permeability and capillary pressure (courtesy of Schlumberger).

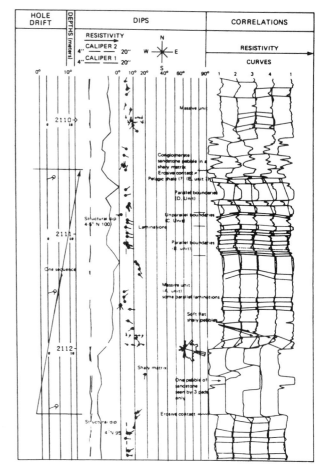

Fig. 1-32. An example of a sequence of homogeneous bedded and heterogeneous strata as distinguished by the dipmeter (From Payre and Serra, 1979).

of the measuring device (see 2.2.2.2). If this is not the case, and the bed thickness can not be reliably determined, a short-spacing device is required (Fig. 1-32).

(b) A homogeneous bed is characterized by featureless log curves (Fig. 1-32).

(c) A heterogeneous, conglomeratic or burrowed bed produces fluctuations in log response. Peaks having poor or no correlation will be recorded by the dipmeter resistivity tool, and no dip trend will be discernible (Fig. 1-32).

(d) A continuous trend within an elementary sequence of sufficient thickness will give rise to log responses that also show a regular evolution between two end-values. (Fig. 1-33).

(e) Figure 1-34 shows how the dipmeter arrow-plot indicates oblique lamination or cross-bedding.

(f) A well-bedded sequence with parallel boundaries will be seen as alternating peaks and troughs, (especially on the micro-tools) and a homogeneity of dips.

(g) Meso- and megasequential evolution will be recorded very clearly by well-logs.

(g) Lateral variations are monitored by correlations between adjacent wells.

1.5. CONCLUSION

Interpretation of well-logs can provide a large part of the geological information usually deduced from examination of outcrops, cores and cuttings. This will be developed in volume 2.

Tables 1-4 set out the correspondence between geological factors and the log parameters which describe them.

We will now study the principles of each of the logging services. This will provide an understanding of how formation geology affects tool response, and the uses to which the various log measurements can be put.

1.6. REFERENCES

See end of chapter 2.

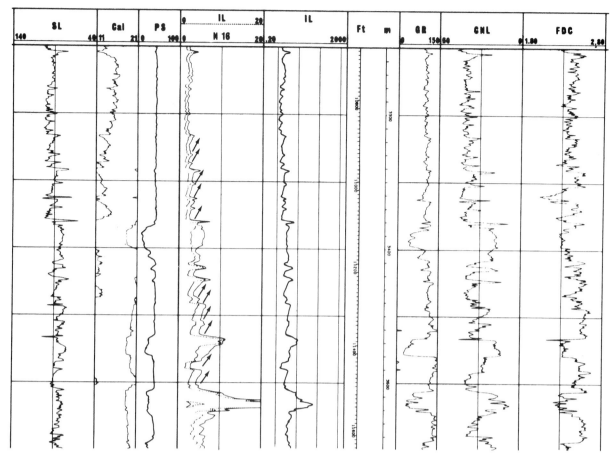

Fig. 1-33. Example of continuous trends within individual beds, showing clearly on well-logs.

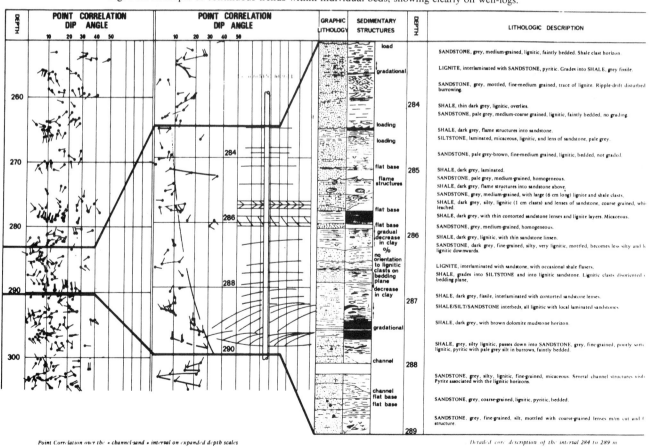

Point Correlation over the « channel-sand » interval on expanded depth scales

Detailed core description of the interval 284 to 289 m

Fig. 1-34. Oblique and cross-bedding as seen by a dipmeter log (courtesy of Schlumberger).

TABLE 1-4

Relations between geological factors and well-logging parameters

GEOLOGICAL FACTORS				WELL LOGGING PARAMETERS AFFECTED	RELEVANT WELL LOGGING TOOLS
ENVIRONMENT OF DEPOSITION	**FACIES OF A BED**	COMPOSITION See Table 1-4a			
		TEXTURE See Table 1-4b			
		SEDIMENTARY STRUCTURES See Table 1-4c PALAEOCURRENTS (after regional dip removal)	TYPE { UNI-MODAL	consistency of dip azimuth	HDT
			BI-MODAL	two main dip azimuths	
			MULTI-MODAL	spread out of azimuths	
			DIRECTION	dip magnitude evolution	
		GEOMETRY	THICKNESS { Apparent	h_a	all logs
			true	h_t	HDT (GEODIP)
			LATERAL EXTENT (by using correlations, dip data and mapping techniques)		all logs
	SEQUENCE OF BEDS	VERTICAL FACIES EVOLUTION	TYPE OF EVOLUTION { Sequences Rhythms Cycles	curve shapes and bed thickness changes	all logs but especially HDT
			EVOLUTION OF BED THICKNESS (upwards)		
		LATERAL FACIES EVOLUTION (detected using correlation and mapping techniques)			all logs
GEOLOGICAL HISTORY	**STRATIGRAPHY**	Normal (relative age determination)			all logs
		Reverse (detected by correlations and dip data)			HDT
	BREAK OF SEDIMENTATION	NON DEPOSITION		no change in regional dip	HDT
		EROSION (detected by using correlations, HDT data (curves & dips), mapping technique)			all logs
		ANGULAR UNCONFORMITY		change in regional dip	HDT
	DIAGENESIS	RECRYSTALLIZATION		$\emptyset$, $[\rho_b', I_H, \Delta t, R']$ P_e, $\rho_{ma}, I_{Hma}, \Delta t_{ma}$ = CONSTANT	
		CEMENTATION { Same cement than grains			
		{ other cement than grains		$\emptyset$, P_e, $\rho_{ma}, I_{Hma}, \Delta t_{ma}$ variable vertically	
		DOLOMITIZATION		$\emptyset$; P_e, $\rho_{ma}, I_{Hma}, \Delta t_{ma}$ variable vertically	
		DISSOLUTION		$\emptyset$, $P_e, \rho_{ma}, I_{Hma}, \Delta t_{ma}$ constant	
		COMPACTION		$\emptyset$, $[\rho_b', \Delta t, I_H, Th/K]$, P	
	TECTONIC STRESSES	STRUCTURAL REGIONAL DIP		constant dip in low energy environment	HDT
		FOLD (Tilt and plunge are defined by stereographic projection)		dip patterns	HDT
		FRACTURE		± all	all logs
		STYLOLITES			NGS, GR, HDT
		FAULT { normal (lack of series seen by correlations) with or without drag reverse (repetition of series seen by correlations) growth with or without rollover and drag		dip patterns	HDT & all logs for correlations
		THRUST (detected by repetition of series, correlations, dip data)		dip patterns	
	TEMPERATURE (depends on temperature gradient, rock composition, flow.....)			$R_f, \rho_f, I_{Hf}, \Delta t_f, \Sigma_f$	Resistivity logs FDC, CNL, TDT, EPT, GST
	PRESSURE (depends on density of overburden, and on composition)			d_i, P	
	BOREHOLE INFLUENCE	DIAMETER		ΔP, d_i, ρ_m, P_e	Resistivity log, BHC, FDC,....
		DRILLING FLUID { Density		d_i, h_{mc}	Resistivity logs, FDC, CNL, TDT, EPT, GST, NGS
		{ Nature { Fluid loss		$R_{mf}, \rho_{mf}, I_{Hmf}, K_{mf}, \Sigma_{mf}$	
		{ Filtrate (dissolved salts)		$h_{mc}, R_{mc}, \rho_{mc}, K_{mc}, P_e$	
		{ mud-cake (composition)		Pad contact with borehole wall	ML, MLL, MSFL, PL, FDC, SNP, LDT, EPT
		RUGOSITY			

TABLE 1-4 a

Relations between composition of a rock and well-logging parameters

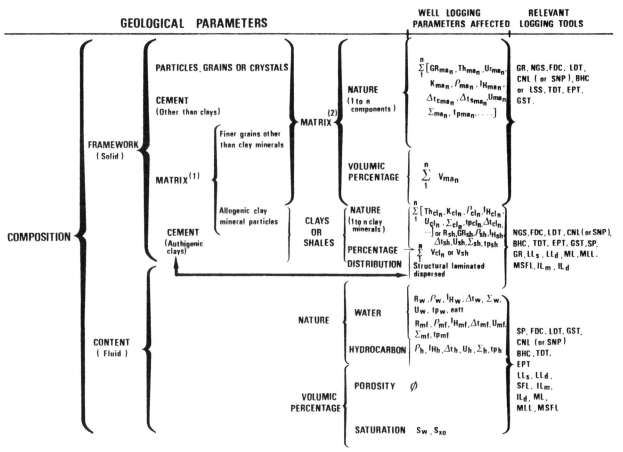

TABLE 1-4 b

Relations between rock texture and well-logging parameters

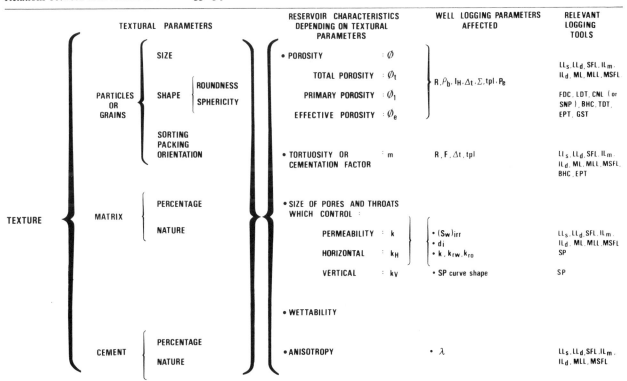

TABLE 1-4 c

Relations between rock structure and well-logging parameters

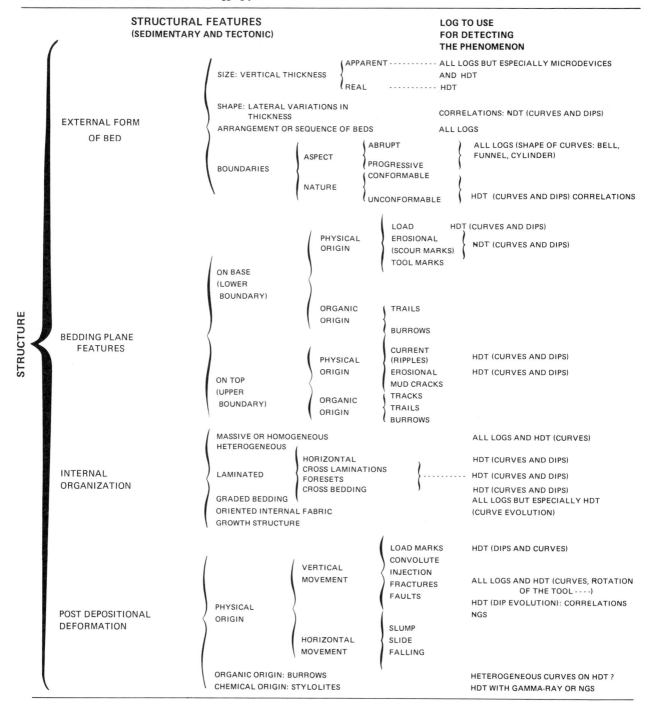

2. LOGGING TECHNIQUES AND MEASUREMENTS

2.1. CLASSIFICATION OF LOG MEASUREMENTS

In this book we are confining our interest to logs that are used for the evaluation of the rock and its fluid content. This will include both open-hole and cased-hole measurements.

Table 2-1 shows these measurements grouped into two broad categories; those arising from *natural* (or spontaneous) phenomena, and those arising from *induced* phenomena. The first group simply employs a suitable detector to obtain the measurement; the second group requires an appropriate type of emitter to "excite" a particular response in the formation, in addition to a detection system.

2.1.1. Natural phenomena

(a) Natural gamma radioactivity, which can be measured: (1) as a total gamma-ray count-rate, as in the classical gamma-ray log; and (2) as count-rates corresponding to selected energy bands, as in the natural gamma-ray spectrometry log (NGS * or Spectralog).

(b) Spontaneous potential: the S.P. log.

(c) Formation temperature: the temperature log. We should also include in this category:

(d) Hole-diameter: the caliper log, which in fact is a measurement strongly related to the mechanical or chemical properties of the rock.

(e) Inclination of the hole: the deviation log, which measures both the angle of the hole from the vertical, and its direction (or azimuth).

2.1.2. Physical properties measured by inducing responses from the formation

(A)   *Electrical* measurements, by the emission of an electrical signal:

---
* Mark of Schlumberger.

TABLE 2-1

A classification of log measurements

1. *Natural or spontaneous phenomena*			
Basic equipment—Single detector (passive system)			
Spontaneous potential	SP	Temperature	T
Natural gamma-ray activity		Hole Diameter	CAL
Total	GR	Deviation	DEV
Spectrometry	NGS * (spectralog)		

2. *Physical properties measured by inducing a response from the formation*
Basic equipment—source (or emitter)+detector (s).

Resistivity:	
(a) Long-spacing devices:	
non-focused:	N(ormal), L(ateral), ES
focused	LL *, SFL *
(b) Micro-devices:	
non-focused	ML *
focused	MLL *, PL *, MSFL *
(c) Ultra long-spacing devices	ULSEL
Conductivity	IL
Dielectric constant (electromagnetic propagation)	EPT *
Hydrogen index (using neutron bombardment)	N, NE, NT *, SN(P) *
	CN(L) *
Neutron capture cross-section, or thermal neutron decay time (neutron lifetime)	TDT *, NLL. GST *
Photoelectric absorption cross-section	LDT *
Electron density	FDC *, LDT *, D, CD
Relaxation time of proton spin (nuclear magnetic resonance)	NML
Elemental composition (induced gamma-ray spectroscopy)	IGT *, GST *, HRS *
Acoustic velocity	SV, SL, LSS *, BHC *, WST *
Formation dip	DM, CDM *, HDT *, SHDT *
Mechanical properties (amplitude of acoustic waves)	A, VDL *

---
* Mark of Schlumberger

(1) Resistivity or conductivity:
   (a) using an electrode system: electrical survey (ES), laterolog (LL) *, microlog (ML) *, microlaterolog (MLL) *, spherically focused log (SFL) *, micro-spherically focused log (MSFL) *, high-resolution dipmeter (HDT) *, (SDT) *.
   (b) using inductive coils: induction log (IL).
(2) Dielectric constant, using inductive coils: electromagnetic propagation tool (EPT) *.
(B) *Nuclear* measurements, by the irradiation of the formation with gamma rays or neutrons:
   (1) Density. Gamma rays are emitted from a source. The Compton scattered gamma rays returning from the formation are detected: formation density or gamma-gamma log. (FDC *, D, CD, LDT *).
   (2) Photo-electric absorption coefficient (related to the mean Atomic Number). This is a low energy gamma-ray phenomenon and is measured in addition to density in the lithodensity log (LDT) *.
   (3) The hydrogen index. The formation is continuously bombarded by high energy neutrons, which are slowed by successive elastic collisions with atomic nuclei, particularly those of hydrogen.
   There are several techniques in use, involving the detection of:
   (a) Thermal neutrons, i.e. those neutrons that have been slowed down to thermal energy: neutron-thermal neutron logging (CNL *, NT).
   (b) Gamma rays emitted when these thermal neutrons are captured by atomic nuclei: neutron-gamma logging (N).
   (c) Epithermal neutrons i.e. those neutrons not yet slowed down to thermal energy:
   neutron-epithermal neutron logging NE, SNP *, CNL * (epithermal).
   (4) Macroscopic thermal neutron capture cross-section (Σ). High-energy neutrons are emitted in short bursts. The rate of decay of the thermal neutron population in the formation is measured between bursts. This is a neutron capture phenomenon: thermal neutron decay time (TDT) * or neutron lifetime logging (NLL).

(5) Elemental composition. Gamma rays emitted from interactions between high energy neutrons and certain atomic nuclei are analyzed spectroscopically. There are three types of interaction important for induced gamma ray spectroscopy:
   (a) Fast neutron or inelastic interactions: inelastic gamma ray spectrometry (IGT, GST), carbon-oxygen logging.
   (b) Neutron capture: capture gamma ray spectrometry (GST, IGT), chlorine logging.
   (c) Activation and subsequent decay of radio-isotopes: activation logging, high resolution spectrometry (HRS).
(6) Proton spin relaxation time. A pulsed DC magnetic field momentarily aligns the nuclear magnetic moments of the protons. After the pulse, the time required for the protons of the formation to stop precessing about the Earth's magnetic field is measured. This spin relaxation time can be used to evaluate residual oil. Nuclear magnetic resonance log (NML).
(C) Acoustic measurements—an acoustic signal is sent into the formation. We may measure:
   (1) The velocity of a compressional wave, from the transit time between two receivers: sonic log (SV, SL, BHC *). The shear-wave velocity can also be measured.
   (2) The transit time from a surface gun to a downhole geophone: well seismic (WST) *.
   (3) The amplitude of a selected peak or trough in the acoustic wave-train arriving at a receiver. The compressional or shear-wave arrivals may be of interest: amplitude logging (A).
   (4) The relative amplitudes of the various components of the wave-train, the configuration of the wave-train; variable density logging (VDL) *, sonic waveform photography, well seismic (WST), borehole televiewer (BHTV).

2.2. PROBLEMS SPECIFIC TO WELL-LOG MEASUREMENTS

Although we would like logs to be direct measurements of the formation, log responses are invariably affected by the presence of the well-bore, certain near-hole phenomena associated with the drilling of the well, and the geometry of the logging tool itself. Operational problems may be posed by temperature and pressure in the well.

* Mark of Schlumberger.

2.2.1. Borehole effects, invasion

2.2.1.1. Drilling mud

The influence of the drilling mud on a log response depends on several factors:

2.2.1.1.1. Hole diameter

The larger the hole, the greater the volume of fluid around the logging tool, and the stronger its effect on the log reading. Above a certain hole-size, there may be very little or no signal from the formation. Well-logging companies always specify a recommended maximum hole-size, in addition to the minimum size for safe passage of the tool.

2.2.1.1.2. Mud-type and mud-density

Whether or not a certain log can or should be run depends on the type of drilling mud in the hole. For instance, acoustic signals are poorly transmitted in an air-filled hole; oil or air will not conduct current; a salt-saturated mud, because of its high conductivity, will contribute a large borehole signal to the induction log; (on the other hand, it improves conduction between electrodes and formation, which is advantageous for the focussed resistivity devices).

Mud-salinity affects conductivity, resistivity, hydrogen index measurements, among others. The density of the mud influences the absorption of gamma rays.

Charts are available to correct for these borehole effects

2.2.1.2. Invasion

Ignoring for the moment wells drilled with air or emulsion, the functions of the drilling mud are:
(a) cooling of the drill-bit.
(b) preventing the hole from collapsing inwards.
(c) preventing flow of formation fluids (an extreme case of which is the "blow-out").
(d) bringing the cuttings up to surface.

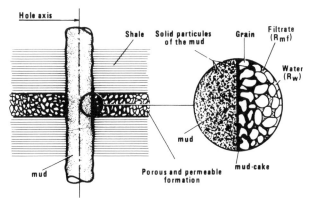

Fig. 2-1. The formation of mud-cake on the hole-wall.

In principle, the mud is kept at a slightly higher pressure than the formation pore-fluids, by careful control of the mud-density. Because of this pressure difference, there is a tendency for the mud to infiltrate porous, permeable beds. This is known as invasion. The solid particles in the mud are usually larger than the pores, and only the liquid content can invade the formation. So during drilling, there is a build-up of mud-cake on the wall of the hole wherever the mud-filtrate infiltrates. Eventually the mud-cake forms an almost impermeable membrane which impedes further invasion (Fig. 2-1).

The mud-filtrate displaces some of the formation fluid. The depth of invasion depends on the porosity and permeability of the rock, the "water-loss" factor of the mud (the quantity of water which is liberated from the mud), and the pressure difference between well-bore and formation.

For a given mud-type, in contact with a formation of a certain permeability and wettability, under a given pressure differential, the depth of invasion is larger the smaller the porosity—we consider the volume of filtrate to be constant for the prescribed set of conditions—the total volume of rock necessary to contain this filtrate increases as its porosity decreases.

Figure 2-2 is a schematic representation of the invasion profile. The reservoir rock near the hole does not contain the same fluids as before invasion. Changes have occurred both in the nature of the fluids and their proportions. Since the logging tools will always read at least some of the invaded zone signal, these changes must be taken into account when attempting to evaluate the fluid saturation of the virgin zone, which represents the reservoir at large.

There is no perfect solution to this problem. A satisfactory approach models the invasion fluid distribution as a step-profile (shown in Fig. 2-3) between the flushed zone of resistivity, R_{xo}, and the virgin reservoir, of resistivity R_t. The resistivities measured by tools of different depths of investigation are combined to solve for the saturations of the two zones (S_{xo}, S_w respectively).

2.2.1.3. Casing and cement

The presence of casing and cement precludes certain logging measurements (resistivity for instance). Generally, only nuclear (and some acoustic) measurements can be made through casing.

2.2.1.4. Fluid mobility

Although a troublesome phenomenon from a reservoir evaluation point of view, invasion can be used as an indication of the mobility of the reservoir

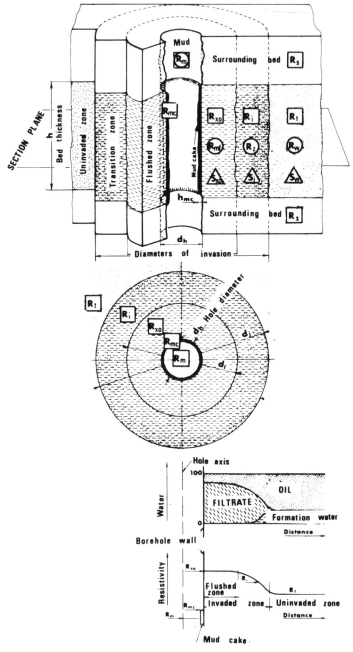

Fig. 2-2. The distribution of fluids and resistivities near the well-bore.

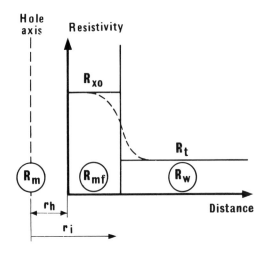

fluids. Hydrocarbon mobility is its ease of displacement.

The difference between the saturations S_{xo} and S_w calculated in the invasion and virgin zones is the quantity of hydrocarbon displaced by the filtrate.

The producible oil index, POI, (also called the movable oil index) is defined as:

$$POI = \phi(S_{xo} - S_w) \qquad (2\text{-}1)$$

If we are justified in assuming that the displacement of hydrocarbon during filtrate invasion is representa-

Fig. 2-3. Simplified model of the fluid and resistivity distribution of Fig. 2-2.

tive of what will occur during subsequent production by water-drive, then the POI is a useful index of probable recoverability. Certainly, if the difference $(S_{xo} - S_w)$ is small, it is likely that the hydrocarbon mobility is poor, and recoverability will be low *. Conversely, a large difference promises good recoverability.

The recoverability factor, f, is defined as:

$$f = (S_{xo} - S_w)/(1 - S_w) \qquad (2\text{-}2)$$

which is simply the recoverable fraction of the initial hydrocarbons in place.

2.2.2. The effect of tool geometry

2.2.2.1. Tool diameter, excentralization

There is a minimum hole diameter (or casing inner diameter) through which a tool of a certain size may safely pass. For most logging services, there exists a range of tool diameters appropriate to the common hole-sizes, including special "slim-hole" equipment.

The logging tool may take any of three positions relative to the hole axis (Fig. 2-4); centralized, excentralized against the wall ($\delta = 0$), and stood-off from the wall by a small amount ($\delta = $ constant). Correct tool positioning is mandatory for some measurements, and is ensured by mechanical means; one or several multi-armed centralizers (BHC); one-armed excentralizer (CNL, FDC); rubber stand-off (induction logging). This equipment becomes even more important in deviated wells, where the logging sonde would otherwise tend to lie along the low side of the hole. (In fact, wells are rarely perfectly vertical.)

The coefficient of excentralization is defined as:

$$\epsilon = 2\delta/(d_h - d_{tool}) \qquad (2\text{-}3)$$

ϵ is 1.0 for a perfectly centred tool sonde, and 0 when it is against the hole wall.

2.2.2.2. Spacing of sensors, depth of investigation

Logging tools do not take point readings, the signals they measure come from a finite volume of

---

* The fluid mobility is essentially a function of its relative permeability. If this is zero, the mobility is zero. In the case of high permeability (and therefore high mobility), the displaced hydrocarbon will often return rapidly once mud-circulation has stopped. If invasion is extremely deep, the deep-reading resistivity measurement may be very close to R_{xo} rather than R_t. The invaded zone saturation will appear to differ very little from that of the virgin zone, and we might conclude, erroneously, that mobility is poor. A similar conclusion could be drawn in a situation where the pressure differential is small, and invasion insignificant despite a high hydrocarbon mobility. (A zero or reversed pressure differential would have the same effect.) One should always carefully consider the factors which may influence invasion (porosity, rock-type, mud-density, formation pressure, etc) before computing saturations (S_{xo} and S_w) and drawing conclusions about good or poor mobility.

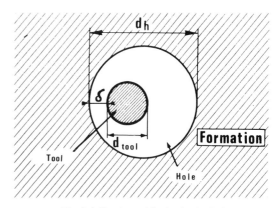

Fig. 2-4. Sonde positioning in the hole.

formation (and borehole) surrounding the sensor.

Single-detector devices (the S.P. electrode, the natural gamma-ray detector, for instance) respond to a volume which would be a sphere, centred on the sensor, if the surrounding medium were homogeneous.

Logging tools which use a source or emitter (current electrodes array, induction coils, neutron source, acoustic wave transmitter, etc) and one detector (measuring electrode, receiver coil, neutron or gamma-ray detector, acoustic receiver, etc) receive a signal from a volume of formation whose height is of the same order as the source-detector spacing.

Where two detectors are used (e.g. compensated neutron or sonic tools), the difference between the received signals is a measure of the formation over a distance roughly equal to the detector-detector spacing.

Thus the log measurements are average values, integrated over a volume of formation whose dimensions and shape depend on the tool geometry, as well as the nature of the measurement.

We categorize logging devices according to their "depth of investigation":

(a) Micro-tools, whose sensors and emitters are usually mounted on a pad which is pressed against the hole-wall. The volume investigated is small; for a compensated density tool it is a hemisphere of radius less than 10 cm; for the microlog * (ML) the hemisphere is only a few centimetres radius; for a micro-laterolog * (MLL) it is trumpet shaped, extending perhaps 10 cm into the formation; for the micro-spherically focussed log (MSFL) the volume is bean-shaped rather than spherical (Fig. 2-5).

(b) Macro-tools measure over a volume of, say, 0.5 to 5 m^3, which may be spherical, cylindrical or even almost a disc (Fig. 2-5), with the tool sonde as axis. (Although not strictly true, this is usually assumed to coincide with the hole axis.)

---

* Mark of Schlumberger.

34

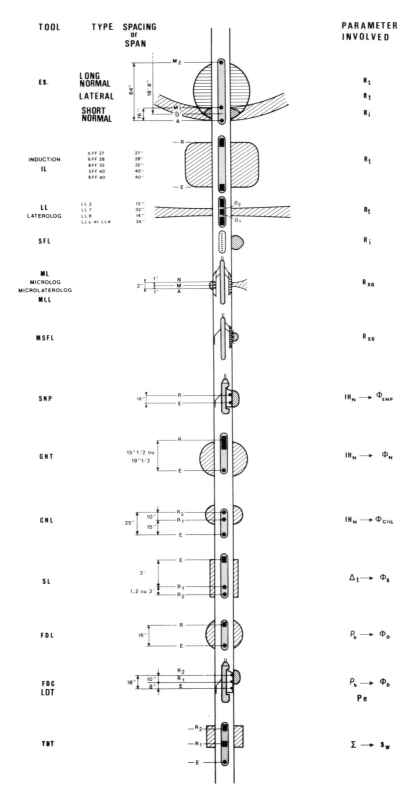

Fig. 2-5. Zones of investigation of the principal logging tools (Desbrandes, 1968; revised by the author). N.B. Not drawn to scale.

Among the resistivity tools, there exist deep and shallow reading devices. The so-called shallow depth of investigation lies somewhere between the deep and micro measurements (LLs, LL8, SFL, for instance).

As a general rule, the depth of investigation increases with the sensor spacing, while vertical defini-tion deteriorates. For instance, the microlog provides the finest vertical bed resolution, while the deep induction (ILd) and laterolog (LLd) have the largest depths of investigation in most conditions. Figure 2-6 shows how these factors vary with tool configuration.

2.2.2.3. *Vertical resolution*

A sedimentary series consists of a sequence of beds of various thicknesses, with differing lithological and petrophysical properties. In theory, each bed should be distinguished from its neighbours by its own particular reading on each of the logs. This is indeed seen in practice when the beds are thick.

Thin beds present a different picture—on certain logs it will be barely possible to pick out the bed boundaries, on others not at all. Two factors must be taken into account, one related to the logging tool design, the other to the fact that the measurement is made while the tool is moving.

The bed resolution on a log run across a sedimentary series depends on a number of factors, and will be different for each tool: (a) the thickness of the beds; (b) the tool geometry and type of measurement being made, inasmuch as they affect the volume of investigation (relative to the bed-thickness); (c) the contrast between readings in the bed in question and its immediate neighbours; and (d) in a few cases, ancillary or parasite tool responses, (such as the blind-zone and inverse deflections characteristic of the conventional resistivity measurements, the normal and lateral) which may mask or distort bed thickness, and give erroneous values.

The bed resolution on radio-active logs is affected by the "time constant" and logging speed discussed in 2.2.3.

Bed boundaries on logs are not perfectly sharp, but appear as a more or less gradual transition between a lower and a higher reading. The steepness of the transition depends on the bed resolution of the logging tool, discussed above. Often, the boundaries appear displaced slightly upwards (Fig. 2-8).

"Thin" beds, (that is, beds thinner than the spacing of the logging tool) may still be discerned on the log. The measured signal, however, is an average of contributions from all the beds within the volume of investigation, and the true log value in a thin bed is rarely obtained, even after correction.

At this point we introduce the concept of "electro-bed", which correspond to the beds we discern by eye on a log. Electro-beds have the following features: (a) they are at least as thick as the effective spacing of the logging tool, and may well be made up of several thinner lithological strata which are being "averaged" over the volume of investigation; (b) their lithology is a volume for volume mixture of each of these substrata; and (c) their apparently homogeneous log response is in fact an average of the contributions of any substrata, according to their thicknesses and log characteristics (Fig. 2-7).

An electro-bed corresponds to an interval of depth in which log response is constant (within certain limits).

Logging devices with fine bed resolution which may be used for distinguishing the detailed structure of a sedimentary series are the electromagnetic propagation tool (EPT), microproximity (MPL), micro-

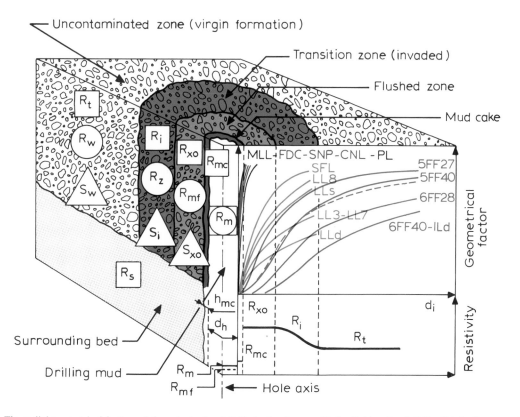

Fig. 2-6. The radial geometrical factors of the principal resistivity tools, shown with the fluid and resistivity distributions near the well-bore.

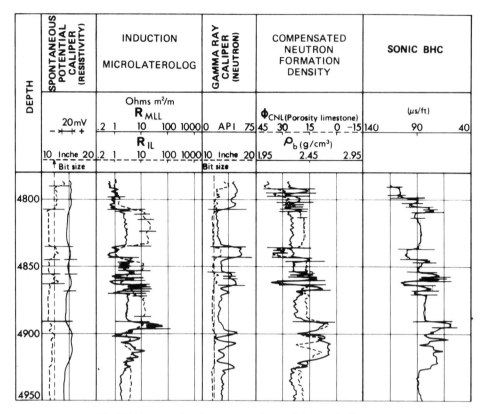

Fig. 2-7. An example of a composite log, and "electrobed" zoning.

laterolog (MLL), or better still, the microlog (ML) and high-resolution dipmeters (HDT and SDT).

Since the electro-bed definition of the longer spacing tools is inferior to the micro-tools, it is customary to apply "depth-smoothing" or vertical averaging to the latter so as to render the bed resolution of all logs more or less compatible, before an evaluation is attempted.

2.2.3. Logging speed

Logging speeds are by no means the same for all types of log.

Since natural and induced radio-active phenomena are random by nature, it is necessary to accumulate count data over a period of time and compute the mean in order to obtain a representative reading. This accumulation or sampling period corresponds to the "Time constant" of conventional (capacitative-type) measuring equipment. The time constant is chosen according to the count-rate level and measurement precision desired; the logging speed is then usually adjusted such that the tool travels 1 ft (0.30 m) in one time constant period, as shown in Table 2-2. Vertical resolution of 2–3 ft (0.60–0.90 m) may be attained under these conditions.

Other factors which limit logging speed are galvanometer inertia (they must have sufficient time to deflect to the full value, which precludes very high logging speeds where high contrasts in readings are to be encountered), and various safety considerations, particularly cable tension and the risk of damage to pad-type equipment.

On log films, a speed marker provides a record of the logging speed. This is usually a small break or pip

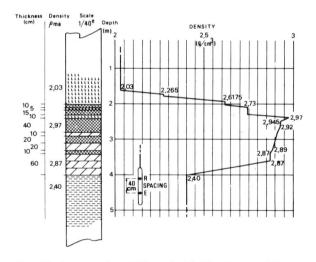

Fig. 2-8. An example of "electrobeds". The theoretical log response of a single-detector formation density tool (40 cm spacing), across a series of fine alternating beds of dolomite, anhydrite and salt. Note that the true density reading is not attained in beds significantly thinner than the tools spacing; and the upward displacement of the apparent bed boundaries.

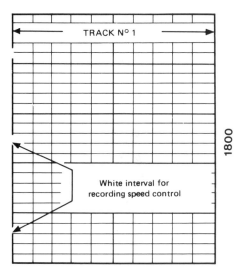

Fig. 2-9. The time-marker, a record of distance moved by the cable per minute.

1800

Track N° 1

White interval for recording speed control

occurring every minute on the vertical grid-line at the extreme left or right edge of the log (Fig. 2-9). Tool speed (and acceleration) may also be recorded on magnetic tape.

Table 2-2 shows the maximum logging speeds recommended for some of the logging services. Obviously, for combinations of several tools, the maximum speed of the slowest survey must be respected.

2.2.4. Hostile environments

Well-bore temperature and pressure increase with depth as a function of the geothermal gradient, and mud density, respectively (Figs. 2-10 and 2-11). Logging tools must be able to withstand extreme hole conditions which might be encountered. (Freshly circulated drilling mud may initially be considerably cooler than the formations with which it is in contact. The "warming-up" period may provide a short safe period for logging very hot wells.)

The presence of corrosive gases, such as hydrogen

TABLE 2-2

Recommended maximum logging speeds.

Survey		Maximum logging speed	
		(ft/min)	(m/min)
SP		100	30
Induction		100	30
Laterolog		50	15
Microlaterolog		35	10
Neutron	TC = 2 sec	30	9
GR	TC = 3 sec	20	6
Density	TC = 4 sec	15	4.5
TDT			
Sonic		70	20
Amplitude		35	10

sulphide, may require special precautions, and resistant equipment (H_2S-proof cable, for instance).

Logging companies provide the operating limits for each tool. Outside these limits, there is the risk of breakdown or destruction of the equipment by temperature failure of electronic components, leaking of mud past pressure seals, collapse of the pressure housing, and so on.

2.3. LOGGING EQUIPMENT—SURFACE AND DOWNHOLE

Log measurements are made using a measuring sonde (with electronic cartridge) lowered on a cable from a winch, which is mounted on a logging truck or offshore unit. The truck and unit are laboratories containing the recording equipment (optical and tape), control panels, and perhaps a computer (Figs. 2-12 to 2-18) or micro-processor.

2.3.1. Logging truck and offshore units (Figs. 2-13 and 2-14)

Figure 2-15 shows the equipment contained in a conventional logging truck:
– The main winch (E), which may hold as much as 26,000 ft (8000 m) of multi-conductor, steel-armoured cable, with a pulling capacity of several tons (Fig. 2-16).
– auxillary winch (G) containing thinner mono-conductor cable, generally for use when there is well-head pressure (production or work-over operations, mainly).
– the winchman's control panel (C).
– the surface logging panels (A) which power and control the downhole tool, process the incoming information and transmit the information to:
– the recording equipment: (a) camera (B and Fig. 2-22) and (b) magnetic tape recorder (J and Fig. 2-23).
– depth-measuring system (F).
– electrical generator (H).
– dark-room for development of the film (D).
– a printer (not shown).

Recent years have seen the introduction of fully computerized logging units, which not only handle the data acquisition, but permit well-site evaluations to be made. (Fig.2-18 shows Schlumberger's CSU * (Cyber Service Unit). Similar units are Gearhart's DDL (Direct Digital Logging), Dresser Atlas's CLS (Computerized Logging Service) and Welex's DLS (Digital Logging System).

---

* Mark of Schlumberger.

38

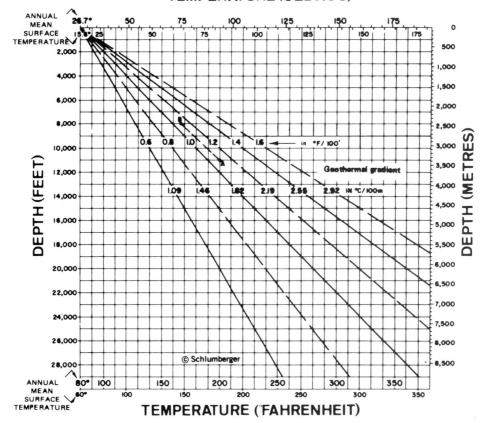

Fig. 2-10. The variation of temperature with depth. (courtesy of Schlumberger).

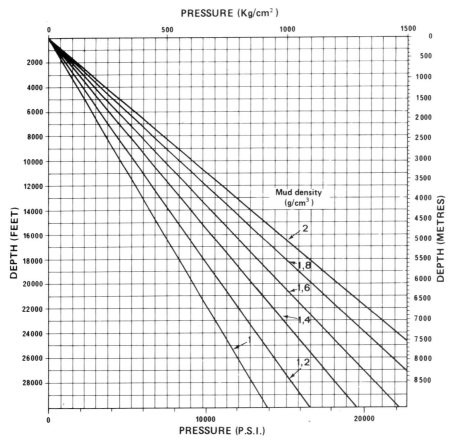

Fig. 2-11. The variation of hydrostatic pressure with depth.

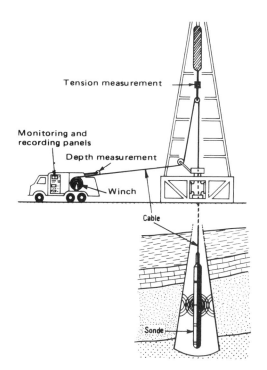

Fig. 2-12. Schematic representation of a logging "set-up" on a land-rig.

2.3.2. Cable

The logging cable fulfills three functions: (a) running-in and pulling out the tool, and control of tool speed; (b) electrical interface between the downhole logging tool and the surface processing and recording equipment; and (c) depth measurement.

2.3.2.1. Tool transport

The cable has an external, load-carrying armour consisting of two layers of helical steel winding, wound in opposite senses. A multicable is typically 15/32″ (11.8 mm) in diameter, and a monocable 7/32″ (5.5 mm). The cable has the following physical characteristics (quantities refer to multi-cable):

(a) It can support as much as 8 tons of tension at surface. However, non-elastic deformation (and rupture of the conductors) will occur at considerably lower tensions. In practice, tensions higher than 3–4 tons (for a logging tool weight of 1500 lb) are rarely encountered.

(b) It is able to pass round relatively small pulleys or sheave-wheels, typically 4 ft in diameter, mounted in the rig-structure (Fig. 2-12).

(c) The insulation of the conductor wires must not be able to extrude between armour windings, even at several thousand psi of pressure (several tens of megapascals).

(d) It must resist abrasion by rocks, and corrosion (saline muds, gas, etc.).

The life of a cable generally runs to two or three hundred descents in moderate operating conditions.

2.3.2.2. Conductor of electrical signals

Monocable: control signals and power from surface panels must share the single conductor wire with perhaps several channels of measured data from downhole, requiring special transmission modes.

Multiconductor cable (Hepta-cable) (Fig. 2-19): the seven copper conductors are each insulated by polypropylene, teflon, rubber, etc, jackets. The insulation, better than 10 MΩ at surface, must remain effective at high temperatures. Power, commands and data are transmitted on combinations of conductors in such a way as to minimize signal attenuation and "crosstalk" among channels.

Signal distortion is kept to a minimum despite the considerable distance of transmission (as much as 26,000 ft or 8000 m) and high temperatures involved (175–200°C). Signal frequencies up to 200 kHz are used.

The following extract from "The Electrical Logging Cable" by B. Chauve (Said, Paris 1977) lists some of the electrical properties of a rubber insulated cable:
- Resistance of conductor wire at 20°C, 36 ± 1.5 ohm/km.
- Minimum resistance with respect to armour:
  300 MΩ/km at 20°C
  2   MΩ/km at 100°C
  0.1   MΩ/km at 150°C
  0.04 MΩ/km at 175°C
- Capacitance to armour (rubber insulation):
  0.263 ± 0.05 μF/km at 1000 Hz.
- Capacitance between two adjacent conductors $\leqslant$ 0.03 μF/km.
- Dielectric breakdown potential $\geqslant$ 1000 V at 50 Hz for 2 minutes.
- Attenuation of signal for a standard cable length (5 km):
  7 db at   1 kHz
  30 db at   10 kHz
  55 db at 100 kHz
  90 db at 200 kHz

2.3.2.3. Depth measurement

Film and tape movement are governed by cable motion, which is transmitted by a calibrated Spooler wheel (F in Fig. 2-15) to mechanical or electrical drive systems, which permit a choice of tape speeds or depth scales on film.

The cable is marked magnetically every 100′ or 25 m under constant tension. These marks are detected as the cable passes the spooler device (F, Fig. 2-15) and serve as the reference for precise depth

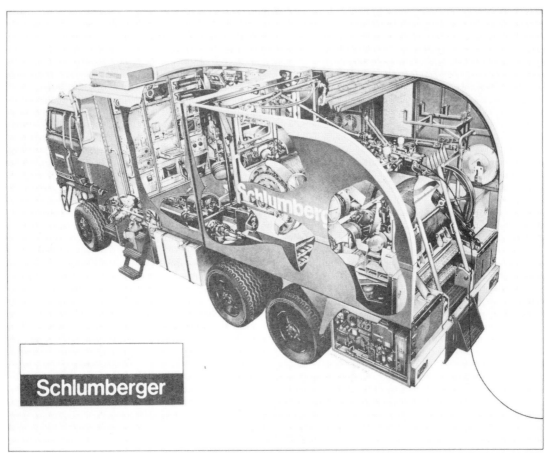

Fig. 2-13. A logging truck (courtesy of Schlumberger).

OSU-F WITH WINCH DRUM ATTACHMENT

Fig. 2-14. An offshore logging unit (courtesy of Schlumberger).

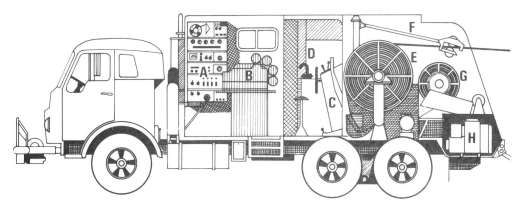

Fig. 2-15. Sectional view of a logging truck (reproduced by courtesy of Schlumberger).

control, permitting corrections to be made for spooler wheel inaccuracy. The cable has a very low stretch coefficient in its normal operating range; elastic stretch under logging tensions is small, and is corrected automatically, or manually by a hand-crank adjustment.

2.3.2.4. *The "bridle"*

Certain electrode devices require the use of a "bridle", which is a length of deca-cable, insulated on the outside, with two lead electrodes which serve as remote returns (Fig. 2-20a). The bridle is connected to the cable by a quick-connection consisting of two rope-sockets clamped in a split-shell, the "torpedo". Figure 2-20b shows the bridle head, which connects electrically and mechanically to the logging tool.

Fig. 2-16. Rear view of a logging truck, showing the cable and spooler (courtesy of Schlumberger).

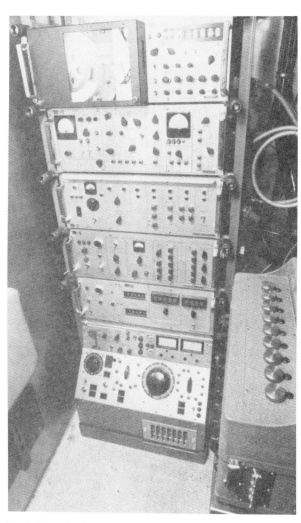

Fig. 2-17. A view inside the truck, showing the logging panels, tape recorder (top) and part of the camera (right) (courtesy of Schlumberger).

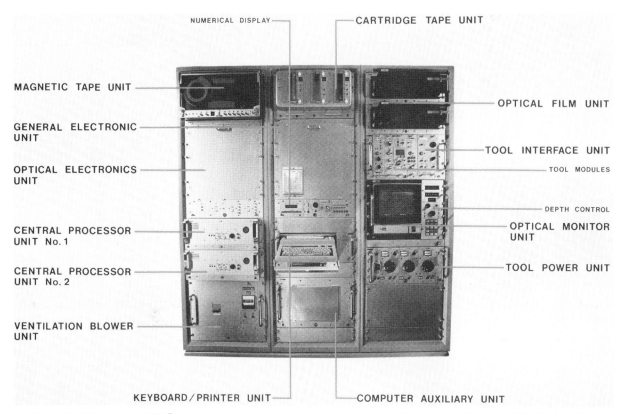

NUMERICAL DISPLAY

CARTRIDGE TAPE UNIT

MAGNETIC TAPE UNIT

GENERAL ELECTRONIC UNIT

OPTICAL ELECTRONICS UNIT

CENTRAL PROCESSOR UNIT No. 1

CENTRAL PROCESSOR UNIT No. 2

VENTILATION BLOWER UNIT

OPTICAL FILM UNIT

TOOL INTERFACE UNIT

TOOL MODULES

DEPTH CONTROL

OPTICAL MONITOR UNIT

TOOL POWER UNIT

KEYBOARD/PRINTER UNIT

COMPUTER AUXILIARY UNIT

Fig. 2-18. Schlumberger's CSU© —a view of the computers, tape units and optical display section (courtesy of Schlumberger).

Most logging tools dispense with the bridle. In this case, a head similar to the one in Fig. 2-20b, is connected directly to the cable, using a torpedo quick-connection.

The logging head contains the important weak-point, which permits the cable to be pulled off a stuck tool prior to a fishing job.

2.3.2.5. *The "fish"*

This is a remote electrode made out of lead, connected to the logging panel by a long insulated wire, via a truck installation. It is usually buried in damp earth, the mud-pit, or lowered into the sea, suffi-

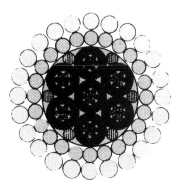

Fig. 2-19. Section of a multi-conductor cable (reproduced by courtesy of Schlumberger).

ciently far from electrical sources to be at zero potential. It is needed as the reference electrode-at-infinity for the SP, and for the current return of the LLD (deep laterolog). The fish is not strictly a common ground return, and is insulated from the truck chassis and cable armour.

2.3.3. The logging tool

Logging tools vary in complexity from a simple electrode-carrying mandrel, to a sophisticated system of electronic circuits, enclosed in a pressure-resistant metal housing and capable of operating at high temperatures.

All tools consist at least of a detector, receiver or sensor. For measurements of induced phenomena (2.1.2) there will also be a source or emitter. These components are mounted in the logging sonde, which may also contain a hydraulic or mechanical system (for opening and closing the arms of a pad-type tool, for instance), and some electronics (preamplifier, power-transformer Fig. 2-21a).

The sonde is generally attached below an electronic cartridge, which carries in a protective housing the electronic modules or hardware for the downhole instrument, and which has a multiple role:

(a) Power supply to the sonde emitter system (if necessary), and control of the timing and characteristics of the emitted signal.

Where several tools are being run in combination, each of the sondes and cartridges in the tool-string has a pass-through facility for the signals to or from tools lower in the string.

Each sonde-cartridge set can be connected electrically and mechanically to the bridle or cable head by a quick-connect system consisting of pins and sockets, and a threaded ring (Fig. 2-21b).

As we discussed in 2.2.2.1, (ex)centralizers or stand-offs may be attached to the sonde and cartridge.

2.3.4. Recording equipment

2.3.4.1. *Photographic recorder* (Fig. 2-22)

This is usually a nine-galvanometer system, recording on two films. Each galvanometer mirror deflects a light-beam, assigned to a log measurement, onto both films. The deflection of the beam varies with the magnitude of the logging signal; up to nine measurements can be recorded.

Each film is drawn from its light-tight supply can, across an aperture, where it is exposed to the galvanometer beams, and into its receiving can. The spooler wheel and a drive mechanism transmit the cable motion to the two films. As the cable moves, the rate at which the "upper" and "lower" films advance is governed by a gear-reduction which provides depth-scale options of 1/1000, 1/500, 1/200, 1/100, 1/40 and 1/20. For most purposes, the upper film is run on a small scale like 1/1000, and the lower at 1/200.

Although the same traces are played onto both films, certain of them can be suppressed from either film if desired.

Cathode-ray and electronic camera systems are gradually replacing these galvanometer cameras.

2.3.4.2. *Magnetic tape recorder* (Fig. 2-23)

With the advent of multiple tool combinations containing several sensors, more data can be sent to surface than a 9-trace optical recorder can handle.

Magnetic tapes permit many more channels of information to be recorded, and offer several other advantages:

(1) Both raw and functioned log data (such as raw count-rates, and computed porosity from the CNL) are recorded, in addition to various tool operational data such as temperature and current level.

(2) Playback capabilities permit changes in optical log presentation to be made, recalibration, functioning of raw data, and so on. This can be performed at the well-site with the computerized logging unit, or at the log computing centre.

(3) Computerized interpretation programs can be run on the logs, permitting rapid and continuous

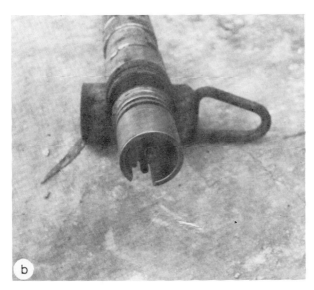

Fig. 2-20. a. The bridle (note the electrodes, head, and part of the quick connection). b. The bridle-head.

(b) Power supply to the detector or receiver system if necessary.

(c) Filtering and amplification of the incoming signal, analog-to-digital conversion, scaling and so on.

(d) Transmission of the signal up the cable, control of telemetry system, etc.

(e) Power supply and control of any sonde mechanical operations (e.g. opening or retraction of the dipmeter arms (HDT) shown in Fig. 19-4a).

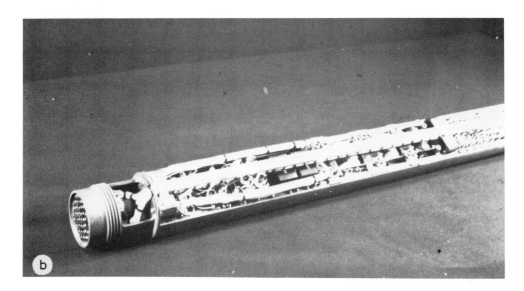

Fig. 2-21. a. Part of the complex electronic section of a logging sonde (courtesy of Schlumberger). b. An electronic cartridge removed from the pressure-housing, showing the circuitry and an intermediate multi-pin connecting head (courtesy of Schlumberger).

evaluation of data. The computerized logging unit permits well-site interpretations to be made immediately after the survey.

(4) Data can be transmitted by telephone or radio-telephone from the well-site to a distant office or computing centre.

Fig. 2-22. The photographic recorder (courtesy of Schlumberger).

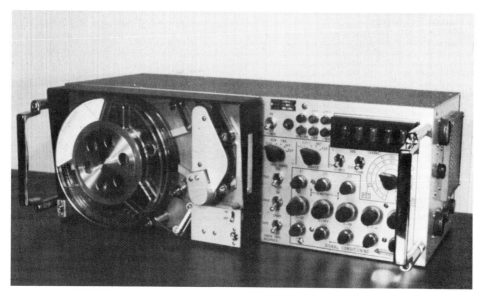

Fig. 2-23. Magnetic tape-recorder (courtesy of Schlumberger).

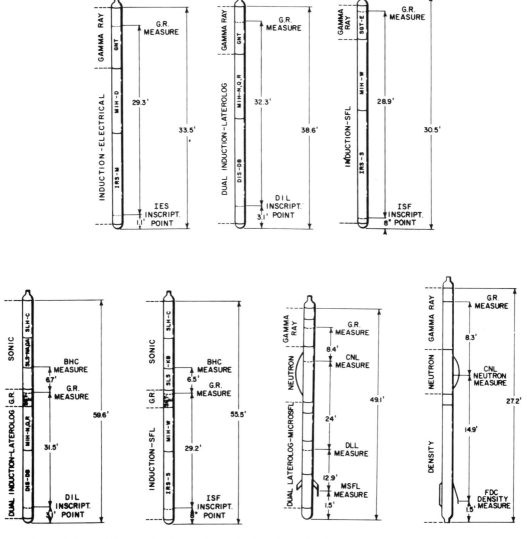

Fig. 2-24. Some of the combination logging tool-strings. (reproduced by courtesy of Schlumberger, 1974).

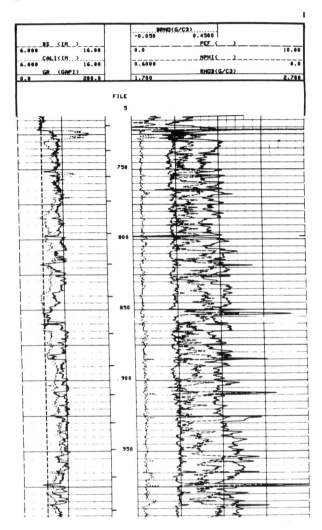

Fig. 2-25. A well-log.

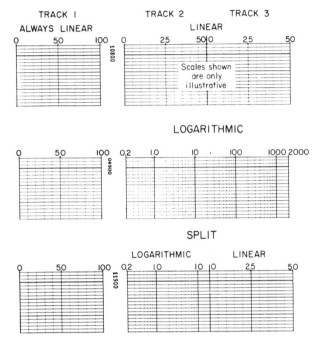

Fig. 2-26. The three most common track scalings.

2.3.5. Tool combinations

Some of the combination services presently offered by Schlumberger are shown in Fig. 2-24. Appendix 1 contains a more general list of the services of Schlumberger and the other logging companies.

Schlumberger is at present developing the possibility of combining any desired number and types of tools, subject only to operational considerations such

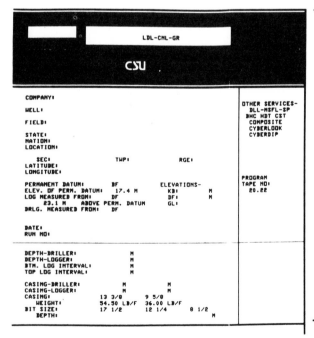

Fig. 2-27. A log heading.

as length and weight. It should be noted that in very long tool combinations, the upper tools may not be able to enter a zone of interest too near the bottom of the well. In such a situation, it will be necessary either to drill deeper, or run a shorter tool-string.

2.3.6. Memorization

Where several tools are in combination, it is necessary to store in a memory the readings of each tool as it passes a given depth, and to discharge this memory only when the last sensor reaches that depth. In this way all filmed and taped data are recorded on depth. This is performed by a multi-channel memorizer.

2.4. LOG PRESENTATION

The variations in parameters being measured are recorded on film and tape as a function of depth. The film is developed, or the tape played back, to produce the log, which may be an opaque or transparent print. Fig. 2-25 is an example.

The API * standard grid presentation used by all logging companies consists of a left-hand track (Track 1) 6.4 cms wide, a depth-track 1.9 cms wide, and two right-hand tracks, each 6.4 cms wide (Tracks 2 and 3).

Track 1 is always scaled linearly into 10 "large divisions". (A "small division" is 1/10 of a "large division", i.e. 1/100 track-width). Tracks 2 and 3 may be both linear (Fig. 2-26 upper), both logarithmic, covering two decades each (Fig. 2-26 middle), or a mixture of logarithmic and linear (Fig. 2-26 lower) with Track 2 serving for resistivity measurement, Track 3 for sonic, for example.

Certain old resistivity logs employed a "hybrid" grid which was linear in resistivity over the left half of one track (0–50 Ωm, for instance), and linear in conductivity over the right half (20 mmho–0 mmho, corresponding to 50–∞ Ωm, in this case).

The depth scaling is chosen according to the purpose of the log. 1/1000 and 1/500 scales (respectively 1.2" per 100 ft, and 2.4" per foot) are used for a quick overview of the entire logged interval, and for correlation (with a lithological record for instance). 1/200 and 1/100 are the conventional scales used for reservoir evaluation. The 1/40 is common for the micro-focussed resistivity logs to take advantage of their fine bed resolution.

Optical correlation of dipmeter curves requires a 1/20 (60" per 100 ft) scaling.

Note that in North America, a scaling of 1/1200 (1" per 100 ft) replaces 1/1000, and so on; thus the

1/200 scale (6" per 100 ft) becomes 1/240 (5" per 100 ft). This is called the "decimal foot" system, as opposed to the "duodecimal foot".

The log heading (Fig. 2-27) displays all relevant information about the well and the logging operation: well-name, company, field, well coordinates, bit-size, mud data, data logged, type of equipment

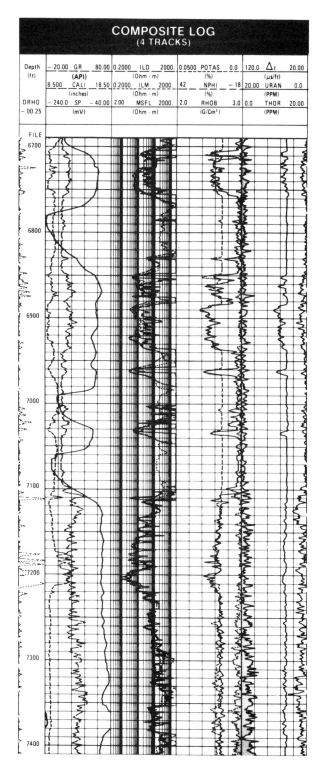

Fig. 2-28. A composite log presentation obtained with CSU.

---

* American Petroleum Institute.

used, calibrations, plus any special remarks concerning the job.

Figure 2-28 is a composite log, made up of several logs side by side and on depth. With the CSU it is now also possible to play back a number of logs onto a single film, with the desired coding of traces, shading of separations between traces, and so on, at the well site.

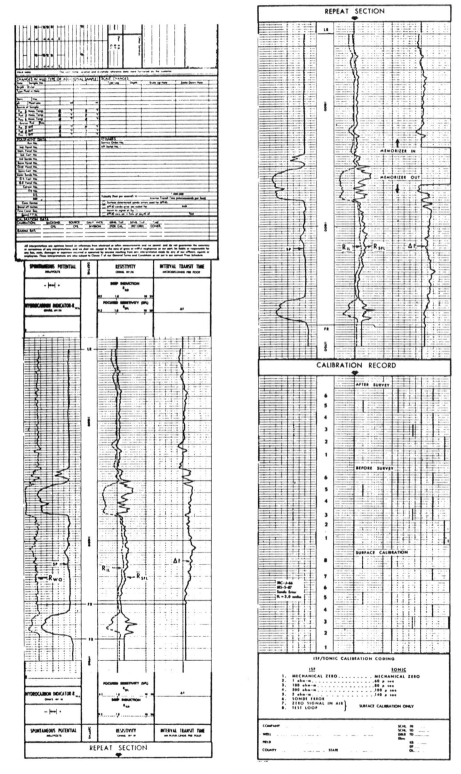

Fig. 2-29. Example of a log with repeat section and calibration tails.

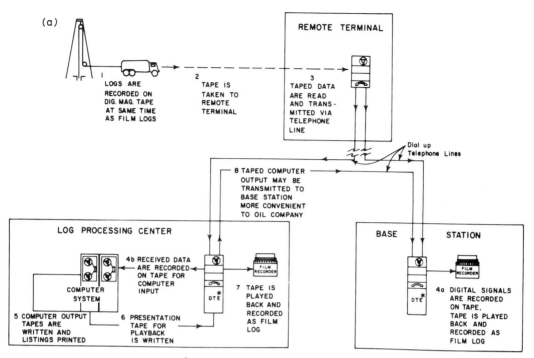

*DTE IS DIGITAL TRANSMIT–RECEIVE–PLAYBACK EQUIPMENT

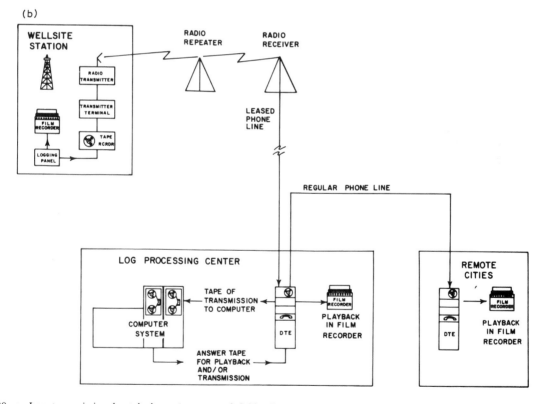

Fig. 2-30. a. Log transmission by telephone (courtesy of Schlumberger). b. Log transmission by radio and telephone (courtesy of Schlumberger).

2.5. REPEATABILITY AND CALIBRATIONS (see Appendix 5 for more details)

At the end of a log, a repeat section and calibration tail are usually attached (Fig. 2-29, right). The repeat section verifies that the tool is functioning consistently. For radio-active logs, several repeat sections may be run for subsequent stacking (averaging) to reduce statistical variations (e.g. NGS and TDT).

A calibration record is made on film and tape before and after the survey, to show that the equipment was correctly adjusted, and that no drift in adjustment has occurred during the log. For certain tools, a "master calibration" is made at the workshop prior to the logging operation, using laboratory equipment. The well-site calibration repeats this adjustment using portable calibration "jigs" (FDC, CNL, GR... Appendix 5). Other tools have their own internal calibrators (IL, LL,...), and may be adjusted while down-hole.

2.6. DATA TRANSMISSION

Taped data can be transmitted by telephone, or radio and telephone (Fig. 2-30) to a log computing centre or company office if quick decisions, or rapid interpretation, are to be made away from the well-site.

2.7. REFERENCES

Allaud, L. and Martin, M., 1976. Schlumberger: Histoire d'une Technique. Berger-Levrault, Paris.

Arps, J.J., 1964. Engineering concepts useful in oil finding. Bull. Am. Assoc. Pet. Geol., vol. 48 (2).

Chilingar, G.V., 1969. Deltaic and Shallow Marine Deposits. L.M.J.U. van Straaten, ed., Elsevier, Amsterdam.

Chombart, L.G., 1959. Reconnaissance et évaluation des formations par diagraphies electriques et nucléaires. Technip, Paris.

Choquette, P.W. and Pray, L.C., 1970. Geologic nomenclature and classification of porosity in sedimentary carbonates. Bull. Am. Assoc. Pet. Geol., 54 (2).

Clavier, C., Coates, G. and Dumanoir, J., 1977. The theoretical and experimental basis for the "dual water" model for the interpretation of shaly sands. Soc. Pet. Eng., AIME, Pap. No. 6859.

Dadone, R., 1958. Diagraphies différées. Tome 1 des "Techniques d'exploration profonde dans la recherche du pétrole". Technip, Paris.

Desbrandes, R., 1968. Théorie et interprétation des diagraphies. Technip, Paris.

De Witte, L., 1950. Relations between resistivities and fluid contents of porous rocks. Oil Gas J., August 24.

Flandrin, J. and Chapelle, J., 1961. Le pétrole. Technip, Paris.

Guillemot, J., 1964. Cours de Géologie du Pétrole. Technip, Paris.

Guyod, H., 1952. Electrical Well Logging. Fundamentals, Houston.

Guyod, H., 1969. Geophysical Well Logging. Guyod, Houston.

Lynch, E.J., 1962. Formation Evaluation. Harper & Row, New York.

Martin, M., 1957. Possibilités actuelles des méthodes de diagraphies électriques et nucléaires. Technip, Paris.

Millot, G., 1969. Géologie des argiles. Masson, Paris.

Patchett, J.G., 1975. An investigation of shale conductivity. SPWLA, 16th Am. Log. Symp. Trans., paper V.

Perrodon, A., 1966. Géologie du pétrole. Presses Univ. Franç., Paris.

Pirson, S.J., 1963. Handbook of Well Log Analysis for Oil and Gas, Formation Evaluation. Prentice Hall, New York.

Schlumberger, 1959. Etudes des sondages par les méthodes Schlumberger, document 8. Log interpretation: Vol. I. Principles 1972; Vol. II. Applications 1974; Charts 1979. Soc. Pet. Eng. AIME, 1958. Petroleum Transactions, vol. I: Well logging.

Timur, A., 1968. An investigation of permeability, porosity and residual water saturation relationship for sandstone reservoirs. Log Analyst, 9 (4).

Waxman, M.H. and Smits, L.J.M., 1968. Electrical conductivities in oil-bearing shaly sands. J. Soc. Pet. Eng., 8 (2).

Wyllie, M.R.J., 1957. The Fundamentals of Electric Log Interpretation. Academic Press, New York.

Wyllie, M.R.J. and Rose, W.D., 1950. Some theoretical considerations related to the quantitative evaluation of the physical characteristics of reservoir rock from electrical log data. J. Pet. Technol., 189.

3. THE MEASUREMENT OF RESISTIVITY

3.1. INTRODUCTION

The measurement of formation resistivity is funda-mental to the evaluation of hydrocarbon saturation. There are several measuring techniques in use, all variations of a common basic system: an emitter (electrode or coil) sends a signal (electrical current, electromagnetic field) into the formation. A receiver (electrode or coil) measures the response of the for-mation at a certain distance from the emitter.

Generally, an increase in the spacing of the system results in an improved depth of investigation (and a reading nearer to R_t), at the expense of vertical resolution.

Long-spacing devices which are medium to deep reading include:
ES —the conventional electrical survey, with normal and lateral (or inverse) electrode arrays;
IL —the induction;
LL —the laterologs;
SFL—the spherically focused log.
The last three, being focused devices, are less sensi-tive to the borehole. Depending on the spacing and the nature of the focusing, either R_{x0} (SFL, LL8) or R_t (LLd, IL) may make the predominant contribu-tion to the measured signal, under average conditions of invasion.

The micro-tools, mounted on pads which are ap-plied against the hole wall, are designed to read R_{x0}, by virtue of their very shallow depths of investiga-tion. There is very little borehole fluid effect, but the mud-cake contributes a small signal. The micro-tools include:
ML —the microlog (normal and lateral);
MLL —the microlaterolog (not to be confused with the microlateral of the ML);
PL —the microproximity log;
MSFL—the microspherically focused log;
HDT —the high-resolution dipmeter tool (not strictly used for a quantitative measure of R_{x0}-see chapter 19).
The vertical definition obtained with these electrode tools is much finer than with the longer spacings. Like the ES, the ML is the only non-focused system in the group.

A combination of deep-, medium- and shallow-reading tools enables us to evaluate R_t, R_{x0}, and d_i. The following sections will discuss in detail the oper-ating principle and measurement characteristics (borehole sensitivity, zone of investigation, geometri-cal factor, etc) of each tool. After this we will see how the geology of the formation affects resistivity, and finally the applications of the measurement.

3.2. NON-FOCUSED LONG-SPACING TOOLS

3.2.1. Measuring principle

In Fig. 3-1, a single point electrode, A, sends current in an infinite, homogeneous medium, to a remote return B.

The current will radiate uniformly in all direc-tions, and the equipotential surfaces will be con-centric spheres centred on A. If the potential at distance r from A is $V(r)$, then the difference dV between two equipotentials dr apart is:

$$-dV = \frac{RI}{4\pi r^2} dr \qquad (3-1)$$

where I is the total current flowing, and R is the resistivity of the medium (so that $Rdr/4\pi r^2$ is the resistance between the two surfaces).

Integrating dV between r and infinity (zero poten-tial):

$$V = \int_r^\infty RI \frac{dr}{4\pi r^2} = \frac{RI}{4\pi} \left[-\frac{1}{r} \right]_r^\infty = \frac{RI}{4\pi} \left[\frac{1}{r} - \frac{1}{\infty} \right] = \frac{RI}{4\pi r}$$

$$(3-2)$$

The field H is given by:

$$H = -\frac{dV}{dr} = \frac{RI}{4\pi r^2} \qquad (3-3)$$

and from these we obtain R:

$$R = \frac{V}{I} 4\pi r \qquad (3-4)$$

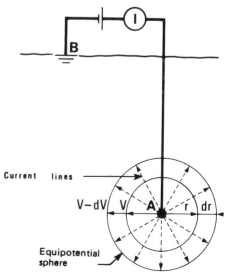

Fig. 3-1. The basic principle of resistivity measurement.

$$R = -\frac{4\pi r^2}{I}\frac{dV}{dr} \qquad (3\text{-}5)$$

Equations 3-4 and 3-5 suggest two ways of measuring the resistivity and these are discussed next.

3.2.1.1. *The normal configuration*

The measuring electrode M is situated close to the current electrode A, (Fig. 3-2a). A constant current I flows from A to the remote return B. The potential V_M of M is measured with respect to a reference electrode N (at zero potential) by means of a volt-meter. Although, theoretically, N should be on surface (at "infinity"), inductive phenomena necessitate placing it downhole, but at a distance from M considerably greater than is A (one of the bridle electrodes, for instance) (Fig. 3-2b). From eq. 3-2:

$$V_M = \frac{RI}{4\pi\overline{AM}} \qquad (3\text{-}6)$$

neglecting borehole effects, i.e. we assume an infinite homogeneous medium.

Since I is held constant, V_M is proportional to R. $4\pi\overline{AM}$ is the coefficient K_N of the normal device:

$$R = K_N\frac{\Delta V}{I} \qquad (3\text{-}7)$$

Thus a continuous recording of V_M on an appropriate scale is a log of the resistivity. There are two conventional $\overline{AM}$ spacings for the normal:

$\overline{AM} = 16''$, the "short normal";

$\overline{AM} = 64''$, the "medium normal".

3.2.1.2. *Lateral and inverse configurations*

In the lateral configuration (Fig. 3-3a), two measuring electrodes, M and N, are placed close together below A. The difference ΔV between the spherical equipotential surfaces on which M and N lie, is derived as follows:

$$V_M = \frac{RI}{4\pi\overline{AM}}$$

V_M is the potential at electrode M, from eq. 3-2.

$$V_N = \frac{RI}{4\pi\overline{AN}}$$

V_N is the potential at electrode N, from eq. 3-2.

$$\Delta V = V_M - V_N = \frac{RI}{4\pi}\left[\frac{1}{\overline{AM}} - \frac{1}{\overline{AN}}\right]$$

$$\Delta V = RI\frac{\overline{MN}}{4\pi\overline{AMAN}} \qquad (3\text{-}8)$$

The constant $4\pi\overline{AMAN}/\overline{MN}$ is the coefficient K_L for the lateral device. So:

$$R = K_L\frac{\Delta V}{I} \qquad (3\text{-}9)$$

The formation resistivity R is proportional to ΔV if the current is constant.

In practice, return electrode B is placed downhole, and a modified, but equivalent "inverse" configuration is used (Fig. 3-3b). The electrode pair A-B takes the place of M-N; however, by the principle of reciprocity the resistivity relationship derived above still holds, with measure return electrode N remote from M.

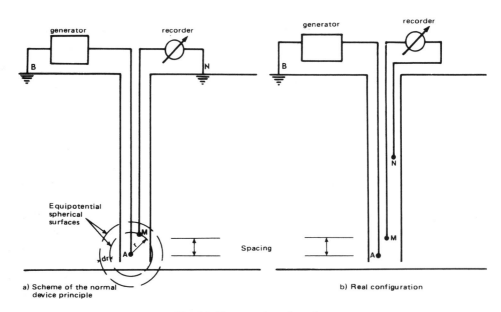

a) Scheme of the normal device principle

b) Real configuration

Fig. 3-2. The normal configuration.

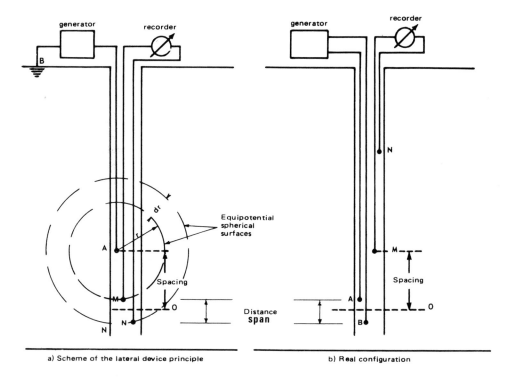

a) Scheme of the lateral device principle   b) Real configuration

Fig. 3-3. The lateral configuration.

For the lateral, M and N are very close relative to their distances from A. If O is the mid-point of $\overline{MN}$, then since $\overline{MN} \ll \overline{AM}$ and $\overline{AN}$ (from eq. 3-8):

$$R = \frac{4\pi \overline{AO}^2}{I} \frac{dV}{dl}$$

dV/dl is the intensity of the electric field (H, eq. 3-3) at O, and R is proportional to this intensity. For the inverse, $\overline{AO}$ is replaced by $\overline{MO}$. The spacing $\overline{AO}$ (or $\overline{MO}$ for the inverse array) is usually 18′8″.

3.2.2. The current path

The resistivity measured is not exactly that of the virgin formation because it is neither infinite in extent, nor homogeneous. Heterogeneity is introduced by the presence of a fluid-filled borehole, the invaded zone (Fig. 2-6), and adjacent formations in the sedimentary series.

The current traverses these zones, the equipotential surfaces are no longer spheres, and eqs. 3-6 and 3-8 are no longer strictly valid.

A general relationship $R_a = K(V'/I)$ still holds, where: R_a is the apparent (i.e. measured) resistivity; V' is equal to V for the normal array, dV for the lateral and inverse; and K is the geometrical coefficient, for the spacing and configuration used.

After correction for the effects of the borehole (Fig. 3-7 for instance), a good approximation to R_t can be made using the concept of the pseudo-geomet-rical factor (J), considering the invaded and virgin zones to be in series electrically:

$$(R_a)_c = J_i R_{x0} + (1 - J_i) R_t \qquad (3-10)$$

$(R_a)_c$ is the borehole-corrected measured resistivity. J_i is the pseudo-geometrical factor corresponding to the depth of invasion.

To determine R_t, it is necessary to know R_{x0}. The ES sonde has three different electrode arrays to facilitate the solution of the problem: the 16″ and 64″ normals, and the 18′8″ inverse.

N.B. The above equation is only valid if we can consider there to be no parallel current paths, such as a borehole fluid, mud-cake or adjacent strata which are highly conductive relative to the formation.

3.2.3. Measuring point (depth zero)

3.2.3.1. The normal

The depth reference point is taken at the middle of the spacing. For the normal, this corresponds to the mid-point of $\overline{AM}$. Thus for the 16″ and 64″ normals, which share a common A electrode, the measuring points differ. This is corrected optically on the film.

3.2.3.2. The lateral and inverse

The measuring point is at O, the mid-point of $\overline{MN}$ (lateral) or $\overline{AB}$ (inverse).

3.2.4. Radius of investigation

3.2.4.1. The normal

In a homogeneous isotropic medium, referring to eq. 3-6, it is easy to show that 50% of the potential drop towards zero occurs within a sphere of radius $\overline{AM}$, 90% within a radius 10 $\overline{AM}$. So for the 16″-normal, only 10% of the signal comes from the formation beyond 160″ from A.

The radius of investigation (at which an arbitrary 50% of the potential drop has occurred) is therefore equal to twice the spacing $\overline{AM}$ (Fig. 3-4). Similarly, the vertical resolution is 2 $\overline{AM}$.

3.2.4.2. Lateral and inverse

The radius of investigation of the lateral is approximately equal to $\overline{AO}$, with most of the signal deriving from the farthest part of the sphere (Fig. 3-5). For the inverse, it is $\overline{MO}$.

3.2.4.3.

In reality, where the surrounding medium is certainly not homogeneous, the situation cannot be described in terms of simple spheres. The volume of material contributing to the signal depends very much on the geometry and resistivity of the zones of inhomogeneity—the borehole, mud-cake, invaded and virgin formation. However, we can make the follow-

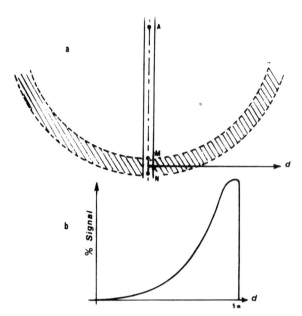

Fig. 3-5. Radius of investigation of the lateral. a. The electrode lay-out in the well-bore. b. The contributions from the various zones, as a function of distance from the sonde axis (from Desbrandes, 1968).

ing general observations about the radius of investigation:

(a) All other factors being equal, the radius of investigation is greater the larger the spacing of the electrode array.

(b) For a given electrode configuration, it will be smaller when the formation is more resistive than the mud.

(c) For a given spacing, a normal device investigates more deeply than a lateral or inverse.

3.2.5. Environmental corrections

We can write:

$$R_a = f(R_m, d_h, R_{mc}, h_{mc}, R_{x0}, d_i, R_t, h, R_s) \quad (3\text{-}11)$$

To obtain R_t, it is necessary to correct the measured R_a for the effects of all the other parameters describing the environment.

This is done using charts, which are available for "borehole correction", "invasion correction", "bed thickness (or shoulder-bed) correction", and so on. To permit covering a wide range of cases on a single chart, resistivity is normalized with respect to R_m (R_t/R_m for instance). Figure 3-6 shows three correction charts for the lateral and normal devices; the upper two for a bed of infinite thickness, the lower one for a thickness $h = 50d_h$. Invasion is assumed to be zero. Note the horizontal axis is scaled in units of $\overline{AO}/d_h$ (inverse) or $\overline{AM}/d_h$ (normal) so that the charts may be used for any electrode spacing.

Figure 3-7 is the simplified correction chart specific to the 18′8″ lateral or inverse, and the 16″ short

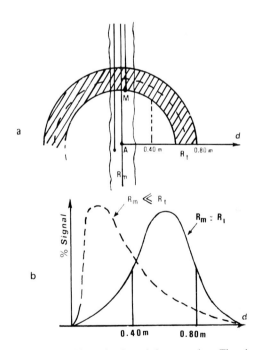

Fig. 3-4. Radius of investigation of the normal. a. The electrode lay-out in the well-bore. b. The contributions from the various zones, as a function of distance from the sonde axis (from Desbrandes, 1968).

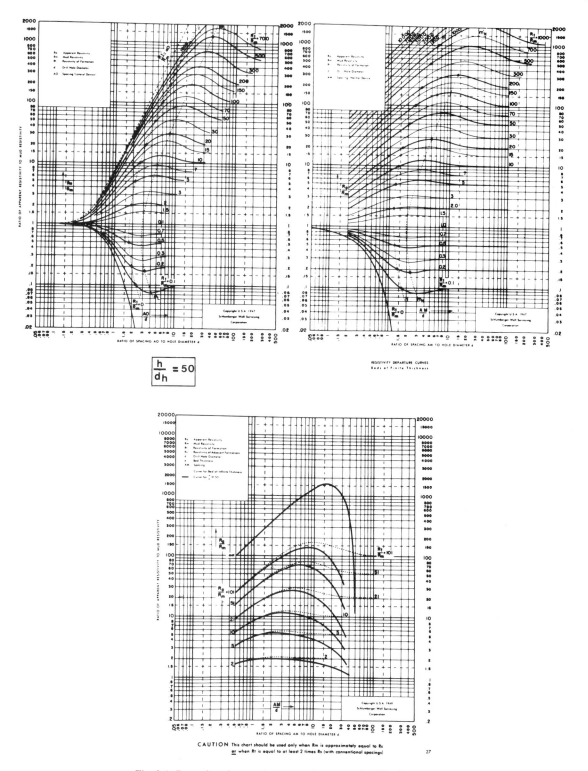

Fig. 3-6. Examples of correction curves (courtesy of Schlumberger).

normal. Invasion is assumed zero, and bed-thickness infinite, i.e. these are solely borehole corrections.

3.2.6. The shape of the apparent resistivity curve

As well as influencing the magnitude of the apparent resistivity, the environmental factors discussed above affect the shape of the log response to a bed of finite thickness.

3.2.6.1. *The normal*

3.2.6.1.1. *Thick resistive beds (h > $\overline{AM}$)*
The curve is symmetric about the middle of the

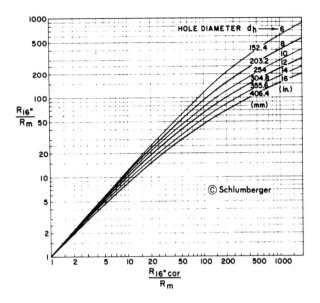

Fig. 3-7. Simplified correction curves (courtesy of Schlumberger).

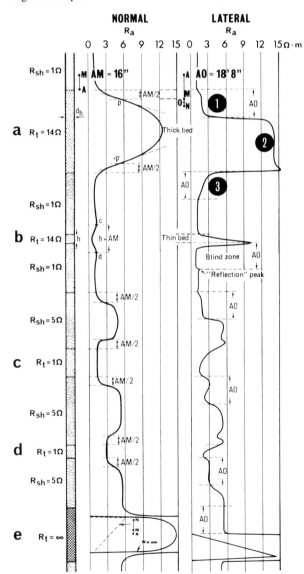

Fig. 3-8. The influence of bed-thickness and resistivities on the shapes of the lateral and normal responses.

bed (Fig. 3-8a). However, the points of inflection (p and p') on the lower and upper slopes give an apparent bed-thickness shorter than true by a distance equal to $\overline{AM}$. The peak value of R_a will depend on borehole and invasion effects (and, to a lesser extent, bed-thickness unless $h \gg \overline{AM}$).

3.2.6.1.2. Thin resistive beds ($h < \overline{AM}$)

The response resembles an apparently conductive bed, with two small resistive shoulders (c and d), Fig. 3-8b. The shoulders are spaced $h + \overline{AM}$ apart. This inverted response is stronger for higher resistivities.

3.2.6.1.3. Very resistive thick beds ($h > \overline{AM}$)

This is the case of anhydrite, salt or tight carbonate beds. If reference electrode N is on the surface, the response is a symmetrical bell-shape (Fig. 3-8e). If N is downhole (on the bridle, 18'8" from M) the curve is triangular in form (dashed line in Fig. 3-8e).

3.2.6.1.4. Conductive beds

Apparent bed thickness (points of inflection) is $h + \overline{AM}$. The response is symmetrical (Fig. 3-8c and d). The thinner the bed, the harder it becomes to distinguish it from the adjacent beds.

3.2.6.2. The lateral and inverse

3.2.6.2.1. Thick resistive beds ($h > \overline{AO}$)

The curve is not at all symmetrical with respect to the bed and can take one of several rather complex forms (Fig. 3-8a is one example). When M and N (lateral) enter the bed (zone I), only a small difference of potential is measured because most of the current is reflected into the adjacent shale. As A enters the bed, a resistivity reading quite close to true value (allowing for borehole and invasion effects) is obtained (zone 2). As M and N leave the bed, the potential difference falls rapidly (after a small increase) to a value a little higher than that of the adjacent bed (zone 3), until A has also passed through.

3.2.6.2.2. Thin resistive beds ($h < \overline{AO}$)

The response is not symmetrical, and the true resistivity is never attained. A maximum is reached as N approaches the lower boundary, because current is flowing in the upper shale and not in the lower, and the potential difference between M and N is at a maximum.

When both measure electrodes are in the lower shale, the thin resistive bed acts like a screen between A and M-N. Consequently, only a weak potential difference exists and $R_a < R_s$ in this "blind zone" (Fig. 3-8b). The small "reflection peak" occurs when A reaches the bottom of the bed and the current is reflected down into shale.

3.2.6.2.3. *Very resistive thick beds (h ≫ $\overline{AM}$)*

Figure 3-8e shows the unsymmetrical triangular response. As in 3-8a, the upper boundary appears displaced downwards by a distance equal to $\overline{AO}$.

3.2.6.2.4. *Conductive beds*

The unsymmetrical response produces an apparent bed thickness too large by a distance $\overline{AO}$ (Fig. 3-8c and d).

Comments: You will appreciate how difficult it can be to delineate bed boundaries with any precision using the non-focused conventional resistivity tool, especially in a sedimentary series containing a sequence of closely spaced resistive and conductive strata.

3.3. FOCUSED LONG-SPACING TOOLS

Focused resistivity tools overcome, to a greater or lesser extent, the following short-comings of the ES-type tools:

(a) In thin beds (*h* of the same order as the spacing) the apparent resistivity is a poor estimate of the true value, because of the influence of the adjacent beds.

(b) The borehole (mud-column) and invaded zone signals are often appreciable.

(c) The available correction charts go some way towards correction for these environmental effects, but are rarely 100% effective.

(d) Bed boundaries are difficult to define precisely.

The current path of a focused tool is constrained to flow in a desired direction. This is achieved using:
–electrodes: laterologs (LL), spherically focused log (SFL);
–coils: induction (IL).

N.B. Do not confuse the laterolog (LL) with the lateral log of the ES!

3.3.1. **Induction—IL**

3.3.1.1. *Principle (Fig. 3-9)*

An oscillator produces an alternating high frequency current in the emitter coil. The resulting electromagnetic field induces co-axial horizontal current loops in the surrounding formation (Foucault effect). Each of these loops generates its own electromagnetic field. The total field is detected by a receiver coil, and the resulting e.m.f. is proportional to the magnetic flux.

Since the emitter coil current is of constant frequency and amplitude, the intensity of the Foucault currents induced in the formation is proportional to

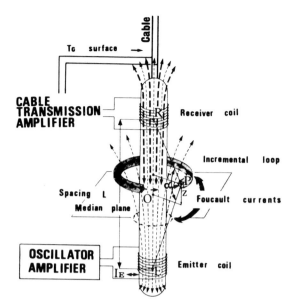

Fig. 3-9. The principle of the induction tool.

the conductivity, and so the resulting e.m.f. induced at the receiver coil.

N.B. The receiver coil current (the "R-signal") is 180° out of phase with the emitter current. Any direct inductive coupling between the two coils will produce an additional receiver current (the "X-signal") that is 90° out of phase. A phase-sensitive detector is required in the down-hole tool to separate the R-signal, coming via the formation, from the unwanted direct X-signal.

Figure 3-10 shows the relative phases of the emitter current, Foucault formation current loops, receiver measure current (R-signal), and direct coupled X-signal.

Induction sondes in fact contain several additional focusing coils (both emitting and receiving). Their purpose is to minimise the effects of the borehole mud and adjacent beds (see 3.3.1.4.).

3.3.1.2. *The geometrical factor*

A first approximation to the response of the tool can be made by considering the idealized case of a perfectly centred tool, using a two-coil system, opposite a homogeneous, isotropic formation of infinite extent (Fig. 3-9).

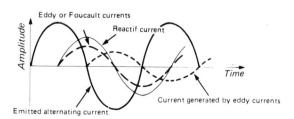

Fig. 3-10. The phase differences between the various induced currents relative to the emitter current.

If we ignore the mutual and self-inductances of the coaxial current loops, the signal from each elemental loop will be proportional to its conductivity, weighted according to a geometrical factor which depends on the position of the loop relative to the tool.

$$e = KgC \qquad (3\text{-}12)$$

where:

e = e.m.f., in volts, at the receiver coil;
g = geometrical factor for the elemental loop;
C = conductivity of the loop;
K = constant for the sonde configuration.

$$K = \frac{\pi \mu^2 f^2 I_E S_E S_R}{L} \qquad (3\text{-}13)$$

where:

f = emitter current frequency (cps);
I = emitter current intensity (amp);
L = coil spacing (m);
S_E, S_R = surface area of one emitter (receiver) coil winding (m^2) = $S_{\text{winding}} \times n_E$ (or n_R);
n_E, n_R = number of windings in the emitter (receiver) coil;
μ = formation magnetic permeability (henry/m).

g can be expressed either as a function of an angle α (Fig. 3-10) describing the position of each loop:

$$g = \frac{\sin^3 \alpha}{2L^2} \qquad (3\text{-}14)$$

or in terms of radial (r) and vertical (Z) distance relative to a plane passing through O (the mid-point of $\overline{ER}$), perpendicular to the sonde axis.

$$g = \frac{L}{2} \frac{r^3}{\left[r^2 + \left(\frac{L}{2} - z \right)^2 \right]^{3/2} \left[r^2 + \left(\frac{L}{2} + z \right)^2 \right]^{3/2}} \qquad (3\text{-}15)$$

N.B. The multiplier $L/2$ is introduced to normalize the integration of g from $Z = -\infty$ to $+\infty$ to a dimensionless unity.

Thus, for the total receiver signal:

$$E = K \iint gC \, dr \, dz \qquad (3\text{-}16)$$

integrating r from 0 to ∞ and Z from $-\infty$ to ∞. We can extend this equation to cover the case of several coaxial cylindrical zones of different conductivities (borehole, invaded and virgin formation...):

$$E = K \left[C_A \iint_A g \, dr \, dz + C_B \iint_B g \, dr \, dz + \ldots \right] \qquad (3\text{-}17)$$

where A, B,... are the coaxial zones.

This becomes:

$$E = K \left[C_A G_A + C_B G_B + \ldots \right] \qquad (3.18)$$

where:

$$G_A = \iint_A g \, dr \, dz$$

$$G_B = \iint_B g \, dr \, dz \qquad (3\text{-}19)$$

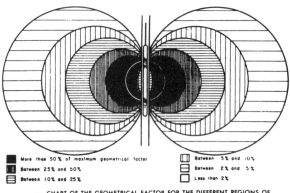

■ More than 50% of maximum geometrical factor	▦ Between 5% and 10%
▨ Between 25% and 50%	▤ Between 2% and 5%
▤ Between 10% and 25%	☐ Less than 2%

— CHART OF THE GEOMETRICAL FACTOR FOR THE DIFFERENT REGIONS OF GROUND AROUND THE COIL SYSTEM.

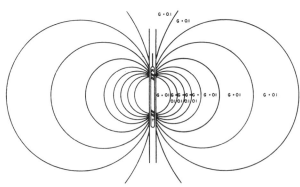

— CHART ILLUSTRATING SECTIONS OF EQUAL GEOMETRICAL FACTOR, 2 COIL SONDE.

Fig. 3-11. Geometrical factors (courtesy of Schlumberger).

and so on.

G_A, G_B,... are the integrated geometrical factors of zones A, B,... respectively. Their magnitudes depend on the dimensions and positions of the respective zones (Fig. 3-11), and they represent the fraction that each zone contributes to the total signal (assuming a uniform conductivity within each zone).

The induction log is a measurement of E/K, which is the apparent conductivity of the formation:

$$C_a = C_A G_A + C_B G_B + \ldots \qquad (3\text{-}20)$$

Note that:

$$G_A + G_B + \ldots = 1$$

From eq. (3-20) you will appreciate that the higher the conductivity of a zone, the stronger will be its contribution to C_a, for a given geometrical factor.

3.3.1.3. *Measuring point—radius of investigation—vertical definition*

The measuring point is the mid-point of the emitter–receiver spacing.

3.3.1.3.1. *Radial characteristics*

The geometrical factor of the homogeneous medium contained between two cylinders (whose axis is that of the sonde) of radius r and $r + dr$, extending

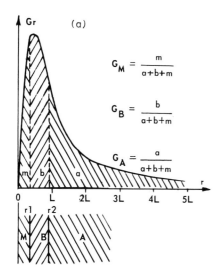

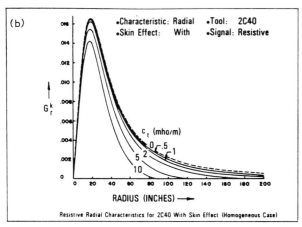

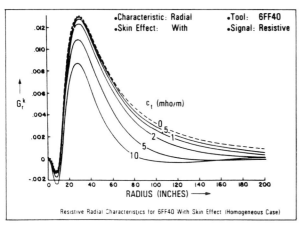

Fig. 3-12. Radial geometrical factors (two-coil sonde).

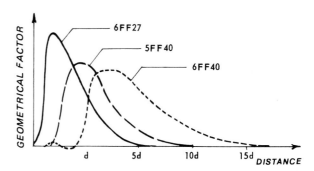

Fig. 3-13. Radial geometrical factors for different induction sondes.

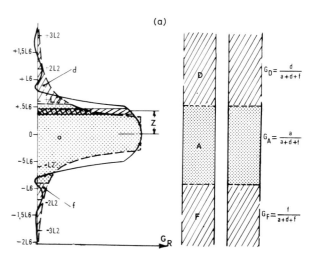

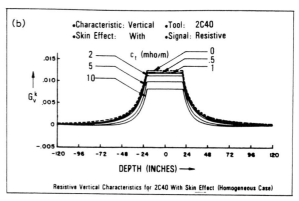

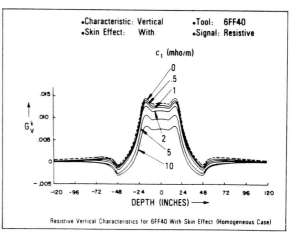

Fig. 3-14. Vertical geometrical factor.

vertically to $+\infty$ and $-\infty$, is written as:

$$G_r = \int_{z=-\infty}^{z=+\infty} g\, dz \qquad (3\text{-}21)$$

G_r is shown in Fig. 3-12 as a function of r for a two-coil system. Such a geometrical factor would correspond to a very thick homogeneous formation, so that shoulder-bed effects were negligible. The major part of the electromagnetic field propagates within a radius $L/4$ to L, L being the coil-spacing. Figure 3-13 demonstrates how a longer spacing produces a deeper investigation (the 6FF40 has 40″ between main emitter and receiver coils).

3.3.1.3.2. *Vertical characteristics*

A vertical geometrical factor, at vertical distance Z from the sonde measure point (Fig. 3-14), is defined as:

$$G_z = \int_{r=0}^{r=+\infty} g \, dr \qquad (3\text{-}22)$$

from which follows:

$$G_z = \frac{1}{2L} \quad \text{when} \quad -\frac{L}{2} < z < +\frac{L}{2}$$

$$G_z = \frac{L}{8z^2} \quad \text{when} \quad z < -\frac{L}{2} \text{ or } z > +\frac{L}{2}$$

A bed of thickness greater than L is therefore hardly affected by the adjacent formations, and the vertical definition is of the same order of magnitude as the spacing.

3.3.1.4. *Focusing*

The sondes represented in Fig. 3-13 are in fact multi-coil systems, the 6 in "6FF40" indicating a total of six coils, for instance. The extra coils are used to focus the signal, i.e. to enhance the response from the virgin formation at the expense of the invaded zone.

The response can be estimated by considering each transmitter-receiver coil pair in turn, and adding their responses algebraically, taking into account polarity and position relative to the measure point.

Focusing achieves several improvements in signal response: (a) better vertical resolution by stronger suppression of the adjacent bed signals; (b) smaller borehole effect, and deeper investigation, by suppression of the unwanted signals (hole and invaded zone); and (c) reduced X-signal component.

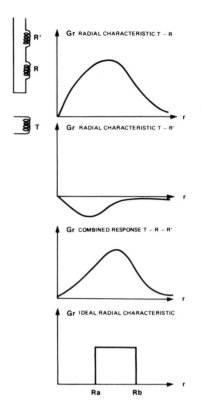

Fig. 3-16. Radial investigation characteristics of a 3-coil sonde (courtesy of Schlumberger).

Figures 3-15 and 3-16 demonstrate the improvements obtained by adding a second receiver coil R′. The response T-R-R′ is sharper than that of T-R at the top of the bed (Fig. 3-15); additional coils would be used to eliminate the undesirable negative response shaded in Fig. 3-15c. In Fig. 3-16 you will see that the near-tool response (invaded zone) has already been considerably reduced.

3.3.1.5. *Deconvolution (6FF40 only)*

Even the six-coil array does not give perfect vertical definition. It is improved further by a computation made automatically on surface while logging,

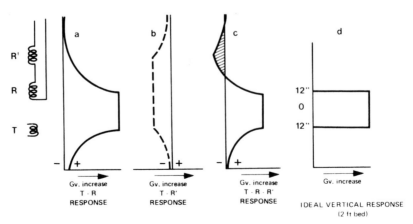

Fig. 3-15. The vertical investigation characteristics of a 3-coil sonde (courtesy of Schlumberger).

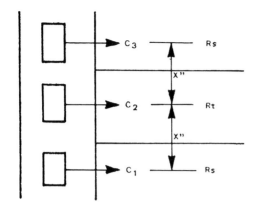

Fig. 3-17. The principle of the shoulder-bed correction.

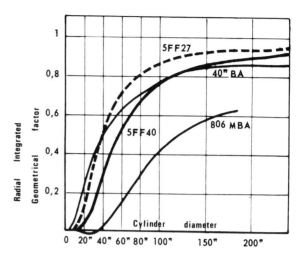

Fig. 3-18. Integrated radial geometrical factors.

called deconvolution, or shoulder-bed correction.

Referring to Fig. 3-17, the apparent conductivity C_2, measured in the middle of the bed, contains some contributions from the adjacent formations. C_1 and C_3 are the measured values at a distance X on either side of level 2 at which C_2 was recorded. (The value of X depends on the type of sonde.) C_2 is corrected to a new value, M, which is a better measurement of the true conductivity at level 2:

$$M = \theta_2 C_2 + \theta_1 C_1 + \theta_3 C_3 \qquad (3\text{-}23)$$

The weights θ_1, θ_2, θ_3 depend on the average "shoulder bed resistivity", SBR. For SBR = 1Ωm, θ_2 is 110%, and θ_1, θ_3 are both minus 5%. The readings C_1 and C_2 are stored in memory until the sonde reaches level 3. At this point, the three readings are combined to compute M, which is then recorded at level 2, after a skin-effect correction (described in 3.3.1.7).

3.3.1.6. *Integrated geometrical factors*

These are shown in Figs. 3-18 and 3-19. They are exactly what their name implies, and they represent the percentage of the total signal coming from within a certain diameter or vertical distance from the sonde. Based on these, correction charts for mud, invasion and bed thickness have been derived.

3.3.1.7. *Skin effect*

This effect, stronger for higher conductivities, is the result of the interactions between the current loops, and is such that the measured conductivity is too low. Each Foucault current loop is not independent of the others, as we had assumed in the simplified treatment in 3.3.1.2. E.m.f.'s are induced as a result of the self- and mutual inductances of the loops, which alter both the amplitude and the phase of the signal reaching the receiver coil. The loops nearer the tool tend, in fact, to reduce the electromagnetic field reaching the deeper formation:

(a) energy is dissipated by current flow in the loops, diminishing the energy available for transmission to the remoter formation.

(b) The out-of-phase electromagnetic fields produced by the Foucault currents nearer the tool destructively interfere with the emitter field propagating farther out.

The formation actually contributing to the received signal is therefore a skin of a certain depth, defined as:

$$\delta = \sqrt{\frac{2}{\omega \mu C}} \qquad (3\text{-}24)$$

where

δ = the depth (m) at which 63% of the emitter signal amplitude has been attenuated. This is the skin depth. (63% corresponds to $(1 - 1/e)$, where $e = 2.718$);

$\omega = 2\pi f$ and f = signal frequency (Hz);

μ = magnetic permeability ($4\pi \times 10^{-7}$ for free space);

C = conductivity (mmho/m).

Figure 3-20 shows how the skin depth is reduced

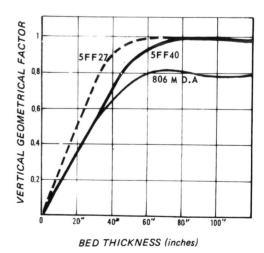

Fig. 3-19. Integrated vertical geometrical factors.

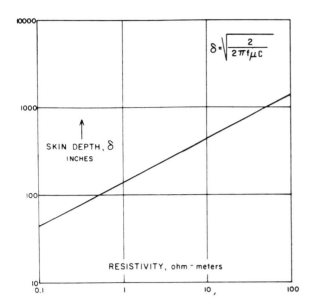

Fig. 3-20. Attenuation of the transmitted signal.

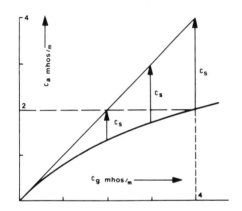

Fig. 3-21. The skin-effect correction (courtesy of Schlumberger).

by increasing formation conductivity.

If C_g is the true conductivity, and C_a is what we measure ($= M$ in 3.3.1.5.), then the skin effect, C_s is:

$$C_s = C_g - C_a$$

C_a always being too small. The magnitude of C_s depends on C_g, (Fig. 3-21). It increases with the conductivity.

A skin effect correction is automatically applied to the conductivity while logging, except in the case of the older tools (5FF27 or 40) where a correction chart (Fig. 3-22) is needed.

3.3.1.8. *Environmental corrections*

Refer to Fig. 3-11 for the general schema of the investigative characteristics of the induction tools.

If we ignore skin effect, we can write:

$$C_a = C_m G_m + C_{mc} G_{mc} + C_{x0} G_{x0} + C_i G_i + C_t G_t + C_s G_s \qquad (3-25)$$

and:

$$G_m + G_{mc} + G_{x0} + G_i + G_t + G_s = 1 \qquad (3-26)$$

The effect of the mud-cake is considered small enough to be ignored. We need to correct C_a for the other contributions in order to obtain C_t.

3.3.1.8.1. *Borehole effect and tool stand-off*

The contribution to C_a from the mud column is $C_m G_m$. In holes drilled with fresh or oil-based mud, or air, C_m is zero or very small and the borehole effect is almost nil. If the hole is extremely large ($> 60''$ diameter) G_m becomes important and the mud-column term may not be negligible, even when C_m is small.

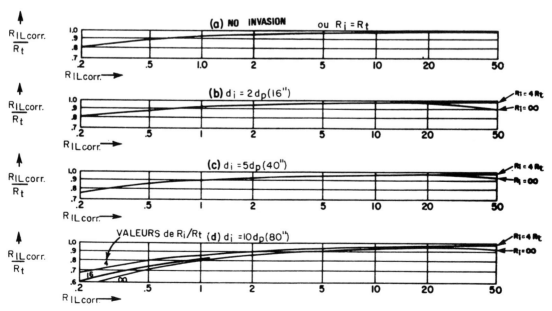

Fig. 3-22. Skin effect correction charts for the 5FF40 (thick beds) (courtesy of Schlumberger).

INDUCTION LOG BOREHOLE·CORRECTION

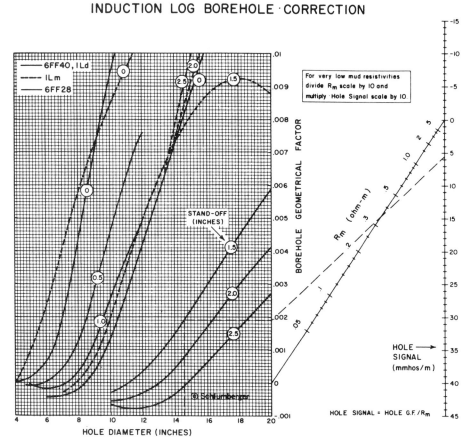

Fig. 3-23. Borehole correction chart for the 6FF40, ILD, ILm, 6FF28 (courtesy of Schlumberger).

The chart of Fig. 3-23 corrects the induction measurement for borehole effect for a range of mud resistivities and hole sizes. Note how even a small borehole signal of a few mmho/m can become significant in highly resistive formations, say above 100 Ωm (= 10mmho/m).

The tool position, or stand-off, is important because if the sonde is not in the centre of the well-bore, the current loops are not axial with the hole. In practice, $1\frac{1}{2}''$ rubber-finned stand-offs are used in holes larger than 8″ diameter.

3.3.1.8.2. Bed thickness and adjacent beds

Figure 3-24 corrects the induction for bed thickness where adjacent bed resistivities are 1Ωm and 4Ωm. From the left-hand chart you can see how relatively conductive adjacent beds strongly perturb the induction reading, especially when h approaches the coil spacing. If the bed is thin and very conductive, correction is necessary because the measured resistivity is too low (Fig. 3-25).

3.3.1.8.3. Invasion

The invaded and flushed zone signals are important when d_i (and therefore G_{x0} and G_0) is large, or when C_{x0} and C_i are high. The problem of correct-

ing for invasion is simplified in the step-profile model, since G_i becomes zero, and we need only to know d_i and C_{x0} (or R_{x0}) in order to make the correction. This will be discussed in more detail in the section on R_t determination in volume 2.

3.3.1.8.4. The annulus

In an oil-bearing formation of high permeability, with very low water saturation and high oil mobility, it is possible for an annulus of high formation water saturation to form between the invaded and virgin zones (Fig. 3-26). This has the effect of reducing the induction resistivity so that an erroneously low R_t is obtained after applying the standard corrections.

3.3.1.9. Summary

The induction is the preferred R_t tool in fresh muds ($R_t < 2.5R_{x0}$, and $d_i > 100''$, but $R_t < 500 R_m$). The induction is the only resistivity tool that can be run in oil-based muds, or air-drilled holes.

It is less effective in saline muds, and very high resistivities ($R_t > 100$ Ωm), because of the large borehole signal in the former case, and the zero-reading uncertainty of ± 2 mmho/m in the latter.

Bed thickness effects become significant in beds

64

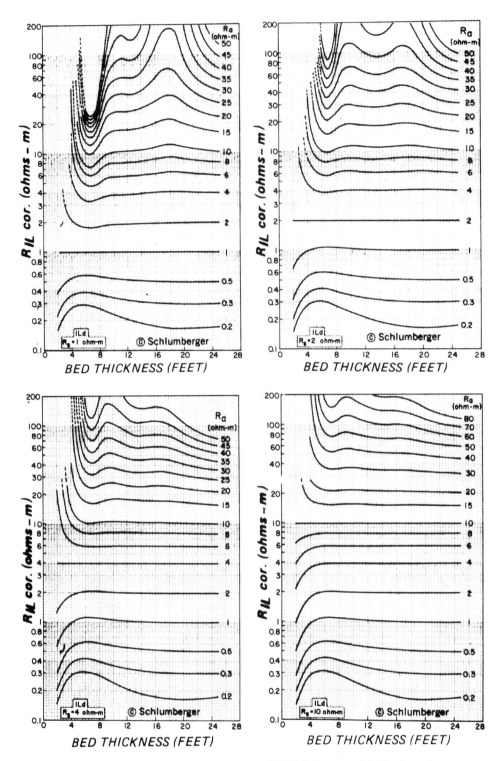

Fig. 3-24. Bed-thickness corrections (6FF40 and 6FF28) (courtesy of Schlumberger).

thinner than 5' (6FF40) or $3\frac{1}{2}'$ (6FF28).

The Dual Induction (DIL) combines a deep reading 6FF40 (ILd) with a medium investigating ILm, and a shallower electrode device (either the LL8 or SFL described later) in one sonde.

3.3.2. The laterologs—LL

3.3.2.1. Principle

The laterolog configuration uses a current electrode A_0 (Fig. 3-27) with remote return. Two symmetrical guard electrodes, A_1 and A'_1 emit focusing

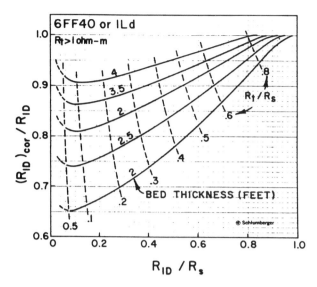

Fig. 3-25. Correction chart for thin conductive beds (courtesy of Schlumberger).

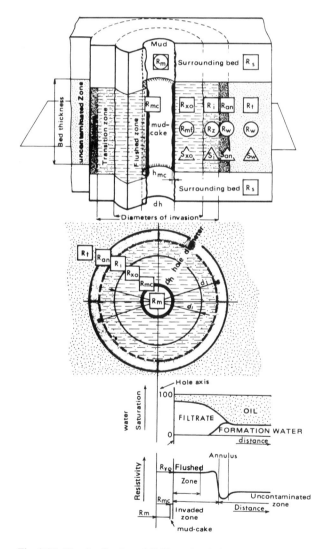

Fig. 3-26. The distribution of fluids and resistivities in an invaded formation with an annulus. □ = formation resistivity; ○ = formation fluid resistivity; and △ = saturation of water. N.B. If there is no transition zone or annulus, $d_i = d_j$.

currents which constrain the A_0 current beam I_0 to flow perpendicularly out into the formation (the guard electrodes in fact establish equipotential surfaces coaxial with the tool, forcing the A_0 current to radiate perpendicularly to the axis). As a result, the borehole and adjacent bed signals are considerably less than those of the normals and laterals. There are several types of laterolog:

3.3.2.1.1. Laterolog—3, LL3 (Fig. 3-27)

A_0 is the main current electrode, which emits a variable current I_0 to a remote return. A_1 and A'_1 are two long guard electrodes, connected together. Their potential V_g is maintained equal to an internal reference potential V_R by a self-adjusting bucking current I_g which flows from A_1-A'_1. The potential V_0 of A_0 is held equal to V_g by varying the main current I_0. A_1-A_0-A'_1 is thus an equipotential surface, and current I_0 can only flow out perpendicularly, as a disc of thickness OO'. The magnitude of I_0 is measured; it is proportional to the formation conductivity, since:

$$KV_0 = RI_0 \qquad (3-27)$$

so that: $I_0 = KV_0C$, where C is the conductivity. The LL3 is thus better suited to conductive formations.

3.3.2.1.2. Laterolog—7, LL7 (Fig. 3-28)

A constant current I_0 is sent from central electrode A_0 to a remote return. A bucking current I_B flows from the symmetrical guard electrode pair A_1-A'_1, so as to maintain the monitor electrode pairs M_1-M'_1 and M_2-M'_2 at the same potential. Equipotential surfaces, shown dotted in Fig. 3-28, are thus created. The current from A_0 is constrained by these equipotentials to flow as shown, in a disc of thickness 00' perpendicular to the sonde. The potential of one of the M electrodes is measured; since I_0 is constant,

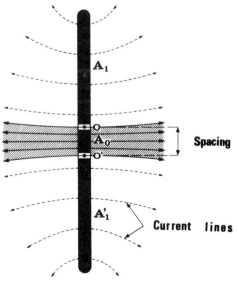

Fig. 3-27. The laterolog-3 (courtesy of Schlumberger).

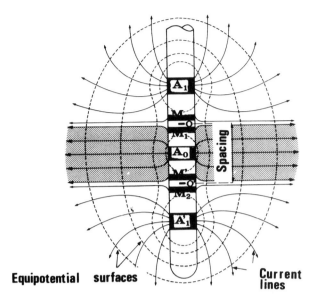

Fig. 3-28. The laterolog-7 (courtesy of Schlumberger).

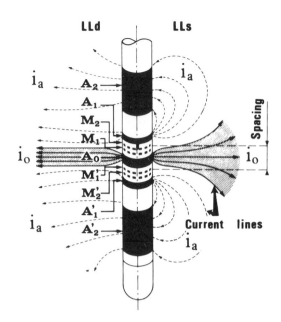

Fig. 3-30. The dual laterolog-9 (courtesy of Schlumberger).

V_m is proportional to the formation resistivity.

The LL7 is more suited to resistive beds than the LL3.

N.B. $\overline{O_1O_2}$ is 00′ in Fig. 3-28. The degree of focusing of I_0 depends on the ratio $\overline{A_1A_1'}/\overline{O_1O_2}$. (See Fig. 3-29, $n = \overline{A_1A_1'}/\overline{O_1O_2}$).

3.3.2.1.3. *Laterolog—8, LL8*

This is a shallow-reading laterolog, similar in design to the laterolog 7, except that the spacings are smaller, and the current returns are much closer.

3.3.2.1.4. *The dual laterolog–DLL (Fig. 3-30)*

This 9-electrode tool makes simultaneous deep and shallow readings (LLd, LLs). Earlier versions were sequential (LL9). The principle of focusing and measurement is similar to that of the LL7. Symmetrically placed electrodes are connected in pairs A_2-A_2', A_1-A_1', etc. The LLd system uses remote (surface) returns

for the main and bucking currents. The LLs system uses the A_2-A_2' electrode pair as the return for the bucking current from A_1-A_1', which reduces the effectiveness of the focusing of I_0.

3.3.2.2. *Measuring point*

For all the laterologs, depth zero is at the centre of A_0.

3.3.2.3. *Vertical resolution*

It is easy to appreciate that bed-thickness definition will be equal to the distance 00′ (refer to Figs. 3-27 to 3-30). This corresponds to the thickness of the I_0 current disc: LL3—12″; LL7—32″; LL8—14″; DLL—24″.

3.3.2.4. *Radius of investigation*

The investigation characteristics of a laterolog depend on the lengths of the guard electrodes, and the resistivity contrast between invaded and virgin zones. The integrated pseudo-geometrical factor, J, defined on the same principles as that of the induction, is shown in Fig. 3-31.

The deep-reading LLd measures closest to R_t, followed by the LL3 and LL7. The shallowest readings, closer to R_{x0}, are obtained with the LLs and LL8.

While the LL3 and LL7 are run alone, the LLd and LLs are comprised in the DLL. The LL8 is combined with the ILd and ILm in the earlier DIL sondes.

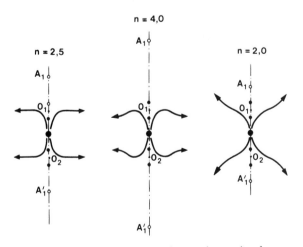

Fig. 3-29. The dependence of focusing on the spacing factor n. (courtesy of Schlumberger).

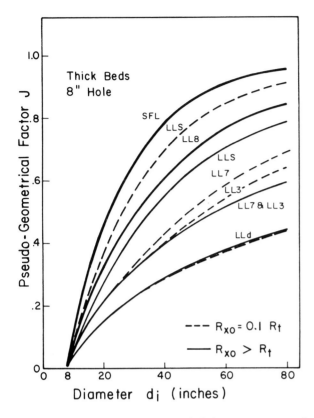

Fig. 3-31. Integrated pseudo-geometrical factors (courtesy of Schlumberger).

3.3.2.5. *Environmental corrections*

The measured apparent resistivity R_a is a function of a number of different parameters:

$$R_a = f(R_m, d_h, R_{x0}, d_i, R_s, h, R_t) \qquad (3-28)$$

The apparent resistivity $(R_a)_c$ obtained by correcting for the borehole (Fig. 3-32) and adjacent beds (Fig. 3-33), is related to the true R_t by:

$$(R_a)_c = R_{x0}J_{x0} + R_tJ_t \qquad (3-29)$$

where $J_{x0} + J_t = 1$.

3.3.2.5.1. *The borehole*

Borehole signal is larger the bigger the hole size, and the more resistant the mud. The laterologs cannot be run in air or oil-based mud having infinite resistivity—here the induction must be used. In conductive muds, Fig. 3-32 shows that the hole signal varies with the ratio R_{LL}/R_m, and hole size.

3.3.2.5.2. *Bed thickness and adjacent beds*

In beds thicker than the spacing of the tool (3.3.2.3.) neighbouring beds exert very little influence on R_a, when R_a/R_s is reasonably close to 1.0. Otherwise, the charts of Fig. 3-33 must be used.

3.3.2.5.3. *Invasion*

When invasion is zero or very shallow (J_{x0} and $J_i = 0$), $(R_a)_c$ is very close to R_t. To correct $(R_a)_c$ for the invaded zone signal, we need to know R_{x0} and J_{x0}. J_i is assumed to be 0 (step profile). This requires two other measurements of resistivity having different depths of investigation. For instance, the LLs and MSFL along with the LLd. This will be covered in detail in the section on the determination of R_t.

3.3.2.5.4. *Delaware and anti-Delaware effects*

When the B and N return electrodes are located downhole rather than on surface (Fig. 3-34), the so-called Delaware effect will be observed below a thick highly resistive bed. An erroneously high, and steadily increasing resistivity is recorded, masking the true resistivity of the formations over a distance of 80ft below the resistive bed. When electrode B enters the bed, the current can only flow in the borehole mud to reach B. Once N enters the bed, it is no longer at zero potential because of the effect of this large current flux. The further N travels into the resistive section, the more negative its potential becomes; this means that V_0 increases relative to N, so the measured R_a increases.

The use of a surface current return reduces this effect (LL7, LLd), but does not always eliminate it completely. It has been found that mutual inductance phenomena can create a Delaware-type effect at moderate current frequencies, even with B on surface. For this reason, low frequencies are used.

Finally, there remains an anti-Delaware effect; an increase in the potential of N, caused again by the current from the sonde. R_a is seen to decrease. This is minimized by using the lower part of the cable armour as N electrode (LLd of the simultaneous DLL).

3.3.2.5.5. *Groningen effect*

This resembles a rather weak Delaware effect, and occurs below a resistive bed through which casing has been run. The sonde currents are effectively shorted directly to surface once they reach the casing, which is equivalent to moving B down to the casing-shoe.

3.3.3. **The spherically focused log—SFL**

This relatively new configuration gives a shallow measurement, with minimal borehole effect even in highly resistive formations, and a good vertical definition.

3.3.3.1. *Principle*

Referring to Fig. 3-35, the tool consists of a central current electrode A_0 and eight symmetrically placed electrodes, connected in pairs M_0-M_0', A_1-A_1', etc.

LATEROLOG 7 BOREHOLE CORRECTION

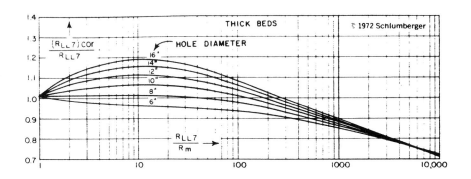

LATEROLOG 3 BOREHOLE CORRECTION

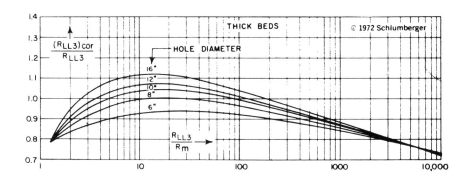

DEEP LATEROLOG* BOREHOLE CORRECTION

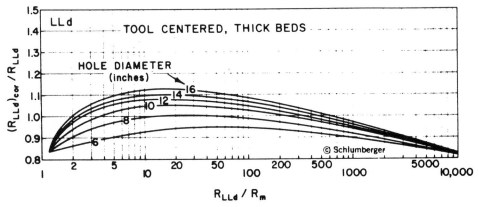

Fig. 3-32. Borehole corrections for the laterologs.

A variable current I_0 is sent from A_0 so as to maintain M_1-M_1' and M_2-M_2' at the same potential. A focusing current I_a flows between A_0 and the A_1-A_1' pair, to hold the potential difference between M_0-M_0' and M_1-M_1' equal to a fixed reference level V_{ref}. The path of I_a is shown in Fig. 3-35. It has the effect of forcing I_0 out into the formation. The equipotential surfaces are roughly spherical. In the figure, the surface B passing from M_0 to M_0', and C from the mid-point O of M_1-M_2 to O' of M_1'-M_2', are at a constant potential difference equal to V_{ref}. The I_0 current intensity is therefore inversely proportional to the resistivity of the formation contained between the two equipotentials, and almost all of the measured signal comes from this volume.

Bringing M_1-M_1' and M_2-M_2' nearer to A_0 will

SHALLOW LATEROLOG BOREHOLE CORRECTION

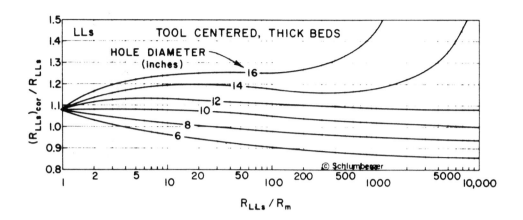

LATEROLOG*-8 BOREHOLE CORRECTION

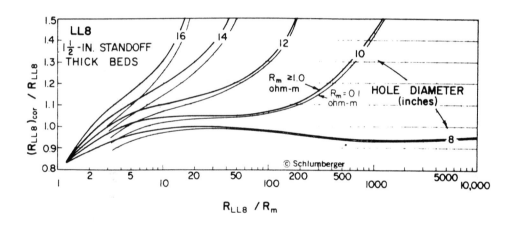

SFL* BOREHOLE CORRECTION

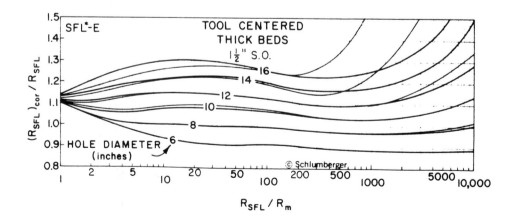

Fig. 3-32. Borehole corrections for the laterologs and SFL (courtesy of Schlumberger).

70

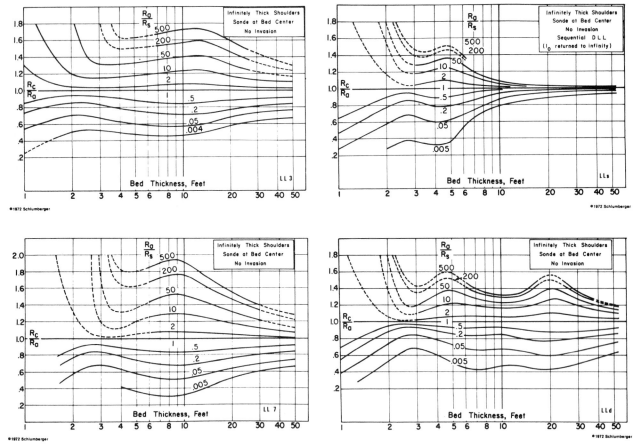

Fig. 3-33. Shoulder-bed corrections (courtesy of Schlumberger).

reduce the depth of investigation, at the expense of an increased borehole effect. The SFL is essentially an R_{x0} measurement.

3.3.3.2. *Measuring point*—at the centre of electrode A.

3.3.3.3. *Vertical resolution*—equal to distance OO', which is 30".

3.3.3.4. *Radius of investigation*—is dependent on the spacing. For the 30" SFL, the pseudo-geometrical factor is shown in Fig. 3-36.

3.3.3.5. *Environmental corrections*

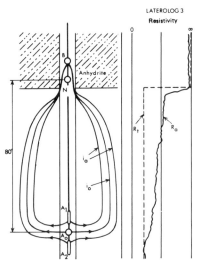

Fig. 3-34. The Delaware effect (courtesy of Schlumberger). N is affected by the currents flowing towards B.

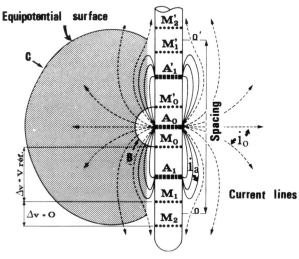

Fig. 3-35. The SFL (courtesy of Schlumberger).

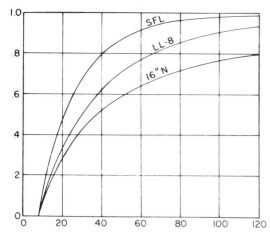

Fig. 3-36. Integrated pseudo-geometrical factor (courtesy of Schlumberger).

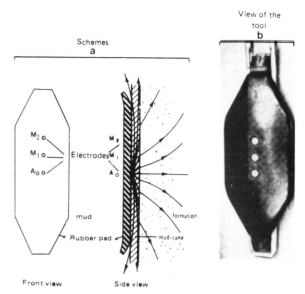

Fig. 3-38. The microlog (courtesy of Schlumberger). a. Measurement principle. b. The hydraulic pad.

3.3.3.5.1. *The borehole*

For hole diameters less than about 10″, the spherical focusing renders the borehole effect almost negligible. Figure 3-37 is a comparison of the influence of the borehole size and mud resistivity on the SFL and 16″ Normal—the SFL is a significant improvement.

3.3.3.5.2. *Adjacent beds*

Their effect on the SFL is much less than on the 16″ Normal.

3.3.3.5.3. *Invasion*

You will appreciate from Fig. 3-36 that the SFL is strongly affected by the invaded zone. It generally reads between R_{x0} and R_t, and is used in combination with deep and micro-resistivity measurements to evaluate R_t.

The induction tool in current use, the ISF, contains an SFL electrode system, which replaces the 16″ Normal system of the earlier IES.

3.4. NON-FOCUSED MICROTOOLS: THE MICROLOG (ML)

3.4.1. Principle

Three electrode buttons are mounted in line on the face of an oil-filled rubber pad (Fig. 3-38). The pad-face is pressed against the borehole wall by a hydraulically controlled spring pressure system.

The electrodes are 1″ apart, and are combined electrically in two configurations which operate

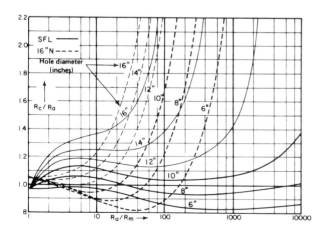

Fig. 3-37. Comparison of borehole effects on the 16″N and SFL (courtesy of Schlumberger).

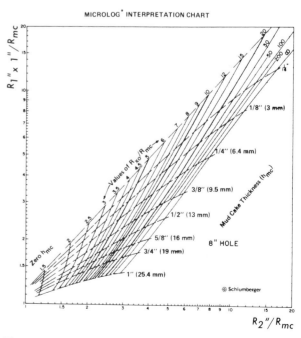

Fig. 3-39. Microlog interpretation chart (courtesy of Schlumberger).

simultaneously—a micro-normal (2″) and a micro-inverse (1″ × 1″), whose principles have already been covered.

3.4.2. Environmental effects

If the pad is in perfect contact with the hole wall, the borehole fluid has no effect on the reading. However, these very shallow measurements are very sensitive to the mud-cake, across which the current must flow opposite permeable, porous beds. Figure 3-39 is a chart which combines the micro-inverse and micronormal measurements to obtain R_{x0}, correcting for the thickness and resistivity of the mud-cake. Note how sensitive the readings are to the thickness of the cake.

In rugose holes, the pad makes irregular contact. There may be mud between the electrodes and the uneven wall of the hole, causing erroneously low measurements.

The vertical definition is very fine, and adjacent beds will only affect the readings when bed thickness is less than a few inches.

3.4.3. Tool response

3.4.3.1. Porous, permeable formations

When mud-cake is present, and $R_{mc} < R_{x0}$, there is a distinctive "positive" separation ($R_{1″ \times 1″} < R_{2″}$) on the micro-log. This is because the depths of investigation of the two tools (and therefore the mud-cake effect) differ. However, negative separations will be observed when $R_{x0} < R_{mc}$, or when invasion is very shallow (with $R_{x0} > R_t$).

3.4.3.2. Shaly formations

There is generally no mud-cake, and the separation is negative.

3.4.3.3. Tight formations

In the absence of mudcake and invasion, the two systems are, in principle, measuring the high R_t. However, the readings are very sensitive to the presence of mud-filled fissures, fine conductive strata, and poor application, and any such anomalies will show up very clearly.

A "mud-log" is usually recorded while running into the hole, with the pad arms retracted. Both measurements will jump erratically as the tool makes intermittent contact with the hole wall on the way down. The minima on the shallow-reading micro-inverse (corresponding to when the tool was farthest from the wall) are taken as indicative of R_m.

3.5. FOCUSED MICROTOOLS

3.5.1. The microlaterolog—MLL

3.5.1.1. Principle

The principle of measurement is the same as that of the LL-7. The electrode array is mounted on an oil-filled rubber pad, as shown in Fig. 3-40. The central A_0 electrode is surrounded by three concentric rings of buttons, constituting the M_1, M_2 and A_1 electrodes respectively. (This amounts to a circular LL-7 array, in fact.)

A constant current I_0 flows from A_0 to the return electrode in the bridle. The variable bucking current holds M_1 at the same potential as M_2, thus forcing I_0 to flow out perpendicularly to the pad-face, at least for a short distance. Since I_0 is constant, the potential V_0 of M_1 and M_2 is proportional to the resistivity of the formation.

3.5.1.2. Measuring point

The measuring point is at A_0. The spacing is the diameter of a circle passing mid-way between M_1 and M_2. Vertical definition is about 1.7″. Depth of investigation is 1–2″.

3.5.1.3. Environmental effects

We can ignore the borehole and mud provided pad contact is good.

3.5.1.3.1. Mud-cake influence
The mud-cake, however, cannot be ignored. The chart of Fig. 3-41 is used to correct R_{MLL} for mud-cake thickness and resistivity. $(R_{MLL})_C$ should be close to R_{x0}.

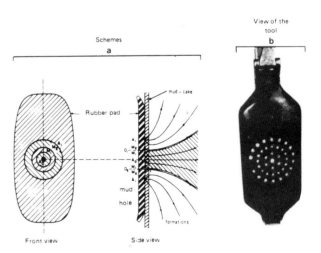

Fig. 3-40. The microlaterolog (courtesy of Schlumberger). a. Measurement principle. b. The hydraulic pad.

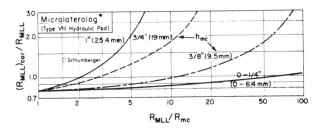

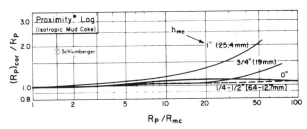

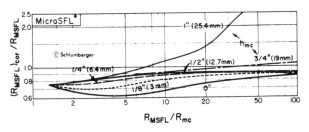

Fig. 3-41. Mud-cake correction charts (courtesy of Schlumberger).

3.5.1.3.2. *Bed thickness influence*

For beds thicker than 1.7″, adjacent beds have no appreciable effect on R_{MLL}.

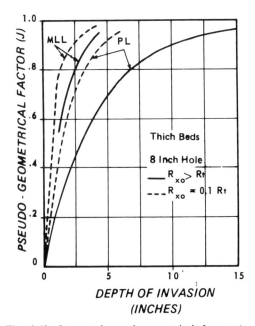

Fig. 3-42. Integrated pseudo-geometrical factors (courtesy of Schlumberger).

3.5.1.3.3. *Invasion influence*

The virgin zone does not affect R_{MLL} if invasion is deeper than a few inches (Fig. 3-42). Very shallow invasion will produce a reading somewhere between R_{x0} and R_t. In this case, we can compute R_{x0} by combining several different resistivity measurements, and using the step-profile model.

3.5.2. The micro-proximity log (PL)

3.5.2.1. *Principle*

This device works on the same principle as the LL-3, but uses rectangular electrodes with a common centre (Fig. 3-43), mounted on a solid rubber pad.

3.5.2.2. *Environmental effects*

The mud-cake signal is weaker (Fig. 3-41) by virtue of the larger depth of investigation (Fig. 3-42). The measurement is therefore, except in the case of deep invasion (> 10″), influenced to some extent by R_t. Computation of R_{x0} requires other measurements to fix d_i and R_t.

3.5.3. The micro-SFL (MSFL)

This is a small-scale SFL array, mounted on a flexible rubber pad (Figs. 3-44 and 3-45). It has two advantages over the MLL and PL: (a) it is less sensitive to the mud-cake than the MLL, and reads shallower than the PL (Fig. 3-41); and (b) it can be combined with other tools, such as the DLL and DIL, while the MLL or PL require a separate run (and therefore more rig-time, and risk of sticking).

3.5.4. The high-resolution dipmeter (HDT)

This is dealt with in chapter 19, but is included here for completeness.

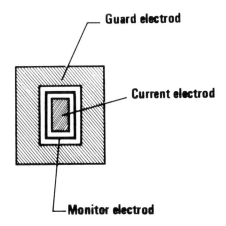

Fig. 3-43. The microproximity electrode configuration.

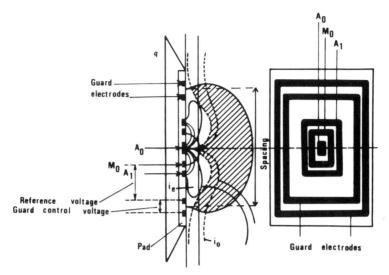

Fig. 3-44. The micro-SFL (courtesy of Schlumberger). a. The electrode configuration. b. The measurement principle.

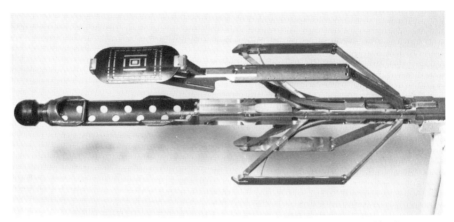

Fig. 3-45. The MSFL sonde, showing the pad and caliper arms (courtesy of Schlumberger).

3.6. CONCLUSIONS

3.6.1. Geological factors which influence resistivity

We have so far considered "environmental" effects–unwanted signals which distort the measurement. We will now look at the significance of some of the geological aspects of the formation.

3.6.1.1. *Rock composition*

(a) The nature of the solid part of the rock: as a general rule, the solid grains and cement are infinitely resistive. However, some clays are more or less conductive, and there exist some conductive minerals (graphite, hematite, metals, sulphides, etc).

(b) The nature of the fluids in the pores: hydrocarbons are infinitely resistive; the resistivity of the water will depend on the dissolved salts present.

(c) The porosity and saturation.

3.6.1.2. *Rock texture*

The shape, size, sorting, orientation and arrangement of the grains determine the porosity, its distribution, the sizes of the pores and channels, which in turn influence the resistivity by:

(a) Tortuosity; which means the "a" and "m" terms in the relationship between F and ϕ.

(b) Permeability; variations will alter the invasion profile and saturations in both the virgin and invaded zones.

(c) Microscopic anisotropy; by which tools employing horizontal current flow will read differently from those using predominantly vertical flow. The manner in which clay or other conductive minerals are distributed is also important (laminar, dispersed, etc).

Fissures or fractures, if filled with conductive mud or water, present preferential current paths to some of the tools, and each will be affected differently. The "a" and "m" factors will therefore change.

3.6.1.3. *Dips*

In beds which are not in a plane perpendicular to the axis of the hole, the apparent resistivity may be in error. This is a form of anisotropy.

3.6.1.4. *Sedimentary structure, depositional environment, the geological sequence*

The thickness of the beds, their internal organisation, and the nature of the adjacent strata, (that is, the anisotropy on a macroscopic scale) is dependent on the depositional history.

3.6.1.5. *Temperature*

The effect of temperature on the resistivity of the fluids was discussed in 1.3.3.2.2. A bottom-hole temperature measurement must accompany a resistivity log, if it is not already known.

3.6.1.6. *Pressure-compaction*

Pore-pressure is a function of several factors—including tectonic forces, overburden, compaction. Compaction is also important in its effect on the arrangement of the grains (packing), and the porosity.

The invasion profile is determined in part by the difference between hydrostatic pressure in the wellbore, and formation pore pressure.

3.6.1.7. *Summary*

To summarize, the resistivity of a rock depends principally on the water saturation, and porosity. Further geological information can be deduced, using the measurement in conjunction with other logs, and making certain reasonable hypotheses.

3.6.2. **Applications**

They will be developed in volume 2 but can be summarized as follows.

3.6.2.1. *Hydrocarbon saturation*

The major application of resistivity logging is in the determination of water (and hence hydrocarbon) saturation, in the flushed and virgin zones.

R_{x0} can be measured fairly directly with micro-tools.

R_t can rarely be measured directly because of the omni-present borehole and invaded zone signals, and the effects of adjacent beds. Under favourable conditions, the focused deep-reading tools (LLd, ILd) can be read very closely to R_t even without correction.

We have already seen the various correction charts available to deal with borehole and adjacent beds. Correction for the flushed zone usually requires at least three resistivity measurements of different depths of investigation to solve for the three unknowns; d_i, R_{x0}, R_t (we assume the step-profile of section 2.2.1.2.). We have:

$$(R)_{a\,cor} = R_{x0}G_{x0} + R_iG_i + R_tG_t$$

$$G_{x0}(d_i) + G_i(d_i) + G_t(d_i) = 1$$

which becomes:

$$(R_a)_{cor} = R_{x0}G_{x0} + R_tG_t$$

if we eliminate the transition zone, so that:

$$G_{x0}(d_i) + G_t(d_i) = 1$$

G_{x0}, G_t correspond to the invasion diameter, d_i. (G is used for induction, J is reserved for laterolog device.)

Given three different resistivity measurements, or by making assumptions about R_{x0} or d_i, R_t can be estimated by calculation or charts. This is covered in the next volume on quantitative interpretation.

3.6.2.2. *Porosity*

If we assume a formation to be clean and 100% water-bearing, its porosity can be calculated, since:

$$F = R_0/R_w = R_{x0}/R_{mf}$$

and:

$$F = a/\phi^m$$

3.6.2.3. *Water resistivity*

Given the porosity, we may obtain R_w or R_{mf} in a clean, wet formation:

$$R_{wa} = R_t/F$$

$$R_{mfa} = R_{x0}/F$$

R_{wa} and R_{mfa} in clean, 100% water-wet zones correspond to R_w and R_{mf}. Hydrocarbons will cause $R_{wa} > R_w$ and $R_{mfa} > R_{mf}$. A good estimate of R_w and R_{mf} can therefore be obtained by picking the minima on R_{wa}, R_{mfa} (in clean beds) over a log interval.

If, in addition, S_w is known by another means (such as TDT), fluid resistivity may be calculated from Archie's equation in hydrocarbon-bearing zones.

3.6.2.4. *Correlation between wells*

This is the earliest application of resistivity logging. Similarity between the resistivity profiles in two wells may often quite reasonably be assumed to indicate the occurrence of the same geological sequence. It must be borne in mind that depths will certainly not be exactly the same (dips, faults between the wells), and that some strata appearing in one well may "pinch-out" before reaching the other.

3.6.2.5. *Compaction, fracturing*

The degree of compaction with depth in sand-shale series has been studied with resistivity, density and acoustic logs. As already mentioned, both deep and micro-tools can serve as fracture indicators, though confirmation is needed from other logs.

3.6.2.6. *Sedimentology, lithology*

The resistivity can shed light on geological properties such as shaliness, tightness, texture, sequential evolution, etc.

3.6.2.7. *Bed thickness*

The micro-tools give fine definition of thin boundaries, down to a few inches thickness. (This resolution is surpassed by the HDT.) Large depth scales are often used with micro-tools (1/40), to bring out the fine detailing. Of the deep-reading tools, the laterologs have the best vertical definition.

3.7. REFERENCES

Doll, H.G., 1949. Introduction to induction logging. J. Pet. Technol., 1 (6).

Doll, H.G., 1950. The microlog. Trans. AIME, 189.

Doll, H.G., 1951. The laterolog. J. Pet. Technol., 3, 11.

Doll, H.G., 1953. The microlaterol. J. Pet. Technol. 5, 1.

Duesterhoeft, Jr. W.C., 1961. Propagation effects in induction logging. Geophysics, 26 (2).

Duesterhoeft, Jr. W.C., Hartline, R.E. and Sandoe Thomsen, H., 1961. The effect of coil design on the performance of the induction log. J. Pet. Technol., 13 (11).

Dumanoir, J.L., Tixier, M.P. and Martin, M., 1957. Interpretation of the induction electrical log in fresh mud. Trans. AIME, 210.

Gianzero, S. and Anderson, B., 1981. A new look at skin effect. Trans. SPWLA, 22d Annu. Log. Symp., paper I.

Guyod, H., 1944–1945. Electrical well logging: fundamentals. Oil Weekly, 114 (7/8/44) and 115 (25/12/45).

Guyod, H., 1964. Factors affecting the responses of laterolog-type logging system (LL3 and LL7). J. Pet. Technol. 16(2).

Hamilton, R.G., 1960. Application of the proximity log. Trans. SPWLA, 1st Ann. Log. Symp.

Moran, J.H. and Kunz, K.S., 1962. Basic theory of induction logging and application to study of two-coil sondes. Geophysics, 27, 6.

Owen, J.E. and Greer, W.J., 1951. The guard electrode logging system. Petrol. Trans. AIME, 192.

Schlumberger, C.M., and Leonardon, E., 1934. Some observations concerning electrical measurements in anisotropic media, and their interpretation, p. 159. Electrical coring: a method of determining bottom-hole data by electrical measurements, p. 237. A new contribution to subsurface studies by means of electrical measurements in drill holes, p. 273. Petrol. Trans. AIME, 110.

Schlumberger Well Surveying Co., 1947–1949. Resistivity departure curves, document 3.

Schlumberger Well Surveying Co., 1950. Interpretation Handbook for resistivity logs. Document 4.

Schlumberger Well Surveying Co., 1951. Interpretation of induction logs.

Schlumberger Well Surveying Co., 1962. Induction log correction charts.

Schlumberger Ltd. 1970. The dual laterolog. Technical report.

Schlumberger Ltd, 1971. The ISF/SONIC, Technical report.

Tixier, M.P., 1956. Fundamentals of electrical logging - Microlog and microlaterolog.- In: Fundamentals of logging. Univ. Kansas, Petroleum Eng. Conf. 2 and 3, April, 1956.

Tixier, M.P., Alger, R.P. and Tanguy, D.R., 1959. New developments in induction and sonic logging. Soc. Petrol. Eng. of AIME, Pap. 1300G, October 1959.

Tixier, M.P., Alger, R.P., Biggs, W.P. and Carpenter, B.N., 1963. Dual induction-laterolog: A new tool for resistivity analysis. Soc. Petrol. Eng. of AIME, Pap. 713, 6–9 October 1963.

Tixier, M.P., Alger, R.P., Biggs, W.P. and Carpenter, B.N., 1965. Combined logs pinpoint reservoir resistivity. Petrol. Eng., Feb-March.

Witte, L. de, 1950. Relations between resistivities and fluid contents of porous rocks. Oil Gas J., Aug. 24.

Witte, A.J. de and Lowitz, D.A., 1961. Theory of Induction Log. Trans. SPWLA, 2nd. Annu. Log. Symp.

Woodhouse, R. and Taylor, P.A., 1974. The varying radial geometrical factors of the induction log. Paper no. SPE4823 presented at the SPE-European Spring Meeting.

4. THE SPONTANEOUS POTENTIAL—SP

Schlumberger discovered this phenomenon in 1928. An electrical potential difference exists, spontaneously, between an electrode in the borehole, and a remote reference electrode on surface. This potential varies from formation to formation, usually within the range of a few tens or hundreds of millivolts (mV), measured relative to the level in shales (Fig. 4-1).

This phenomenon has been studied by several authors: Doll (1948 and 1950), Wyllie (1949, 1951), Gondouin et al., (1957), Gondouin and Scala (1958), Hill and Anderson (1959), Gondouin et al., (1962). The following text summarizes their work and conclusions.

The spontaneous potential opposite a formation can be attributed to two processes involving the movement of ions:

(a) Electrokinetic (electrofiltration or streaming) potentials, symbol E_k, develop while an electrolyte penetrates a porous, non-metallic medium.

(b) Electrochemical potential, symbol E_c, is present when two fluids of different salinities are either in direct contact, or separated by a semi-permeable membrane (such as a shale).

4.1. THE ORIGIN OF THE ELECTROKINETIC POTENTIAL

Electrokinetic potential appears when filtrate, from the drilling mud, is forced into the formation under the differential pressure between the mud column and the formation.

Filtrate flow takes place and an electrokinetic potential is produced: (a) across the mud-cake in front of the permeable formation; (b) across the permeable formation being invaded; and (c) across the shale beds.

Hill and Anderson (1959) have studied the streaming potential across mud-cake (Fig. 4-2), and several years before, Wyllie (1951) proposed the following relation for E across a mud-cake:

$$E_{kmc} = K_1(\Delta p)^y \qquad \text{,in mV} \qquad (4-1)$$

y ranging from 0.57 to 0.90.

The existence of electrofiltration potentials across shales has been demonstrated in laboratory experiments by Gondouin and Scala (1958) and by Hill and Anderson (1959) (Fig. 4-3). Field data reported by Gondouin et al. (1962), also confirm the existence of shale electrokinetic potentials. Gondouin and Scala (1958) give for E across shales:

$$E_{ksh} = K_2 \Delta p \qquad \text{,in mV} \qquad (4-2)$$

with $K_2 \simeq -0.018 \, (R_{mf})^{1/3}$.

The magnitude of streaming potentials depends on several factors:
– differential pressure across the medium, Δp;
– effective resistivity of the moving filtrate, R_{mf};
– dielectric constant of the filtrate, D;
– zeta potential, ζ;
– viscosity of the filtrate, μ.

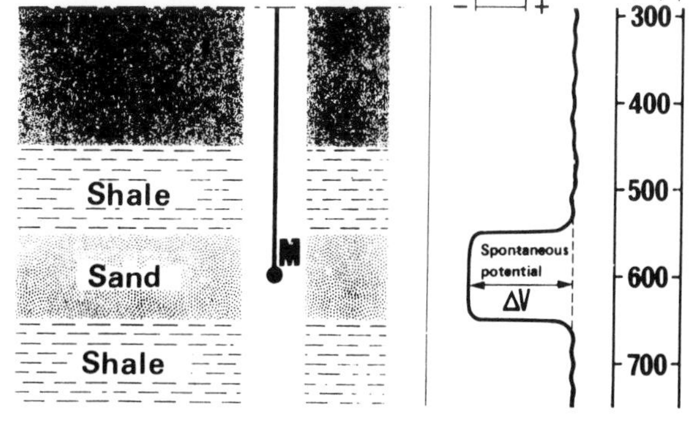

Fig. 4-1. The principle of measurement of the SP.

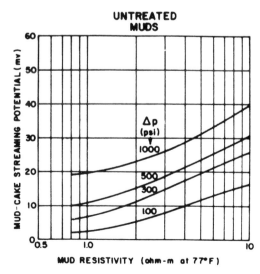

Fig. 4-2. Streaming potential across mud-cake (from Hill and Anderson, 1959).

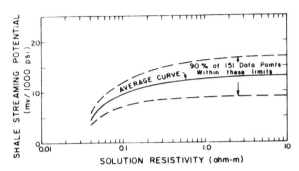

Fig. 4-3. Laboratory values of shale streaming potential as found by Hill and Anderson, 1959.

Because of the very low permeabilities of mud-cakes (10^{-2} to 10^{-4} md), most of the pressure drop between the mud column and the formation occurs across them. The remaining differential pressure across the formation is very small. Usually, then, practically all the electrokinetic potential in a permeable bed is that produced across the mud-cake. However, an electrokinetic potential of the same polarity * is also produced across the shale beds.

The SP deflection opposite a permeable bed being measured from the shale base-line, any contribution of the electrokinetic potential to the SP will be the difference between the contributions of mud-cake E_{kmc} and shale E_{ksh}. In most cases no significant residual electrokinetic potential appears, the two electrokinetic potentials mentioned being very close. There are some exceptional cases:

(a) Depleted formations, in which the pressure on the formation fluids is less than the original pressure, whereas the pressures within the shales may still be close to the original value. So the balance between the two electrokinetic potentials is no longer present.

(b) Special muds. Some muds have been reported to produce systematically some residual streaming potential (Hill and Anderson, 1959; Gondouin et al., 1962).

(c) Heavy muds. They produce larger pressure differentials than normally encoutered (Gondouin et al., 1962; Althaus, 1967).

(d) Low permeability formations.

When the formation is of very low permeability ($k < 5$ md) mud-cake does not build up at all rapidly, and the mud is in direct contact with the rock. So, all the differential pressure is applied to the formation.

If the mud-filtrate is fresh, the movement of fluid into the formation (allbeit a very slow one) can produce a negative SP of some tens of mV. Should pore pressure be greater than mud pressure, the electrokinetic SP will be positive corresponding to an outflow of formation fluid.

When a solid (rock components) and a solution (pore fluid) are in contact, a potential is set up at the interface. This potential may result from:

(a) Preferential adsorption of specific ions from the solution onto the surface of the solid.

(b) Ionization of the surface molecules of the solid; for instance, in clays, light cations (Na^+, K^+, H^+, ...) tend to pass into solution leaving a negative charge on the clay platelets.

(c) The solid lattice itself may have a charge. In clays a negative charge results, owing to substitution of Mg^{2+} for Al^{3+}, or by imperfections in the lattice (absence of an atom).

The polarity of the resultant potential depends on the nature of the solid. For sandstones and limestones the surface of the solid becomes negative. Clay surfaces become strongly negative.

The charges on the solids are compensated by the accumulation of an equivalent number of counter ions of opposite sign in the adjacent solution. Negative charge on the solid attracts positive ions from the solution to its surface (Fig. 4-4).

The first few molecular layers may constitute a hydrodynamically immobile zone termed the "fixed layer". A little farther from the surface is the "diffuse layer" in which there is also an excess of positive ions, but in which the liquid is movable. Beyond this, the bulk solution is finally reached where the liquid is at electrical neutrality.

The potential profile from the interface into the liquid is shown in Fig. 4-4, assuming the potential of the neutral bulk liquid as zero. The potential at the inner edge of the "diffuse layer" is known as the zeta potential, ζ. Because of the greater ionic density near the pore wall the liquid there has a greater electrical conductivity than the bulk liquid in the pore, even though ionic mobility may be less. The resulting

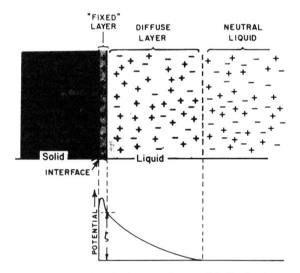

Fig. 4-4. Schematic of charge and potential distribution near solid-liquid interface.

---

* Negative on the hole (high-pressure) side, positive on the formation (low-pressure) side.

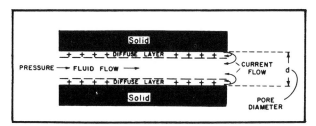

Fig. 4-5. Generation of electrokinetic potential in a cylindrical pore by application of differential pressure across pore.

increase of conductance across the pore over that expected from the conductivity of the bulk liquid is defined as the surface conductance. It is considered to be relatively independent of pore-water conductivity.

Figure 4-5 shows how the streaming potential is produced in a cylindrical pore. A differential pressure applied across the pore causes fluid flow through it. Liquid also flows in the diffuse layer carrying along with it its excess positive ions. This movement of charge creates a potential across the length of the pore. This potential results in current flow back through the pore, and a resultant potential drop through the liquid in the pore.

Lynch (1962) proposed the relation:

$$E_k = \frac{\zeta D \Delta p R_w}{4 \pi \mu} \qquad (4\text{-}3)$$

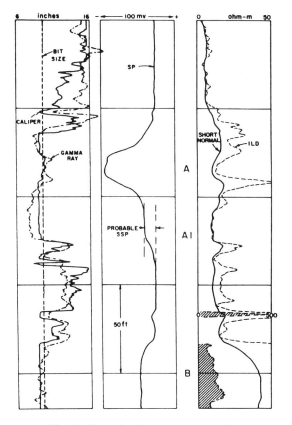

Fig. 4-6. Example of streaming potential.

where:

ζ is the zeta potential, mentioned above;
D is the dielectric constant of the water;
Δp is the pressure differential;
R_w is the water resistivity;
μ is water viscosity.

Figure 4-6 gives an example of streaming potential appearing in front of a low-permeability formation.

Generally, the electrokinetic potential across the mud-cake is compensated by that across the shale. So one neglects the residual streaming potential and one assumes that the spontaneous potential measured is only related to the electrochemical potential. However, the streaming potential must be recognized before interpreting the SP deflection for R computation. Several features by which one can recognize it are summarized below:

SP values are abnormally large, resulting in erroneously low values of R_w (compare with R_{wa} or known values of R_w).

Often SP deflections change quickly with depth giving a "peaky" appearance of the curve.

There may be diminished or no mud-cake, due to low formation permeability (see caliper).

The mud is fresh and formation water is fresh or brackish.

The formation is relatively clean, since shaliness drastically decreases streaming potential (see gamma ray or natural gamma ray spectrometry log).

There must be some porosity (see porosity logs: density, neutron, sonic).

A large differential pressure exists (check mud density).

In these cases it is good to have other methods to estimate R_w.

4.2. THE ORIGIN OF THE ELECTROCHEMICAL POTENTIAL

The electrochemical potential, E_c, is the sum of two potentials described below.

4.2.1. Membrane potential

An e.m.f. develops when two electrolytes of different ionic concentrations (mud and formation water) are separated by shale (Figs. 4-7 and 4-9). The clay minerals in shale are made up of lattices of atomic Al, Si, O etc. O^{2-} ions occupy the outer extremities, and there is a net negative charge on the lattice. As a result, Na^+ ions from solution are attracted and allowed to pass through the shale, while Cl^- ions are repelled. Na^+ ions thus migrate between the two solutions, with a net influx from the more saline to the less. The resulting imbalance of ions on either side of the shale constitutes an e.m.f., the membrane

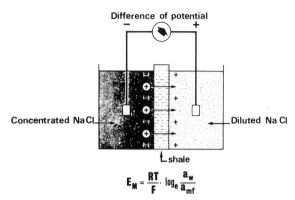

Fig. 4-7. The membrane potential (after Desbrandes, 1968).

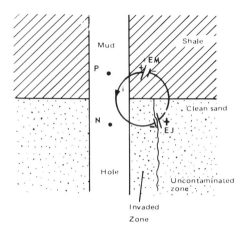

Fig. 4-9. The principle of the electrochemical SP. $E_C = E_M + E_J$.

potential (E_m).

$$E_m = K_3 \log(a_w/a_{mf}) \tag{4-4}$$

where:

$$K_3 = 2.3 \, RT/F \tag{4-5}$$

(the term 2.3 takes care of the conversion from naperian to base-10 logarithms);

R = ideal gas constant (8.314 joules/°K-mol);
T = absolute temperature (in degrees Kelvin);
F = Faraday constant (96, 489 coulombs/mol);
a_w = the ionic activity of the formation water;
a_{mf} = the ionic activity of the mud filtrate;
K_3 = 59.1 mV at 25°C (77°F).

For all intents and purposes, E_m can be considered constant with time, the actual rate of ionic diffusion being very small (a few mA of current) and the volumes of electrolytes very large.

N.B. A small membrane potential also occurs across the mud cake, between the filtrate and the mud. This is usually negligible.

4.2.2. Liquid junction or diffusion potential

This is an e.m.f. established at the contact of the mud filtrate and connate water in an invaded formation (Figs. 4-8 and 4-9). Ions Na$^+$ and Cl$^-$ have different mobilities, and therefore do not diffuse at the same rates between the two electrolytes. Na$^+$ tends to be less mobile because of its larger size and an affinity for water molecules. Consequently, a build-up of Cl$^-$ ions occurs in the weaker solution, and of Na$^+$ ions in the more saline solution. This imbalance causes an e.m.f., E_j, described by:

$$E_j = K_4 \log(a_w/a_{mf}) \tag{4-6}$$

The coefficient $K_4 = 11.6$ mV at 25°C (77°F). It is approximated by:

$$K_4 = 2.3 \frac{v-u}{v+u} \frac{RT}{F} \tag{4-7}$$

where:
v = mobility of Cl (67.6×10^{-5} cm^2/sV);
u = mobility of Na (45.6×10^{-5} cm^2/sV).

4.2.3. The electrochemical potential, E_c

This is the sum of the membrane and liquid junction potentials:

$$E_c = E_m + E_j = K \log \frac{a_w}{a_{mf}} \tag{4-8}$$

where:

$K = K_3 + K_4 = 60 + 0.133T \, (T°F)$.

E_c is arbitrarily defined as zero opposite a shale.

4.3. IONIC ACTIVITY, CONCENTRATION, AND RESISTIVITY

The ionic activity depends on the valency of the cations. Formation waters are usually predominantly solutions of sodium, calcium or magnesium salts. The activity of such a solution is:

$$a_f = \left(a_{Na} + \sqrt{a_{Ca} + a_{Mg}} \right)_f \tag{4-9}$$

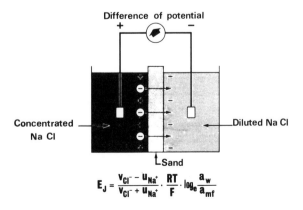

Fig. 4-8. The liquid junction potential (after Desbrandes, 1968).

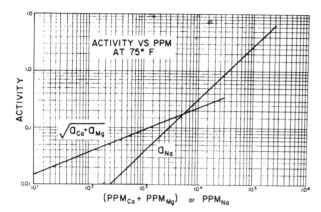

Fig. 4-10. The relationship between ionic activity and concentration (courtesy of Schlumberger).

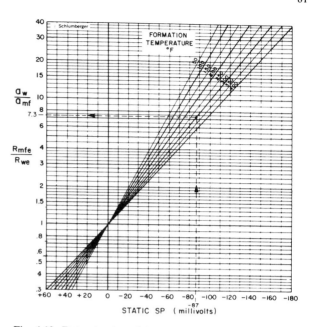

Fig. 4-12. Determination of R_{we} from the SP (clean formation) (courtesy of Schlumberger).

The chart in Fig. 4-10 can be used to determine the activity from the concentrations. In Fig. 4-11 the resistivity of the solution is drawn as a function of ionic activity. Note that it is inversely proportional to the activity below about 80 Kppm total salinity, so that:

$$E_c = K \log \frac{R_{mf}}{R_w} \qquad (4\text{-}10)$$

At higher concentrations (or more precisely, at higher activities, such as might result from divalent ions), the relationship is more complicated. We simplify the problem by defining equivalent resistivities such that:

$$E_c = K \log \frac{(R_{mf})_e}{(R_w)_e} \qquad (4\text{-}11)$$

Equivalent and true resistivities are related as in Fig.

4-13. The SP (E_c) is now a simple logarithmic function of $(R_{mf})_e/(R_w)_e$ as shown in Fig. 4-12.

4.4. THE STATIC SP

In Fig. 4-14a, the spontaneous potential is drawn for the idealized case where no current is allowed to flow in the borehole. The SP is constant across the permeable bed, dropping sharply to zero in the shales (dashed curve). This is the static SP, or SSP, given

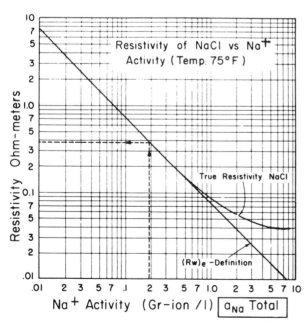

Fig. 4-11. The relationship between ionic activity and resistivity (courtesy of Schlumberger).

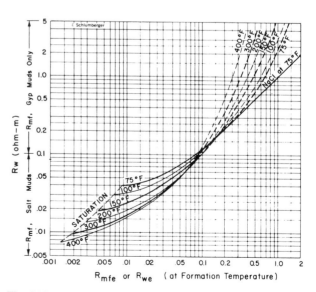

Fig. 4-13. Determination of R_w from R_{we} (courtesy of Schlumberger).

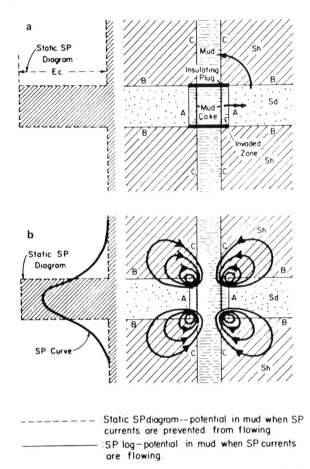

Static SP diagram--potential in mud when SP currents are prevented from flowing.

SP log--potential in mud when SP currents are flowing.

Fig. 4-14. Schematic representation of potential and current distribution in and around a permeable bed (courtesy of Schlumberger).

by:

$$SSP = -K \log \frac{(R_{mf})_e}{(R_w)_e} = -E_c \qquad (4-12)$$

The SSP, being defined as zero in the shale, all deflections are measured relative to the SHALE BASE-LINE. Consequently, it is of the utmost importance to correctly define the shale base-line. In reality, a current loop is completed through the mud, and an SP profile similar to that of Fig. 4-14b is produced.

Note that the SP builds up rather gradually from the base-line and that the peak value is less than the SSP. This is referred to as the pseudo-static SP or PSP, and is strongly dependent on bed thickness, the resistivities of the invaded and virgin zones, depth of invasion, shaliness, and so on. These factors also affect the shape of the peak.

4.5. AMPLITUDE AND SHAPE OF SP PEAKS

4.5.1. Hole diameter

The SP is reduced by an increase in hole size, all other factors remaining unchanged.

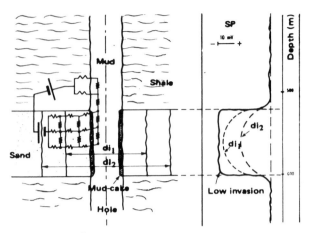

Fig. 4-15. The effect of invasion on the SP (after Desbrandes, 1968).

4.5.2. Depth of invasion

As Fig. 4-15 shows, the SP decreases as the invasion deepens.

4.5.3. Bed thickness

The SP is in fact a measurement of the rise and fall of electrical potential produced by current flow in

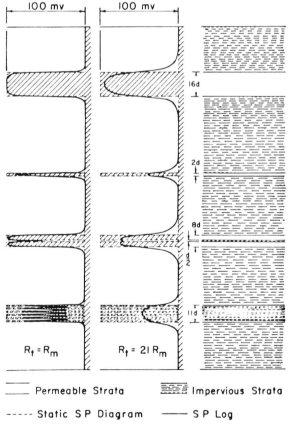

Fig. 4-16. The effect of bed thickness and resistivity on the shape of the SP (from Doll, 1948).

the mud. Its amplitude approaches the SSP value only when the resistance to current flow offered by the formation and adjacent beds is negligible compared with that of the mud (Fig. 4-16). This condition is met only when the bed is thick. So, in general, the SP is reduced in thin beds.

4.5.4. Formation resistivities

The virgin zone—as R_t/R_m increases, the SP deflection decreases—and the bed boundaries are less sharply defined (Fig. 4-16). The presence of hydrocarbons therefore attenuates the SP. Neighbouring beds—the SP increases with R_s/R_m; invaded zone—the SP increases with R_{xo}/R_m.

Charts of the form of Fig. 4-17 are necessary to correct the PSP for these variables, which can have significant effects on the log reading. "SP_{corr}" in the figure is a good estimate of the PSP. More precise correction charts have been proposed by Segesman (1962).

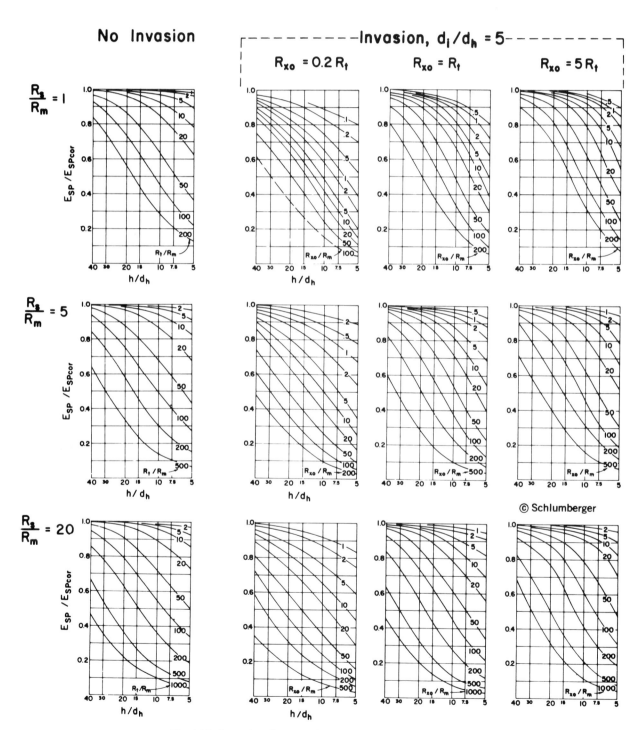

Fig. 4-17. Correction charts for the SP (courtesy of Schlumberger).

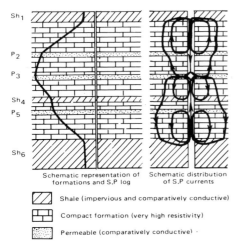

Schematic representation of formations and S.P log   Schematic distribution of S.P currents

[/////] Shale (impervious and comparatively conductive)

[▭▭] Compact formation (very high resistivity)

[▭▭] Permeable (comparatively conductive) ·

Fig. 4-18. The SP in highly resistive formation (courtesy of Schlumberger).

4.5.5. Tight formations

Because of the high resistivity, the SP current tends to flow deeply in a tight formation (that is, over a large cross-sectional area). The shale bed provides the only conductive path back to the mud and thence to the permeable strata (Fig. 4-18) (note that current return to the mud directly along a permeable bed is prevented by an opposing e.m.f.).

Opposite the impermeable section, the current flowing in the mud is constant, so the potential gradient is uniform. This produces an SP of constant slope, (the straight-line sections in the figure). The equivalent circuit of Fig. 4-19 depicts the mud as a potentiometer, and the SP electrode as the slider measuring the potential at each point along it.

The boundaries of tight beds are difficult to locate precisely, but are always associated with a change of slope, or curvature, on the SP log, with a concave side towards the shale base-line (Fig. 4-19), except where the boundary is with a shale, in which case the curvature is convex.

4.5.6. Shale base-line shifts, and drift

An SP base-line shift is not a common occurrence (excepting the mechanical or electrical shift occasionally imposed by the logging engineer to keep the SP curve from going off-scale!). It occurs where two zones of different connate water salinities are separated by a shale that is not a "perfect" cationic membrane, or where the salinity changes within a single bed.

Figure 4-20 is a hypothetical example of this. Some care is necessary in estimating the amplitude of the PSP, since it must be measured relative to a base-line. Shale C, for instance, being an imperfect membrane, should not be used as the base-line for the SP in D. Rather use shale E, which indicates that the SP in D is at least + 44 mV (E may not be perfect either), and in F at least − 23 mV.

Figure 4-21, from Doll (1948), shows two shale base-line shifts. In these cases the types of shales are sufficiently different to shift the base-line. The level where the shift occurs constitutes a horizon marker which can correspond to an unconformity.

SP base-line drift occurs on most logs, and is the result of a gradual electrode polarization in the mud. It is seen as a gentle and fairly steady creeping of the shale base-line towards more negative values with the passage of time.

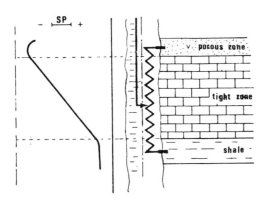

Fig. 4-19. The SP in a thick, infinitely resistive bed.

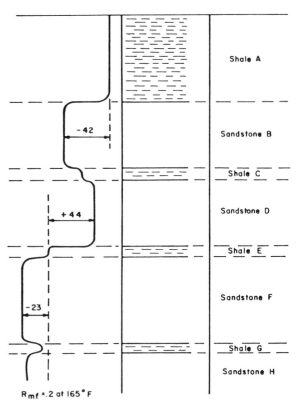

Fig. 4-20. SP baseline shifts (courtesy of Schlumberger).

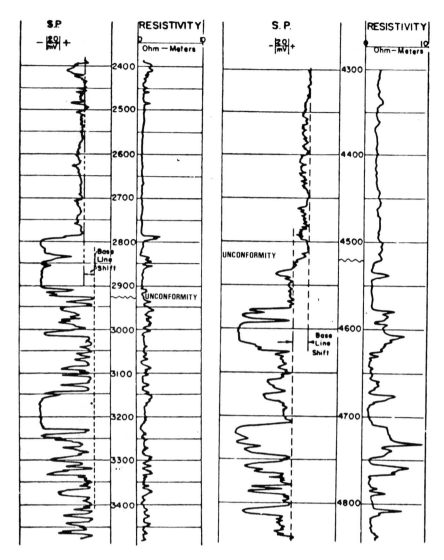

Fig. 4-21. Field examples of SP base line shift and unconformity (from Doll, 1948).

4.5.7. Irregular invasion profile

In a reservoir of high vertical permeability, where the density of the filtrate is lower than that of formation water, there is an upward gravity segregation of filtrate as illustrated in Fig. 4-22. There is an extensive lateral spreading at the top, and a total absence of filtrate may result at the bottom of the zone (Fig. 4-23).

At the bottom, the liquid junction potential is non-existent. However, there is a membrane potential (E_{mc}) across the mud-cake, between the (saline) connate water and the (fresher) mud, in opposition to the membrane potential set up by the shale (E_m). The two may even cancel out if the mud-cake is particularly "active". A slight loss of SP deflection may occur at the top of the bed, where invasion is very deep.

The characteristic "saw-tooth" SP is produced where thin impermeable layers cross a thick permeable bed (Fig. 4-22).

4.5.8. SP anomalies

4.5.8.1. Bimetallism

Exposed metallic parts of logging tools are potential sources of weak DC currents (current will flow between two metal surfaces in contact with an electrolyte). Where a degree of electrical continuity exists inside the tool between these metals, the electrical circuit is completed via the mud column and a bimetallic current will flow. This current can affect the SP electrode (one of the bimetallic sources can be the electrode itself). In highly resistive formations, these currents are confined to the well-bore and may drastically alter the SP (Fig. 4-24). The use of a remote bridle electrode avoids this problem.

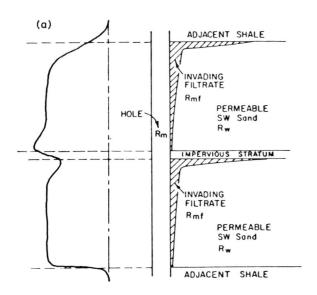

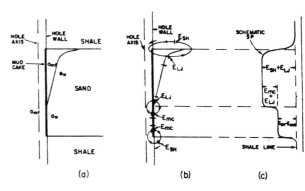

Fig. 4-23.a. Distribution of liquids in a sand having no invasion in bottom. b. The e.m.f.'s driving of the SP current. c. The schematic SP curve (courtesy of Schlumberger).

Fig. 4-22. Segregation of mud-filtrate in a permeable reservoir—the saw-tooth SP. a. Theory. b. Field examples.

4.5.8.2. *Crosstalk*

An imbalance in the measuring currents of certain resistivity tools may affect the SP in a similar manner to bimetallism. If this is suspected, the SP should be recorded with the measuring currents shut off.

4.5.8.3. *Magnetism*

The appearance of a sinusoidal or periodic noise on the SP curve while the tool is moving, probably arises from the inadvertant magnetisation of part of the logging cable drum, sprocket, drive chain, etc. (This often arises from perforating jobs where high DC currents are used). The SP deflections are correct but noisy (Fig. 4-25). Solutions are filtering of the log, or demagnetisation of the truck.

4.5.8.4. *Stray currents, erratic noises*

The remoteness of the SP fish is important. It must be at zero potential. Stray currents may arise from the fish being too close to the drilling rig or platform or an overhead power cable. Movement of the sea water around the SP fish (offshore logging), or telluric currents can also be troublesome.

4.6. GEOLOGY AND THE SP

4.6.1. Composition of the rock

4.6.1.1. *Matrix minerals*

Excepting coal, metallic sulphides and conductive minerals, the matrix composition has no effect on the SP. (A coal seam often produces an SP similar to that of a permeable formation.)

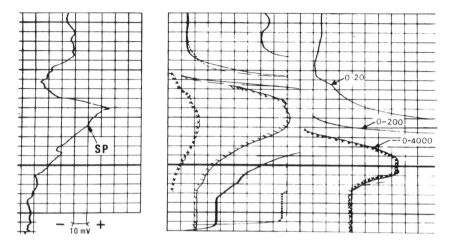

Fig. 4-24. Example of bimetallism (courtesy of Desbrandes, 1968).

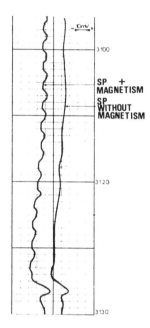

Fig. 4-25. Example of magnetism (courtesy of Desbrandes, 1968).

4.6.1.2. *Shales*

The influence of a shale on the SP depends not only on how much is present, but on its manner of repartition.

4.6.1.2.1. *Laminar shales*

The important factors in this case are the relative thicknesses of the shales and permeable beds, and the resistivities R_t, R_s and R_m (Fig. 4-16, lower). We can write:

$$PSP = SSP(1 - V_{sh}) \qquad (4-13)$$

Thus the PSP can serve as an indicator of shaliness, since:

$$V_{sh} = 1 - \alpha$$

where $\alpha = PSP/SSP$ is the SP reduction coefficient.

4.6.1.2.2. *Dispersed shales*

Dispersed shales impede the movement of Cl^- ions, and strongly attenuate the SP. This attenuation is a function of the amount of shale present in the pores. The diffusion of anions can be reduced to zero at a certain shale percentage. Beyond this the process is reversed, creating an opposing e.m.f.

4.6.1.2.3. *Structural shales*

As long as the sand grains constitute a continuous "phase", the effect of structural shale particles will be rather similar to that of dispersed shales. Once the continuity is lost (sand grains surrounded by shale), no SP can develop.

4.6.1.3. *The fluids*

The predominant factor here is the difference in salinities of the filtrate and the connate water:
– $R_w < R_{mf}$ (more saline formation water); in this case we have:

$$R_{mf}/R_w > 1$$

so: $\log(R_{mf}/R_w) > 0$
and: the SSP < 0, i.e. a negative deflection (towards the left of the shale base-line).
– $R_w > R_{mf}$ (more saline mud filtrate), we have by the same reasoning:
SSP > 0, i.e. a positive deflection.

N.B. We assume the dissolved salts are of the same nature in both solutions. Where this is not the case, a negative SP can occur even though R_{mf} is less than R_w. Refer to the dashed curves in Fig. 4-13 for R_w, and the solid curves for R_{mf}.

– $R_w = R_{mf}$, then the SP = 0.
– An increase in hydrocarbon saturation has the effect of attenuating the (positive or negative) SP, as described in 4.5.4.

4.6.2. Rock texture

There is no known direct correlation between permeability, porosity, and the amplitude of the SP. Certainly, both must be present for an SP to develop, and in that they influence invasion (4.5.2.) and fluid segregation (4.5.7), we can say that the sedimentology of the formation is one of the factors affecting the SP.

Shale distribution is discussed in 4.6.1.2.

4.6.3. Temperature

K is temperature-dependent, and is approximated by:

$$K = 60 + 0.133T \; (°F) \tag{4-14}$$

4.6.4. Pressure

The pressure difference between the well-bore and the formation influences: (a) the diameter of invasion; and (b) a possible electrokinetic SP.

4.6.5. Depositional environment, sequential evolution

These factors determine the bed thicknesses, and for a given bed, the nature of its "environment", i.e. its adjacent (4.5.3 and 4.5.4).

The actual shape of the SP curve (Shell geologists have devised a classification including bell-, funnel- and egg-shaped, etc.) depends on the sequential evolution within the bed, since this will manifest itself as an "evolution" of spontaneous potential (upward coarsening, upward fining, and so on).

4.7. APPLICATIONS

(a) Identification of porous, permeable beds (provided the other conditions discussed are met).

(b) Determination of R_w.

(c) A lithological indicator: shales, coal-seams, shale fraction estimation from: $V_{sh} \leqslant 1 - PSP/SSP$.

(d) An analysis of the facies and grading, based on the curve shape.

(e) Correlation.

(f) An indication of possible hydrocarbon saturation in shaly sands, including the presence of a gas-oil contact.

These applications will be discussed in volume 2.

4.8. REFERENCES

Althaus, V.E., 1967. Electrokinetic potentials in South Louisiana Tertiary sediments. The Log Analyst, 8 (1).

Chombart, L.G., 1957. Reconnaissance d'évaluation des formations par diagraphies éléctriques et nucléaires. Technip, Paris.

Dickey, P.A., 1944. Natural potentials in sedimentary rocks. Trans. AIME, 155.

Doll, H.G., 1948. The S.P. Log: theoretical analysis and principles of interpretation. Trans. AIME, 179 (11).

Doll, H.G., 1950. The S.P. log in shaly sands. Trans. AIME, 189.

Gondouin, M. and Scala, C., 1958. Streaming potential and the S.P. log. J. Pet. Technol., 10 (8).

Gondouin, M., Tixier, M.P. and Simard, G.L. (1957). An experimental study on the influence of the chemical composition of electrolytes on the S.P. curve. J. Pet. Technol., 9 (2).

Gondouin, M., Hill, H.J. and Waxman, M.H., 1962. A field streaming potential experiment. J. Pet. Technol., 14 (3).

Hill, H.J. and Anderson, A.E., 1959. Streaming potential phenomena in S.P. log interpretation. Trans. AIME, 216.

Kerver, J.K. and Prokop, C.L., 1957. Effect of the presence of hydrocarbons on well logging potential. Soc. Petrol. Eng. of AIME, Los Angeles Meeting, October 1957.

Lynch, E.J., 1962. Formation Evaluation. Harper's Geoscience Series, Harper and Row, New-York.

McCardell, W.M., Winsauer, E.O. and Williams, M., 1953. Origin of the electric potential observed in wells. J. Pet. Technol., 5 (3).

Mallenburg, J.K., 1973. Interpretation of gamma ray logs. The Log Analyst, 14(6).

Mounce, W.D. and Rust, W.M., Jr., 1943. Natural potentials in well logging. J. Pet. Technol., 6.

Pied, B. and Poupon, A., 1966. SP base-line shift in Algeria. 7th Ann. Symp. S.P.W.L.A.

Pirson, S.J. (1947) A study of the self potential curve. Oil Gas J. Oct. 4.

Poupon, A. and Lebreton, F. Diagraphies nucléaires. Inst. franç. Pétr. ENSPM, ref. 2432.

Schlumberger, C.M. Schlumberger, C. and Leonardon, E.G., 1934. A new contribution to subsurface studies by means of electrical measurements in drill holes. Trans. AIME, 110.

Segesman, F., 1962. New SP correction charts. Geophysics, vol. 27, n 6, part 1.

Segesman, F. and Tixier, M.P., 1959. Some effects of invasion on the S.P. curve. Trans. AIME, 216.

White, C.C., 1946. A study of the self potential curve. Oil Gas. J., 14 Dec.

Witte, L. de 1950. Experimental studies on the characteristics of the electrochemical potential encountered in drill holes. AIME, Los Angeles, 12-13 October 1950.

Worthington, A.E. and Meldau, R.F., 1958. Departure curves for the self-potential log. J. Pet. Technol., 10 (1).

Wyllie, M.R.J., 1949. A quantitative analysis of the electrochemical component of the S.P. curve. J. Pet. Technol., 1 (1).

Wyllie, M.R.J., 1951. An investigation of the electrokinetic component of the S.P. curve. J. Pet. Technol., 3 (1).

Wyllie, M.R.J., Witte, A.J. de, and Warren, J.E., 1958. On the streaming potential. Trans. AIME, 213.

5. AN INTRODUCTION TO NUCLEAR LOGS

5.1. DEFINITION

Measurements classified as nuclear include:
- Natural gamma radio-activity;
- Compton scattered gamma radiation and photo-electric effect by gamma absorption when the formation is bombarded with gamma rays from a source;
- the flux of thermal or epithermal neutrons resulting from bombardment of the formation with high-energy neutrons;
- gamma ray emission resulting from this bombardment.

Table 5-1 summarizes the principles of the various nuclear techniques, and how the radiation is measured.

In the following sections are discussed the characteristics common to all these techniques.

5.2. RECORDING CAPABILITY

Since both neutrons and gamma rays can pass through steel, radioactive logs can all, in principle, be run in cased-hole as well as open-hole. However, the quality of the measurement of more shallow-reading tools is in some cases poorer through casing.

5.3. STATISTICAL VARIATIONS

All radioactive phenomena are random in nature. Count-rates vary about a mean value, and counts

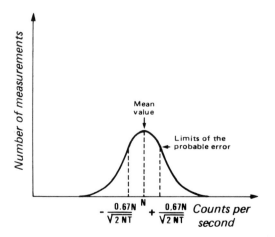

Fig. 5-1. Distribution of the measurements about the mean value (from Dewan, 1954).

TABLE 5-1

Principle of the different nuclear logging techniques

Measurement principle	Schlumberger tool
Natural radio-activity	
Natural γ-radioactivity (total)	SGT(GR)
Natural γ-ray spectroscopy	NGT
Radiation induced by neutron bombardment	
Spectroscopy of γ-rays emitted by the activation of oxygen:	GST IGT
$^{16}O(n,p)^{16}N(\beta -)$ $^{16}O*(\gamma 6.13MeV)^{16}O$	
or the activation of a wide range of elements	HRS
Spectroscopy of γ-rays emitted from fast neutron interactions (mainly inelastic):	GST IGT
eg $^{16}O(n,n'\gamma 6.13MeV)^{16}O$, $t_{1/2} = 1.7 \times 10^{-11}$ sec	
Spectroscopy of γ-rays emitted by thermal neutron capture	GST IGT
Epithermal neutron density at a fixed distance from a high-energy neutron source	SNP CNT
Thermal neutron density at a fixed distance from a high-energy neutron source, by:	
(a) detection of neutrons themselves	CNT
(b) detection of gamma-rays arising from capture of the thermal neutrons	GNT
Decay rate of the thermal neutron population. The thermal neutron density is sampled at two different times between neutron bursts, by detecting the capture gamma-rays.	TDT
Compton scatter of gamma rays	
Gamma rays emitted from a source are scattered by the formation. The count-rate of those reaching the detector is measured.	FDC LDT*
Photo-electric absorption	
Low-energy gamma absorption by the photo-electric effect	LDT *
Decay of proton spin precession induced by a strong magnetic field	
Protons are caused to precess about the local Earth magnetic field by a strong DC magnetic pulse. This precession decays with a time characteristic of the formation fluids.	NML

* Note that the LDT makes both measurements.

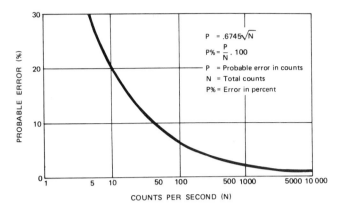

Fig. 5-2. Propable error as a function of a count-rate.

must be accumulated over a period of time and averaged in order to obtain a reasonable estimate of the mean. The accumulation period is the time constant. The estimate of the mean will be more precise the higher the count-rate and the longer the time-constant.

Figure 5-1 shows that for random statistical variations, 50% of the sampled data will lie within $\pm 0.67 N/\sqrt{2NT}$ (the probable error) of the mean, where T is the time constant, and N is the mean count-rate. This would correspond, for instance, to taking a stationary measurement at one depth, and inspecting the data over a long period of time. T is typically 1–6 seconds for most logging purposes, and N is of the order of $10-10^3$ cps. Figure 5-2 shows how the relative probable error, expressed as a percentage of mean counts, decreases as the count-rate increases ($T = \frac{1}{2}$ s).

The time constant is chosen to keep the standard deviation less than 2.5% of the mean reading wherever practical, which implies that NT should be over 400. T must be increased where count-rate level is low.

With the CSU system, the measurement time is fixed and samples are taken every six inches of tool travel. The sampling time chosen is sufficient to maintain the standard deviation at less than 2.5% when the standard logging speed of 900 ft/hr (= 0.25 ft/s) is used.

If lower statistical variations are required, a slower logging speed may be adopted; this will increase N, hence the product NT in the above equation, reducing the probable error.

A small anomaly on a radioactive log may correspond to a physical occurrence, or simply to statistical variation. The standard deviation expresses the probability that the anomaly is correlatable.

Log analysts commonly speak of the uncertainty or repeatability of radio-active measurements. These terms are not rigorously defined, but allude to the probable error.

Remarks. The repeatability between several measurements over the same interval cannot be perfect, owing to the statistical fluctuations which are always present, and are especially important at low count rates. The stacking (averaging) of several runs on the same interval will improve the measurement by decreasing the statistical uncertainty.

5.4. DEAD-TIME

The "dead-time" of a detector system is the short period following the detection of a count or pulse, during which no other counts can be detected. It is a recovery period for the crystal or tube, and the measuring circuitry.

For example, in a scintillation detector system, the incoming gamma ray produces a photon in the crystal, which is detected and amplified by a photo-multiplier. The output voltage pulse is generally some 2 μs wide. However, the width of this pulse by the time it reaches the surface panel depends very much on the down-hole processing circuitry and the logging cable. As an extreme example, consider that the (negative) P-M output pulse creates a drop in the cable driver supply voltage from 125 V to 120 V, that the voltage source has a resistive impedance of 200 ohm, and the logging cable a capacitance of 14 μF. Then the line voltage will require a time $T = RC = 200 \times 14.10^{-6} = 28$ μs to recharge, even though the initial pulse was only 2 μs wide. (With the introduction of CSU the dead-time related to the logging cable is eliminated.)

Figure 5-3 shows the probable error in recorded count-rates for a variety of dead-times.

The probability that the phenomenon will occur is:

$$P = \Delta N/N = nt \qquad (5-1)$$

where:

ΔN is the number of counts lost due to the dead-time (cps)

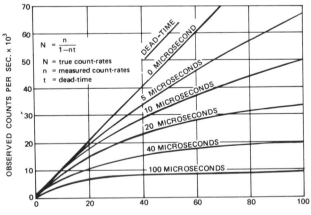

Fig. 5-3. Correction for dead-time.

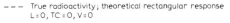

--- True radioactivity; theoretical rectangular response
  L = 0, TC = 0, V = 0

——— Response with L = 4', TC = 0, V ≠ 0 (effect of length)

—·—· Response with L = 0, TC = 7, V = 720'/h (effect of the
  time constant and the velocity)

········ Response with L = 4', TC = 7, V = 720'/h (combined effect
  of length, time constant and velocity)

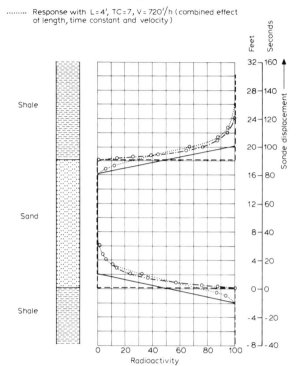

Fig. 5-4. Gamma ray responses in the case of an infinitely slow recording speed.

N is the true count-rate (cps)

t is the dead-time (secs)

n is the measure count-rate (cps) = $N - \Delta N$

From eq. 5-1 we have:

$$\frac{\Delta N}{N} = \frac{N - n}{N} = nt \qquad (5\text{-}2)$$

From eqs. 5-1 and 5-2 we obtain:

$$N = n/(1 - nt) \qquad (5\text{-}3)$$

which means that n is closest to N when t is very small, or when n itself is small (see Fig. 5-3). The dead-time of counting circuitry is optimized according to expected count-rates. The dead-time effect decreases when the count-rate decreases.

5.5. LOGGING SPEED

The time-constant introduces a lag in the response of the recorded log with respect to bed-boundaries, because averaging the circuitry requires a finite period of time to respond fully to a change in count-rates. After traversing a bed boundary, the measured response attains 97% of the new value only after a time equal to $5T$ (Fig. 5-5).

You will appreciate that, were the time-constant

zero, or the speed infinitely slow, the response to the shale-sand-shale sequence of Fig. 5-4 would be very nearly symmetrical (ignoring the minor effects of source-detector disposition which introduce a slight dissymmetry in induced radioactivity logs).

Figure 5-5 summarizes the effects of finite detector length (L) (which tends to smooth the sharp bed boundaries) and time-constant (whereby the time-log in response introduces a depth displacement to the apparent boundary). The longer the time-constant, or the faster the logging speed, the farther up the bed boundary will appear to be displaced. Furthermore, the less accurate will be the peak reading obtained in a thin bed.

Faced with the trade-off between statistical precision (= long time constant) and bed resolution (= short time constant), a satisfactory compromise is usually employed whereby logging speed and time constant are such that:

Logging speed (ft/s) × time constant T (s) = 1 ft

where T has, of course, been chosen to suit the count-rate levels expected.

Returning to Fig. 5-4, you will see that T seconds after crossing the lower boundary, the measurement has reached about 50% of the level in the sand, and the converse for the upper interface. Whether one picks the (apparent) bed boundaries on the dotted curve at the points of inflection, or 50% level, the apparent depth displacement is quite close to the distance travelled in T seconds.

Provided the speed × time-constant relationship is respected, than all beds will be displaced upwards by 1 ft. This small shift is simply corrected on the log using a memorizer.

Logs are usually run at 1800 ft/h (1/2 ft/s) with $T = 2$ s; or 900 ft/h (1/4 ft/s) with $T = 4$ s.

N.B. With the advent of the CSU, a different averaging technique is employed, whereby data sampled at one depth is averaged with data sampled just before and just after, before being recorded on film. The capacitor-type time-constant described above is not used. This gives a symmetric curve with no lag related to logging speed.

5.6. BED THICKNESS

For a hypothetical point detector, the radiation detected will originate from within a certain sphere of influence (natural radiation) or within a volume of influence dependent on source-detector spacing and the nature of the measuring technique (induced radiation).

Taking into consideration the finite length of the detector, the effects of time-constant smoothing (or vertical depth smoothing in the CSU) and logging speed, the responses in beds of various thicknesses are illustrated in Fig. 5-5 (the Natural Gamma Ray). Values of L for several detectors are:

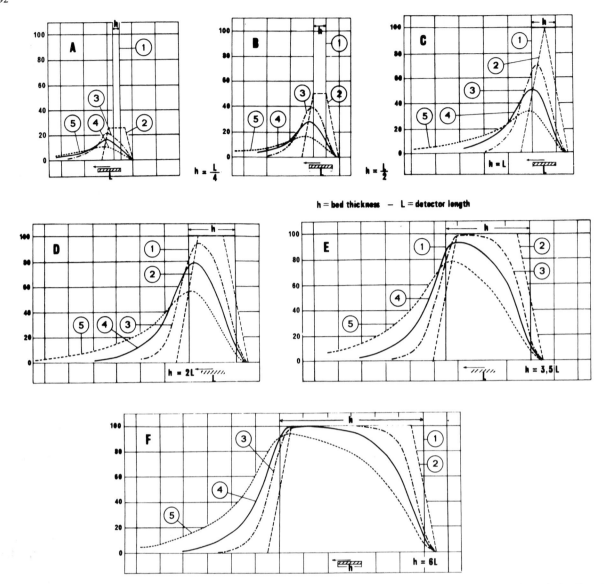

Fig. 5-5. Variations in response of the gamma ray tool as a function of the bed thickness, and the recording speed. *1* = Theoretical radioactive anomaly (equivalent to the log response of a point detector at an infinitely slow logging speed). *2* = Response of a detector of length *L*, at infinitely slow speed. *3* = Response for a logging speed such that the detector of length *L* moves a distance *L*/2 during one time constant (*t*). *4* = Response for a logging speed such that the detector of length *L* moves a distance *L* during one time constant (*t*). *5* = Response for a logging speed such that the detector of length *L* moves a distance 2*L* during one time constant (*t*).

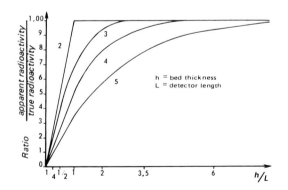

Fig. 5-6. Determination of the true radioactivity as a function of the bed thickness (relative to the detector length) and the recording speed.

Geiger-Mueller for capture gamma rays (GNT): 6″
Geiger-Mueller for natural gamma rays (GNT): 3 ft
NaI crystal for natural gamma rays (SGT): 8″ or 4″
NaI crystal for natural gamma-ray spectrometry (NGT): 12″
^{3}He tube for thermal neutron detection (CNT-A): 12″
G-M detectors are less efficient than NaI crystals, and require a greater active length (the GNT uses a bundle of 5 G-M tubes).

In thin beds, the measured response is an average of all contributions within the volume of influence, defining an electro-bed (2.2.2.3.). The net response may be considered as the sum of the parameters appropriate to each sub-stratum, weighted by the contributing volume of each. Thus the thinner the

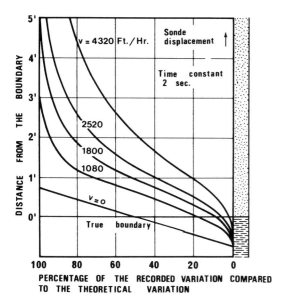

Fig. 5-7. Variation of response of a GNAM neutron at a bed boundary as a function of the recording speed (courtesy of Schlumberger).

bed, the weaker the response, until it becomes indistinguishable from statistical variations. Referring to Fig. 5-5, curve D, a bed-resolution of about 2 ft appears a reasonable minimum for a 6″ detector.

Where spherical thin-bed resolution is required, logs can, of course, be run at lower speeds or shorter time constants than usual.

The chart in Fig. 5-6, which follows from Fig. 5-5, corrects the gamma-ray reading for bed-thickness and detector length, at several different speeds and time-constants.

Figure 5-7 is a correction chart for the early GNAM gamma-ray neutron tool (fore-runner of the GNT). For the given logging speeds at $T = 2$ secs, it provides the depth log and reading correction factor, as a function of bed thickness.

With the exception of the NGS * and certain specialized versions of the SGT, modern nuclear tools use relatively short detectors of about 4″–6″ length. The influence of L in the preceding discussion becomes negligible in beds more than a few feet thick.

5.7. MEASURING POINT

Natural radiation—the measuring point is at the centre of the active part of the detector.

Induced radiation—single detector: mid-way between the source and the detector; two detectors: mid-way between the detectors.

N.B. These are rule-of-thumb definitions of measuring points. Their exact location varies somewhat with the nature of the phenomena being measured.

5.8. REFERENCES

See chapters 6 to 12.

---

* Mark of Schlumberger.

6. MEASUREMENT OF THE NATURAL GAMMA RADIOACTIVITY

6.1. DEFINITION NATURAL RADIOACTIVITY

Natural radioactivity is the spontaneous decay of the atoms of certain isotopes into other isotopes. The products of decay, or daughters, may be stable or may undergo further decay until a stable isotope is finally created. A radioactive series refers to a group of isotopes including the initial radioisotope and all its stable or unstable daughters. Radioactivity is accompanied by the emission of alpha or beta particles, gamma rays and by the generation of heat.

Natural gamma radiation is one form of spontaneous radiation emitted by certain unstable atomic nuclei. Gamma ray frequencies lie, by definition, between 10^{19} and 10^{21} Hz.

6.2. BASIC CONCEPTS

Certain unstable elemental isotopes are in the process of decaying by the spontaneous emission of radiation, regardless of what chemical combination the element may be in. This natural radiation is of three types, α, β and γ. Figure 6-1 shows the decay series of three such isotopes; ^{238}U, ^{232}Th and ^{235}U.

6.2.1. α-radiation

An α-particle may be emitted from an atomic nucleus during radioactive decay. It is positively charged and has two protons and two neutrons. It is physically identical to the nucleus of the helium atom *.

By α-emission the element of atomic number Z is transformed into an element of atomic number $Z - 2$ and the number of nucleons decreases from A to $A - 4$.

$$^{A}_{Z}E \rightarrow {}^{A-4}_{Z-2}E' + {}^{4}_{2}He^{+}\,(=\alpha) + \gamma + Q \qquad (6\text{-}1)$$

The factor γ is symbolic only. In some transitions there will be no gamma emission, in others several photons may follow the particle emission. The Q term represents the additional energy released in the transition.

The alpha-particles have a very low penetration

---

* Helium atom $^{4}_{2}$He has an atomic mass, A, (= sum of the number of protons and neutrons) equal to 4 and an atomic number, Z, (= number of protons, also equal to number of electrons if the nucleus is in neutral, unionized state) equal to 2.

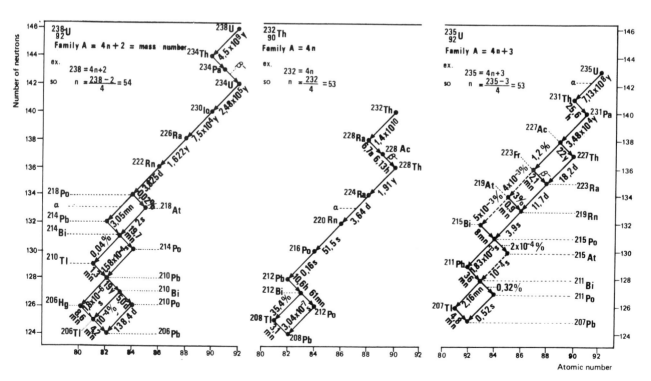

Fig. 6-1. Radioactive families and their decay scheme.

TABLE 6-1

Penetration of α-particles in different materials as a function of their energy

Energy of α-Particles (MeV)	1	2	3	4	5	6	7
Air (mm)	5	10	17	25	35	47	59
Aluminium (μm)	4	7	11	16	22	29	
Iron (μm)	2	4	6	8	11	14	
Water, paper Plastic (μm)		10	19	29	41	54	71

power and at normal temperature and pressure can only penetrate most materials to only a few tens of μm (Table 6-1).

6.2.2. β-radiation, β⁺ or β⁻

A β-particle may be emitted from an atomic nucleus during radioactive decay. It is physically identical to the electron. It may be either positively charged (positron) or negatively charged (electron).

During β^--emission the element of atomic number Z becomes the element of atomic number $Z + 1$.

$$_Z E \rightarrow _{Z+1} E' + \beta^- + n_0 + \gamma + Q \qquad (6-2)$$

γ and Q as above, n_0 represents the neutrino.

During β^+-emission the element of atomic number Z becomes the element of atomic number $Z - 1$.

$$_Z E \rightarrow _{Z-1} E' + \beta^+ + n_0 + \gamma + Q \qquad (6-3)$$

β-particles have a higher penetration power than α. It depends on the kinetic energy of the electron/positron. For instance, at 1 MeV energy, penetration is about 4 m in air, and 2 mm in aluminium (Table 6-2).

Like α-radiation, β-radiation is generally too weak to be of direct interest as a logging measurement.

6.2.3. γ-radiation

γ-radiation may be considered as an electromagnetic wave similar to visible light or X-rays, or as a particle or photon. Gamma rays are electromagnetic radiations emitted from an atomic nucleus during radioactive decay. These radiations are characterized by wave lengths in the range of 10^{-9}–10^{-11} cm, equivalent to frequencies ranging from 10^{19} to 10^{21} sec^{-1}.

The gamma-emission corresponds, in a nucleus, to the transition from one state to another of lower energy with emission of a photon of energy $h\nu$, equal to the difference between the energies of the two states. The energy is related to the wavelength, λ, or frequency ν, by:

$$E = h\nu = hc/\lambda \qquad (6-4)$$

where c is the velocity of light, and h is Planck's constant (6.626×10^{-27} erg s or 6.626×10^{-34} joule s).

The energy is expressed in electron-volts (eV) *. The energies of gamma rays are of the order of the KeV (10^3 eV) or the MeV (10^6 eV).

Usually, the α-, β- and γ-emissions are simultaneous. α- and β-particles do not penetrate far enough to be detected by logging techniques. The gamma rays have a very high power of penetration (Table 6-3) and can be detected and recorded in present-day hole condictions. For that reason, they are the basis of several important logging techniques.

6.2.4. Radioactive decay

The *decay* or *disintegration* of an atom involves the emission of α, β, or γ radiations, and heat, resulting in the transformation of that atom into another, lower down the decay series (Fig. 6-1). The final atomic state at the end of the series is stable. The radioactive decay is a pure statistical process. At a

---

* One electron-volt is the energy acquired by a charged particle carrying unit electronic charge when it is accelerated by a potential difference of one volt. One eV is equivalent to 1.602×10^{-12} erg or 1.602×10^{-19} joule. One eV is associated through the Planck constant with a photon of wavelength 1.239 μm. For an energy of 1 MeV, the wave-length is equal to 0.0124 Å.

TABLE 6-2

Penetration of electrons in different materials as a function of their energy (all values in mm)

Energy in Mev Material (density)	0.01	0.03	0.1	0.3	0.7	1	2	3
Air (0.0013)	1.9	15	120	730	2440	3800	8350	12800
Water (1.0)	0.002	0.017	0.14	0.8	2.7	4.3	9.6	14.9
Vinyl polychloride	0.002	0.015	0.12	0.74	2.4	3.8	8.5	13.1
Flexiglass (1.19)	0.002	0.014	0.12	0.71	2.4	3.7	8.4	13.1
Aluminium (2.7)	0.001	0.009	0.07	0.40	1.3	2.1	4.5	6.9
Iron (7.8)	0.4×10^{-3}	3.4×10^{-3}	0.027	0.16	0.50	0.78	1.7	2.6
Lead (11.35)	0.52×10^{-3}	3.7×10^{-3}	0.028	0.14	0.44	0.67	1.4	2.1

TABLE 6-3

Half-value thickness for a narrow beam of gamma rays in different materials as a function of their energy (all values in mm)

Energy of gamma ray (MeV) Material (density)	0.1	0.2	0.5	1	2	5
Air (0.0003)	36	44	61	83	121	
Water (1.0)	42	51	72	98	142	230
Vinyl polychloride (1.3)	27			78		
Concrete (2.3)	$\simeq 19.2$	$\simeq 26$	$\simeq 37$	$\simeq 50.4$	$\simeq 72$	
Limestone (tight) (2.7)		$\simeq 21$		$\simeq 46$		$\simeq 100$
Aluminium (2.7)	16	21.4	30.4	42	60	
Iron (7.8)	2.6	6.4	10.6	14.7	20.6	$\simeq 28$
Lead (11.3)	0.11	0.65	4.0	8.8	13.6	14.7?
Uranium (18.7)	0.029			4.7		
Plutonium 239 (19.8)	0.025			4.2		

certain instant of its life a radioisotope (the unstable form of an element) emits a radiation. After a radiation the initial number of atoms, N, of the radio-isotope decreases by one:

$N \rightarrow N - 1 +$ radiation

The probability of emission of a radiation depends on the number of atoms of the radioisotope: the higher this number, the higher is the probability of emission.

Assume that at some observation time t we have N atoms of some radioisotope. Then the law of constant fractional decay requires that over a short interval of time, dt, the number of atoms decaying, dN, will be:

$$dN = -\lambda_d N dt \qquad (6-5)$$

where λ_d is the *decay* or *disintegration constant*. The minus sign is required because dN represents a decrease in N. An integration of eq. 6-5 gives the exponential relation between N and t:

$$N = N_0 e^{-\lambda_d t} \qquad (6-6)$$

where N_0 is the number of atoms present at $t = 0$. By taking logarithms of both sides we obtain:

$$\ln(N_0/N) = \lambda_d t \qquad (6-7)$$

This can be illustrated by Fig. 6-2. Early studies of radioactive materials showed that the activity of each species decreased at its own characteristic rate. Each rate of decrease was a constant fraction of the activity present. Because of the constancy of this fractional decrease, it was convenient to define the *half-life*, $T_{1/2}$, of any radioactive nuclide as the time required for any amount of it to decay to one-half of its original activity. Thus, if the activity of a sample was originally 100 units, it would decay to 50 units in one half-life. At the end of two half-lives, one-half of the 50 would have decayed, 25 would be left; and so on.

Since the decay process is based on probabilities, the above relationship is true only on average. Constant fractional decay also requires that each decay

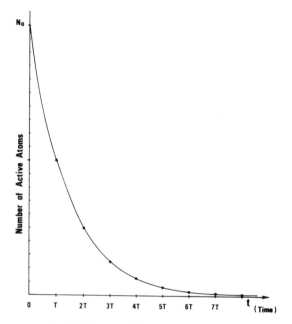

Fig. 6-2. Theoretical law of disintegration.

be an independent event. And for a given nucleus, the probability of decay must be independent of nuclear age. So on the decay scheme of Fig. 6-2 we must add a noise function which on average is equal to zero but varies in time. This can be represented by Fig. 6-3.

Measurements show that when N is large enough to minimize for statistics, eq. 6-6 is obeyed over an enormous range of decay rates. In natural radioactivity half-life between 10^{-6} sec and 10^{10} years are known.

An important relation between decay constant λ_d and half-life, $T_{1/2}$, is obtained by putting $N = N_0/2$ in eq. 6-7. Then:

$$\ln\left(\frac{N_0}{N_0/2}\right) = \ln 2 = \lambda_d T_{1/2}$$

and

$$\lambda_d T_{1/2} = 0.693 \qquad (6-8)$$

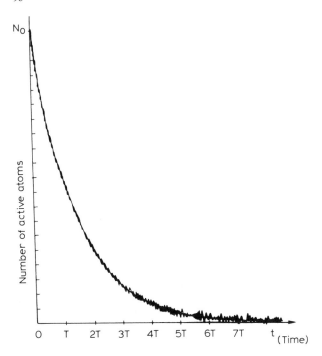

Fig. 6-3. Real law of disintegration.

$T_{1/2}$ is usually given in either years, days, hours, minutes, seconds or even μsec. Each value of λ_d will be the corresponding reciprocal y^{-1}, d^{-1}, h^{-1}, m^{-1}, s^{-1}, μs^{-1} respectively.

Mean lifetime, $\bar{T}$, is another term that is sometimes useful in describing radioactive decay. From eq. 6-5, the number of nuclei decaying during the interval from t to $(t + dt)$ is $\lambda_d N dt$ and, since each of these nuclei had a lifetime t, the total lifetime associated with this interval is $\lambda_d N t dt$. An integration of this factor over all values of t gives the total lifetime of all N_0 nuclei, and a division by N_0 gives the mean life time $\bar{T}$. The integral is:

$$\bar{T} = \frac{1}{N_0} \int_0^\infty \lambda_d N t dt = \int_0^\infty \lambda_d t \, e^{-\lambda_d t} \, dt = \frac{1}{\lambda_d} \qquad (6-9)$$

$$\bar{T} = \frac{T_{1/2}}{0.693} = 1.44 T_{1/2} \qquad (6-10)$$

Half-life has a unique unchangeable value for each radioactive species, and geological evidence indicates that the presently observed values have obtained for a geologically long time. No change in the decay rates of particle emission has been observed over extreme variations of conditions such as temperature, pressure, chemical state, or physical environment.

6.2.5. Radioactive equilibrium

A material is in radioactive secular equilibrium, when the disintegration rate is the same for all the members of the series, the ratio of the number of nuclei of each member to that of any other member present in the system being inversely proportional to the ratio of the respective decay constants. This can be written as follows:

$$\lambda_{d_1} N_1 = \lambda_{d_2} N_2 = \lambda_{d_n} N_n \qquad (6-11)$$

We generally consider rocks older than a million years to be in secular equilibrium.

6.2.6. The units of radioactivity

Equation 6-6 shows the time course of the number of radioactive nuclei of a species characterized by a decay constant λ_d. Since at every instant the rate of decay is $\lambda_d N$, this rate, known as the *activity A*, will also follow an exponential decay law. That is:

$$A = \lambda_d N = \lambda_d N_0 \, e^{-\lambda_d t} \qquad (6-12)$$

Originally, the unit of activity, the curie, was defined as equal to the number of disintegrations per second occurring in 1 gram of ^{226}Ra.

As originally defined, the curie was a changing standard, since it depended upon experimentally determined quantities. In 1950, the curie (Ci) was redefined as exactly 3.7×10^{10} disintegrations per second (dps). The curie is a relatively large unit; for most purposes the millicurie, mCi, and the microcurie, μCi, are more applicable.

The curie is defined in terms of the number of decaying nuclei and not in the number of emissions. Details of the decay scheme must be known in order to calculate a source strength from an emission measurement.

Each radioactive species has an *intrinsic specific activity* (ISA) which is the activity of a unit mass of the pure material. When the radioactive isotope is diluted with a stable isotope of the same species, the corresponding calculation gives the specific activity. From Adams and Weaver (1958) the relative gamma activity of K, Th and U is respectively 1, 1300 and 3600.

The gamma radioactivity is expressed either in μg Ra equivalent/metric tonne, or in A.P.I. units (American Petroleum Institute). The API unit is defined as the one two-hundredth of the difference in curve deflection between zones of low and high radiation in the API gamma ray calibration pit in Houston, Texas (see 6.11.). Generally,

1 μg = 16.5 API

except for the Schlumberger 1 11/16" gamma ray tools GNT-J/K, SGT-G where:

1 μg = 13.5 API

For the older version of the GNT-K, with G.M. detector:

1 μg = 11.5 API

6.3. THE ORIGIN OF NATURAL RADIOACTIVITY IN ROCKS

Elements other than hydrogen, whether stable or unstable (radioactive), have been formed in very hot and high pressure environments (stars, supernovae). Natural conditions on the Earth are not suitable for the formation of these elements and the only elements or isotopes found on the Earth are those which are stable or which have a decay time comparable with, or larger than the age of the Earth (about 5×10 years). The isotopes with shorter life-times disappeared long ago, unless they result from the decay of longer lived radioactive elements.

The radioisotopes with a sufficiently long life, and whose decay produces an appreciable amount of gamma rays (Table 6-4) are:

(a) Potassium ^{40}K with a half-life of 1.3×10^9 years, which emits 1 β^-, 1 γ and zero α.

(b) Thorium ^{232}Th with half a life of 1.4×10^{10} years, which emits 7 α, 5 β^- and numerous γ of different energies.

(c) Uranium ^{238}U with a half-life of 4.4×10^9 years, which emits 8 α, 6 β^- and numerous γ of different energies.

N.B: The radioactive ^{40}K isotope constitutes a fairly constant 0.0118% of the total potassium (^{38}K, etc.) present. The isotope $^{235}_{92}U$, also radioactive, is much less abundant than $^{238}_{92}U$ (0.71% compared with 99.28%) and can be ignored.

Potassium ^{40}K disintegrates to give argon ^{40}A, which is stable (Fig. 6-4). The spectrum consists of a single peak. The process is more complex for uranium and thorium which give a series of isotopes, some of which are gamma ray emitters.

The detailed decay schemes of uranium and thorium are indicated on Fig. 6-4. Figure 6-5 shows the theoretical gamma ray emission spectra of the potassium, thorium and uranium families.

Note in Table 6-5 (from Adams and Weaver, 1958) that the relative contributions of these isotopes to the total mean radiation is quite closely linked to the relative abundances of the elements in the lithosphere and hydrosphere (Clarke's numbers).

6.4. MINERALS AND ROCKS CONTAINING RADIOACTIVE ELEMENTS

6.4.1. Potassium-bearing minerals and rocks

The source minerals of potassium are the alkali potassic feldspars and micas plus a large number of minerals of minor importance.

During alteration, some silicates such as the feldspars are completely dissolved; the potassium is thus liberated in ionic form and transported in solution (Fig. 6-6).

Micas, on the other hand, may lose only part of their potassium during alteration, the remainder staying in the crystal lattice. The minerals resulting from the alteration of micas therefore contain potassium in varying amounts (illite, montmorillonite, mixed layers I-M). However, under temperate, humid conditions muscovite is at an unstable phase and undergoes incongruent dissolution to form kaolinite:

$$2\ KAl_3Si_3O_{10}(OH)_2 + 2\ H^+ + 3\ H_2O$$
(muscovite)

$$\rightarrow 3\ Al_2Si_2O_5(OH)_4 + 2\ K^+$$
(kaolinite)

The liberated potassium is transported by rivers to the sea. In arid regions it may remain with the altered minerals.

Potassium ions may be adsorbed onto the surfaces of clay particles, or absorbed by plants, to reappear in relatively insoluble complex organic compounds. This explains why, in sea-water, potassium is only present at 380 ppm, a small fraction of the original concentration.

The concentration of potassium in sea-water is affected by: (a) dilution by river-water; (b) biological

TABLE 6-4

Main radioactive elements

Element	Isotope	Percentage of the total element %	Emissions			Half life (year)	Relative abundance in Earth's crust (ppm)
			α	β	γ		
Primeval natural gamma-ray emitters							
Potassium	^{40}K	0.0118		1	1	1.3×10^9	2.5 [c]
Uranium [d]	^{235}U	0.72	8 [d]	5 [d]	[a]	7.1×10^8	0.02
Secondary gamma-ray emitters by their daughters							
Uranium series	^{238}U	99.27	8 [d]	6 [d]	[b]	4.5×10^9	3
Thorium series	^{232}Th		7 [d]	5 [d]	[b]	1.4×10^{10}	12

[a] Gamma ray emitter by itself and its daughters. [b] For more detail see Fig. 6-4. [c] From Krauskopf (1967): Introduction to Geochemistry. McGraw Hill, New York. [d] From Adams and Gasparini (1970).

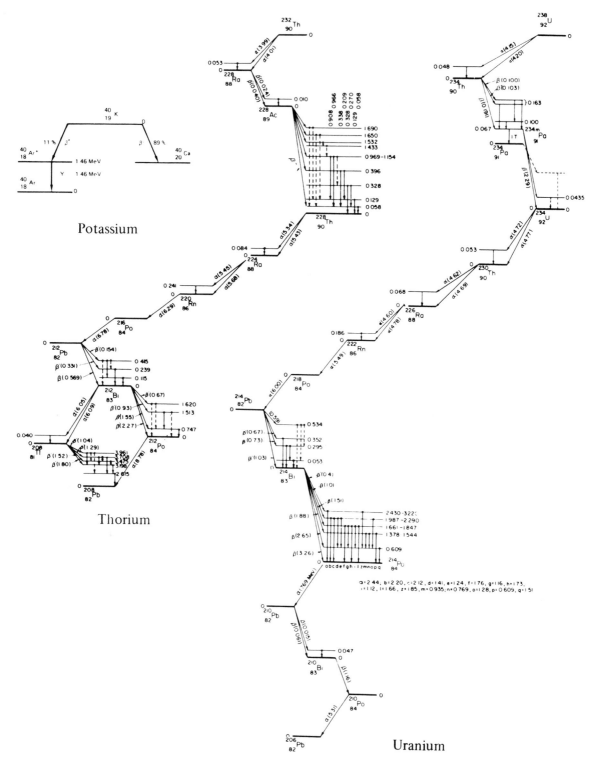

Fig. 6-4. Detailed schemes of disintegration of ^{40}K, ^{232}Th and ^{238}U (from Adams and Gasparini, 1970).

activity; certain algae, for instance, accumulate potassium; and (c) interactions between sea-water and detrital minerals, colloidal particles and clay minerals. Potassium minerals may also crystallize out of solution to form evaporites.

Summarizing, there follows a list of the more common potassium minerals:

Evaporites: these are listed in Table 6-6.

Clay minerals: potassium is present in the crystal lattice (refer to Fig. 1-1) in the micas or clay minerals from the mica group. (See Table 6-7).

Potassium is also absorbed onto the negatively charged clay platelets. Figure 6-7, from Blatt et al. (1979), gives the average composition of the most

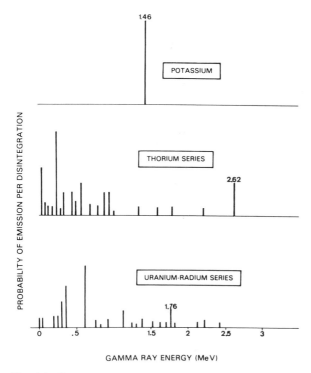

Fig. 6-5. Gamma-ray emission spectra of radioactive elements (courtesy of Schlumberger).

common micas and clay minerals. Considering all the clay minerals, average K content is 2.7%.

Feldspars such as microcline and orthoclase (see Table 6-7). Sandstones contain on average 12% of feldspar, which corresponds to 1.5% potassium. In extreme cases, sandstones (arkose) may contain up to 50% of feldspar which means 7–8% potassium.

Feldspathoids such as leucite and nephelite (see Table 6-7).

Carbonates which may contain between 0 and 7% potassium, averaging at 0.3%.

6.4.2. Uranium-bearing minerals and rocks (see Table 6-9)

The source minerals are in igneous rocks of acid origin. Table 6-8 lists the average content (in ppm) of uranium and thorium in several rock types. These

TABLE 6-5

Relative contribution of the three radioactive elements to the gamma ray flux (from Adams and Weaver, 1958)

	K	Th	U
Relative abundance on the Earth (Clarke's number)	2.35%	12 ppm	3 ppm
Relative gamma-ray activity per unit weight of element	1	1300	3600

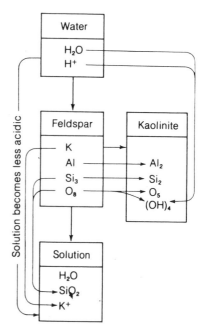

The flow of material in the alteration of feldspar to kaolinite by the dissolving action of water. Some water and some hydrogen ions (H+) are absorbed into the kaolinite structure, but all of the potassium (K+) and some of the silica (SiO2) ends up in solution.

Fig. 6-6. Alteration of feldspar giving kaolinite (from Press and Siever, 1978).

are averages; actual values can vary considerably.

Uranium is very soluble, and is transported mainly in solution, rarely in suspension. It is dissolved out during the alteration or leaching of source minerals. Leaching is predominant in the presence of water rich in organic acids.

6.4.2.1. *Solubility of uranium*

Uranium ions exist with two valencies: 4+ and 6+.

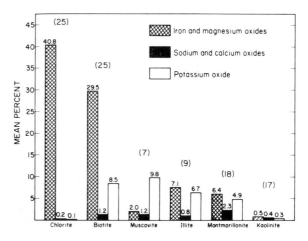

Fig. 6-7. Mean amounts of metallic cations (except aluminium) in the most abundant micas and clay minerals. Based on 101 analyses (in parenthesis) from various sources (from Blatt et al., 1979).

TABLE 6-6

The potassium-bearing minerals of evaporites

Name	Composition	K (% weight)	ρFDC (g/cm^3)	P_e (b/e)	ϕ_{CNL} (%)	Δt (μs/ft)
Sylvite	KCl	52.44	1.86	8.51	-3	74
Langbeinite	K_2SO_4, $(MgSO_4)_2$	18.84	2.82	3.56	-2	52
Kainite	$MgSO_4$, KCl, $(H_2O)_3$	15.7	2.12	3.5	> 60	
Carnallite	MgCl, KCl, $(H_2O)_6$	14.07	1.57	4.09	> 60	83
Polyhalite	K_2SO_4, $MgSO_4$, $(CaSO_4)_2$, $(H_2O)_2$	13.37	2.79	4.32	25	57.5
Glaserite	$(K,Na)_2$, SO_4	24.7	2.7			
Halite	NaCl		2.04	4.65	-3	67
Anhydrite	$CaSO_4$		2.98	5.05	-2	50
Gypsum	$CaSO_4$, $2(H_2O)_2$		2.35	3.99	> 60	52
Sulfur	S		2.02	5.43	-2	122
Tachydrite	$CaCl_2$, $(MgCl_2)_2$, $(H_2O)_{12}$		1.66	3.84	> 60	92
Kieserite	$MgSO_4$, H_2O		2.59	1.83	43	
Epsomite	$MgSO_4$, $(H_2O)_7$		1.71	1.15	> 60	
Bischofite	$MgCl_2$, $(H_2O)_6$		1.54	2.59	> 60	100
Trona	Na_2CO_3, $NaHCO_3$, H_2O		2.08	0.71	35	65

TABLE 6-7

Potassium-bearing minerals in sand-shale series

Name	Chemical formula	K content (% weight)
Feldspars		
Alkali		
Microcline	$KAlSi_3O_8$ Triclinic	16 (ideal) to 10.9 [a]
Orthoclase	$KAlSi_3O_8$ Monoclinic	14 (ideal) to 11.8 [a]
Anorthoclase	$(Na.K)AlSi_3O_8$	
Plagioclases		0.54
Micas		
Muscovite	$KAl_2(AlSi_3O_{10})(OH.F)_2$	9.8 (ideal) to 7.9 [a]
Biotite	$K(Mg, Fe)_3(AlSi_3O_{10})(OH, F)_2$	6.2–10.1 (av.: 8.5)
Illite	$K_{1-1.5}Al_4(Si_{7-6.5}, Al_{1-1.5})O_{20}(OH)_4$	3.51–8.31 (av.: 6.7)
Glauconite	$K_2(Mg.Fe)_2Al_6(Si_4O_{10})_3(OH)_{12}$	3.2–5.8 (av.: 4.5)
Phlogopite	$KMg_3(AlSi_3O_{10})(F.OH)_2$	6.2–10.1 (av.: 8.5)
Feldspathoids		
Metasilicates		
Leucite	$KAl(SiO_3)_2$	17.9 (ideal)
Orthosilicates		
Nephelite	$(Na.K)AlSiO_4$	4 to 8
Kaliophilite	$KAlSiO_4$	24.7 (ideal)
Other clay minerals *		
Montmorillonite		0–4.9 [b] (av.: 1.6)
Chlorite		0–0.35 (av.: 0.1)
Kaolinite [c]		0–0.6 (av.: 0.35)

* Potassium is fixed by adsorption.

[a] Corresponds to a beginning of alteration.

[b] Some montmorillonites might correspond to imperfectly degraded muscovite, or to an incomplete transform in illite by diagenesis.

[c] Kaolinite contents sometimes more potassium due to imperfectly degraded feldspars.

(a) The ion U^{4+} has a tendency to oxidize and assume the valency U^{6+}, forming UO_4^{2-} and $U_2O_7^{2-}$. Consequently, the U^{4+} form is found only in certain environments such as sulphurous hot-spring water (pH < 4, Eh * < 0). Where the pH exceeds 4, U^{4+} dissociates and an insoluble uranium oxide, UO_2, or uraninite is formed by dissociation.

(b) U^{6+} only exists in solution as the complex uranyl ion UO_2^{2+}. If Eh > 0 and pH < 2, UO_2^{2+} is stable. If 2 < pH < 5, the uranyl ion goes to $U_2O_5^{2+}$ or $U_3O_8^{2+}$ or forms more complex ions such as $(UO_2(OH)_2UO_3)_n$. At pH = 5, hydrolysis occurs and schoepite, 4 $UO_3 \cdot$ 9 H_2O precipitates. UO_2^{2+} co-precipitates with phosphates which may contain as much

---

* Eh is the Redox or oxidation–reduction potential. Oxidation is equivalent to a loss of electrons, reduction a gain of electrons, to the substance under reaction. Eh (in mV) expresses the capacity to give or receive electrons under oxidation or reduction.

TABLE 6-8

Thorium, uranium and potassium contents of some igneous rocks

Igneous Rocks	Th (ppm)	U (ppm)	K (%)
Acid intrusive			
Granite	19–20	3.6–4.7	2.75–4.26
Rhode Island [1]	21.5–26.6 (25.2)	1.32–3.4 (1.99)	3.92–4.8 (4.51)
Rhode Island [1]	6.5–80 (52)	1.3–4.7 (4)	5.06–7.4 (5.48)
New Hampshire	50–62	12–16	3.5–5
Precambrian	14–27	3.2–4.6	2–6
Average for granitic rocks	15.2	4.35	4.11
Syenite [1]	1338	2500	2.63
Acid extrusive			
Rhyolite	6–15	2.5–5	2–4
Trachyte	9–25	2–7	5.7
Basic intrusive			
Gabbro	27–3.85	0.84–0.9	0.46–0.58
Granodiorite	9.3–11	2.6	2–2.5
Colorado [a]	99–125 (110.6)	0.19–2.68 (1.98)	2.62–5.6 (5.48)
Diorite	8.5	2.0	1.1
Basic extrusive			
Basalt			
Alkali basalt	4.6	0.99	0.61
Plateau basalt	1.96	0.53	0.61
Alkali olivine basalt	3.9	1.4	1.4
in Oregon [1]	5.5–15 (6.81)	1.2–2.2 (1.73)	1.4–3.23 (1.68)
Andesite	1.9	0.8	1.7
in Oregon [1]	5–10 (6.96)	1.4–2.6 (1.94)	2.4–4.28 (2.89)
UltraBasic			
Dunite	0.01	0.01	0.02
Peridotite	0.05	0.01	0.2
in California [1]	0.0108	0.0048	0.019

[1] From U.S.G.S. Geochemical standards, in Adams & Gasparini, 1970.

TABLE 6-9

Uranium and uranium-bearing minerals (from Roubault, 1958)

Name	Composition	U Content %
Uranium minerals		
Autunite *	$Ca(UO_2)_2(PO_4)_2$, 10–12 H_2O	
Baltwoodite	U-silicate high in K	
Becquerelite	CaO, 6 UO_3, 11 H_2O	70 to 76
Carnotite *	$K_2(UO_2)_2(VO_4)_2$, 1–3 H_2O	52.8 to 55
Gummite		70
Ianthinite	UO_2, 5 UO_3, 12 H_2O	70 to 71.5
Pechblende *	U_3O_8 to UO_2 amorphous	
Schoepite	4 UO_3, 9 H_2O	68 to 74
Soddyite	5 UO_2, 2 SiO_2, 6 H_2O	69 to 71
Tyuyamunite *	$Ca(UO_2)_2(VO_4)_2$ 5–8 H_2O	54.4 to 56.7
Uraninite	UO_2 cubic	
Uranopilite	$(UO_2)_6(SO_4)(OH)_{10}$ 12 H_2O	
Weeksite	U-silicate high in Ca	
Uranium-bearing minerals		
Betafite	$(U.Ca)(Nb, Ta, Ti)_3O_9 n\ H_2O$	16 to 25
Brannerite	$(U, Ca, Fe, Y, Th)_3Ti_5O_{16}$	40
Chalcolite	CuO, 2 UO_3, P_2O_5, 8–12 H_2O	47 to 51
Euxenite	$(Y, Ca, Ce, U, Th) (Nb, Ta, Ti)_2O_6$	3 to 18
Fergusonite	$(Y, Er, Ca, Fe) (Nb, Ta, Ti)O_4$	0 to 7
Microlite	$(Na, Ca)_2 (Ta, Nb)_2O_6(O, OH, F)$	0 to 15
Parsonite	2 PbO, UO_3, P_2O_5, H_2O	18 to 25
Polycrase	$(Y, Ca, Ce, U, Th) (Nb, Ta, Ti, Fe)_2O_6$	3 to 18
Pyrochlore	$(Na, Ca)_2 (Nb, Ta)_2O_6F$	0 to 1.4
Renardite	PbO, 4 UO_3, P_2O_5, 9 H_2O	52 to 55
Uranotile	CaO, 2 UO_3, 2 SiO_2, 6 H_2O	53 to 56

* Most important commercial ore of uranium.

as 0.1% uranium. Peat, wood and cellulose are the best reducing agents over a wide range of pH. So if a small part of the uranium is absorbed during plant growth most of it may be directly precipitated or absorbed in a stable form as disseminated uraninite or as a uranium-organic compound on products of plant decomposition or disintegration (Swanson, 1960; Fig. 6-8).

(c) The amphoteric hydroxide which forms from UO_2^{2+} (above) is weakly acidic and can react with minerals to produce soluble compounds:

(1) Sulphate-rich environment: $[UO_2(SO_4)_3]^{4-}$; stable at pH ⩽ 2.5.

(2) Carbonate-rich environment: $[UO_2(CO_3)_2]^{2-}$; stable at 7.5 < pH < 8.5.

(3) Polyuranates precipitate if pH > 8.5.

(4) Insoluble schoepite if pH < 7.5.

6.4.2.2. Transport mechanisms

6.4.2.2.1. Solution

Over short distances, uranium is carried by surface and subterranean rivers as complex carbonates (excess of CO_2) or sulphates. In sulphurous water the quadrivalent ionic form exists in solution.

The stability of the various uranium compounds is precarious, being highly dependent on the environment. Only complex organic compounds can be transported over large distances. Organic material is therefore conducive to stability.

6.4.2.2.2. Suspension

Metallic elements are replaced by uranium in the crystal lattices of certain minerals, such as zircon, monazite, xenotime, sphene, allanite, biotite etc. These are resistant minerals which can be transported great distances.

Isomorphic replacement of calcium in phosphates produces carbonate fluorapatite.

Schoepite and polyuranates have already been mentioned. Transport is also possible in clay particles containing organic matter.

6.4.2.2.3. Living organisms

Uranium and thorium accumulate in the thyroid glands of animals, in plant leaves, and especially in the bones of fish, where concentrations as high as 4500 ppm of uranium and 2700 ppm of thorium are possible.

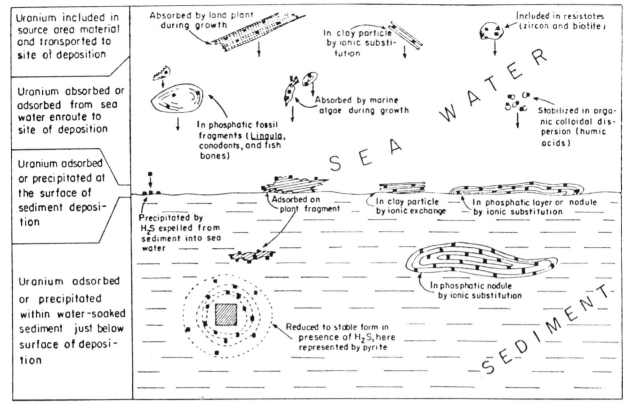

Fig. 6-8. Diagramatic sketch showing possible association and time of emplacement of uranium with common constituents of marine black shales. Uranium is represented by black squares (from Swanson, 1960).

6.4.2.3. *Precipitation of uranium*

Uranium will precipitate and accumulate in sediments under the following conditions:

(a) *The presence of organic matter*: (1) uranium is irreversibly adsorbed from uranyl solutions in the presence of bacteria and humic fractions; (2) in acid pH environments, humic and fulvic acids, ether, alcohol, aldehydes, favour the precipitation of uranium by the reduction of U^{6+} to U^{4+}, forming urano-organic complexes or chelates; and (3) organic matter acts as a reducing agent and UO_2^{2+} is converted to the insoluble quadrivalent ion.

(b) *Platey minerals*: Platey minerals reduce the solubility of organic material, and the uranium is adsorbed within the resulting flocculate. Clays also encourage the formation of schoepite by the hydrolysis of the uranyl ions.

(c) *Sulphur*: Uranium is reduced and precipitated as uraninite from uranyl solution, by bacterial action in an anaerobic, reducing environment.

(d) *Phosphates*: A natural phosphate will extract 63% of the uranium in a uranyl solution; uranium replaces calcium in carbonate-fluorapatite.

(e) *pH*: An acid environment (2.5 < pH < 4) is favourable to the precipitation of uranium.

(f) *Eh, Redox potential*: Precipitation is most likely in a reducing environment with $0 > Eh > -400$ mV.

(g) *Adsorptive material*: Amorphous silica, alumina, alumino-silicates, coals, promote the hydrolysis of the uranyl ion to schoepite in an acid medium.

From this it is apparent that uranium will tend to precipitate or concentrate in confined, reducing environments, which are themselves favourable to the accumulation and preservation of organic matter.

6.4.2.4. *Diagenetic migration*

The soluble ions of uranium can be recirculated during diagenesis by the flushing action of subterranean or hydrothermal waters. This is enhanced by the presence of fissures in the rock. Such migration is comparable to that of the hydrocarbons with which the uranium may well be associated, but generally precedes it by virtue of its solubility. The uranium, therefore, will tend to accumulate down-dip, while the hydrocarbons will later migrate up-dip.

Figure 6-8 explains the sedimentary environments where uranium may accumulate:

(a) clay particles, adsorbed onto their surfaces;

(b) phosphates;

(c) very resistant minerals such as zircon, sphene, monazite, allanite, biotite, xenotime in detrital, fluvial, lacustral or deltaic sediments;

(d) organic matter, that of vegetable (humic) origin being much richer in uranium than that of algal

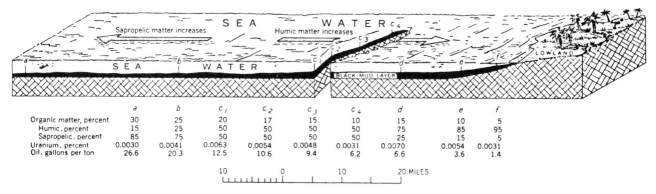

	a	*b*	*c₁*	*c₂*	*c₃*	*c₄*	*d*	*e*	*f*
Organic matter, percent	30	25	20	17	15	10	15	10	5
Humic, percent	15	25	50	50	50	50	75	85	95
Sapropelic, percent	85	75	50	50	50	50	25	15	5
Uranium, percent	0.0030	0.0041	0.0063	0.0054	0.0048	0.0031	0.0070	0.0054	0.0031
Oil, gallons per ton	26.6	20.3	12.5	10.6	9.4	6.2	6.6	3.6	1.4

Fig. 6-9. Sketch showing theoretical distribution of humic and sapropelic materials in a shallow sea in which black muds are accumulating, and the estimated uranium content and oil yield of the resulting black shale. Increase of total organic matter seaward due chiefly to seaward decrease in amount of detrital sediment; proportional increase of sapropelic matter seaward due to decrease of humic land-plant debris and predominance of planktonic matter (from Swanson, 1960).

(sapropelic) origin. Figure 6-9 (after Swanson) shows the occurrence of uranium in organic matter and other sediments in different environments.

6.4.3. Thorium-bearing minerals and rocks

Thorium originates from igneous rocks of the acid and acido-basic types (granites, pegmatites, syenites, nepheline syenites).

Table 6-8 lists the average concentrations (in ppm) encountered.

During the alteration and leaching of rocks, up to 90% of the thorium present can be removed. How-ever, thorium is fairly insoluble, and thorium-bearing minerals are stable; consequently, almost all thorium is transported in suspension, and is a common constituent of the detrital fraction of sediments. The small amount of thorium that passes into solution is readily adsorbed onto clay minerals, or forms secondary products of hydrolysis with the resistant thorium-bearing minerals. Unlike uranium, thorium does not migrate during diagenesis.

Thorium is found principally:

(a) in clays of detrital origin, adsorbed onto the platelets;

(b) in certain altered igneous rocks of the acidic or

TABLE 6-10

Thorium and thorium-bearing minerals (after Frondel et al., 1956, and Roubault, 1958)

Name	Composition	ThO_2 Content, %
Thorium minerals		
Cheralite	$(Th, Ca, Ce) (PO_4 SiO_4)$	30, variable
Huttonite	$ThSiO_4$	81.5 (ideal)
Pilbarite	$ThO_2, UO_3, PbO, 2 SiO_2, 4 H_2O$	31, variable
Thorianite	ThO_2	Isomorphous series to UO_2
Thorite **	$ThSiO_4$	25 to 63-81.5 (ideal)
Thorogummite **	$Th(SiO_4)_{1-z}(OH)_{4-ai} x < 0.25$	24 to 58 or more
Thorium-bearing minerals		
Allanite	$(Ca, Ce, Th)_2 (Al, Fe, Mg)_3 Si_3 O_{12}(OH)$	0 to about 3
Bastnaesite	$(Ce, La)Co_3 F$	less than 1
Betafite	about $(U, Ca)(Nb, Ta, Ti)_3 O_9, n H_2O$	0 to about 1
Brannerite	about $(U, Ca, Fe, Th, Y)_3 Ti_5 O_{16}$	0 to 12
Euxenite	$(Y, Ca, Ce, U, Th)(Nb, Ta, Ti)_2 O_6$	0 to about 5
Eschynite	$(Ce, Ca, Fe, Th)(Ti, Nb)_2 O_6$	0 to 17
Fergusonite	$(Y, Er, Ce, U, Th) (Nb, Ta, Ti)O_4$	0 to about 5
Monazite *	$(Ce, Y, La, Th)PO_4$	0 to about 30; usually 4 to 12
Samarskite	$(Y, Er, Ce, U, Fe, Th) (Nb, Ta)_2 O_6$	0 to about 4
Thucholite	Hydrocarbon mixture containing U.Th. rare earth elements	
Uraninite	UO_2 (ideally) with Ce, Y, Pb, Th, etc.	0 to 14
Yitrocrasite	About $(Y, Th, U, Ca) (2(Ti, Fe, W)_4 O_{11}$	0 to 9
Zircon	$ZrSiO_4$	usually less than 1

* Most important commercal ore of thorium. Deposits are found in Brazil, India, USSR, Scandinavia, South Africa, and U.S.A.

** Potential thorium ore minerals.

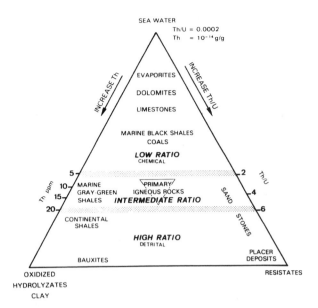

Fig. 6-10. Distribution of thorium and Th/U ratio in sediments (from Adams and Weaver, 1958).

acidobasic type (cinerites, bentonites),

(c) in certain beach-sands and placers, associated with resistive heavy minerals such as monazite, zircon, xenotime, allanite, where thorium is present by substitution;

(d) in chemical compounds of thorium.

Table 6-10 gives a list of the main thorium-bearing minerals.

Figure 6-10 is a schematic presentation of the distribution of thorium in sediments, and its abundance relative to uranium (from Adams and Weaver, 1958).

6.4.4. Summary

The principal radioactive rocks, then, are:

(a) Clays and shales, which by one means or another "fix" K, Th and U. In addition, they may contain significant quantities of phosphates or organic matter, rich in uranium, and radioactive minerals (feldspars, micas, uranium and thorium-bearing heavy minerals, Table 6-11).

TABLE 6-11

Thorium and uranium contents of accessory minerals (from Clark, 1966)

Name	Th (ppm)	U (ppm)
Allanite	500–20,000	30–700
Apatite	20–150	5–150
Epidote	50–500	20–50
Magnetite	0.3–20	1–30
Monazite	$2.5 \times 10^4 - 20 \times 10^4$	500–3000
Sphene	100–600	100–700
Xenotime	low	$500 - 3.5 \times 10^4$
Zircon	50–4000	100–6000

(b) Potassium salts.

(c) Bituminous and alunitic schists.

(d) Phosphates.

(e) Certain arkosic or graywacke sands, silts, sandstones, siltstones, or conglomerates, rich in minerals like feldspars, micas, or in uranium- or thorium-bearing minerals such as zircon, monazite, allanite, sphene, xenotime, or phosphates.

(f) Certain carbonates, originating from algae that have fixed potassium and uranium, or are rich in organic matter, or in phosphates.

(g) Some coals that have adsorbed or accumulated uranium that has been leached out of uranium-rich rocks.

(h) Acid or acido-basic igneous rocks such as granite, syenite, rhyolite.

6.5. MEASUREMENT OF GAMMA RADIATION

Gamma radiation can be measured by Geiger-Mueller counters, ionization chambers, and scintillation counters (Fig. 6-11).

6.5.1. Geiger-Mueller counter

This is composed of a metal chamber with a central wire maintained at a positive potential relative to the cylindrical chamber wall (900–1000 V). The chamber contains gas (argon, helium, or neon) at a low pressure. Incident gamma rays cause the ejection of electrons from the detector wall into the gas. As the ejected electron is drawn towards the highly charged central wire, other collisions occur between each electron and gas atoms, thus producing additional electrons which in turn cause additional ionization by collision. This results in a multiplication of the ionization events. An avalanche of electrons arrives at the central wire for each incident gamma-ray interaction with the cylindrical wall of the chamber. The dead-time is of the order of 0.1–0.2 ms.

6.5.2. Ionization chamber

It consists of a gas-filled cylindrical metal chamber containing a central rod maintained at about 100 V positive to the cylinder wall. The gas is maintained at high pressure (100 atmospheres). Incident gamma rays interact with the wall material and cause the ejection of electrons. The electrons drawn to the central rod produce additional electrons in the collisions with gas atoms. The electrons moving to the central rod constitute a minute flow of electrical current, proportional to the number of gamma-ray interactions.

6.5.3. Scintillation counter

This consists of both a crystal detector, often called phosphor, and a photomultiplier to produce countable pulses. Gamma rays entering the crystal produce light flashes. The flash produced strikes the sensitive surface of a photocathode in the photomultiplier, causing the emission of a number of primary electrons by photoelectric effect. These electrons are attracted to the first of a series of anodes, each of which is maintained at a successively higher positive potential by a voltage source and a potential divider. Each anode surface emits several electrons (4 to 8) when struck with a single electron of sufficient energy. The anodes are shaped and arranged so that each secondary electron moves toward the next anode, accelerated by the higher positive potential. Thus a 10-anode tube with a gain of 6 per stage has an overall gain of 6^{10}. The last anode is connected to the positive voltage supply through a series of resistances. A flash of light on the photocathode will result in the appearance of a pulse of electrons at the anode, which will produce a negative output pulse because of the voltage drop across the load resistor. This negative pulse can be transmitted to the recording circuits through capacitance C. The resulting pulse is proportional to the energy of the incident gamma ray.

This last type has an efficiency of some 50–60% (compared with 1–5% for the others) and crystals some 4–8 inches in length can be used (compared with 30 inches or more), giving good vertical resolution (Fig. 6-11).

6.5.4. Response of the tool

The response of the tool is a function of the concentration by weight of the radioactive mineral in the rock, and the density of the rock.

$$GR = \frac{\rho V}{\rho_b} A \qquad (6\text{-}13)$$

where:
GR is the total measurable γ-radiation
ρ is the density of the radioactive mineral or element
V is its percentage by volume
ρ_b is the bulk density of the formation
A is a constant of proportionality which characterizes the radioactivity of the mineral or element.

A certain volume fraction of a radioactive mineral present in a dense rock would produce, therefore, a lower radiation count than the same fraction present in a lighter rock. This is in part due to the fact that gamma-ray absorption by the formation increases with density (Table 6-3).

Where there are several radioactive minerals present, we have:

$$GR = \frac{\rho_1 V_1}{\rho_b} A_1 + \frac{\rho_2 V_2}{\rho_b} A_2 + \ldots + \frac{\rho_n V_n}{\rho_b} A_n \qquad (6\text{-}14)$$

i.e.

$$\rho_b GR = B_1 V_1 + B_2 V_2 + \ldots + B_n V_n \qquad (6\text{-}15)$$

$B_1 = \rho_1 A_1, B_2 = \rho_2 A_2$, etc., and are constant for a given mineral type. The GR response is in a formation of density ρ_b is normalized by taking the product $\rho_b GR$. (ρ_b can be obtained from a density log).

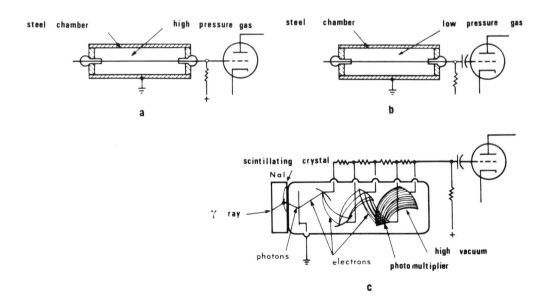

Fig. 6-11. The three types of gamma-ray detectors and their principle of measurement. a. Ionization chamber. b. Geiger-Mueller counter. c. Scintillation counter.

TABLE 6-12

Relative contribution of an average shale to gamma-ray flux

	K (total)	Th	U
Relative gamma-ray activity per unit weight of element	1	1300	3600
Relative abundance in average shale	27.000	12 ppm [a]	3.7 ppm [a]
Relative contribution to the gamma ray flux of average shale	48% (2.7)	28% (1.56)	24% (1.33)

[a] from Adams and Weaver, 1958.

Considering the three radioactive elements; K, U and Th, Adams and Weaver (1958) found that, weight for weight, the γ-activity of uranium and thorium was, respectively, 3600 and 1300 times stronger than that of potassium. (The A terms in eq. 6-14. The B terms of eq. 6-15 are derived from this relationship (Table 6-5).

Potassium is, therefore, considerably less active than uranium and thorium. However, because it is more abundant, its contribution to the *GR* response is as significant as that of the other two isotopes (Table 6-12).

Measurement of the total γ-radiation (Fig. 6-12) follows the overall proportion of radioactive materials present. But in general, it is not possible to evaluate either the nature of the active elements, or their relative proportions. [Potassium salts would be an exception. In this case, we can reasonably assume that all the gamma activity is related to potassium and can be calibrated directly in percentage of potassium (Fig. 6-13).] To achieve this, the technique of natural gamma-ray spectrometry must be used.

Shales are the most common radioactive rocks (if

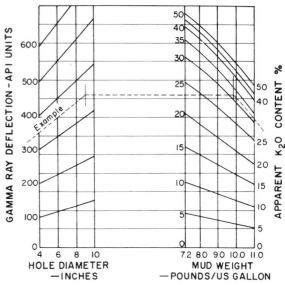

VALID FOR: SGD-F DETECTOR
SGC-G OR GNC-H CARTRIDGE
WITH SCALER OF 2
TIME CONSTANT = I SECOND
LOGGING SPEED = IO FT/MIN
JAN.'66   BEDS THICKER THAN 2 FT

Fig. 6-13. Nomogram for computation of K_2O content (from Crain and Anderson, 1966).

we ignore potassium salts), with the radiation arising primarily from the clay fraction. To a reasonable approximation we can consider that the *GR* level is related to shaliness by:

$$V_{sh} \leqslant (V_{sh})_{GR} = \frac{GR - GR_{min}}{GR_{sh} - GR_{min}} \qquad (6-16)$$

It must be understood that *GR* may include radioactivity from sources other than shale. For this reason, the right side of the equation is an upper limit to V_{sh}.

6.6. MEASURING POINT

It corresponds to the mid-point of the active part of the detector.

6.7. RADIUS OF INVESTIGATION

As we have seen, gamma rays are absorbed or attenuated by the medium through which they travel, particularly when their energy is low or the medium dense. Consequently, a natural gamma-ray tool only detects radiation originating from a relatively small volume surrounding the detector.

Figure 6-14 is the geometrical factor appropriate for a limestone formation.

In a homogeneous formation, the volume of investigation is approximately a sphere centred on the

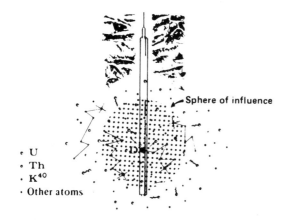

Fig. 6-12. Scheme of the sphere of influence contributing to the gamma-ray measurement.

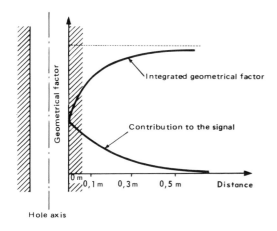

Fig. 6-14. Radius of investigation and pseudo geometrical factor (from Desbrandes, 1968).

detector (Fig. 6-15). Some deformation from the spherical will be introduced by the finite size of the detector, and the presence of the borehole. The radius r of this "sphere" depends on the gamma-ray energy and the densities of both the formation and the mud, and becomes smaller as the energy decreases or the densities increase.

As a consequence of the finite volume of investigation, bed boundaries are not defined sharply; pro-

vided the sphere (or deformed ovoid, to be more exact) is not completely clear of the radioactive bed, some gamma activity will be measured (left side of Fig. 6-15).

6.8. VERTICAL DEFINITION

Vertical definition is equal to the diameter of the "sphere" of investigation, and varies accordingly with formation and densities, and gamma ray energies.

6.9. FACTORS AFFECTING THE GAMMA-RAY RESPONSE

6.9.1. Statistical variations

These are discussed in chapter 5.3.

6.9.2. Logging speed

See chapter 5.5.

6.9.3. Hole conditions

Gamma radiation is attenuated to different degrees by the hole-fluid and any tubing, casing, ce-

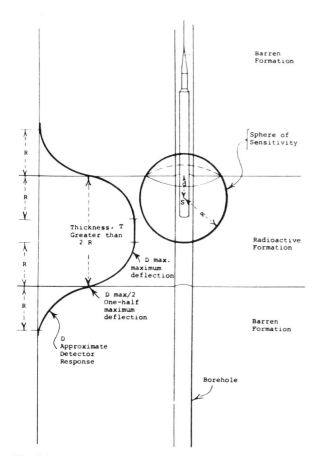

Fig. 6-15. Sphere of influence for a detector compared to the bed thickness, and shape of the curve (from Hallenburg, 1973).

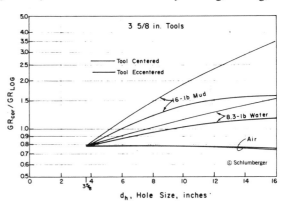

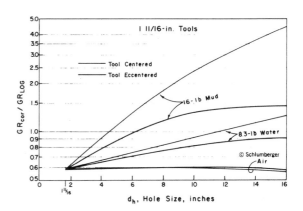

"GR$_{cor}$" is defined as the response of a 3⅝-in. tool eccentered in an 8-in. hole with 10-lb mud.

Fig. 6-16. Correction charts for borehole influence.

ment, etc. that may be present. Correction charts for Schlumberger's 3 5/8″ and 1 1/16″ tools are shown in Fig. 6-16 (which corresponds to Chart POR-7 of the 1979 Schlumberger Chart Book). Note that the position of the sonde in the hole affects the log reading to some extent.

6.9.3.1. *The hole-fluid*

The influence of the hole-fluid depends on:

(a) its volume, i.e. the hole or casing size,

(b) the position of the tool (centered or ex-centered),

(c) its density (air-, water- or oil-based, solids content, etc.),

(d) its composition (the nature of materials dissolved or in suspension; NaCl, KCl, bentonite, barite, etc). Bentonite is radioactive, and will introduce a shift in reading levels, roughly constant if the borehole is uniform. KCl solution, on the other hand, will invade permeable sections. The next result will be an overall increase in gamma radioactivity because of the mud-column, and an additional increase wherever filtrate has penetrated the formation (and this varying with the degree of flushing).

The significance of the inequality sign in eq. 6-16 of Section 6.5 is easily appreciated in these conditions.

6.9.3.2. *Tubing, casing, etc.*

The effects of tubing, casing, packers, and the like, depend on the thickness, density and nature of the materials (steel, aluminium, etc). All steel reduces the gamma-ray level (Eq. 6-16).

6.9.3.3. *Cement*

Deciding factors here are the cement type, nature of additives, density and thickness.

6.9.4. **Bed thickness**

The gamma-ray curve will not attain the correct value in a bed whose thickness is less than the diameter of the "sphere" of investigation. Once again, in a series of thin beds, the log reading will be a volume average of the contributions within the sphere. Following Hallenburg (1973) a good approximation of the *GR* reading D_a can be written as:

$$D_a = V_1 D_1 + V_2 D_2 + \ldots + V_n D_n^{(1)} \qquad (6\text{-}17)$$

where V_i represents the volume of the ith substratum, and D_i is its *GR* level (were it thick enough to be measured) *.

---

* In fact, each point within the volume of investigation does not have the same influence at the detector, the energy of the gamma rays decreasing with distance. Referring to Fig. 6-13, any definition of "radius" must be arbitrary—say the 80% point).

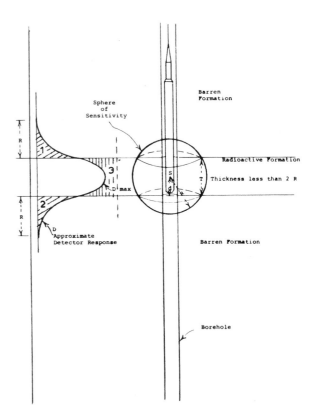

Fig. 6-17. Response in a thin bed (from Hallenburg, 1973).

In Fig. 6-17, the detector is centered in a thin bed. Assuming the adjacent strata to have zero gamma activity, it is possible to approximate the maximum *GR* level, D_{max}, that the thin bed can have. If the log reading is D_a:

$$D_{max} = \frac{16r^3 D_a}{h(12r^2 - h^2)} \qquad (6\text{-}18)$$

Here, h is the thickness of the bed, and r is the radius of the sphere of investigation.

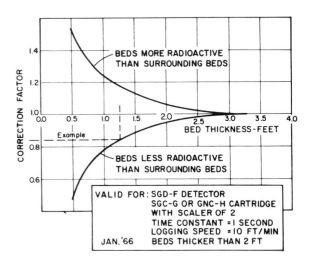

Fig. 6-18. Gamma-ray bed-thickness correction chart (from Crain and Anderson, 1966).

Equation 6-18 applies to a stationary tool. The effect of logging at a finite speed is to produce a response as shown in Fig. 5-5 whose shape depends on the logging speed.

If A is the total area under the peak, then we can approximate:

$$A = GR_t h \qquad (6-19)$$

where h is the bed thickness and GR_t is its true radioactive level. This says, effectively, that shaded areas 1 and 2 in Fig. 6-17 equal area 3.

Figure 6-18 is a correction chart for bed thickness, for one particular set of conditions.

6.10. APPLICATIONS

(a) Evaluation of lithology: shales, evaporites, uranium.

(b) Estimation of shale fraction of reservoir rocks:

$$V_{sh} \leqslant (V_{sh})_{GR} = \frac{GR - GR_{min}}{GR_{sh} - GR_{min}}$$

(c) Well-to-well correlations. Detection of unconformities.

(d) Sedimentology: the gamma-ray evolution with depth gives information on the grain-size evolution.

(e) Depth control of perforating and testing equipment (FIT, RFT, CST, various shaped; charge guns) in open hole where the SP is poorly developed, or through tubing or casing.

(f) The evaluation of injection profiles, behind-casing leaks, etc using radioactive tracer materials.

(g) An approach to estimating permeability.
All these applications will be developed in volume 2.

6.11. CALIBRATION

The API unit of gamma radiation corresponds to 1/200 of the deflection measured between two reference levels of gamma-ray activity in a test pit at the University of Houston (Fig. 6-19). The test pit consists of three zones, two of low activity, one of high activity. Mixtures of thorium, uranium and potassium are used to obtain these different levels. All commercial logging tools are calibrated here.

Schlumberger ensures that its gamma-ray equipment operates within specification by adjusting the tool response before each job against a portable 266Ra source (0.1 mCi) clamped 53″ (1.35 m) from the sonde. This produces a calibration level of:
165 API for the 3 5/8″ diameter tools (GNT-F/G, SGT-B/C/E)
135 API for the 1 11/16″ diameter GNT-K, SGT-g
155 API for the 2″ diameter high-pressure GNT-K
115 API for the old GNT-K with G-M detector

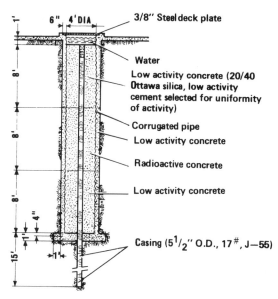

Fig. 6-19. The gamma-ray test pit at Houston (from Belknap, 1959).

These levels are equivalent to approximately 10 μg Ra equivalent/tonne (see paragraph 6.2.6.).

6.12. REFERENCES (see also end of chapter 7.)

Adams, J.S. and Gasparini, P., 1970. Gamma-ray spectrometry of rocks. Elsevier, Amsterdam.

Adams, J.A. and Weaver, C.E., 1958. Thorium to uranium ratios as indicators of sedimentary processes; example of concept of geochemical facies. Bull. Am. Assoc. Pet. Geol., 42 (2).

American Petroleum Institute, 1959. Recommended practice for standard calibration and for nuclear logs. API. R.P. 33, September 1959.

Belknap, W.B., 1959. Standardization and calibration of nuclear logs. Pet. Eng., December.

Blanchard, A. and Dewan, J.T., 1953. The calibration of gamma ray logs. Petrol. Eng., 25.

Blatt, H., Middleton, G. and Murray, R., 1980. Origin of sedimentary rocks. Prentice-Hall, Inc., Englewood Cliffs, N.J.

Crain, E.R. and Anderson, W.B., 1966. Quantitative log evaluation of the Prairie evaporite formation in Saskatchewan. J. Can. Pet. Technol., July-Sept.

Green, W.G. and Fearon, R.E., 1940. Well logging by radioactivity. Geophysics, 5 (3).

Hallenburg, J.K., 1973. Interpretation of gamma ray logs. Log Analyst, 14 (6).

Howel, L.G. and Frosch, A., 1939. Gamma ray well logging. Geophysics, 4 (2).

Hyman, S.C. et al., 1955. How drill-hole diameter affect gamma ray intensity. Nucleonics, Feb. 1955, p. 49.

Kokesh, F.P., 1951. Gamma ray logging. Oil Gas J., July 26, 1951.

Press, F. and Siever, R., 1978. Earth. W.H. Freeman & Co. San Francisco, Calif.

Rabe, C.L., 1957. A relation between gamma radiation and permeability, Denver Julesburg Basin. J. Pet. Technol., 9 (2).

Russel, W.L., 1944. The total gamma ray activity of sedimentary rocks as indicated by Geiger-Counter determination. Geophysics, 9 (2).

Scott, J.H. et al., 1961. Quantitative interpretation of gamma ray logs. Geophysics, 26 (2).

Swanson, V.E., 1960. Oil yield and uranium content of black shales. U.S. Geol. Surv., Prof. Pap., 356-A.

Tittman, J., 1956. Radiation logging: physical principles. In: Univ. Kansas Petroleum Eng. Conf., April 2 and 3, 1956.

Tixier, M.P. and Alger, R.P., 1967. Log evaluation of non metallic mineral deposits. SPWLA, 8th Annu. Symp. Trans., paper R.

Toelke, L.W., 1952. Scintillometer used in radiation logging. World Oil, 135 (3).

Westby, G.H. and Scherbatskay, S.A., 1940. Well logging by radioactivity. Oil Gas. J., 38 (41).

7. NATURAL GAMMA-RAY SPECTROMETRY

7.1. PRINCIPLES

For the conventional gamma-ray log we record the total natural gamma radiation. As we have seen, this radiation is in fact emitted from three main types of source elements: ^{40}K, ^{232}Th or ^{238}U, (and their decay products),

The gamma rays emitted by the three decay series have a number of discrete energies. In Fig. 7-1 are shown the three corresponding gamma-ray emission spectra. Each spectrum characterizes a decay series, each series has a spectral "signature" that enables its presence to be discerned.

As seen previously (Chapter 6) ^{40}K is characterized by a single gamma-ray emission at 1.46 MeV, corresponding to the gamma ray emitted by the $^{40}Ar^*$ isotope (Fig. 6-4) *. In the ^{232}Th series, the most distinctive gamma-ray peak is at 2.62 MeV from ^{208}Tl, Fig. 6-5), but there are several lower energy peaks. The ^{214}Bi peak at 1.76 MeV is used to distinguish the ^{238}U series; again, there are numerous other peaks.

But gamma rays emitted with discrete energies can be degraded by: (a) pair production in the formation and detector crystal, if the gamma energy is higher than 1.02 MeV; and (b) Compton scattering in the formation between the point of emission and the detector.

The observed spectrum takes a continuous rather than a discrete form (Fig. 7-2). Such degradation is worsened by logging conditions (compare Fig. 7-2 and 7-1). It is due to: (a) limitation of detector size (poor efficiency); (b) volume of investigation (important, with large contribution of scattered gamma rays); and (c) duration of the measurement (limited in order to achieve a reasonable logging speed). The result is a very large background in the spectrum, so that its fine structure, given by about 50 peaks, is almost entirely lost (Fig. 7-2).

The relative amplitudes of the three spectra will depend on the proportions of the radioactive components present, so that a quantitative evaluation of the presence of thorium, uranium and potassium can be obtained by breaking down the total spectrum into the three characteristic spectra.

---

* 11 percent of the decays of ^{40}K involve electron capture and $^{40}Ar^*$ is the product. The $^{40}Ar^*$ is, however, in an unstable excited state, and upon returning to its ground-state, emits the gamma photon at 1.46 MeV (Fig. 6–4).

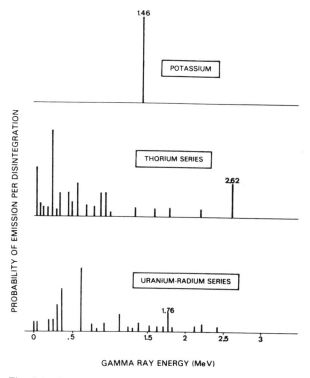

Fig. 7-1. Gamma-ray emission spectra of radioactive elements (courtesy of Schlumberger).

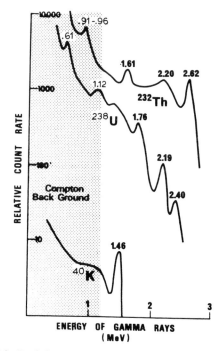

Fig. 7-2. Real Spectrograms of K, Th and U obtained from NaI(Tl) crystal detector (from Doig, 1968).

This analysis will only be valid if we can assume that the spectra of the three radioactive series always have the same energy distribution. As thorium and uranium series are normally in a state of secular equilibrium, each series has the characteristic spectrum observed on Fig. 7-1 with the different elements occurring in fairly well-defined proportions.

To obtain a quantitative evaluation of thorium, uranium and potassium from an analysis of the total energy distribution, it is helpful to divide the spectrum into two regions:

(1) The high-energy region, with three main peaks:
Thallium ^{208}Tl at 2.62 MeV (from the ^{232}Th family)
Bismuth ^{214}Bi at 1.76 MeV (from the ^{238}U family)
Potassium ^{40}K at 1.46 MeV

(2) The low-energy region, covering the energy range of the gamma rays resulting from Compton scattering in the formation, plus lower-energy emissions from the thorium and uranium series.

7.2. TOOL DESCRIPTION

The Dresser Atlas Spectralog® tool measures the counting rates in a number of "windows", each of which spans a certain energy band. As can be seen from Fig. 7-3 the high-energy region is divided into three windows (W 3, W 4 and W 5), centered on the three characteristic high-energy peaks. Thus, broadly speaking, we can see that these three windows will have responses corresponding to the amounts of thorium, uranium and potassium respectively. There will be a certain amount of correlation (for example, it is clear that window 3 will respond not only to the presence of potassium, but also to some extent to the presence of uranium and thorium). However, with three windows it is possible to compute the exact solution:

$$^{232}\text{Th} = m_{13}\text{W}_3 + m_{14}\text{W}_4 + m_{15}\text{W}_5 \qquad (7\text{-}1)$$

$$^{238}\text{U} = m_{23}\text{W}_3 + m_{24}\text{W}_4 + m_{25}\text{W}_5 \qquad (7\text{-}2)$$

$$^{40}\text{K} = m_{33}\text{W}_3 + m_{34}\text{W}_4 + m_{35}\text{W}_5 \qquad (7\text{-}3)$$

where m_{ij} is the current element of the so-called measurement matrix.

The Schlumberger Natural Gamma ray Spectrometry tool (NGS *) uses five windows (Fig. 7-4), making fuller use of the information in the spectrum so as to reduce the statistical uncertainty on the analysis of Th, U and K. Indeed the high-energy region represents only ten percent of the spectrum in terms of counting rates. As a result, the above computations are subject to a very substantial statistical error.

Much-improved results can be obtained if we take into consideration the remaining (low-energy) portion of the detectable spectrum. This region contains meaningful information that is pertinent to the spectrometry measurement.

Figure 7-5, from Marett et al. (1976), compares the result of a three-window analysis with one incorporating low-energy information. The reduction in statistics is clearly seen in the second case: the standard deviation of the statistical noise for Th, U and K is reduced substantially (Fig. 7-6).

To illustrate the importance of the low-energy windows an artifical log was created. By adding statistical variations to a theoretical model (Fig. 7-7), which assumes a succession of levels with different thorium, uranium and potassium content, artificial window count-rates were created (Fig. 7-8). As we observe, small peaks on thorium, uranium or potassium are only clearly seen on the low-energy windows. On the high-energy windows, these peaks are difficult to separate from the statistical noise. To define the sensitivity of the tool another artificial log was created (Fig. 7-9) in which a sequence of thorium, uranium and potassium peaks was created. Even if the spectrum is degraded by Compton scattering it appears that the low-energy windows are the most sensitive: compared to a constant background, peaks of 1 ppm uranium, 2 ppm thorium and 0.3% potassium are detected only on the two first low-energy windows and do not appear on the three high-energy windows where they are masked by statistics.

7.3. DETECTOR

The detector consists of a sodium iodide crystal doped with thallium—referred to as NaI (Tl). It is optically coupled to a photo-multiplier (Fig. 7-10), and an amplifier and a multi-channel analyser com-

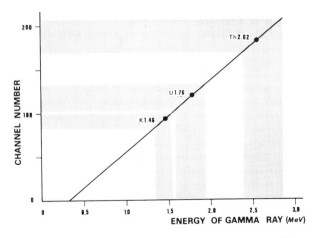

Fig. 7-3. Window repartition in the Dresser Atlas' Spectralog.

---

* Trade Mark of Schlumberger. Although the NGS spectrum is degraded, mainly by Compton scattering, it can be retrieved by statistical analysis.

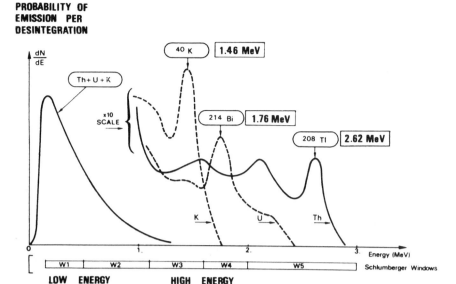

Fig. 7-4. Potassium, Thorium and Uranium response curve (NaI crystal detector) (from Serra et al., 1980).

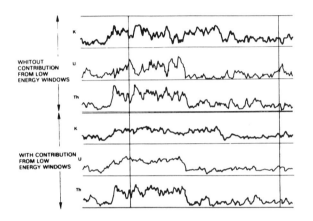

Fig. 7-5. Gamma-ray spectrometry with and without low-energy windows (from Marett et al., 1976).

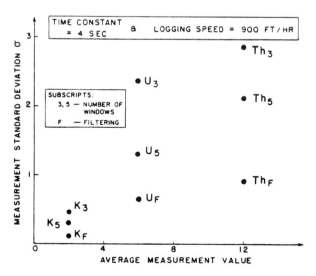

Fig. 7-6. NGS statistical error (from Serra et al., 1980).

plete the measuring circuitry. Unlike previous gamma-ray tools, which only counted pulses, the energy determination of the gamma rays entering the crystal is of the utmost importance.

Figure 7-11 shows how the proportionality between the energy E of the gamma ray entering the crystal and the output voltage V at the preamplifier is achieved (see figure caption).

The pulse amplitude V is analyzed by a set of comparators each of which has a fixed reference voltage. The window count rates obtained by these comparators are stored in a buffer and sent to the surface through the telemetry. The purpose of the feedback loop, shown on the downhole block diagram of Fig. 7-10, is to stabilize the gamma spectrum in energy. It uses a reference gamma-ray source and can be considered as a continuous energy calibration of the tool. Since the behaviour of the detector versus temperature is unpredictable, this stabilization is essential.

At the surface, the transmitted information is decoded and made available for further digital processing in the CSU mode. For optical display the data are depth-averaged. In this averaging technique the data sampled at one depth are averaged with data sampled just before (past) and just after (future), before being recorded on film. This gives a symmetric curve in time with no lag related to logging speed. A further reduction in the statistical noise is achieved using the filtering technique described later. The standard gamma ray, GR, along with the filtered Th, U, K and "uranium-free" gamma-ray measurement, CGR, are available for playback at the well-site with the CSU system.

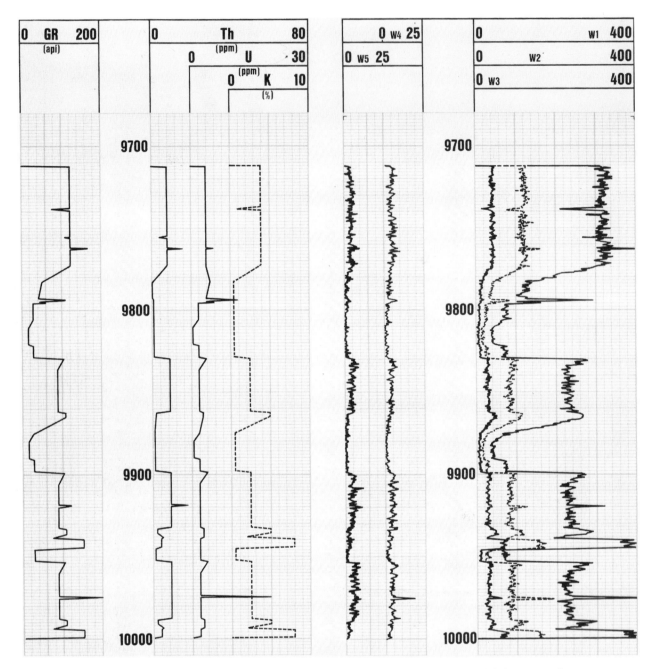

Fig. 7-7. Main log. Model.

Fig. 7-8. Main log. Window counts (in cps).

A time-constant of 4 seconds was used for logging the NGS-tool at 900 ft/hr with panel. A time-constant of 6 seconds and a recording speed of 600 ft/hr are required for the Spectralog. With the CSU system, the measurement time is fixed and samples are taken every six inches of tool travel. The sampling time chosen is sufficient to maintain the standard deviation at less than 2.5%, when the standard logging speed is used (900 ft/hr).

7.4. CALIBRATION

For Schlumberger's tools, the base of the master calibration is a special test pit (TUK pit) which has been built in Clamart (Fig. 7-12).

The basic structure of this pit consists of four zones. Each of the top three zones contains the three radioactive elements, thorium, uranium and potassium, with the greatest possible contrast between the zones.

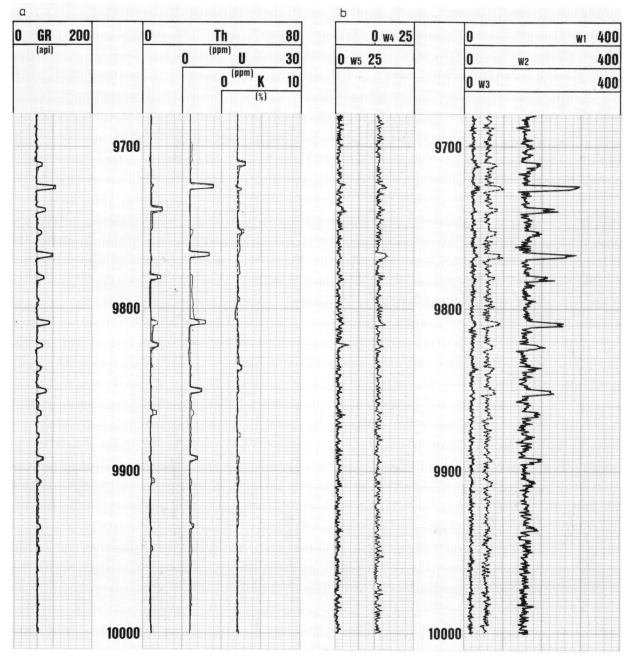

Fig. 7-9. a. Model for testing the sensitivity. Th, U, K computation using a Kalman filter. b. Model for testing the sensitivity. Window counts (in cps).

A fourth zone, at the bottom of the pit, enables the descent of long tools and the evaluation of the radioactive contribution of the concrete that is the main component of the lower zone. The size of the pit was dictated by the most powerful radiation emitted by the components, i.e. the 2.62 MeV gamma ray of the thorium series.

The NGS calibration involves the measuring of the counting rates in the five windows when the tool is placed in the centre of the three zones.

The counting rate in window i when the tool is placed in zone j is called W_{ij}. There are 15 W_{ij} values.

These values are related to the Th_j, U_j, K_j contents by the linear relations:

$$W_{ij} = A_i Th_j + B_i U_j + C_i K_j \qquad (7\text{-}4)$$

where Th_j, U_j, K_j are known and W_{ij} measured.

The 15 coefficients A_i, B_i, C_i are computed by resolving the 3 equations system:

$$W_{i1} = A_i Th_1 + B_i U_1 + C_i K_1 \qquad \text{(zone 1)}$$
$$W_{i2} = A_i Th_2 + B_i U_2 + C_i K_2 \qquad \text{(zone 2)}$$
$$W_{i3} = A_i Th_3 + B_i U_3 + C_i K_3 \qquad \text{(zone 3)}$$

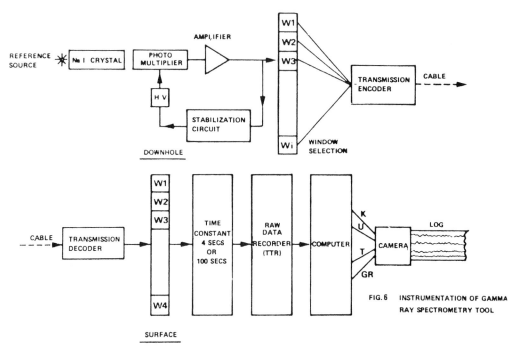

Fig. 7-10. Schematic of the NGS tool (from Marett et al., 1976).

The determination of these 15 coefficients can be considered as the tool master calibration. However, as there are no TUK pits in the field (each zone corresponds to about 6 tonnes of concrete), a unique matrix is applied to all tools, because the instrumental dispersion from tool to tool is small. The field is provided with secondary calibrations in order to check the operation of the tool and its stability.

The tool response can be considered as the product of two parameters:

(a) The detector efficiency ϵ: number of gamma ray detected for 1 ppm of contents in the formation. This is essentially stable.

(b) The energy response of the detector: resolution

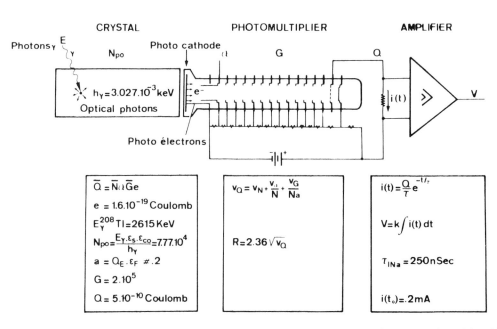

Fig. 7-11. Detector assembly illustrating how the proportionality between the energy E and the output voltage V is achieved.

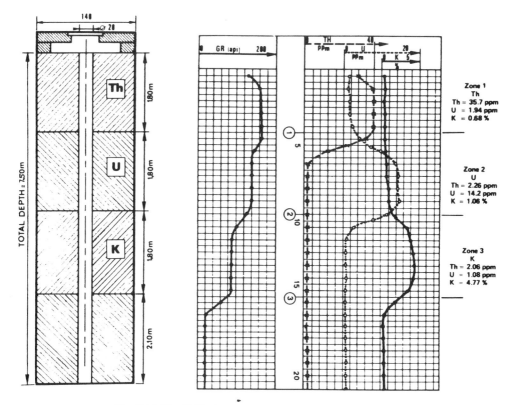

Fig. 7-12. The TUK pit in Clamart (courtesy of Schlumberger).

and conversion slope in volts output per MeV input. This slope is basically controlled by the Americium loop and can be re-adjusted in the lab following the test and inspection procedure.

Calibration at the well-site is achieved with a flexible calibrator containing monazite. This is carefully positioned around the sonde, and the count-rates in each window channel are checked.

7.5. RADIUS OF INVESTIGATION

The radius of investigation depends not only on hole size, mud-density and formation bulk density, but on the energies of the gamma rays themselves (Fig. 7-13). Higher-energy radiation can reach the detector from deeper in the formation.

7.6. FUNDAMENTAL FACTORS INFLUENCING THE MEASUREMENT

The measurement responds essentially to the concentration of Th, U, and K in the formation. There are, however, several minor perturbing factors; the occurrence of interfering peaks close to the principle peak in each window; the existence of two "escape peaks" associated with each principle high-energy peak, which result in Th interfering in the U window, and Th and U in the K window. (The escape peaks are the result of electron and positron pair production in the crystal, producing a characteristic triplet effect in the NaI (Tl) crystal.)

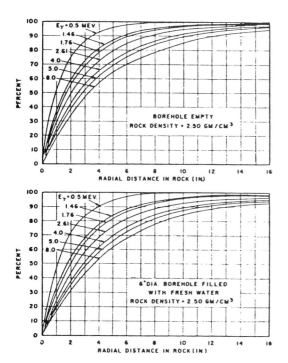

Fig. 7-13. Radius of investigation, expressed as a percentage of the infinite volume response, (pseudo geometrical factor, J), as a function of the gamma ray energy (from Rhodes et al., 1976).

120

7.7 COMPUTATION OF Th, U AND K CONTENT

An equation can be written for each window, relating its count-rate to the concentrations of the three radioactive elements. In theory three equations would be sufficient to compute the Th, U and K contents of the rock. But as seen previously the three high-energy peaks represent only 10% of the spectrum in terms of counting rates.

Although the low-energy part of the spectrum is degraded, mainly by Compton scattering, it can be retrieved by statistical analysis, and the information it carries, used for more accurate computation of the Th, U and K contents. To accomplish this, a set of equations is written as follows:

$$W_i = A_i\text{Th} + B_i\text{U} + C_i\text{K} + r_i \qquad (7\text{-}5)$$

where r_i is a factor representing the statistical errors, W_i is the count-rate from window i, and A_i, B_i and C_i are the calibration coefficients (for windows i) obtained using a special calibration pit (called the TUK pit). It is then possible to solve by the least squares method, obtaining an equation:

$$\sum_{i=1}^{n} r_i^2 = \sum_{i=1}^{n} (W_i - A_i\text{Th} - B_i\text{U} - C_i\text{K})^2 = r^2 \qquad (7\text{-}6)$$

where coefficients are optimized in such a way as to minimize r^2. The results can be further improved by weighting the r terms according to their expected standard deviations (according to Poisson's law, the expected standard deviation is proportional to the average count-rate).

7.8. FILTERING

As seen before, the Th, U, K computation results are affected by statistical variations that are more important in the three high-energy windows than in the two low ones. This creates anti-correlation especially between Th and U, and negative readings (Fig. 7-14).

To decrease these effects and to improve the log appearance a Kalman filtering technique has been used and introduced for both CSU and computing centre (FLIC) processing.

The estimations of the thorium, uranium and potassium contents are performed via a Kalman filtering type algorithm. The algorithm implemented on CSU works in real time, whereas the one designed for FLIC provides more reliable estimations by using a more sophisticated processing.

To illustrate the usefulness of a Kalman filtering and its efficiency, artificial logs have been created. The choice of artificial logs was made for the following reasons: (a) the theoretical model can simulate all the particular events which can occur in reality; and

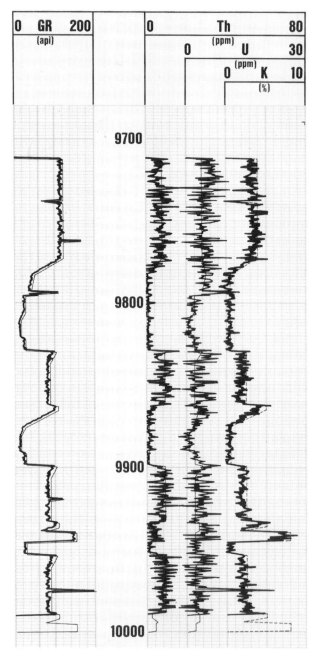

Fig. 7-14. Example of natural gamma ray spectrometry log. Computation of Th, U and K content using the matrix of standards, without filtering.

(b) the values are very well known and one can easily verify which filtering technique gives the most reliable results both in bed value and in bed boundary definition.

By adding statistical variations to a theoretical model (Fig. 7-7) which assumes a succession of levels with different Th, U and K composition, artificial window count-rates were created (Fig. 7-8) from which an artificial Th, U, K log was computed by applying the matrix of standards without filtering (Fig. 7-14).

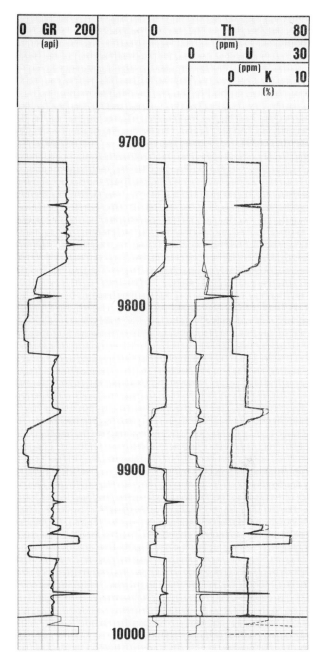

Fig. 7-15. The Th, U, K computation using a Kalman filtering technique. Compare with the model.

These artificial Th, U, K logs show all that we generally observe on a non-filtered log: anticorrelations, negative readings, high statistics.

After applying filtering anticorrelations, negative readings, and high statistics have practically disappeared, and the filtered logs appear now very close to the model (Fig. 7-15).

7.9. APPLICATIONS

7.9.1. Lithology determination

7.9.1.1. *Evaporitic environment*

In this case the NGS alone can:

(a) Differentiate between shales and potassium salts; these last minerals having a much higher potassium content than the clay minerals, and no thorium content since thorium is insoluble and can be considered as an indicator of detrital origin.

So in front of potassium evaporites, the Th curve will be flat and near zero while the K curve will show a high percentage of potassium and a shape generally very similar to that of the total gamma ray, at least if at the same time the uranium curve is flat and near zero (showing little organic material in the rock) (Fig. 7-16).

(b) Recognize the potassium evaporite mineral, through its potassium content (Table 6-6), if this mineral forms a sufficiently thick bed compared to the NGS vertical resolution. If this is not the case a combination of the NGS with other logs is necessary for a complete and accurate mineralogy determination in evaporite series;

(c) Recognize the mineral types present in the rocks and evaluate their percentages. Either cross-plot techniques or computation from a set of equations can be used. In cross-plot techniques, combination of K (from NGS) with U_{ma}, ρ_{ma}, Δt_{ma}, or even ρ_b, ϕ_N, Δt can be sufficient to solve a three-mineral mixture which corresponds to the majority of situations (Fig. 7-17). This will be developed in volume 2.

7.9.1.2. *Sand-shale series*

Very often pure clean sands or sandstones exhibit very low radioactivity, because their thorium, uranium and potassium contents are very low too. They correspond to orthoquartzites. In that case, we can generally assume a very reworked sand and consequently a high chemical and textural maturity of the detrital deposit (Fig. 7-18), with probably a medium to coarse grain size, very well sorted.

But sometimes sands or sandstones which do not contain significant percentages of clay, are radioactive. In these cases, the NGS recognizes the origin of the radioactivity and permits us in most of the cases:

(a) To compute a better shale percentage by using the shale indicators derived from the thorium or the potassium, or from their sum (CGR):

$$(V_{sh})_{Th} = (Th - Th_{min})/(Th_{sh} - Th_{min})$$

$$(V_{sh})_K = (K - K_{min})/(K_{sh} - K_{min})$$

$$(V_{sh})_{CGR} = (CGR - CGR_{min})/(CGR_{sh} - CGR_{min})$$

It is clear then that $(V_{sh})_{Th}$, $(V_{sh})_K$, or $(V_{sh})_{CGR}$ will serve as better shale indicators than $(V_{sh})_{GR}$ and $(V_{sh})_U$ since the general random associativity of uranium with shale has been eliminated. In addition,

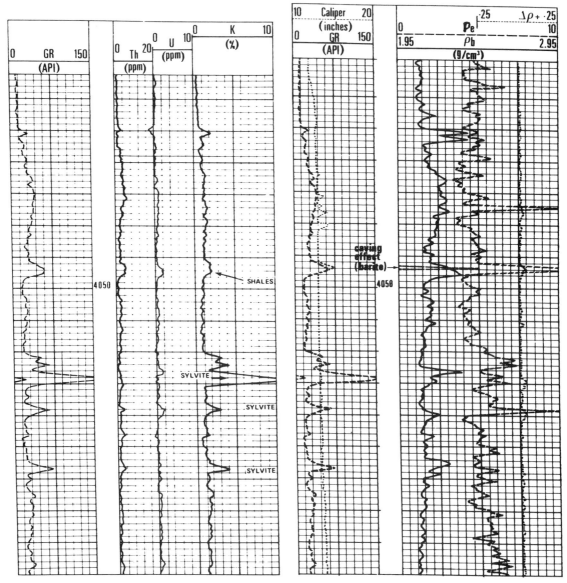

Fig. 7-16. Example of response in evaporite.

for example, in the presence of mica or feldspars, the indicator $(V_{sh})_{Th}$ is preferred as a shale indicator since the amount of mica or feldspars generally does not correlate with the volume of shale.

Figure 7-19 presents a log which displays cumulatively level-by-level counts per second due to: (1) thorium; (2) thorium and potassium; and (3) thorium, uranium, and potassium. The formation is the Langlie Mattix, and for this selected well, is known to consist of "radioactive shale sand" with intermittently distributed dolomite and anhydrite beds. The zones of high thorium and potassium concentration are known to correspond to the radioactive shaly sand, and the zones of high uranium concentration are primarily radioactive dolomite. It is obvious that due to the presence of radioactive dolomite, the gamma ray would not be a viable shale indicator.

(b) To determine the nature and the percentage of the radioactive minerals. Such sands correspond to one of the following groups:

Feldspathic sandstones or arkoses

They will show some potassium content—dependent on the feldspar percentage in the sands—due to the high percentage of potassium in feldspars (Table 6-7). At the same time the apparent matrix density, $(e_{ma})_a$ from the $\phi_N - \rho_b$ cross-plot, will probably be lower than 2.65 since these minerals have lower density (2.52 to 2.53 compared to 2.65 for quartz). The Pe and U_{ma} values will be a little higher than those of a pure sand, due to higher Pe and U_{ma} values of the feldspar: respectively 2.86 and 7.4 against 1.81 and 4.79 for quartz. This type of sandstone will show a very low Th/K ratio, less than 1×10^{-4} due to both the low thorium and the high potassium content of such detritic sands (see Hassan et al., 1975; Fig. 7-20).

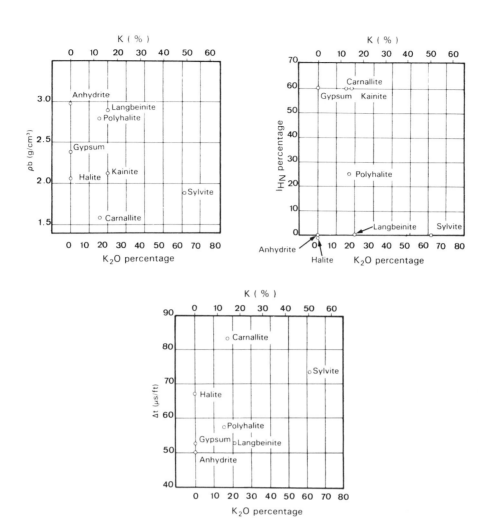

Fig. 7-17. Determination of the evaporite minerals by cross-plot technique.

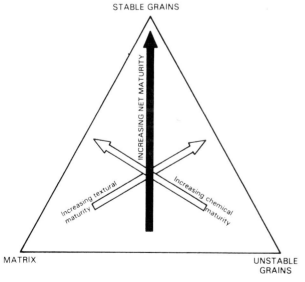

Fig. 7-18. End-member triangle to illustrate how the composition of a sand is a function both of its textural maturity, expressed as a matrix content, and of chemical maturity, expressed as unstable grain content (from Selley, 1976).

Micaceous sandstones

Micas contain potassium (Table 6-7), and consequently the potassium content of micaceous sandstones depends on the mica percentage. For the same percentage of mica or feldspars, the sand will show a lower potassium content in the former case, since the potassium content of mica is lower than feldspars. At the same time the thorium content will be higher and the apparent matrix density $(\rho_{ma})_a$ from the $\phi_N - \rho_b$ cross-plot will probably be higher than 2.65 due to the higher density of mica (2.82 to 3.1). This type of sandstone will show a Th/K ratio close to 2.5×10^{-4} generally due to the heavy thorium bearing minerals associated with micas.

The Pe and U_{ma} values will be higher than those of a pure sand.

Mixed feldspathic-micaceous sandstones or graywackes

Often feldspathic sandstones are micaceous, and vice versa. Of course in that case the ratios Th/K are intermediate as $(\rho_{ma})_a$, Pe and U_{ma}.

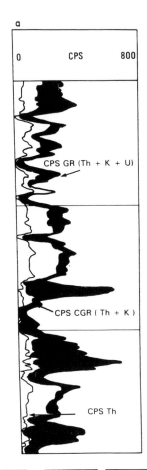

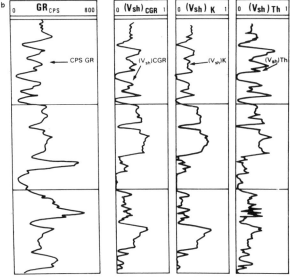

Fig. 7-19. a. Formation response due to thorium; thorium and potassium; thorium, potassium and uranium. b. Comparison of the three shale indicators with the total gamma ray (from Serra et al., 1980).

The cross-plot Th vs K of Fig. 7-21 illustrates of such a case.

It seems clear that most of the points falling between the O-Muscovite and O-Feldspars lines correspond to a mixture of quartz, muscovite and felds-

par, with a very small amount of clay mineral (illite).

In the examples of Fig. 7-22 from the North Sea, Hassan et al. (1976) have better defined the mineralogy of the sandstone by using the NGS data.

Heavy minerals within sandstones

Very often heavy minerals like zircon, allanite, monazite, and sphene are thorium and uranium-bearing, which give rise to some radioactivity in pure sandstones. This case is easy to recognize because the potassium level is generally very low, only the thorium and uranium curves being active. Consequently, this type of sandstone shows a very high Th/K ratio. At the same time ρ_b and the apparent matrix density $(\rho_{ma})_a$ generally increase owing to the denser minerals present.

Figure 7-23 is an example of such a case. We can observe that the thorium, uranium and density increase above 200, without much change in porosity, yet there is a very low potassium content throughout.

Shaly sands and sandstones

The combination of NGS and LDT * data (Pe or U_{ma}) allows the determination of the clay mineral types present within the sands. Figure 7-24 shows for instance the theoretical position of the main potassium-bearing minerals, kaolinitic sands and chloritic sands as distinguished by their Pe values. Montmorillonitic sands are separated from illitic sands by their Pe and K values.

On Fig. 7-25 it is clear that the formation is mainly composed of quartz with variable amounts of mica and feldspar, and with mainly kaolinite and chlorite as clay minerals, and probably traces of illite. Some points with high Pe and a potassium percentage between 1.4 and 1.8% could correspond to glauconitic beds.

"Greensand" or glauconitic sandstone

Glauconite is a dull-green, amorphous and earthy or granular mineral of the mica group with a mixed-layer lattice (by replacement of aluminium by iron). It contains magnesium, iron and potassium. Glauconite occurs as grains which may be mixed in all proportions with ordinary sand. Some greensands contain over fifty percent glauconite. The glauconite may be concentrated in certain laminations or scattered throughout the sand.

7.9.1.3. Carbonate series

In these rocks the standard gamma ray is very often a poor clay indicator, because the observed radioactivity is not related to clay content of the rock, but to the presence of uranium (Fig. 7-26).

In a pure carbonate of a chemical origin, the thorium will be absent, since it is insoluble. So if the

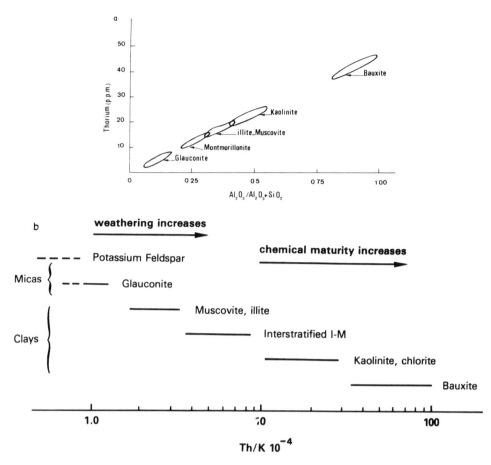

Fig. 20. a. Thorium content of clays as a function of the aluminium content. b. Thorium over potassium ratio for some minerals (from Hassan and Hossin, 1975).

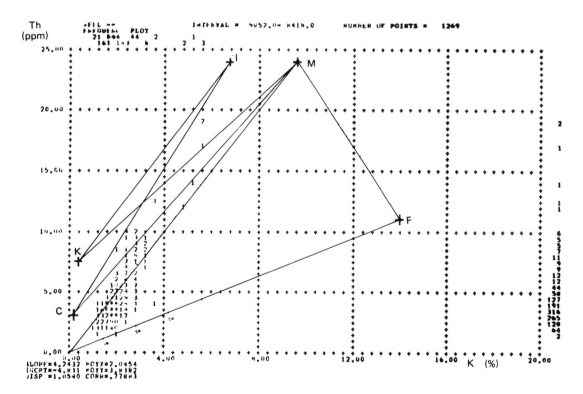

Fig. 7-21. Cross-plot thorium vs potassium and its interpretation.

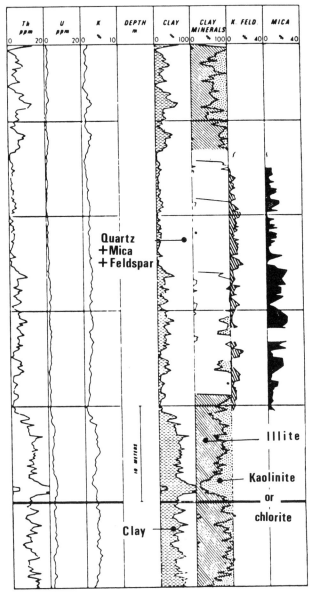

Fig. 7-22. Automatic interpretation of NGS logs (North Sea) (from Hassan et al., 1976).

concentrated. Peaks of uranium can also correspond to phosphate-bearing levels.

If Th and K are present with uranium, this indicates the presence of clays in the carbonate (clayey carbonates to marls).

If K is present with or without uranium it can correspond to a carbonate of algal origin or a carbonate with glauconite.

It is clear that in this type of series the NGS gives a real improvement and permits us to determine the clay percentage with better accuracy. Since, as seen above, the radioactivity is often related to uranium, the clay content determination can be improved by using the new shale indicator derived from the CGR curve. But one must be careful in the use of this relation, and before its use, verify that the thorium is not flat and near zero. If that is the case, it means that potassium is not related to the clay content of the rock but to a carbonate of algal origin or with glauconite. In these last cases, a better estimate of the shale content can be obtained from the thorium curve.

7.9.2. Well-to-well correlations

Peaks on thorium curves are often used for well to well correlations. They correspond generally to volcanic ashes (or bentonitic levels) and, consequently, can be considered as deposited at exactly the same time over a wide area (Lock and Hoyer, 1971).

7.9.3. Detection of unconformities

Abrupt changes in the mean thorium/potassium ratio are generally indicative of important variations in the proportion of radioactive minerals which occur when there are changes in geological conditions of deposition. These correspond to unconformities. Figure 7-27 is a good example of such unconformities. They are very difficult to detect on the other logs.

7.9.4. Fracture and stylolite detection

In reducing conditions the circulation of hydrothermal or underground waters in fractures may cause precipitation of the uranium salt, uraninite. So fractures can be recognized by peaks of uranium.

The presence of fractures must be confirmed by other methods because uranium is often associated with stylolites: during compaction, insoluble impurities (clay minerals, organic matter, iron oxides...) are often concentrated in very thin layers called stylolites, which can also give radioactive peaks.

Do not forget that uranium is often associated with phosphates which are encountered with carbonates.

NGS shows a carbonate level with thorium and potassium near zero this corresponds to a pure carbonate. If at the same time the uranium is zero too, this carbonate was precipitated in an oxidizing environment.

If the levels show a variable percentage of uranium, the corresponding carbonate can either have been deposited in a reducing environment (restricted), generally favourable also to the conservation of organic material and to its transformation into hydrocarbon; or, if it is compact (low porosity) it corresponds to a carbonate with stylolites, in which impurities such as uranium, organic matter and even clay minerals, are

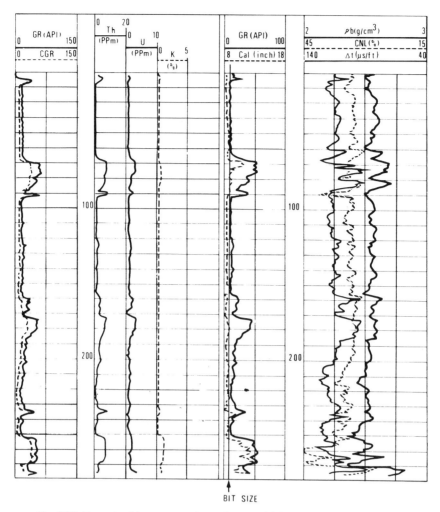

Fig. 7-23. Example of log response in sandstones with heavy radioactive minerals.

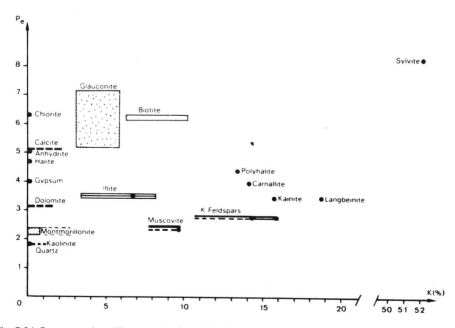

Fig. 7-24. P_e vs potassium (K) cross-plot for radioactive mineral identification (courtesy of Schlumberger).

128

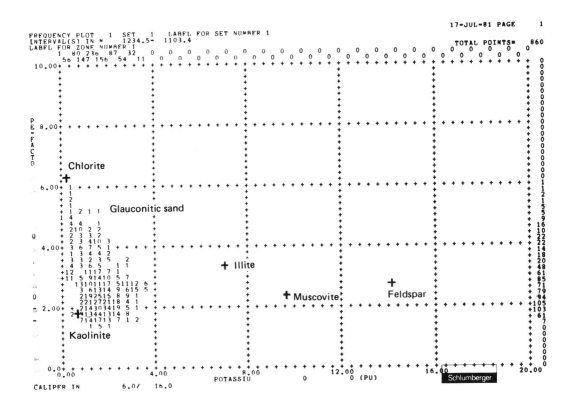

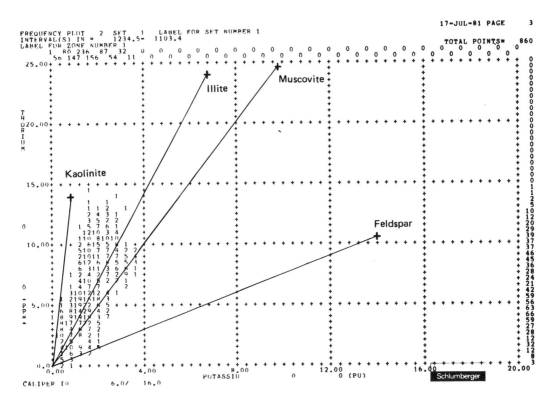

Fig. 7-25. Cross-plots P_e vs potassium (K), and thorium vs potassium for mineral determination.

7.9.5. Hydrocarbon potential

Several authors, Beers and Goodman (1944), (Fig. 7-28), Russel (1945), Swanson (1960), Spackman et al. (1966), Hassan et al. (1977), Supernaw et al. (1978) (Fig. 7-29), have observed a strong correlation between uranium and organic material. So, after calibration with core data, it is possible to evaluate

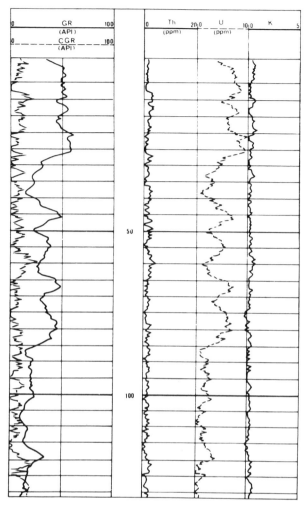

Fig. 7-26. Example of NGS response in a carbonate series showing that the radioactivity is mainly due to uranium.

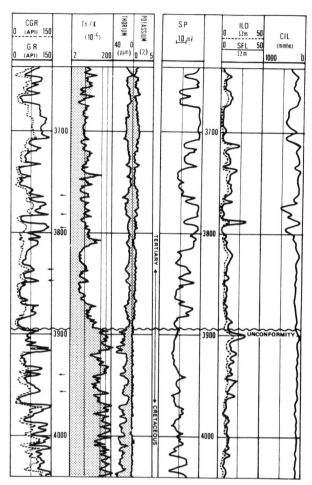

Fig. 7-27. Unconformity detected by the NGS and not easily seen on other logs (from WEC, Venezuela, 1980).

the organic carbon content of source rock from its uranium content and from that its hydrocarbon potential (Fig. 7-30).

7.9.6. Igneous rock recognition

The NGS data can help in recognition of igneous rock type. But this will be more accurate if other well logging information can be added. Density and sonic travel time are the most important.

N.B. Except for syenite, most of the intrusive igneous rocks show a Th/U ratio close to 4 (Fig. 7-31). Deviations from this value seem to indicate weathering effects during which uranium is dissolved and eliminated by rain and running waters; or oxidizing conditions before crystallization of magma; or intrusions of basic igneous rocks.

Adams (1954) (Fig. 7-32), Whitfield et al. (1959) (Figs. 7-33 and 7-34), and Clark et al. (1966) have observed a good correlation between the three radioactive elements in vitreous (lavas), granitic, mafic and intermediate rocks.

By considering the uranium by itself Larsen et al.

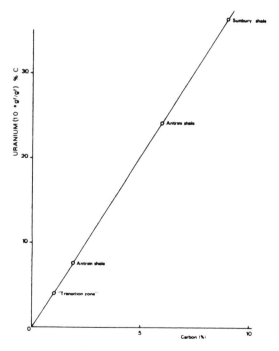

Fig. 7-28. Relation between uranium and organic carbon in sedimentary rocks (from Beers and Goodman, 1944).

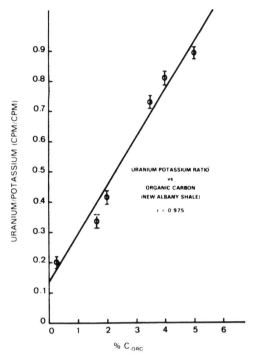

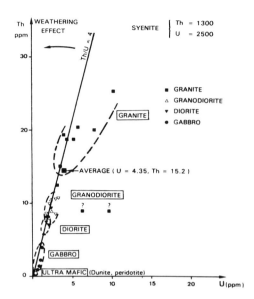

Fig. 7-31. Thorium vs uranium cross-plot for the main igneous rocks.

Fig. 7-29. Relation between uranium and organic carbon (from Supernaw et al., 1978).

(1954) have found a good correlation between the uranium content and the igneous type rocks (Fig. 7-35).

7.9.7. Sedimentology

Through the information on the radioactive elements present in the rocks, the NGS is a very powerful tool for sedimentological applications.

The recognition of radioactive minerals, especially of the clay minerals present in the rocks, and an

understanding of the conditions of deposition through the uranium measurement (oxidizing if free of uranium, reducing if rich in uranium), give a better determination of the mineralogical and grain size vertical distribution, and allow a more accurate reconstruction of the depositional environment. These are often characterized by the presence of certain minerals (Hassan et al., 1976) (Table 7-1):

(a) Glauconite is of marine origin, forming mainly in continental shelf conditions.

(b) Phosphatic deposits occur in similar conditions, but with the added requirements of warm water and a reducing environment.

(c) Feldspars are indicators of the degree of evolution of sand facies. Being unstable, they will only be

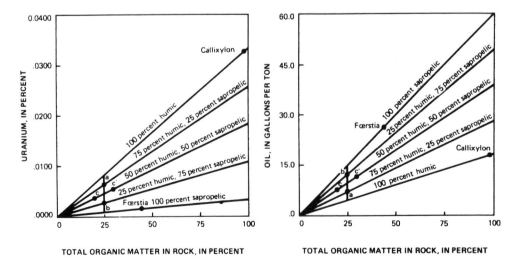

Fig. 7-30. Diagrams showing possible relation of uranium content to oil yield of a marine black shale, as controlled by total organic matter and the proportions of humic and sapropelic material making up the organic matter (from Swanson, 1960).

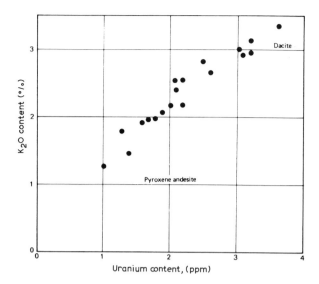

Fig. 7-32. Relation between potassium and uranium content of rocks in the Larsen Peak region, California (from Adams, 1954).

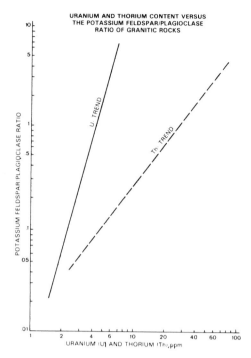

Fig. 7-34. Uranium and thorium content vs Potassium feldspar/plagioclase ratio of granitic rocks (after data from Whitfield et al., 1959).

found relatively close to the source rocks.

(d) Bauxite forms in a warm, humid, well-developed continental environment with good drainage.

(e) Clay typing is widely used in analyzing depositional environments.

(f) Uranium indicates low-energy, reducing conditions.

7.9.8. Diagenesis

Diagenesis causes the alteration of clay minerals (montmorillonite, illite, mixed layer I-M), the disappearance of kaolinite (which is transformed into illite), or a neogenesis in sands (kaolinite). These phenomena can be studied from gamma-ray spectrometry and, in particular, the Th/K ratio.

Under compaction, montmorillonite is trans-

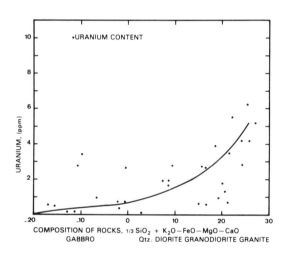

Fig. 7-35. Distribution of uranium in rocks of the southern California Batholith plotted as a function of the composition of the rocks (from Larsen et al., 1954).

TABLE 7-1

Thorium and uranium content of clay minerals (data from Adams and Weaver, 1958; Clark et al., 1966; Hassan et al., 1975)

Minerals	Th (ppm)	U (ppm)
Illite	10– 25	1.5
Montmorillonite	10– 24	2 – 5
Bentonite	4– 55	1 –36
Kaolinite	6– 42	1.5– 9
Chlorite	3– 5	
Glauconite	< 10	
Bauxite	10–132	3 –30

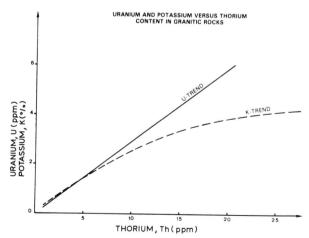

Fig. 7-33. Uranium and potassium vs thorium content in granitic rocks (after data from Whitfield et al., 1959).

formed into illite, passing through an intermediate mixed-layer illite-montmorillonite phase (Hassan et al., 1976). This results in a decrease of the Th/K ratio with depth. In undercompacted shales this trend will be reversed.

In carbonate reservoirs, diagenesis strongly affects the concentration and distribution of uranium (Hassan et al., 1976): (a) uranium can be easily mobilized and migrates during leaching and dissolution; (b) a high uranium concentration is characteristic of material filling or surrounding stylolites (Hassan et al., 1976); and (c) phosphatization in carbonates, too, results in a build-up of uranium (Hassan and Al-Maleh, 1976).

7.9.9. Estimation of the uranium potential

Natural gamma-ray spectrometry allows the direct estimation of the uranium content of the rocks, and consequently the detection of the uranium ores.

7.9.10. An approach to the cation exchange capacity

From the fact that one can define the type and the percentage of the clays present in the rocks, one can compute a parameter related to the cation exchange capacity.

7.9.11. Radioactive scaling

Abnormally high uranium content is frequently observed in front of perforated intervals in old wells. This is due to the precipitation of radioactive salts (radio-barite, $Ba(Ra)SO_4$).

Under dynamic conditions, radium isotopes are transported through permeable reservoirs, during the primary production of a water-flood operation, until final precipitation occurs at perforated (or around unperforated) cased wellbores. This precipitation depends on variations in temperature, pressure, flow and chemical equilibrium.

7.10. ENVIRONMENTAL AND OTHER EFFECTS

7.10.1. Time constant (vertical smoothing), logging speed, dead time

These are discussed in Chapter 5.

7.10.2. The bore-hole

The bore-hole influence is essentially due to absorption and Compton scattering. These are a function of several factors:

(a) Energy of the gamma rays emitted: the absorption is higher when the gamma energy is lower (Fig. 7-36).

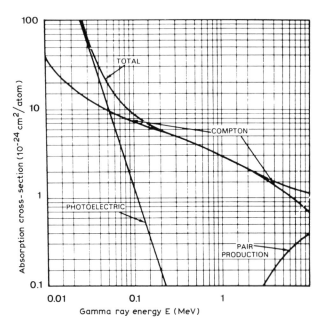

SILICON

Fig. 7-36. The Compton scattering and photoelectric absorption increase when the gamma energy decreases (from Adams and Gasparini, 1970).

(b) Volume of the bore-hole fluid around the tool which depends on the diameters of the hole and the tool.

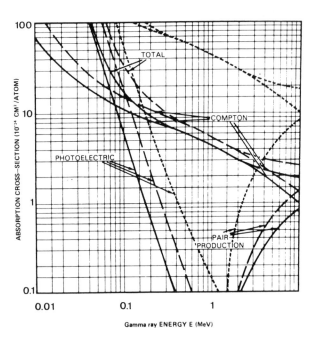

Fig. 7-37. Compton Scattering and photoelectric absorption as a function of density and atomic number (from Adams and Gasparini, 1970).

(c) Nature of the bore-hole fluid.

(1) Its density (air, gas, water, oil): the Compton scattering is higher when the density is higher;

(2) Its effective atomic number Z; the absorption will be higher if its content of strong gamma absorbers like barite is higher (Fig. 7-37).

(3) Its content of radioactive materials: bentonite, potassium salts (e.g. KCl mud), etc.

7.10.3. Tool position

Tool excentering is a departure from the idealized concentrically cylindrical geometry and the log readings are affected accordingly.

7.10.4. Casing

In cased hole the nature of the casing—its thickness, its position in the hole, and the nature and the volume of the cement between the casing and the formation—all contribute to absorption and Compton scattering, preferentially attenuating the low-energy gamma rays. The measured spectrum is thus unfairly weighted towards the high energies.

7.10.5. Bed thickness

As for other radioactive techniques, the log readings in a thin bed depend on the thickness of the bed, and the volume it represents relative to the spheres of influence of the three radioactive elements.

Rhodes et al. (1966) proposed correction charts for these effects. However, they are somewhat difficult to apply.

7.11 REFERENCES

Adams, J.A.S., 1954. Uranium and thorium contents of volcanic rocks. In: Nuclear Geology. J. Wiley & Sons, N.Y.

Adams, J.S. and Gasparini, P., 1970. Gamma ray spectrometry of rocks. Elsevier, Amsterdam.

Adams, J.S. and Weaver, C.E., 1958. Thorium to uranium ratio as indicator of sedimentary processes: examples of concept of geochemical facies. Bull. Am. Assoc. Pet. Geol., 42 (2).

Baleine, O., Charlet, J.M., Dupuis, Ch. and Meys, H. (1976). Dosage par spectrométrie des radioéléments naturels. Application à l'étude de quelques formations du bassin de Mons. Ann. Sci. Dépt. Mines-Géologie, Fac. Polytech. Mons, 2.

Beers, R.F. and Goodman, C., 1944. Distribution of radioactivity in ancient sediments. Bull. Geol. Soc. America, 55.

Bell, K.G., 1954. Uranium and thorium in sedimentary rocks. In: Nuclear Geology. J. Wiley & Sons, N.Y.

Blatt, H., Middleton, G. and Murray, R., 1980. Origin of Sedimentary Rocks (2nd ed.). Prentice-Hall, Englewood Cliffs, N.J.

Brannon, Jr, H.R. and Osoba, J.S., 1956. Spectral gamma ray logging. Trans. AIME, 207.

Caillere, S. and Henin, S., 1963. Minéralogie des Argiles. Masson & Cie., Paris.

Civetta, L., Gasparini, P. and Adams, J.A.S., 1965. Aspetti dell' evoluzione magmatica del vulcano di Roccamonfina attraverso misure di radioattività. Ann. Osserv. Vesuviano, Vol. 7.

Clark, S.P.Jr, Peterman, Z.E. and Heier, K.S., 1966. Abundances of uranium, thorium and potassium. Handbook of Physical Constants, 1966.

Cody, R.D., 1971. Absorption and the Reliability of Trace Elements as Environment Indicators for Shales. J. Sediment. Pet., Vol. 41, No. 2.

Deer, W.A., Howie, R.A. and Zussman, J., 1978. Rock Forming Minerals (2nd Ed.) Halsted Press (6 Volumes).

Doig, 1968. The natural gamma-ray flux: in situ analysis. Geophysics, 33 (2).

Dumesnil, P. and Andrieux, C., 1970. Spectrométrie gamma dans les forages par sonde a semi-conducteur Ge-Li. Revue Industries Atomiques, no 11/12.

Dunoyer De Segonzac, G., 1969. Les minéraux argileux dans la diagenèse. Passage au metamorphisme. Mem. Serv. Cart. Géol. Als.-Lorr., 29.

Evans, R.D., 1955. The Atomic Nucleus. McGraw-Hill, N.Y.

Fertl, W.H., 1979. Gamma ray spectral data assists in complex formation evaluation. The Log Analyst, vol 20, No. 5.

Fertl, W.H., Stapp, W.L., Vaello, D.B., and Vercellino, C., 1978. Spectral Gamma Ray Logging in the Austin Chalk trend. SPE 7431, SPE of AIME, Fall Mtg., Houston, TEXAS, Oct. 1-3.

Fertl, W.H., Welker, D.W. and Hopkinson, E.C., 1978. The Dresser Atlas Spectralog—A look at basic principles, field applications and interpretive concepts of gamma ray spectral logging. Dresser Atlas Publication No. 3332, September.

Frost, E. Jr, and Fertl, W.H., 1979. Integrated Core and Log analysis Concepts in Shaly Clastic Reservoirs. 7th Formation Evaluation Symp. of the Canadian Well Log. Society, Calgary.

Gasparini, P., Luongo, G. and Davia, G., 1963. Misure di radioattivita alla Solfatara di Pozzuoli. Atti Conv. Assoc. Geofis. Ital., 12, Rome, pp. 41-50.

Grim, R.E., 1958. Concept of diagenesis in argillaceous sediments. Bull. Am. Assoc. Pet. Geol., 42 (2).

Grim, R.E., 1968. Clay Mineralogy (2nd ed.). McGraw-Hill, New York, N.Y.

Hassan, M., 1973. The use of radioelements in diagenetic studies of shale and carbonate sediments. Int. Symp. Petrography of Organic Material in Sediment. Centre Nat. Rech. Sci., Paris, Sept. 1973.

Hassan, M. and Al-Maleh, K., 1976. La repartition de l'uranium dans les phosphates Senoniens dans le Nord-Ouest Syrien. C. R. Acad. Sci. (Paris), 282.

Hassan, M. and Hossin, A., 1975. Contribution a l'etude des comportements du thorium et du potassium dans les roches sedimentaires. C.R. Acad. Sci. (Paris), 280.

Hassan, M., Selo, M. and Combaz, A., 1975. Uranium distribution and geochemistry as criteria of diagenesis in carbonate rocks. 9th Int. Sediment. Congr. (Nice, 1975), 7.

Hassan, M., Hossin, A. and Combaz, A., 1976. Fundamentals of the differential gamma-ray log. Interpretation technique. SPWLA. 17th Ann. Log. Symp. Trans., Paper H.

Hodson, G., Fertl, W.H. and Hammack, G.W., 1975. Formation Evaluation in Jurassic sandstones in the northern North Sea area. SPWLA, 16th Ann. Log. Symp. Trans., Paper II.

Institut Français du Pétrole (1963). La spectrométrie du rayonnement naturel des roches sédimentaires. Bureau d'Etudes Géologiques, Rapport no. 9000.

Keller, W.D., 1958. Argillaceous and direct bauxitization. Bull. Amer. Assoc. Petroleum Geol., 42, 2.

Keys, W.S., 1979. Borehole Geophysics in Igneous and Metamorphic Rock. SPWLA, 20th Ann. Log. Symp. Trans., pap. OO.

King, R.L. and Bradley, R.W., 1977. Gamma-ray log finds bypassed oil zones in six Texas oilfields. Oil Gas J., April 4.

Krumbein, W.C. and Sloss, L.L., 1963. Stratigraphy and Sedimentation. W.H. Freeman & Co, San Francisco.

Larsen, E.S., Jr., and Phair, G., 1954. The distribution of uranium and thorium in igneous rocks. In: Nuclear Geology. J. Wiley & Sons, New York.

Lock, G.A. and Hoyer, W.A., 1971. Natural gamma ray spectral Logging. The Log Analyst, 12 (5).

Luongo, G. and Rapolla, A., 1964. Contributo allo studio dell'evoluzione del magma Somma-Vesuviano mediante le determinazione delle concentrazioni in isotopi radioattivi 238U, 232Th, 226Ta, e 40K. Ann. Osserv. Vesuviano, vol. 6.

Marett, G., Chevalier, P., Souhaite, P. and Suau, J., 1976. Shaly sand evaluation using gamma ray spectrometry applied to the North Sea Jurassic. SPWLA, 17th Ann. Log. Symp. Trans., Paper DD.

Millot, G., 1963. Géologie des Argiles. Masson & Cie., Paris.

Millot, G., 1963. Clay. Sci. Am., 240 (4).

Moll, S.T., 1980. Spectral Gamma Logging in the Copper Mountain Uranium District: A case Study in Fractured Quartz Monzonite. SPWLA, 21st Ann. Log. Symp. Trans., Paper O.

Naidu, A.S., Burrel, D.C. and Hood, D.W., 1971. Clay mineral composition and geologic significance of some Beaufort Sea sediments. J. Sediment. Petrol., 41 (3).

Pettijohn, F.J., 1975. Sedimentary Rocks (3rd ed.). Harper & Row, New York.

Pettijohn, F.J., Potter, P.E. and Siever, R., 1972. Sand and Sandstone. Springer-Verlag, Berlin.

Press, F. and Siever, R., 1978. Earth (2nd ed.). W.H. Freeman, San Francisco.

Quirein, J.A., Baldwin, J.L., Terry, R.L. and Hendrieks, M., 1981. Estimation of clay types and volumes from well log data. An extension of the global method. SPWLA, 22nd Ann. Log. Symp. Trans., Paper Q.

Rhodes, D.F. and Mott, W.E., 1966. Quantitative interpretation of gamma-ray spectral log. Geophysics, 31 (2).

Roubault, M., 1958. Géologie de l'uranium. Masson, Paris.

Russell, W.L., 1945. Relation of Radioactivity, Organic content, and sedimentation. Bull. Am. Assoc. Pet. Geol., 29 (10).

Schenewerk, P.A., Sethi, D.K., Fertl, W.H. and Lochmann, M., 1980. Natural Gamma ray Spectral logging aids granite wash reservoir evaluation. SPWLA, 21st Ann. Log. Symp. Trans., Paper BB.

Schlumberger, 1981. Natural Gamma-ray Spectrometry. Technical Book.

Schlumberger, 1982. Essentials of NGS interpretation

Selley, R.C., 1976. An Introduction to Sedimentology. Academic Press, London.

Selley, R.C., 1978. Ancient Sedimentary Environments. (2nd Ed.). Chapman & Hall, London.

Serra, O., Baldwin, J. and Quirein, J., 1980. Theory, interpretation and practical applications of Natural Gamma-ray Spectroscopy. SPWLA, 21st Ann. Log. Symp. Trans., Paper Q.

Stromswold, D.C., 1980. Comparison of Scintillation Detectors for Borehole Gamma-ray logging. SPWLA, 21st Ann. Log. Symp. Trans., Paper EE.

Stromswold, D.C. and Kosanke, K.L., 1979. Spectral Gamma-ray logging I: Energy Stabilization Methods. SPWLA, 20th Ann. Log. Symp. Trans., Paper DD.

Suau, J. and Spurlin, J., 1982. Interpretation of micaceous sandstones in the North Sea. SPWLA, 23rd Ann. Log. Symp. Trans., Paper G.

Supernaw, I.R., McCoy, A.D. and Link, A.J., 1978. Method for in-situ evaluation of the source rock potential of each formation. U.S. PATENT 4,071,744, Jan 31.

Swanson, V.E., 1960. Oil yield and Uranium Content of black shales. Geol. Survey, Prof. Paper 356-A.

West, F.G. and Laughlin, A.W., 1976. Spectral gamma logging in crystalline basement rocks. Geology, 4: 617–618.

Whitfield, J.M., Rogers, J.J.W. and Adams, J.A.S., 1959. The relationship between the petrology and the thorium and uranium contents of some granitic rocks. Geochim. Cosmochim. Acta., 17.

Wichmann, P.A., McWhirter, V.C. and Hopkinson, E.C., 1975. Field results of the natural gamma ray Spectralog. SPWLA, 16th Ann. Log. Symp. Trans., Paper O.

Wilson, R.D., Stromswold, D.C., Evans, M.L., Jain, M. and Close, D.A., 1979. Spectral gamma-ray logging II: Borehole Correction Factors. SPWLA, 20th Ann. Log. Symp. Trans., Paper EE.

Wilson, R.D., Stromswold, D.C., Evans, M.L., Jain, M. and Close, D.A., 1979. Spectral Gamma-ray Logging III: Thin Bed and Formation Effects. SPWLA, 20th Ann. Log. Symp. Trans., Paper EE.

Wilson, R.D., Cosby, M.S. and Stone, J.M., 1980. Field evaluation of direct uranium borehole logging methods. SPWLA, 21st Ann. Log. Symp. Trans., Paper S.

8. NEUTRON LOGS

8.1. GENERAL

When the formation is bombarded by high-energy neutrons, several types of interaction can occur between the neutrons and atomic nuclei. As Table 8-1 shows, several of these interactions lend themselves to logging techniques. They will be covered in this and in the following two chapters.

8.2. MEASUREMENT OF THE APPARENT HYDROGEN INDEX (neutron-gamma, neutron-epithermal neutron, and neutron-thermal neutron logs)

8.2.1. Principles

Neutrons of energies between 4 and 6 MeV are emitted continuously from a chemical source. The neutrons travel initially at some 10,000 km/s and have a high penetrating power. They interact both inelastically and elastically with atomic nuclei in the formation and the borehole surrounding the source. The life of these neutrons can be divided into four phases: fast, slowing down, diffusion, and capture.

8.2.1.1. Fast neutron phase

This will be covered in Chapter 9

TABLE 8-1

Schematic classification of neutron techniques in nuclear geophysics (from Brafman et al., 1977).

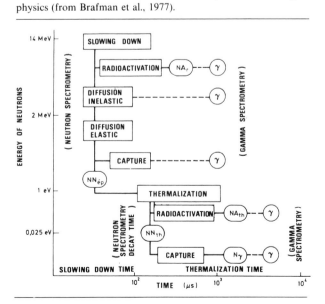

8.2.1.2. The slowing-down phase

Following the fast neutron phase, the neutrons are rapidly slowed down by elastic collisions with nuclei. The energy lost at each encounter depends on the angle of incidence with, and the mass of, the target nucleus. The mechanics of elastic collisions predict that the maximum energy will be lost when the target nucleus has a mass equal to that of the incident neutron. Thus it is that neutron slow-down is most strongly affected by hydrogen atoms (H), the single proton of the nucleus having very nearly the mass of a neutron. (A simple analogy can be made with the collisions of billiard balls: a glancing blow between a moving ball (neutron) and a stationary one (nucleus of hydrogen) will result in almost no energy loss by the moving ball; however, a head-on collision will bring the moving ball to a dead stop, because 100% of its energy is imparted to the target ball. This explains why, in Table 8-2, the *average* energy loss by collisions between neutrons and hydrogen is 50%. Now, if the stationary ball were heavier, it turns out that the maximum energy that the moving ball could lose to the target ball is reduced). The average energy lost in collisions involving carbon nuclei (12), for instance, is only 14%, while for oxygen (16) (which is heavier still) it is 11%. Please refer to the right-hand column of Table 8-2.

The probability of a collision occurring with a particular element depends, obviously, on the number of its atoms present in a given volume of formation, i.e. the atomic concentration per cm. However, another parameter must be considered: the elastic interaction cross-section. This is a characteristic of each type of atom. It has the dimensions of area and can be considered as the effective surface area presented by the nucleus to the on-coming neutron. It is not, however, simply related to the physical size of the nucleus and depends, for instance, on the energy of the neutron. (Fig. 8-1).

We can summarize by saying that the total slowing down power (SDP) of a certain element in the formation is given by the proportionality:

$$SDP = N\sigma_c\xi \tag{8-1}$$

where:
N = concentration of atoms per cm³;
σ_c = average collision cross-section;
ξ = energy lost per collision.
Since, at moderate porosities, hydrogen is relatively

TABLE 8-2

Thermal neutron capture cross-section for the principal elements

Elements	Cross-section in barns ($= 10^{-24}$ cm^2) for neutrons ($v = 2200$ m/s)	Energy of emitted gamma rays (MeV)	Probability of emission	Mean loss of energy by collision
Gadolinium	40,000			
Bore	771			
Lithium	71			
Chlorine	33.4	7.77	0.1	
		7.42	0.08	
		6.12	0.06	
		5.01	0.04	
Potassium	2.2			
Barium	1.2			
Iron	2.56			
Sodium	0.534			
Sulphur	0.49	5.43	0.84	
		4.84	0.20	
Calcium	0.44	6.42	0.83	0.04
		5.89	0.11	
Hydrogen	0.332	2.23	1.0	0.5
Aluminium	0.232	7.72	0.35	
		3.02	0.16	
		2.84	0.13	
Silicon	0.16	6.40	0.19	0.06
		4.95	1.0	
		4.20	0.19	
		3.57	0.94	
		2.69	0.65	
Magnesium	0.064	8.16	0.09	
		3.92	0.83	
		3.45	0.16	
		3.83	0.39	0.14
Carbon	0.0034	4.95	0.49	
		4.05	0.15	
		3.05	0.36	
Oxygen	0.0002			0.11

* From Handbook of Chemistry and Physics, 62 edition (1981–1982). Chemical Rubber Publishing Co., Cleveland, Ohio, U.S.A.

highly abundant, and its atoms are at least a factor of 10 more effective at slowing down neutrons than the other common elements (Fig. 8-2), it follows that the slowing down phase is very dependent on the concentration of hydrogen or the hydrogen index.

Neutrons continue to be slowed down until their

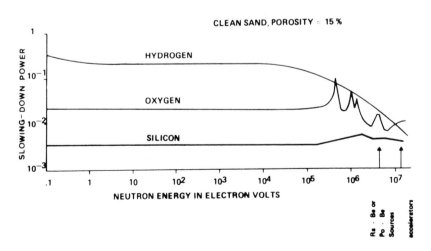

Fig. 8-1. Slowing down power of H, O and Si for different incident neutron energies (courtesy of Schlumberger).

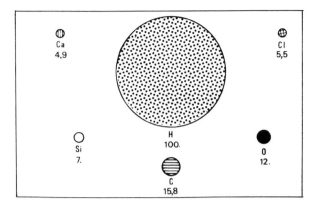

Fig. 8-2. Relative neutron slowing down powers. (from Waddell and Wilson, 1959).

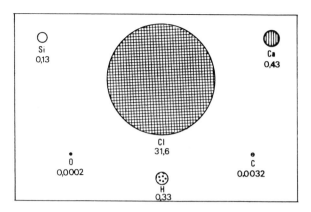

Fig. 8-3. Relative thermal neutron capture cross-sections (from Waddell and Wilson, 1959).

TABLE 8-3

Slowing down power of elements (from I. Kaplan)

Element	Number of collisions needed to reduce the energy of neutrons from 2 MeV to 0.025 eV
Hydrogen	18
Carbon	114
Oxygen	150
Silicon	257
Chlorine	329
Calcium	368

* Nuclear Physics, Irving Kaplan.

mean kinetic energies are equal to the vibration energies of the atoms in thermal equilibrium. Thermal energy is 0.025 eV at 25°C, corresponding to a mean velocity of 2200 m/s. Table 8-3 shows that only 18 collisions are required for hydrogen to slow a neutron down from 2 MeV to thermal energy, while other common elements require several hundred.

The entire slowing down phase requires of the order of 10 to 100 micro seconds, depending on conditions.

The term epithermal is applied to the energy range 100 eV–0.1 eV, representing more or less the final stages of the slowing down phase.

8.2.1.3. *Diffusion* *

A cloud of thermal neutrons forms around the source. It is unevenly distributed in space because of the inhomogeneous nature of the borehole and formation. Collisions between the vibrating neutrons and nuclei continue and there is a general spreading or diffusion of the cloud outwards into the formation, where the concentration of thermal neutrons is low.

---

* For a more detailed discussion refer to chapter 10.

Some diffusion of neutrons back towards the borehole may also occur.

8.2.1.4. *Capture*

Occasionally, during this diffusion phase, a nucleus will capture a neutron, resulting in its total absorption. The nucleus becomes momentarily excited and on returning to its ground-state, emits one or several

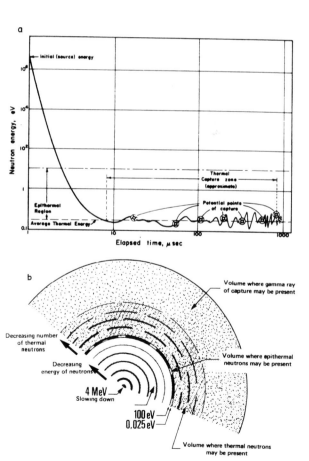

Fig. 8-4. a. Neutron energy versus time after emission (courtesy of Schlumberger). b. Schematic of spatial distribution of neutrons and gamma rays during continuous neutron emission.

138

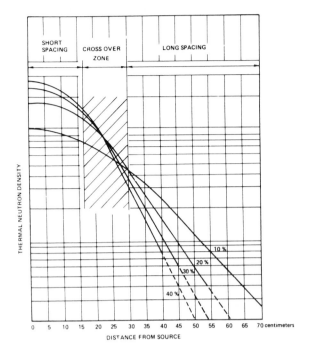

Fig. 8-5. Thermal neutron density with distance from a point source (from Tittman, 1956).

gamma rays, or some other radiation. A number of common elements are compared in Table 8-2. In much the same way as for elastic interactions, each element has its thermal capture cross-section (c in the Table). Note that, for capture, hydrogen is only moderately important, while chlorine ranks as the most effective of the common elements (Fig. 8-3). Gadolinium, boron and lithium occur infrequently as trace elements in formation water, and boron is often found in shales. Its proportion is related to the type of clay mineral and the salinity of the depositional environment (Fertl, 1973). The significance of their presence can be appreciated from Table 8-2.

8.2.2. Spatial distribution of thermal neutrons and capture gamma rays

Consider a point source of neutrons surrounded by an infinite homogeneous medium. High-energy neutrons are continuously emitted in all directions. The lifetime of these neutrons (from emission to capture) averages less than a millisecond in general.

A balance is rapidly established between the influx of "fresh" neutrons from the source and the absorption of thermal neutrons by capture, resulting in a spherical cloud of thermal neutrons whose spatial extent is primarily a function of the hydrogen concentration.

The idealized situation is shown in Fig. 8-4. The thermal neutron density is constant over the surface of any sphere centred about the source, and decreases

with distance from the source according to Fig. 8-5. Note that near to the source (the "short spacing" region) the thermal neutron density increases with hydrogen concentration, while farther out (the "long spacing" region) the opposite occurs. In the intermediate "cross-over" region there is almost no dependence on hydrogen concentration.

Since the number of gamma rays being emitted by capture is proportional to the number of thermal neutrons being captured, the preceding comments apply, with the difference that the gamma rays are able to penetrate further into the formation (Fig. 8-4).

8.2.3. Neutron logs

For practical reasons, all neutron logging tools operate in the "long-spacing" configuration, with a detector to source spacing in excess of 30 cm (12″). Neutron logs are measurements of the apparent concentration of hydrogen atoms per unit volume. There are several types.

8.2.3.1. *The neutron-gamma log (Fig. 8-6) (GNT)*

This measurement is based on the gamma ray emission rate due to thermal neutron capture. The gamma radiation is measured using scintillation or Geiger-Mueller detectors. Gamma activity of natural origin, or arising from the chemical source (Ra-Be, Am-Be, etc.,) is generally relatively weak and can be attenuated with shields and electronic circuitry.

8.2.3.2. *The neutron-thermal neutron log (Fig. 8-7) (CNT)*

This is a measure of the as yet uncaptured thermal neutron density in the formation. ^{3}He detectors are used, the helium having a large capture cross-section. An α-particle is produced in the detector at each neutron detection. Higher energy (epithermal, etc.)

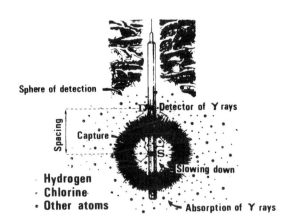

Fig. 8-6. The neutron-gamma (n,γ) logging principle.

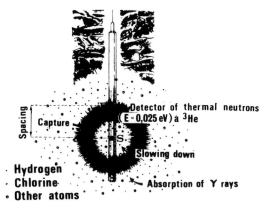

Fig. 8-7. The neutron-thermal (n, n_{th}) logging principle.

neutrons do not interact with the helium, because its capture cross-section becomes negligible above the thermal energy level.

8.2.3.3. The neutron-epithermal neutron log (Fig. 8-8) (SNP, CNT-G)

Here it is the epithermal neutron density (energy between 100 eV and 0.1 eV) in the formation that is measured. Detectors contain activated boron or lithium fluoride crystals.

8.2.4. Neutron sources

A chemical neutron source is an intimate mixture of beryllium and one of the α-emitters described later.

α-particles are produced by radioactive isotopes such as $_{88}^{226}$Radium, $_{94}^{239}$Plutonium or $_{95}^{241}$Americium. They are allowed to bombard beryllium, and neutrons are emitted at high energy:

$$_{4}^{9}Be + _{2}^{4}He \rightarrow _{6}^{12}C + _{0}^{1}n \qquad (8-2)$$

Neutron sources are characterized by their activity in neutrons per second, which depends, in turn, on the alpha activity within the source. The higher the

alpha emission rate, the greater will be the neutron emission rate, and the number of interactions with the formation, and the stronger the measured signal. As we have seen, high count-rates reduce statistical uncertainty on the log measurement. The response in high porosities, shales and coal, which produce low detector count-rates, is therefore enhanced by using a stronger source.

The Ra-Be source has an activity of 300 millicuries, producing 4.5×10^6 neutrons/second, with a mean energy of 4.5 MeV. The half-life is 1620 years. For each neutron, 10,000 gamma rays are emitted.

The ^{239}Pu-Be source has an activity of 5 curies and emits some 8.5×10^6 neutrons/second, of mean energy 4.5 MeV. The half-life is 24,300 years. There is almost no accompanying gamma radiation. The ^{238}Pu-Be combination produces 4×10^6 neutrons/second with an activity of 16 curies. The Am-Be source emits 10^7 neutrons/second, of mean energy 4.5 MeV, with almost no gamma radiation. Its activity is 4 curies, and the half-life is 458 years.

These last two sources are used in Schlumberger's CNL and SNP. The possibility of an electrically controlled accelerator source for continuous emission is under investigation.

N.B. The curie is a unit of total radioactivity, equivalent to 3.7×10^{10} disintegrations/second. This encompasses all particle emissions (see chapter 6, paragraph 6.2.6.).

8.2.5. Calibration and logging units

Initially, each well-logging company had its own system of units for neutron: cps (Schlumberger), standard neutron units (PGAC), environmental units (Lane Wells), etc..

The American Petroleum Institute (API) standardized on the API neutron unit, now used by all service companies. 1000 API units are defined as the difference between neutron tool readings (1) without a source (i.e. background level), and (2) with a source, in a special calibration pit at Houston University.

The pit (Fig. 8-9) contains 3 reference formation blocks, through which a $7^{7/8}''$ hole has been drilled. The hole is filled with fresh water. The 1000 API standard corresponds to when the tool is opposite the middle block of 19% porosity Indiana limestone. 26 percent porosity Austin limestone and 1.9% Carthage marble blocks are used in conjunction with the "100% porosity" water-bath above the blocks to verify the tool response once the main calibration has been made.

Modern neutron logs are scaled directly in porosity or hydrogen index units. The conversion from API units is made with a function-former or software in real time, since the response of the tool is accurately known and can be incorporated in the surface equipment.

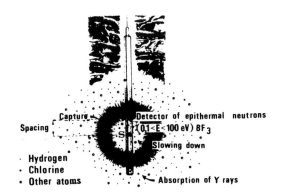

Fig. 8-8. The neutron-epithermal neutron (n, n_{epi}) logging principle.

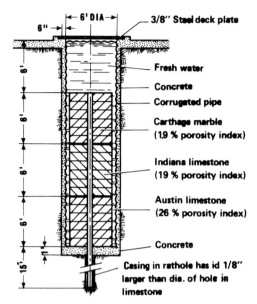

Fig. 8-9. The Houston calibration pit (from Belknap, 1959).

The neutron tool is calibrated in limestones and fresh water. Corrections can be applied in real-time, or later, with the aid of charts, for other lithologies or salinities. (Fig. 8-10 for example).

8.2.6. Schlumberger neutron tools

The earliest Schlumberger neutron logging tool was a neutron-gamma type called the GNAM, scaled in cps.

At present, four neutron porosity tools are widely available.

(a) GNT (Gamma-ray/Neutron Tool): a single detector tool which measures gamma-rays of capture. Cadmium fins, part of the detector, smooth out salinity variations to some extent by capturing thermal

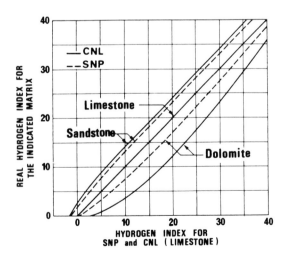

Fig. 8-10. The relationship between real and apparent (measured) hydrogen indices for three lithologies (courtesy of Schlumberger).

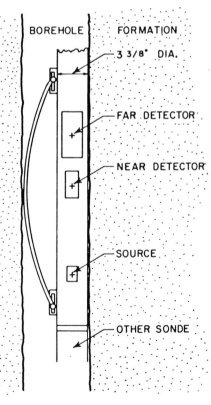

Fig. 8-11. The sidewall neutron porosity sonde (SNP) (courtesy of Schlumberger).

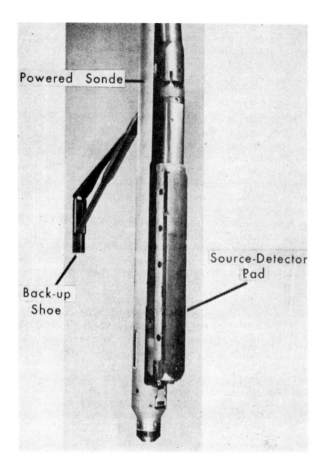

Fig. 8-12. Sketch of the dual-spacing neutron (CNL) tool (courtesy of Schlumberger).

neutrons, and converting them to gamma-rays (see 8.2.10.). The log is in API units.

(b) SNP (Sidewall Neutron Porosity Tool): measures epithermal neutrons. The source and single detector are mounted in a pad which is applied to the hole-wall (Fig. 8-11). The log is scaled directly in limestone porosity units, converted from the detector count-rate.

(c) CNT (Compensated Neutron Tool): The use of a dual-detector system reduces borehole effects. The tool is usually run excentred as shown in Fig. 8-12. The CNT-A detects thermal neutrons. The ratio of the count-rates at the near and far detectors is converted to porosity units by the surface equipment, according to the algorithms shown in Fig. 8-13 (the API test pit readings are also shown). The CNT-G consists of two dual-detector systems, measuring both thermal and epithermal neutrons. The epithermal measurement has the advantage of being insensitive to the presence of strong thermal neutron absorbers which perturb the thermal neutron response (see 8.2.10).

(d) TDT: Although not a conventional neutron tool because it uses pulsed rather than continuous neutron emission, the TDT is included here for completeness. It is dealt with in detail in Chapter 10. This dual-detector pulsed neutron tool measures capture gamma rays. It is not primarily a porosity tool; however the ratio of the count-rates of the two detectors resembles the CNL ratio quite closely. It is converted to porosity subsequent to logging, in the CSU computer unit at the field computing centre, or using charts. The TDT can provide a useful porosity measurement through tubing or casing.

8.2.7. Depth of investigation

The depth of penetration of neutrons into the formation becomes smaller as the hydrogen con-

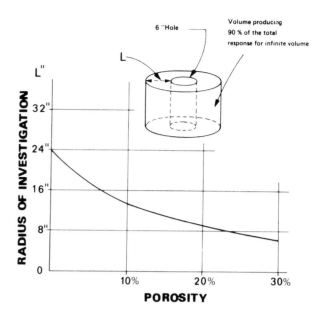

Fig. 8-14. Radius of investigation as a function of hydrogen index (courtesy of Schlumberger).

centration (and hence the slowing down power) increases. The thermalizing and capture processes occur rapidly and close to the source. Figure 8-14 gives an appreciation of the order of magnitudes involved. (The depth of investigation, L, has been arbitrarily defined as the distance from the hole-wall at which 90% of the total (infinite volume) response is obtained).

It depends also, of course, on the source-detector spacing. The integrated pseudo-geometrical factors for the SNP and CNL tools are compared in Fig. 8-15, in terms of distance from the hole-wall, for a 35% porosity formation. The depth of investigation is some 12″ for the CNL, and only 8″ for the SNP (a pad tool).

Capture gamma-ray detection gives a slightly deeper investigation, because the gamma rays can propagate from remote points of capture.

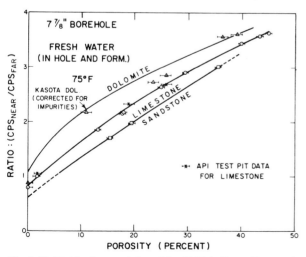

Fig. 8-13. Matrix characteristics of the CNT-A (from Alger et al., 1972).

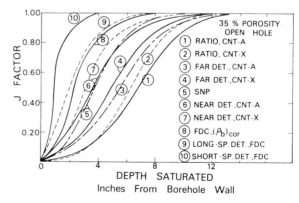

Fig. 8-15. Summary of all J-factor results for 35%-porosity formation, open hole (from courtesy of CWLS and SPWLA).

Neutron porosity tools make essentially invaded zone measurements if porosities and permeabilities are moderate to high.

8.2.8. **Vertical resolution**

It is usually slightly in excess of the source/detector or detector/detector (dual detection) spacings which are:

GNAM : 20″ or 26″ (two spacings were available)
GNT   : 15.5″ or 19.5″ (two spacings were available)
SNP   : 16″
CNL   : 10″ (dual detection)
TDT   : Approx 12″ (dual detection)

Although the logs will respond to the presence of such thin beds, the readings will rarely have time to attain the true values (see 8.2.12.1). Generally, a thickness of at least 3–4 ft is necessary for realistic readings to be obtained.

8.2.9. **Measuring point**

This corresponds to the mid-point of the source-detector (single-detector system), or of the two-detector (dual-detector system) distance.

8.2.10. **Factors influencing the measurement**

8.2.10.1. *Hydrogen*

As we have seen, hydrogen figures predominantly in the slowing down of neutrons. It is also relatively abundant in nature. Neutron logs, therefore, are essentially measurements of hydrogen concentration. The hydrogen index (HI) of a material is defined as the ratio of the concentration of hydrogen atoms per cm in the material, to that of pure water at 75°F. Pure water therefore has an HI of 1.0. Table 8-4 lists the hydrogen indices of a number of reservoir fluids and minerals. The HI of hydrocarbons covers the whole range from nearly zero (low pressure gases) to close to 1.0 (heavier oils), depending on molecular type, temperature and pressure.

The hydrogen content of most pure rock grains (quartz, calcite, etc.) being zero, their slowing down power is weak, and it follows that the neutron log is a porosity measurement, provided the HI of the pore-fluid is equal to 1.0. We will see below how fluid characteristics, and rock type must be taken into account, if we wish to estimate porosity in non-ideal conditions.

8.2.10.2. *Clays, micas, etc.*

Certain minerals contain hydrogen in the crystal lattice (Table 8-4). This may be water of crystalliza-

TABLE 8-4

Hydrogen contents of some substances

Material	Number of hydrogen atoms per cm³ ($\times 10^{23}$)	Hydrogen Index
Water, pure		
60°F, 14.7 psi	0.669	1
200°F, 7000 psi	0.667	1
Water, salted,		
200 000 ppm NaCl		
60°F, 14.7 psi	0.614	0.92
200°F, 7000 psi	0.602	0.90
Methane CH_4		
60°F, 14.7 psi	0.0010	0.0015
200°F, 7000 psi	0.329	0.49
Ethane C_2H_6		
60°F, 14.7 psi	0.0015	0.0023
200°F, 7000 psi	0.493	0.74
Natural gas (mean)		
60°F, 14.7 psi	0.0011	0.0017
200°F, 7000 psi	0.363	0.54
N-octane C_8H_{18}		
68°F, 14.7 psi	0.667	1.00
200°F, 7000 psi	0.639	0.96
N-nonane C_9H_{20}		
68°F, 14.7 psi	0.675	1.01
200°F, 7000 psi	0.645	0.97
N-decane $C_{10}H_{22}$		
68°F, 14.7 psi	0.680	1.02
200°F, 7000 psi	0.653	0.98
N-undecane $C_{11}H_{24}$		
68°F, 14.7 psi	0.684	1.02
200°F, 700 psi	0.662	0.99
Coal, bituminous		
0.8424 (C) 0.0555 (H)	0.442	0.66
Carnalite	0.419	0.63
Limonite	0.369	0.55
Ciment	0.334	0.50 env.
Kernite	0.337	0.50
Gypsum	0.325	0.49
Kainite	0.309	0.46
Trona	0.284	0.42
Potash	0.282	0.42
Anthracite	0.268	0.40
Kaolinite	0.250	0.37
Chlorite	0.213	0.32
Kieserite	0.210	0.31
Serpentine	0.192	0.29
Nahcolite	0.158	0.24
Glauconite	0.127	0.19
Montmorillonite	0.115	0.17
Polyhalite	0.111	0.17
Muscovite	0.089	0.13
Illite	0.059	0.09
Biotite	0.041	0.06

Compared with that of fresh-water (under identical pressure and temperature conditions) in the case of fluids.

tion, or molecularly bound. Although this hydrogen is not associated with porosity, it is nevertheless seen as such by the neutron tool. In addition, clays often retain large amounts of water within their platey structures.

It is common, therefore, to observe high neutron-porosity readings opposite shales, and the presence of shale in a reservoir rock necessitates a correction to the log reading (HI):

$$HI_{Cor\,sh} = HI_{log} - V_{sh}HI_{sh} \qquad (8-3)$$

8.2.10.3. Mineral type

Although the HI's of the common matrix minerals, such as quartz, calcite and dolomite, are zero, the elements present do have some effect on the neutron slowing-down and capture phases, as Fig. 8.10 demonstrates. Measured porosity, usually calibrated in limestone units, must be corrected for matrix lithology (e.g. Fig. 8-13). Apparent neutron porosity readings are well-known for a wide range of minerals (Table 8-4).

8.2.10.4. The presence of neutron absorbers

In the case of neutron tools measuring the thermal phase, the detector count-rates are affected by the presence of strong neutron absorbers, such as chlorine, lithium and boron, because of their influence on the thermal neutron population (and, to a lesser extent, on the slowing-down phase). Chlorine is the most commonly encountered strong absorber, and corrections are routinely made to the measured porosity to account for the salinity of the drilling mud, filtrate, and connate water.

Epithermal neutron measurements are much less sensitive to the presence of these absorbers.

8.2.10.5. Salinity

As well as affecting the thermal neutron count rates as discussed in the section on absorbers, fluid salinity alters the amount of hydrogen present. Dissolved NaCl displaces H, and reduces the hydrogen index of the fluid. Schlumberger has proposed the following relationship:

$$(HI)_w = \rho_w(1 - P) \qquad (8-4)$$

where ρ_w is the fluid density (g/cm^3) and P is the salinity in ppm $\times 10^{-6}$.

This must be taken into account for the fluids in the borehole, flushed and/or virgin zones. For a well drilled with air or oil-based mud, only the connate water salinity needs to be considered. In cased holes, casing fluid and connate water salinities will figure.

8.2.10.6. Hydrocarbons

Most medium to heavy oils have a hydrogen index close to unity, and their presence will have little effect on the neutron measurement. Light oils and gases, on the other hand, can alter the neutron reading markedly, because of their low HI. A light hydrocarbon-bearing zone, therefore, is characterized by an apparent neutron porosity which is less than what would be observed were the zone water-bearing.

Remarks. It has been observed that certain oils have an influence on the very early stages of the neutrons life (high energy fast neutron interactions) which can have a measureable effect on the (epi) thermal neutron measurements, and the simplistic HI concept no longer describes the situation adequately. This phenomena is still under investigation.

The following equation:

$$(HI)_{hy} = (9n/12 + n)\rho_{hy} \qquad (8-5)$$

has been proposed by Schlumberger to determine the HI of a hydrocarbon of the form CH$_n$, density ρ_{hy}.

Figure 8-16 permits the HI of a gas of molecular composition C$_{1.1}$H$_{4.2}$, to be estimated if its density $\rho_{a(gas)}$ temperature and pressure are known.

In its most simplistic form, we can write, that the neutron porosity measured in a clean gas-bearing formation is:

$$\phi_N = \phi\left[HI_{hy}S_{hr} + HI_wS_{xo}\right] \qquad (8-6)$$

where ϕ is the effective porosity.

However, it has been found that this does not entirely explain the neutron response to gas. An "excavation effect" has been postulated which introduces an additional decrease in ϕ_N. It means a reading below that expected on the basis of hydrogen indices of the formation components. Excavation effect results from the presence of a second formation fluid with a hydrogen index lower than that of the

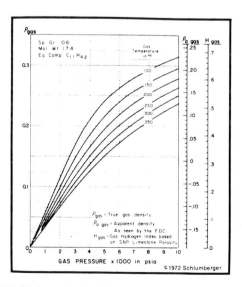

Fig. 8-16. Hydrogen index of a gas as a function of pressure and temperature (courtesy of Schlumberger).

144

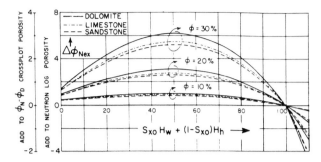

Fig. 8-17. Correction for excavation effect as a function of S_w for three values of porosity and for $H_g = 0$. Effect of limestone, sandstone and dolomite included within the shaded bands (courtesy of Schlumberger).

water. So in fact eq. 8-6 becomes:

$$\phi_N = \phi\left[HI_{gas}(1 - S_{xo}) + HI_f S_{xo}\right] - \Delta\phi_{Nex} \qquad (8\text{-}6b)$$

The chart of Fig. 8-17 can be used to correct the CNL-measured ϕ_N for this.

An approximate correction is given by:

$$\Delta\phi_{N\,ex} = K\left[2\phi^2 S_{wH} + 0.04\phi\right](1 - S_{wH}) \qquad (8\text{-}7)$$

where K is a lithology coefficient ($K = 1.0$ for sandstone, 1.046 for limestone and 1.173 for dolomite), and:

$$S_{wH} = HI_{gas}(1 - S_{xo}) + HI_f S_{xo} \qquad (8\text{-}8)$$

The term excavation effect originates from the comparison of a fully water-saturated formation with another one containing the same water content, but having a larger porosity, the additional pore space being filled with zero-hydrogen index gas. On the basis of hydrogen index both formations should give the same neutron porosity response (Fig. 8-18). However, the second formation differs from the first in that the additional pore space occupied by the gas has been provided by "excavating" some of the rock

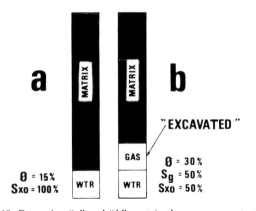

Fig. 8-18. Formation "a" and "b" contain the same amounts of hydrogen. However, the neutron log porosity of formation "b" is reduced because some of its matrix is replaced by gas (excavation effect).

framework. The two formations give neutron log apparent-porosity responses which differ by the amount of the excavation effect for this case.

Excavation effect is greater for larger contrasts between the hydrogen indices of the second fluid and the formation water, for higher formation porosities, and for intermediate water saturations.

8.2.11. Interpretation

8.2.11.1. *Early neutron tools (API logging units)*

(a) Neutron tool response can be related to the apparent hydrogen index of the formation in the following approximate manner:

$$\log N_f = \log(N_a - N_t) = C - K(HI)_N \qquad (8\text{-}9)$$

where:

N_f is the neutron response (API units) due just to the formation;

N_a is the amplitude of the neutron log response (API);

N_t represents the extraneous contributions to Na arising from the borehole, casing, cement, etc.;

C is a response coefficient which is a function of the tool configuration (source, detector, spacing etc.) and of the transmission properties of the formation;

K is a constant representing the transmission characteristics of hydrogen and the forma-

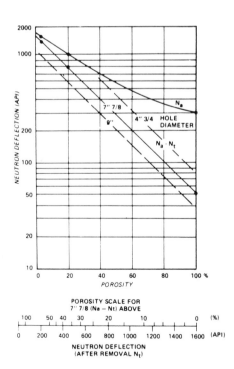

Fig. 8-19. Chart relating the logarithm of the neutron response (in API) to porosity (Eq. 8-9) (courtesy of Dresser Atlas).

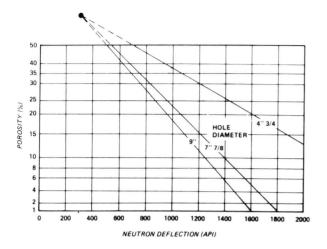

Fig. 8-20. Chart relating neutron response (in API) to the logarithm of the porosity (eq. 8-10) (courtesy of Dresser Atlas).

tion per unit length. It varies also with the source-detector spacing.

This relationship is illustrated in Fig. 8-19 for a range of hole-sizes.

(b) The neutron response equation can be more conveniently written in a different form:

$$e^{-K(HI)}N = C'(N_a - N_t) = C_2 N_f \qquad (8\text{-}10)$$

where C' encompasses the same variables affecting C. N_t is a constant for a given tool and set of hole conditions. This is demonstrated in Fig. 8-20. This equation has the advantage that all lines converge to a common fluid point (100% porosity) provided the hole fluid has the same HI as that in the formation (as is usually the case in a good approximation).

The neutron response can then be easily calibrated for a given hole-size and lithology at a single level of known (or assumed) porosity, since the 100% porosity point is known.

(c) An empirical approach to the problem postulates:

$$\log(HI)_N = C'' - K'N_a \qquad (8\text{-}11)$$

In this case, a plot of the logarithm of the HI versus neutron API reading (N_a) is a straight line (Fig. 8-21) for a given tool, lithology, and hole configuration (this is valid at moderate to high porosities). A two-point calibration is conveniently made using an arbitrary 50% HI shale-point and a near-zero porosity anhydrite or tight zone if available, or a known porous bed.

8.2.11.2. Modern tools

The problem of API neutron unit calibrations, and the errors inherent in the assumptions made, are no longer present in the more recent neutron tools. Tool response is more accurately understood, extraneous signals are reduced or automatically corrected out, and log data is recorded directly in apparent porosity units. Usually only minor corrections are required later to account for residual effects of lithology, salinity, temperature and so on.

8.2.11.3. The neutron response equation

Finally the apparent hydrogen index $(HI)_N$ measured by the tool is related to the porosity as follows:

$$(HI)_N = \phi_e(HI)_{mf}S_{xo} + \phi_e(HI)_{hy}(1 - S_{xo}) + V_{sh}(HI)$$
$$+ \sum_i V_i(HI)_{mai} \qquad (8\text{-}12)$$

where

$$\phi_e + V_{sh} + \Sigma V_i = 1$$

This assumes only the flushed zone is being investigated. V_i are the fractions of minerals ($i = 1, 2, \ldots n$) present, and $(HI)_i$ are their corresponding hydrogen indices; ϕ_e is the effective porosity; $(HI)_{sh}$ includes the HI of the bound and free water in the shale as well as the clay and silt minerals; $(HI)_{mf}$ represents the mud-filtrate; and $(HI)_{hy}$ is the hydrogen index of the hydrocarbons in the invaded zone. $(HI)_N$ is assumed to have been corrected for excavation effect.

8.2.12. Environmental effects

8.2.12.1. Time-constant, logging speed, dead-time, bed-thickness

These have already been dealt with in Chapter 5.

8.2.12.2. Borehole effects

8.2.12.2.1. Mud-type

(a) In air-drilled holes or gas-filled casing the

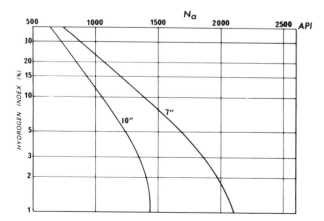

Fig. 8-21. Chart relating neutron response (in API) to the logarithm of the porosity (eq. 8-11).

contributions to neutron slowing-down and capture are negligible. However, tool response is sensitive to empty hole, generally resulting in an increase in detector count-rates, partly from the formation, partly by direct interaction between source and detectors. The GNT, TDT, SNP and epithermal CNT-G can be run in empty hole, but not the thermal CNT (because of count-rate saturation).

(b) In liquid-filled holes, the influence of the fluid on the log reading depends on:

(1) *Mud salinity*. Chlorine being a strong neutron absorber, an increase in mud salinity results in a decrease in thermal neutrons and a consequent increase in capture gamma rays. Neutron-thermal neutron tools will record a lower count-rate, which translates into a higher apparent hydrogen index. The opposite applies to neutron-gamma tools (Fig. 8-22).

(2) *Mud density*. The HI of the mud column (and therefore its slowing down power) is reduced by the addition of weighting additives (barite, NaCl, etc.), increasing the thermal neutron level near the detector. Thermal neutron tools read a lower apparent *HI*.

The response of the neutron-gamma tool has a further dependence on the nature of the additives —particularly their neutron capture and gamma emission/transmission characteristics.

8.2.12.2.2. *Hole diameter*

The larger the hole (and/or casing), the stronger will be the slowing down and capture properties of the well-bore. Formation signal strength becomes weaker with respect to hole signal as the hole-size increases.

8.2.12.2.3. *Tool positioning*

Formation signal is usually stronger when the tool is excentered against the hole or casing wall. Mecha-nical excentralizers can be used (e.g. CNT). In even slightly deviated wells, the tool tends to run along the low side of the hole anyway.

18.2.12.3. *Mud-cake*

Mud-cake effects are important with pad tools such as the SNP, but usually negligible with other tools. Being rich in hydrogen atoms, it tends to increase the apparent *HI*.

8.2.12.4. *Cased hole*

Iron is a strong neutron absorber. It also attenuates gamma-rays. The cement sheath behind a casing has a high *HI* ($\simeq 50\%$). The overall effect of a cemented casing is to increase the apparent *HI* by reducing count-rates (Fig. 8-23).

Tubing causes a similar increase.

8.2.12.5. *Summary*

The pad-tools are fairly insensitive to borehole effects but are strongly influenced by mud-cake. They cannot be run in cased hole.

Other neutron tools are affected to a greater or lesser extent. Ratio methods using dual-detection (CNT, TDT) are less susceptible to environmental effects. In Fig. 8-23, although the effect on the individual detector count-rates is quite marked as cement thickness is increased, for instance, the *ratio* of the count-rates (i.e. the slope of the curves) is only slightly changed for two given detector spacings. This typifies

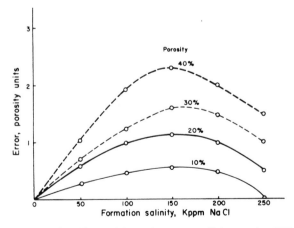

Fig. 8-22. The effects of formation water salinity on the CNL (open hole).

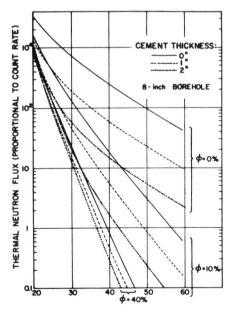

Fig. 8-23. Neutron count-rate versus source-detector distance for various formation porosities and cement thickness (freshwater in hole and formations)—computed data (from Alger et al., 1972).

the improvement gained by using dual detectors. The necessary corrections are applied either automatically or with the aid of charts (Fig. 8-24).

8.2.12.6. *Invasion*

The volume of formation contributing to the measurement depends on the tool configuration, the measuring technique, and the character of the formation itself. The GNT, CNT and TDT have approximately the same depths of investigation, considerably greater than the SNP (Fig. 8-15). Generally, the higher the porosity (and hence the *HI*), the shallower the measurement.

The process of invasion of mud filtrate was discussed in Chapter 2. Invasion depth varies according to mud and formation characteristics and drilling procedure.

Taking all these factors into consideration, we can say as a general rule that neutron tools will read mostly, (and sometimes entirely), the flushed zone. Equation 8-12 may need to be modified to include

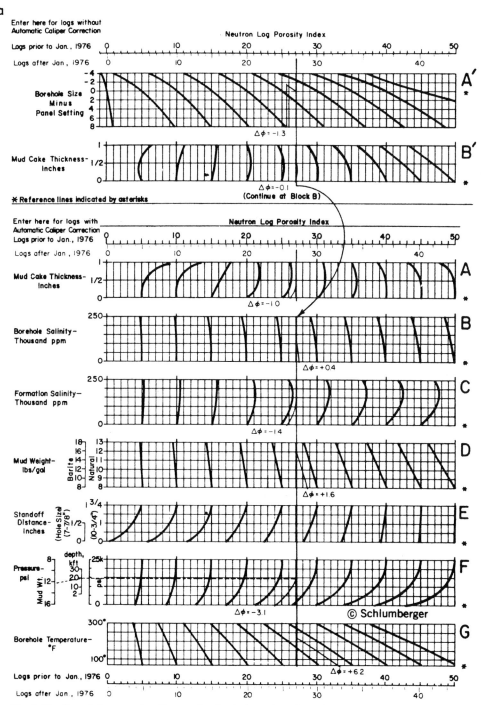

Fig. 8-24a. Open-hole correction diagram for the dual spacing (thermal) CNL (courtesy of Schlumberger).

148

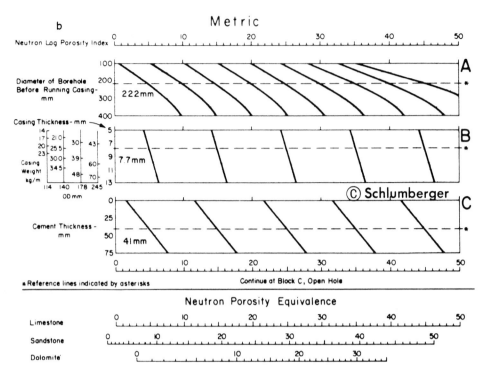

Fig. 8-24b. Cased-hole correction nomogram for the dual spacing (thermal) CNL (courtesy of Schlumberger).

some virgin zone contribution, but this is difficult to quantify and approximations have to be made to model the flushed zone/virgin zone contributions.

8.2.13. Geological factors affecting the hydrogen index

8.2.13.1. *Lithology and pore-fluids*

The measured apparent hydrogen index will be influenced by the slowing down and absorptive properties of the elements present (weighted by their respective concentrations). Strong lithological effects are commonly seen with clay minerals, and volcanic tuffites, or hydrogen-bearing minerals (gypsum, carnallite, etc.), for instance.

The same can be said of the interstitial fluids; neutron response depends on the neutron interactive properties of the fluid constituents, and, of course, porosity and saturation within the volume investigated.

8.2.13.2. *Rock texture*

With the exception of the porosity, which is controlled by sorting, packing, percentage of matrix and cement, and has of course a strong influence, the other rock-textural parameters have no direct effect in neutron and gamma ray interactions.

Indirectly, grain size and sorting can be said to be influential inasmuch as they determine permeability (and hence the invasion process and the nature of fluids in the flushed zone).

8.2.13.3. *Temperature*

As temperature increases, the *HI* of the pore- and well-bore fluids decreases.

Electronic stabilization circuitry ensures that detector response remains within acceptable limits at high temperatures.

8.2.13.4. *Pressure*

Increasing pressure tends to increase the *HI* of the fluids. This is particularly true for gas. The combined effects of temperature and pressure result in a net increase of *HI* with increasing depth. Pressure is also a factor in the invasion process.

8.2.13.5. *Depositional environment, sequential evolution*

These are factors on which depend bed-thickness and lithological sequence. As such, they affect the neutron log response.

8.2.14. Applications

They can be listed as follows and will be developed in volume 2: .

(a) Evaluation of porosity.

(b) Detection of gas or light hydrocarbons.

(c) Evaluation of hydrocarbon density (in conjunction with other logs).

(d) Identification of lithology (in conjunction with other logs).

(e) Correlation (particularly where shales are non-radioactive and give no natural gamma-ray response).

(f) Gravel-pack evaluation (CNT).

8.3. REFERENCES

Alger, R.P., Locke, S., Nagel, W.A. and Sherman, H. The dual spacing neutron log. CNL–SPE of AIME, paper SPE, No. 3565.

American Petroleum Institute, 1959. Recommended practice for standard calibration and form for nuclear logs. API RP 33, September 1959.

Belknap, W.B., 1959. Standardization and calibration of nuclear logs. Pet. Eng.

Brafman, M., Godeau, A. and Laverlochère, J., 1977. Diagraphie neutronique par analyse spectrométrique de l'émission instantanée des rayonnements gamma de diffusion inélastique et de capture. Nuclear Techniques and Mineral Resources, 1977. IAEA-SM-216/17.

Dewan, J.T., 1956. Neutron log correction charts for borehole conditions and bed thickness. J. Pet. Technol., 8 (2).

Dewan, J.T. and Allaud, L.A., 1953. Experimental basis for neutron logging interpretation. Pet. Eng., 25.

Lane Wells, 1966. The Gamma-ray-Neutron log. Lane Wells Techn. Bull., Jan. 1966.

Poupon, A. and Lebreton, F., 1958. Diagraphies nucléaires. Inst. Franç. Pétrole, Réf. 2432.

Stick, J.C., Swift, G. and Hartline, R., 1962. A review of current techniques in gamma-ray and neutron log interpretation. J. Pet. Technol., 14 (3).

Swift, G. and Norelius, R.G., 1956. New nuclear radiation logging method. Oil Gas J., 54 (76).

Tittle, C.W., 1961. The theory of neutron logging. I. Geophysics, 26 (1).

Tittman, J., 1956. Radiation logging. In: Fundamentals of logging. Univ. Kansas.

Tittman, J., Sherman, H., Nagel, W.A. and Alger, R.P., 1966. The sidewall epithermal neutron porosity log. J. Pet. Technol., 18 (10).

9. INDUCED GAMMA-RAY SPECTROMETRY

9.1. EARLY CAPTURE GAMMA-RAY SPECTROMETRY—THE CHLORINE LOG

As we have seen, the capture of thermalized neutrons by atomic nuclei often results in the emission of gamma rays. Now the energies of these gamma rays have discrete values which are characteristic of the element responsible for the capture.

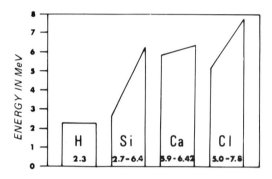

Fig. 9-1. Schematic of the principle gamma ray emission energies for thermal neutron capture by H, Si, Ca and Cl (from Waddell and Wilson, 1959).

Thus each element can be said to have a capture gamma-ray signature (its spectrum), by which, theoretically, its presence can be identified if we were to measure the energies of the gamma rays as well as their count-rate. In addition, the relative count level of each spectrum will be related to the relative proportions of the elements present in the medium surrounding the measuring system. This, then, is the basis for a capture spectroscopic analysis of the reservoir rock, in terms of its constituent elements and their abundances. As we will see later in this chapter, this approach can also be applied to other types of neutron interactions with the formation.

Figure 9-1 shows the main energy peaks for H, Si, Ca and Cl. Figure 9-2 provides a more detailed picture of the individual peaks for these elements in limestone and sandstone. H has one pronounced peak at 2.2 MeV, and Fe at 7.7 MeV, while the others each have several quite prominent emission energies.

Chlorine has by far the largest capture cross-section among the common elements (Fig. 9-3). It is dwarfed by some rarer elements such as boron; however, boron does not emit any detectable gamma rays.

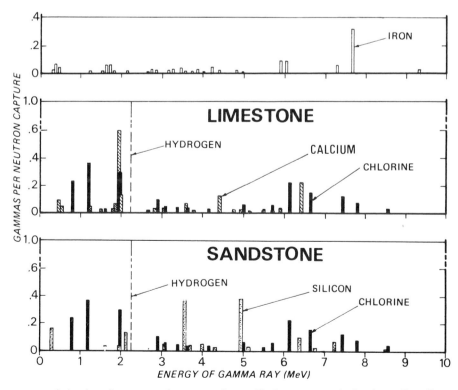

Fig. 9-2. Capture gamma emission from limestone and quartz sandstone. The iron spectrum is also shown (from Dewan et al., 1961).

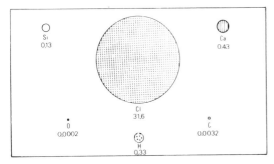

Fig. 9-3. The relative thermal neutron capture cross-sections of some of the common elements (from Waddell and Wilson, 1959).

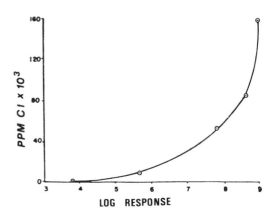

Fig. 9-4. The response of a chlorine-logging tool to water salinity (from Mooring, 1961).

Early spectroscopic techniques concentrated on the detection of chlorine as a means of localizing oil/water contacts and evaluating saturation—tools such as the Dresser Atlas Chlorinilog, Schlumberger's Chlorine Log, and McCullough's Salinity Log. These were, however, soon eclipsed by thermal neutron decay time techniques (Chapter 10), and more sophisticated spectroscopic techniques are in use to-day. The McCullough Shale Compensated Chlorine Log (SCCL) is in present-day use. An outline of the chlorine log is a good introduction to the subject of spectrometry.

9.1.1. Principle of the chlorine log

In its initial conception, it consisted of an americium-beryllium or plutonium-beryllium source emitting some 4×10^6 neutrons/sec at mean energy 4.5 MeV, and a conventional scintillation detector with NaI/thallium crystal. The detector and circuitry were sensitive only to gamma energies over 4 or 5 MeV, so as to respond to the strong chlorine peaks in this range (Fig. 9-2) and, of course, any other elements having peaks in this region.

Source-detector spacing was optimized for the best response. The system was calibrated with a neoprene sleeve which produced a known level of chlorine counts/sec, the neoprene containing some 40% by weight of chlorine.

9.1.2. Measurement characteristics

9.1.2.1. Salinity

The recorded count-rate was, not surprisingly, sensitive to formation water salinity. It was not recommended to run the survey in salinities less than about 25,000 ppm Cl, because the relative importance of other elements' emission peaks became significant (Fig. 9-2). As Fig. 9-4 shows, a saturation effect causes loss of resolution above about 100,000 ppm Cl.

9.1.2.2. Porosity

The effect of changing porosity was small at salinities above 50,000 ppm Cl, and moderate to high porosity. Below about 20 p.u., matrix effects began to interfere, causing a convergence of the iso-salinity curves (Fig. 9-5) and loss of resolution.

9.1.2.3. Formation constituents

Pore fluids: The displacement of water by oil causes a decrease in chlorine yield, i.e. a lower apparent salinity is read in an oil-zone.

In gas, the chlorine yield will also be low. However, the neutron log reading also is decreased because of the low hydrogen index of gas. A typical gas-zone response appears in Fig. 9-6.

Matrix: calcium and silicon have much lower capture cross-sections than chlorine (0.44, 0.16 and 33.4 barns respectively) and their influence is relatively weak at porosities above about 20 pu. Their presence

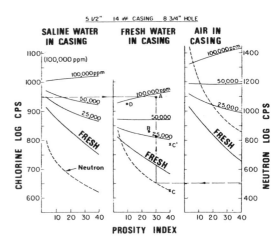

Fig. 9-5. Full response of a chlorine-logging tool to porosity, pore-fluid salinity and casing fluid salinity (from Dewan et al., 1961).

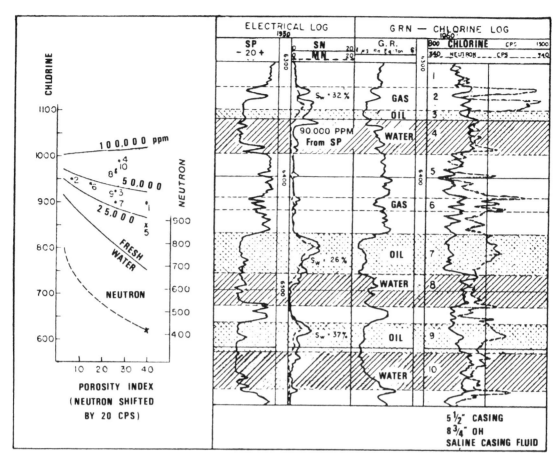

Fig. 9-6. The chlorine log, showing responses to sands and shales, and different formation fluids (from Dewan et al., 1961).

tends to increase the measured apparent salinity. Since calcium will have a stronger effect than silicon, (volume for volume) on neutron capture, it will produce a greater increase in apparent salinity as a result. This increase becomes more noticeable as porosity decreases. In addition, salinity resolution deteriorates.

Dolomite will have an effect intermediate to silicon and calcium, the capture cross-section of magnesium being 0.064 barn).

Shales: Depending on the salinities of the shale bound-water and connate water, shales will cause an increase or decrease in chlorine yield. The relatively fresh bound water will appear as oil if connate water is saline, and reference must be made to the SP or GR for confirmation (Fig. 9-6).

9.1.2.4. *Borehole effects*

Borehole effects originate from:

(a) The diameter and thickness of the casing: steel absorbs some of the neutrons, and scatters the emitted gamma rays. It will therefore contribute iron peaks to the measured spectrum, and perturb the formation spectrum because it degrades the gamma-ray en-

ergies; in addition its scattering cross-section varies with incident energy.

(b) The borehole or casing fluid: the effect of the casing fluid can be seen in Fig. 9-5 for saline and fresh water, and air. The relatively high count-rates in air-filled casing result from the near absence of any neutron scattering in the well-bore; the formation signal is consequently higher.

(c) Hole diameter and the nature of the cement sheath, if any, in the casing-formation annulus (e.g. fresh or saline water cement, channeling, etc.). This is usually impossible to quantify.

9.1.2.5. *Interpretation*

In principle, chlorine log interpretation was quite straightforward if a neutron and gamma-ray log were also available to evaluate porosity and shaliness. A CBL was useful for the quality of the cement sheath.

The log was normalized in a known wet zone. Point A in Fig. 9-5 (middle chart) is the "water point" read off the chlorine and neutron-porosity log; its salinity is about 95,000 ppm. Point B represents the readings in the zone of interest; its apparent salinity is 30,000 ppm. S_w is computed as: apparent salinity (B)/salinity (A) = 30,000/95,000 = 32%.

Alternatively, one could convert the neutron and chlorine log deflections into apparent porosity using appropriate conversion factors, obtained by normalizing the two tool responses in a wet zone, as shown in Fig. 9-6. S_w was obtained via:

$$S_w \simeq \left(\frac{\Delta\phi}{\phi_N} \right) (F)_{Cl} \qquad (9-1)$$

$$\Delta\phi = \phi_N - \phi_{Cl} \qquad (9-2)$$

where: ϕ_N is the apparent neutron porosity suitably scaled; ϕ_{Cl} is the chlorine-log porosity suitably scaled; $(F)_{Cl} \simeq \phi_N S_w / \Delta\phi$ is the chlorine scaling factor, determined in a wet zone ($S_w = 100\%$) from the corresponding values of $\Delta\phi$ and ϕ_N.

9.1.3. The shale compensated chlorine log (SCCL)

Fairly recently, an improved chlorine tool has been made available by McCullough (Fig. 9-7). The chlorine measurement is made using an energy window spanning 3.43–8.0 MeV (Fig. 9-8). In addition, a hydrogen window is set at 1.32–2.92 MeV.

The chlorine count level depends on porosity and shaliness as well as salinity. Since the hydrogen measurement responds to porosity and shaliness like a conventional neutron log, the two measurements can be combined to obtain a salinity estimate corrected for the other two variables.

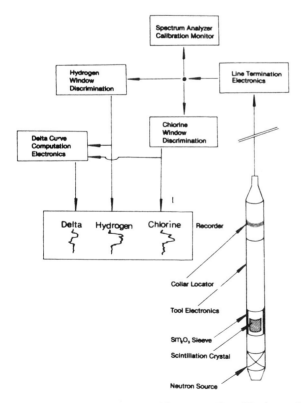

Fig. 9-7. Block diagram of the SCCL system (from Fletcher and Walter, 1978).

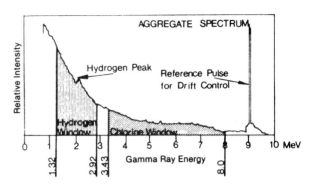

Fig. 9-8. Typical SCCL recorded spectrum with the tool in a 7" casing in 30% porosity sand containing 150‰ (NaCl) water, showing the hydrogen and chlorine windows (from Fletcher and Walter, 1978).

A Delta curve is computed from the Cl and H count-rates. It responds to the concentration of Cl relative to H, i.e. it is a more direct index of S_w.

Figure 9-9b demonstrates the responses of the SCCL curves to shale, water, oil and gas. Note that the Delta curve tracks the SP in wet-sand/shale sequences, but deflects strongly to the right in hydrocarbon zones.

The Delta curve is interpreted relative to its baseline reading in a water sand. For a given water salinity (10^3 ppm), its deflection for saturation S_w is $D_a(S_w)$ relative to this water line. Reference data have been obtained from measurements in a 30% porosity formation at salinity $\delta t = 250 \times 10^3$ ppm, in several different casing configurations (the deflection D_t in Fig. 9-9a). At salinity δ, less than δt, deflections D_a will be smaller than D_t for 100% saturation. $D_a(S_w)$ is calibrated against the reference $D_t(S_w)$ from a scale-factor chart such as Fig. 9-10 for a particular set of hole conditions. Thus at $\delta = 150 \times 10^3$ ppm, the full range of $D_a(S_w)$ from $S_w = 100\%$ to $S_w = 0\%$ is $0.8 \times$ that of D_t. This is shown in Fig. 9-9a. Note that $D_a(S_w)$ is measured from the "water-line" corresponding to wet-zone readings. Since the scale factor is roughly constant above about 100,000 ppm for moderate to high S_w, interpretation of Delta can be made without knowledge of water salinity as long as it is high enough. It is only necessary to establish the wet-sand response line.

In gas zones, a correction can be made for the low hydrogen index of the gas, given reservoir temperature and pressure (see ref.).

The SCCL is run at 12-15 FPM logging speed for the 3" casing tool, and 8 FPM for the 1-11/16" through-tubing tool. This is reduced to 5 FPM for tubing inside casing. Satisfactory statistical accuracy can then be obtained in beds over 6 ft thick. Stationary readings must be made if greater precision is required.

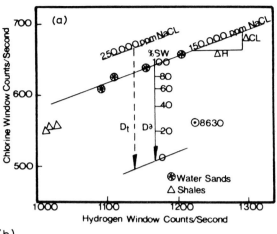

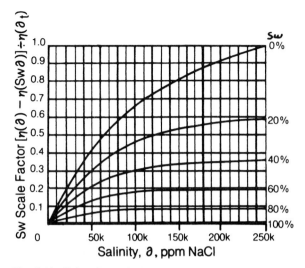

Fig. 9-10. Delta Curve Scale Factor versus Formation Water Salinity (from Fletcher and Walter 1978).

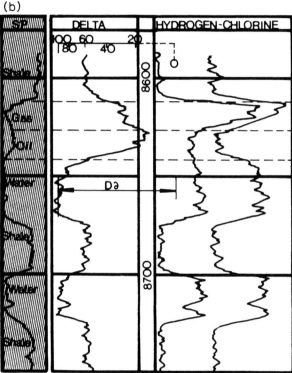

Fig. 9-9.a. The hydrogen vs chlorine cross-plot (from Fletcher and Walter, 1978). b. SCCL log showing the hydrogen, chlorine and delta curves (from Fletcher and Walter, 1978).

9.1.4. Uses

The chlorine-log is used to locate oil/water contacts and estimate saturations behind casing, in moderate to high salinities.

9.2. MODERN INDUCED GAMMA-RAY TECHNIQUES—INELASTIC AND CAPTURE SPECTROMETRY

9.2.1. General

We will now consider the spectroscopic analysis of gamma ray emission from neutron interactions in more detail. Present-day induced gamma-ray spectrometry tools employ accelerator-type pulsed neutron generators emitting neutrons at 14 MeV of energy. (Such a generator is shown in Fig. 10-10 of the Chapter on Thermal Decay Time Measurement.)

At these high energies, a class of interaction known as *fast neutron scattering* can occur between the neutrons and the atomic nuclei of the borehole and formation. (Such interactions are relatively rare when the lower-energy chemical neutron source is used for conventional porosity logging.) The incident neutron has enough energy to excite the nucleus, that is, to raise the energy level of the nucleus to a higher bound-state. This excited nuclear state is usually short-lived, and the nucleus returns to its ground-state with the emission of radiation. In some cases the nucleus is transformed into a different element. The neutron may continue at reduced energy, or may be annihilated. Some fast neutron interactions with certain elements result in the emission of gamma rays (carbon, oxygen, silicon, calcium, etc.).

As a result of multiple fast interactions, and elastic scattering (Chapter 8), the neutrons are rapidly slowed down in the formation. There follows a phase of low-energy interaction, the most significant of which is *thermal neutron capture* with, again, the production of gamma rays from certain of the elements (chlorine, hydrogen, iron, etc.).

In all cases, the gamma emission energy assumes discrete values, functions of the target element, the nature of the interaction, and the incident neutron's energy. Following high-energy neutron bombardment, therefore, many elements produce characteristic gamma ray signatures, or spectra, from fast neutron scatter or thermal neutron capture.

It is quite feasible to record the spectra of a number of significant formation elements in situ. The measurement and analysis of these spectra is the basis of reservoir evaluation by spectrometry logging.

9.2.2. Fast neutron scattering

There are several types of fast neutron interaction, broadly classified as follows:

(a) *Inelastic*: some of the incident neutron energy imparted to the target nucleus excites it to a higher bound-state. The excited state lasts less than a microsecond, and the ensuing prompt return to ground state results in the emission of radiation. Such a reaction involving carbon ^{12}C is shown in Fig. 9-11. Excitation half-life is 3.8×10^{-14}s, and a gamma ray of energy 4.44 MeV is produced. In fact, this is the only emission peak observed from ^{12}C. The interaction is annotated as follows:

$$^{12}C(n, n'\gamma)^{12}C \;^{(1)}$$

It is symbolized in Fig. 9-13, where C * represents the (momentarily) excited nucleus. Note that, in inelastic scatter, the neutron continues at reduced energy.

For oxygen ^{16}O we have (Fig. 9-12, right-hand side):

$$^{16}O(n, n'\gamma)^{16}O$$

with a principle emitted gamma-ray energy of 6.13 MeV and an excitation half of 1.7×10^{11}s. The oxygen spectrum contains other inelastic peaks, at 6.92 and 7.12 MeV for instance, occurring with lower probability.

(b) *Reaction*: an important example is the neutron-induced alpha emission from oxygen, which results in the production of a ^{13}C isotope and the annihilation of the neutron. The ^{13}C nucleus may be already at ground-state, or it may be excited, in which case a gamma ray is promptly emitted at 3.09, 3.68 or 3.86 MeV. Such reactions are written:
$^{16}O (n, \alpha)^{13}C$ in the former case and;
$^{16}O (n, \alpha \gamma)^{13}C$ in the latter (Fig. 9-13).
These peaks form a significant part of the total oxygen spectrum.

(c) *Activation*: the target nucleus is transformed to an unstable intermediate isotope which decays with a relatively long half-life to the final nucleus. If this is in an excited state, a prompt emission of gamma radiation accompanies the return to ground-state. Figure 9-12 illustrates the case of oxygen activation, regarding the figure as a whole:

$$^{16}_{8}O(n, p)^{16}_{7}N(\gamma)^{16}_{8}O$$

---

(1) Reading from left to right, $^{12}C(n, n'\gamma)^{12}C$ indicates: target nucleus, (incident particles, emitted particle(s)), and final nucleus. n' means that the neutron has suffered a non-kinematic energy loss.

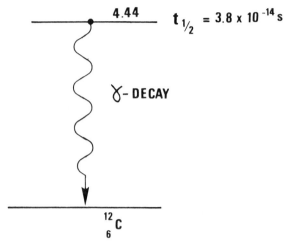

Fig. 9-11. Energy level diagram for the inelastic scatter of fast neutrons by carbon-12, producing gamma ray emission at 4.44 MeV (from Schlumberger publication).

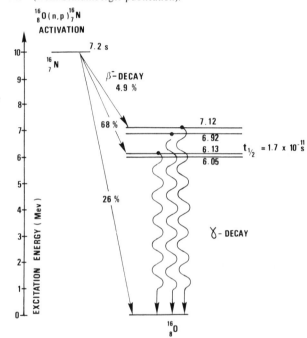

Fig. 9-12. Energy level diagram for the fast neutron interactions of oxygen-16, showing at right some of the possible inelastic gamma emissions and, at left, oxygen activation (from Schlumberger publication).

Fig. 9-13. Schematic of fast neutron bombardment of a formation by an induced gamma ray spectrometry tool (from Hertzog, 1978).

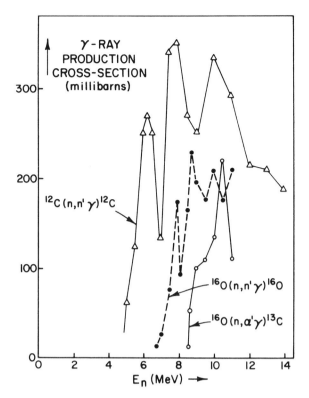

Fig. 9-14. Fast neutron scattering cross-sections of oxygen and carbon nuclei for the interactions shown in Fig. 9-13 (from Hertzog, 1978).

The intermediate unstable nitrogen isotope has a half-life of 7.2 s. It decays by β-emission to one of the excited states of ^{16}O which in turn decays promptly with gamma-ray emission. Figure 9-12 shows that the 68% probable result will be an emission at 6.13 MeV. Other common activations are iron (Fe, half-life 2.6 hours), aluminium (Al, 9.5 min) and iodine (I, 25 min).

Formation and casing activation is sometimes observed during TDT and CNL logging, superposed on the natural gamma-ray curve, if the tool has been stopped for some reason.

In Fig. 9-14 are the cross-sections for gamma-ray production corresponding to three of the interactions just cited. Note that the threshold for oxygen activation is almost 10 MeV of incident neutron energy, while for the inelastic carbon interaction it is some 5 MeV. In the logging context, fast neutron scatter does not occur significantly below a few MeV.

The mean free path of 14 MeV neutrons is relatively short—some 12 cm in a 30 pu porosity formation. Figure 9-15 shows the attenuation of high-energy neutrons at depth 30 cm (12") into the formation, compared with the neutron flux close to the accelerator.

From Figs. 9-14 and 9-15 it is clear that fast-neutron measurements have the following features:

(a) They occur early in the neutron's life, in fact during or very shortly after the accelerator burst.

(b) They are shallow measurements; of the order 5"-10" depth at moderate porosities. Depths of investigation are slightly different for different elements, because of the energy dependence of the associated interactions.

(c) Borehole signal is relatively large.

Fast neutron scattering is commonly referred to as inelastic scattering because of the predominance of that particular type of interaction as far as logging techniques are concerned. The word "inelastic" will be employed from now on in this context, but it should be understood that the term in fact encompasses all significant types of fast neutron interaction.

The common elements currently detected by inelastic spectrometry include: carbon C; oxygen O; silicon Si; calcium Ca; iron Fe; sulphur S: (chlorine Cl, an almost negligible yield).

9.2.3. Thermal neutron capture

The important chlorine capture interaction is:

$$^{35}Cl(n, \gamma)^{36}Cl$$

Its half-life is of the order of 10^{-20} s.

The neutron is absorbed, or captured, and the excited nucleus decays to groundstate with the emission of a gamma ray, in this case at 7.42 or 7.77 MeV among others. Not all elements produce gamma rays, and some are outside the range of detection of the logging equipment. Common elements currently measured by capture spectrometry include: chlorine Cl; hydrogen H; silicon Si; calcium Ca; iron Fe; sulphur S.

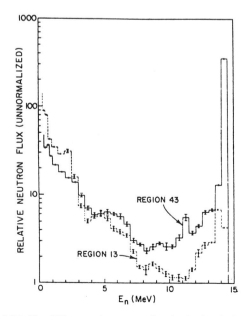

Fig. 9-15. The differences in neutron flux in the borehole (region 43) and formation (region 13) (from Hertzog et al., 1979).

Capture events generally occur at highest probability when neutrons reach thermal energy, and the term "thermal neutron capture" is used commonly for this class of interaction.

The mean free path of 14-MeV neutrons down to thermal energy is, not surprisingly, larger than that of the fast neutron phase.

Neutron capture measurements therefore: (a) tend to be deeper reading; 10″–15″ depth of investigation is a representative range at moderate porosities; and (b) occur during and for some time after the neutron burst. Capture spectral data can thus be measured separately from the inelastic by accumulating counts after the neutron burst. In fact, a certain time delay optimizes formation signal response. (Conversely, some unwanted capture events will be recorded during the inelastic measurement period, both from the current burst output and from the preceding cycles).

9.2.4. Reservoir analysis by spectroscopy

Referring to the list of elements measured from these two classes of interaction, you will see that in many situations a quite full description of the reservoir can be made—both of its elemental composition and fluid content—once the spectral data has been translated into physical concentrations.

The C and O data is of particular significance. It represents a means of detecting and evaluating oil, and, because neither measurement is dependent on the salinity of the water (to a very good approximation in the case of oxygen), a through-casing technique for hydrocarbon logging in very low, mixed or unknown salinities, is available.

In addition, the elemental yield data can be used for mineralogical analysis, and porosity and clay evaluation.

9.2.5. Measured spectra

The total spectrum that is measured by a spectrometer tool will consist of a number of peaks contributed by the various elements present in the borehole and formation.

Figure 9-16 represents the theoretical spectrum that might be accumulated from inelastic neutron interactions in a formation containing C, O, Si and Ca. A peak's amplitude depends both on the relative abundance of the source element, and the probability of emission at that particular energy. There is considerable interference among the peaks (for instance, an oxygen emission at the same energy as the main carbon peak (4.44 MeV), and a silicon peak very close by), and the "carbon region" of the spectrum is quite crowded with non-carbon signals. The same goes for the "oxygen region", and any other for that matter.

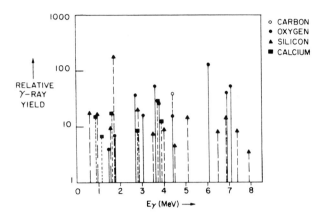

Fig. 9-16. Theoretical fast neutron interaction gamma ray yields, showing the interfering peaks from the different elements (from Hertzog et al., 1979).

The spectrum actually recorded is a considerably less sharp picture than Fig. 9-16. Firstly, the peaks are broadened by Doppler shift caused by recoil and vibration of the target nuclei. Secondly, Compton scattering in the formation, cement, casing, fluid, tool-housing and detector crystal itself, attenuates the energies of many of the gamma rays (Fig. 9-17b and c), producing a significant Compton background signal. Thirdly, the NaI crystal produces a triplet of peaks for each incident gamma energy—the result of electron-positron pair production and annihilation in the crystal. These so-called escape peaks are 0.51 MeV apart. For the 4.44 MeV ^{16}C full energy peak, for example, they are at 3.93 and 3.42 MeV (Fig. 9-17a).

The recorded spectrum, therefore, is like Fig. 9-17d; the peaks become much less distinct bumps superposed on a strong background signal. Figures 9-18 and 9-19 are typical spectra measured from inelastic and capture interactions, respectively, in a cased carbonate laboratory formation.

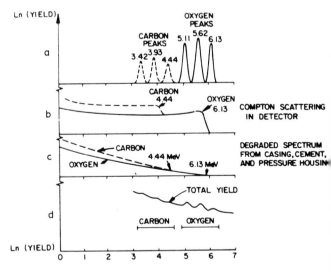

Fig. 9-17. The principal components of the NaI scintillation detector response to gamma rays (from Hertzog, 1978).

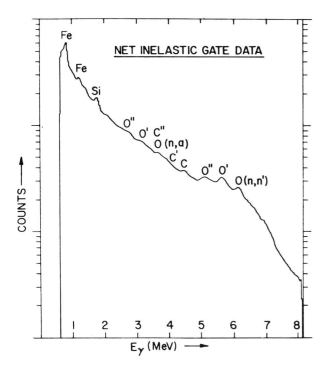

Fig. 9-18. Example of an inelastic spectrum recorded by an induced gamma ray tool (from Hertzog, 1978).

9.2.6. Measuring techniques—The window method

This approach is used by Dresser Atlas, among others, in their C/O tool. The measuring circuit is biased to count the gamma ray yields in two wide

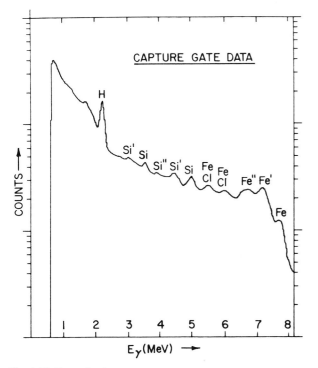

Fig. 9-19. Example of a capture spectrum recorded by an induced gamma ray tool (from Hertzog, 1978).

energy windows straddling the main carbon and oxygen triplets (Fig. 9-17a and d). The ratio of the count-rate in the "carbon window" to that in the "oxygen window" is called the "C/O ratio", and is indicative of the relative abundances of carbon and oxygen surrounding the tool, and hence can be used for hydrocarbon monitoring. Indices related to inelastic and capture Si and Ca are also obtained by this method.

9.2.6.1. *The carbon / oxygen tool*

The Dresser Atlas tool is $3\frac{7}{8}''$ diameter. The accelerator neutron source is pulsed at 20,000 bursts per second, each burst lasting 5–8 μs. Emitted gamma rays are detected by a conventional NaI(Tl) crystal and photomultiplier, thermally insulated in a Dewar flask. The counter output pulses, proportional in amplitude to the incident GR energies, are transmitted to the surface for analysis in a multi-channel analyzer. The fast neutron detection period is coincident with the burst. Between bursts, a 5–8 μs capture detection gate is opened. This is used both to provide capture data and to subtract out any capture events registered during the inelastic gate. Recommended maximum logging speed is 3 FPM.

Figure 9-20 is an example of a continuous C/O log. The capture Si curve is used for correlation, and behaves like a conventional neutron log. The Si/Ca ratio is derived from analysis of the capture spectrum, and the C/O and Ca/Si ratios from the inelastic spectrum. These ratios are indices related to the physical atomic concentrations. They are not absolute atomic ratio values.

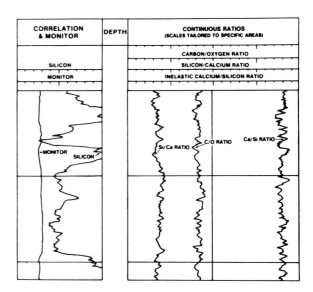

Fig. 9-20. A continuous carbon/oxygen log presentation (from Lawrence, 1981).

9.2.6.2. *Interpretation*

Laboratory test-pit measurements have been made to obtain the tool's response to lithology and fluid-type. Figure 9-21 is a response in $6^{5/8''}$ freshwater-filled casing. Note the increase in the C/O ratio with oil-filled porosity and with carbonate lithology.

Interpretation is handled with a fairly simple model assuming a linear dependence of spectral C/O ratio on oil saturation, for a given lithology and porosity.

$$S_o = (F_{co} - F_{wco})/(F_{oco} - F_{wco}) \qquad (9\text{-}3)$$

where: F_{co} is the C/O log reading, F_{wco} and F_{oco} are the C/O readings in the wet and oil-saturated zones, respectively.

The porosity dependence can be incorporated, assuming that the increase in C/O above the water line can be written as:

$$\Delta_{co} = \alpha \phi^{\beta}$$

where α and β are coefficients derived from test-pit data (Fig. 9-22). Then we can write:

$$F_{co} = F_{wco} + \alpha(\phi S_o)^{\beta} \qquad (9\text{-}4)$$

Returning to Fig. 9-21, the average water line for all lithologies can be approximated by:

$$F_{wco} = m_{wsc} F_{wsc} + b_{wsc} \qquad (9\text{-}5)$$

where m_{wsc} is the slope, b_{wsc} the intercept on the C/O axis, and F_{wsc} is the water-line capture Si/Ca ratio index, which is lithology dependent.

From eq. 9-4 we then have an equation for oil

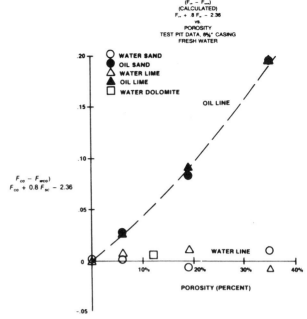

Fig. 9-22. C/O change relative to the water line (from Lawrence, 1981).

saturation which takes into account lithology and porosity:

$$S_o = \left(\frac{F_{co} - m_{wsc} F_{sc} - b_{wsc}}{\alpha \phi^{\beta}} \right)^{1/\beta} \qquad (9\text{-}6)$$

where F_{sc} is the capture Si/Ca ratio. This equation is most effective where formation water is fresh, owing to a dependence of capture Si/Ca on salinity.

A similar treatment can be used to derive a relationship using the inelastic Ca/Si ratio.

Overlay techniques

Overlays are a graphical solution to the response equations to obtain S_o from C/O and capture Si/Ca or inelastic Ca/Si.

This is an approach which blacks out the effect of lithology by displaying the Si/Ca (or Ca/Si) and C/O curves, suitably scaled to overlay in a wet zone (Figs. 9-23 and 9-24). A scaler is used to read oil saturation directly from the curve separation, by referring to the appropriate porosity line (fig. 9-25). Figure 9-26 is the atomic carbon-to-oxygen ratio dependence on porosity and saturation. This calculated data supports the empirically derived spectral ratio response curves of Fig. 9-22. Other interpretation methods are discussed in detail in the references.

9.2.6.3. *Environmental effects*

Readings are sensitive to borehole conditions, casing size, the cement sheath, washouts, well-bore fluid type and salinity. Test-pit data provides calibration coefficients for a limited range of hole configurations.

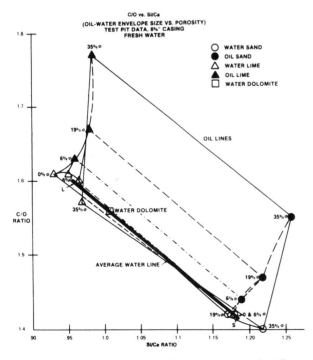

Fig. 9-21. Oil/water calibration envelopes from test-pit data (from Lawrence, 1981).

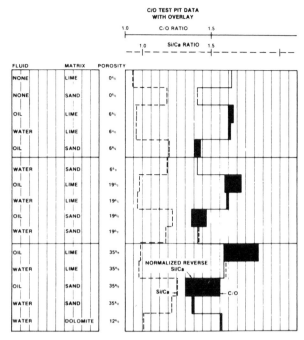

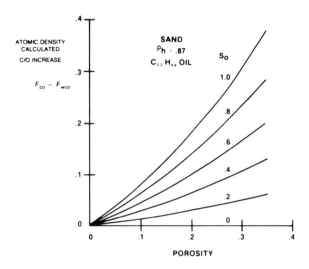

Fig. 9-26. Theoretical C/O dependence on porosity and oil saturation (S_o) in a sandstone formation (from Lawrence, 1981).

Fig. 9-23. The C/O versus capture Si/Ca overlay—test-pit data on log presentation (from Lawrence, 1981).

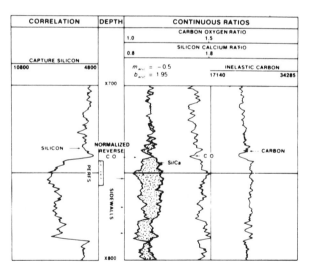

Fig. 9-24. Field example of the overlay method, using C/O and capture Si/Ca (from Lawrence, 1981).

Field cases may require empirical calibration of coefficients by crossplot methods, normalizing in known wet zones, etc. Capture Si/Ca is sensitive to borehole and formation salinity because of the Cl peak in the Ca window. Increasing salinity increases the apparent Ca level, and causes Si/Ca to decrease. Inelastic Ca/Si is not so affected, and is preferred in situations where salinity is high or changes drastically. Statistical uncertainty is, however, higher on the inelastic ratio.

9.2.6.4. Fluid salinity

The C/O ratio is not affected by changes in fluid salinity. The effects on Si/Ca and Ca/Si are mentioned above.

9.2.6.5. Hydrocarbon density

The number of C atoms/cm³ for a particular hydrocarbon is a function of its density. Thus there may be a considerable difference between the sensitivity of the C/O ratio to gas/light oil, and to heavy oil. Test pit data apply to a 30°API gravity oil (0.87 g/cm³) of the formula C_nH_{2n}. The response equations can be modified to include hydrocarbon density and molecular structure.

Gas zones can be distinguished by a low C/O reading (similar to a wet zone) accompanied by a high-capture Si correlation curve (equivalent to a low neutron porosity). Resolution on the C/O ratio is poor in gas zones.

9.2.6.6. Shaliness

The C/O ratio must be corrected for shaliness when the shales are carbonaceous, or where the C/O

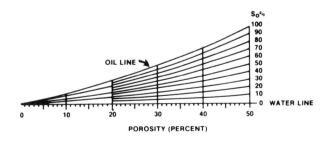

Fig. 9-25. C/O overlay scaler (from Lawrence, 1981).

reading in the shale is not the same as in the wet zones. A simple correction is applied, of the form:

$$F_{eco} = \frac{F_{co} - V_{sh} F_{shco}}{1 - V_{sh}}$$

where F_{eco} is the corrected C/O ratio, F_{co} the log reading, and F_{shco} the value of F_{co} in an adjacent shale bed.

9.2.6.7. Coal

Figure 9-27 shows the response of the continuous C/O log to coal beds—high C/O, low capture Si/Ca, low capture Si correlation. The C/O log has been used to evaluate coal quality (see refs.).

9.2.7. The "weighed least-squares" (WLS) method

9.2.7.1. The gamma spectrometer tool (GST)

One of the drawbacks of the window method is that the apparent spectral ratios are subject to systematic shifts arising from interference from other formation elements, which can not all be accounted for. (This is exemplified by the effect of the capture Cl peak in the Si window, for instance, causing the Si/Ca ratio to be sensitive to salinity.) In addition, the number of elements that can be identified is limited and physical concentrations are not easily quantifiable.

The WLS technique is aimed at an evaluation of all the significant elemental contributions to the spec-

tral measurement. From this it is possible to compute hydrocarbon saturation, salinity, and porosity as well as lithological composition.

9.2.7.2. GST measuring system

The GST is a $3\frac{5}{8}''$ (3.4'' optional) tool operating on multi-conductor cable. The neutron source is a pulsed accelerator, producing 14 MeV neutrons at 3×10^8 per second. A spectrometer is housed in the downhole tool; it consists of a NaI(Tl) detector and multi-channel analyzer, plus telemetry system for data transmission to surface (Fig. 9-28). Spectral analysis is performed in real-time in the CSU computer unit. Downhole circuit stabilization is automatic.

The induced gamma-ray emission is measured in the energy range approximately 1–8 MeV. Inelastic and capture data are sampled independently, and accumulated in the 256-channel spectrometer system. Data are continuously transmitted in digital form to surface where they are processed and stored on tape. The spectra of Figs. 9-18 and 9-19 are typical and were in fact obtained with a laboratory version of the GST.

Two timing modes are used: (a) Inelastic mode, which primarily measures the fast neutron interactions; (b) capture-tau mode, optimized for the capture gamma-ray emission. Each mode requires a separate logging run.

9.2.7.2.1. *Inelastic mode* is a fixed timing cycle (Fig. 9-29) of 100 μs duration. Gamma rays from the fast neutron interactions must be measured during the neutron burst, so the detector gate is coincident with the burst. Both last about 20 μs. Since some capture events also occur during this time, an early "capture background" gate is opened immediately afterwards. This background spectrum is used to clean up the burst gate data (i.e. to subtract out any capture

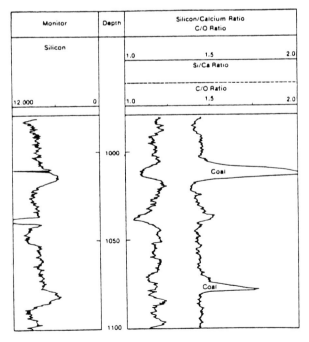

Fig. 9-27. Carbon/oxygen log response to coal-beds. Note the decrease in the silicon correlation curve (from Reike et al., 1980).

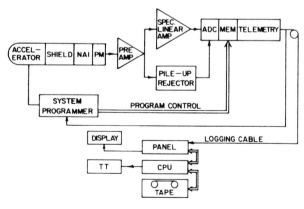

Fig. 9-28. System diagram of the Gamma Spectrometer Tool. TT means teletype terminal; CPU—central processing unit of the CSU; NaI—sodium iodide detector; PM—photomultiplier; ADC—downhole spectrometer; MEM—memory (from Hertzog, 1981).

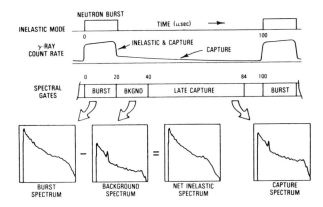

Fig. 9-29. GST inelastic mode timing program (from Westaway and Hertzog, 1980).

events), resulting in the "net inelastic spectrum" in the figure.

A "late capture" gate measures formation (and the inevitable borehole) capture events. All three spectra, plus circuit calibration data, are sent uphole for analysis.

9.2.7.2.2. *Capture-tau mode* allows fast neutron and early borehole-related capture events to die away before the detector gate is opened (Fig. 9-30). The entire gating sequence is dependent on the thermal neutron decay time τ (see Chapter 10). This is measured simultaneously by a self-adjusting tau-loop working along the same lines as the TDT-K except that the background gate is set during the ninth and tenth cycles, which have no neutron bursts. The effect of varying the detector gate delay as formation τ varies, is to enhance the tool response to the formation, at the expense of borehole signal.

This can be further improved by fitting a boron sleeve to the tool housing to displace well-bore fluid. The highly captive boron causes a very rapid depletion of thermal neutrons in the borehole, effectively extinguishing the borehole signal before the detector gates open. Casing or tubing size may preclude the use of this sleeve.

The capture gate, late background and calibration data are transmitted uphole. The formation τ measurement is also recorded as a capture cross-section, Σ, curve.

9.2.7.2.3. *Logging speeds.* On the surface, the spectral data are accumulated over 6″ of tool travel in continuous logging mode. Every 6″ the accumulated data are analyzed, and stored on magnetic tape along with the results of the WLS analysis. Capture-tau mode is usually run at 10 FPM or slower.

Inelastic mode is commonly operated with the tool stationary, in which case data are accumulated over 5 minutes per station. Alternatively, a logging speed of 2-1/2 FPM can be used. To reduce statistical variations, several passes may be averaged.

9.2.7.3. *WLS spectral analysis*

This is a fitting technique. Reference, or standard, spectra for each of the elements it is required to identify, have been derived from laboratory analysis. Each elemental standard represents what the tool would record if surrounded solely by that element. The set in present use for inelastic spectrometry is shown in Fig. 9-31a and the set for capture spectrometry in 9-31b. (Note that the detector crystal's iodine activation yield must be included in the capture standards, since the objective is to analyse ALL significant components of the spectrum).

At each sampling depth, the CSU computer determines which linear combination of these standards best fits the logging spectrum, over almost its entire width. The best fit is assessed according to a least-squares criterion. Rather than using a "trial" and error" approach to the fitting, each standard has its associated linear estimator (Fig. 9-32) which, when multiplied channel for channel into the accumulated log spectrum, produces the corresponding fractional

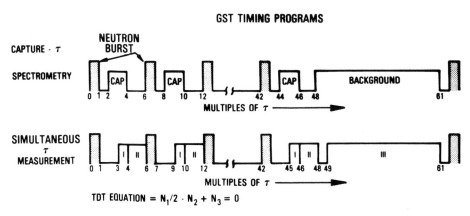

GST TIMING PROGRAMS

TDT EQUATION = $N_1/2 - N_2 + N_3 = 0$

Fig. 9-30. GST capture-tau mode timing diagram (from Westaway and Hertzog, 1980).

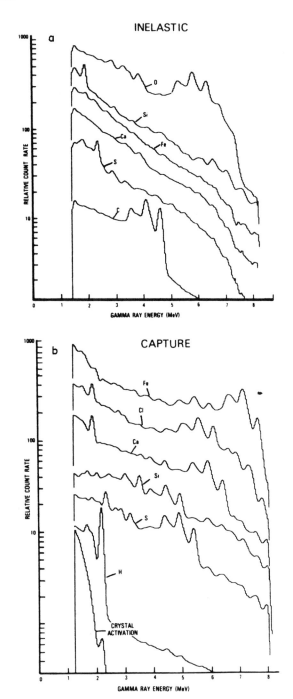

Fig. 9-31. a. The inelastic WLS fitting standards (from Westaway and Hertzog, 1980). b. The capture WLS fitting standards (from Westaway and Hertzog, 1980).

contribution of the element in question.

Inelastic spectra are processed basically with the inelastic standard set, but further cleaning up of residual capture events is possible by incorporation of selected capture standards. Capture spectra are processed with the capture standards. The two analyses are, of course, performed completely independently.

The outputs of the analysis are the elemental yield

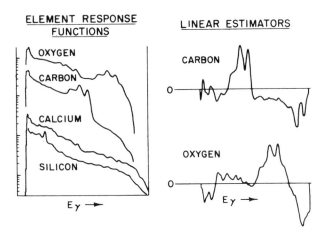

Fig. 9-32. The linear estimators (right) for carbon and oxygen spectral yields, used in the WLS analysis (from Hertzog, 1978).

coefficients. These are the fractions of the total spectrum that are computed to have been contributed by the various elements. The results of a typical inelastic fit in a clean oil-bearing sandstone, with water-filled cased hole, might be:

C : 0.12 (oil)
O : 0.28 (water, quartz, cement)
Si : 0.25 (quartz, cement)
Ca: 0.04 (from the cement)
Fe: 0.32 (casing and tool housing)
S : − 0.01 (none present)

Note that the sum of the yields is, by definition, 1.0. The slightly negative S yield is due to statistical variation about zero. In terms of the fitting, it means that the best fit was obtained by subtracting a small fraction of the S standard rather than adding.

From these coefficients, the spectral yield ratios are computed. They were chosen for correlation with the conventional macroscopic reservoir parameters such as porosity and lithology. These yield ratios are listed in Table 9-1.

Other outputs include gate count-rates, various calibration data concerned with the alignment of the downhole circuitry, crystal resolution (peak width, which deteriorates with crystal age), and a quality indicator for the spectral fitting.

All this data goes onto tape along with the spectra.

TABLE 9-1

The GST spectral yield ratios presently output on CSU (from Westaway and Hertzog, 1980).

Yield ratio	Interaction	Name	Label
C/O	Inelastic	Carbon-oxygen ratio	COR
CI/H	Capture	Salinity-indicator ratio	SIR
H/(Si+Ca)	Capture	Porosity-indicator ratio	PIR
Fe/(Si+Ca)	Capture	Iron-indicator ratio	IIR
Si/(Si+Ca)	Capture and inelastic	Lithology-indicator ratio	LIR

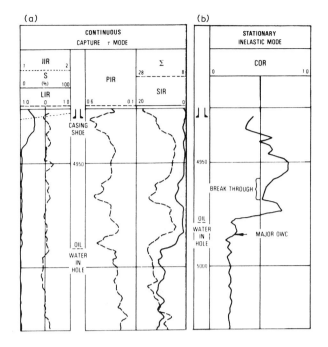

Fig. 9-33. Example of a GST continuous capture-tau mode log (a), and stationary mode inelastic log (b) in a clean carbonate reservoir. Note that the wellbore oil/water interface has moved between runs (adapted from Westaway and Hertzog, 1980).

Selected channels, particularly the yield coefficients, ratios and count-rates, are displayed on film. The spectra can be observed on the viewer of the CSU.

Figure 9-33a is a section of continuous capture-tau log, and Fig. 9-33b the corresponding inelastic mode based on 5-minute stations. The yield ratios are displayed.

9.2.7.4. Reprocessing of data

Since all the downhole spectra are recorded on tape, it is possible to re-analyze them if necessary to improve the results or correct for factors not taken into account in the real-time analysis. Such factors include:

(a) Inclusion of an element not present in the set of standards used, but present in the formation spectrum.

(b) Exclusion from the standards of any element not present in the formation (i.e. less than 2% contribution to the measured spectrum). Minimising the standards used in the W.L.S. fit reduces statistical uncertainty.

(c) The effects of temperature, humidity, age, or damage, on the detector crystal. These tend to degrade the crystal resolution, i.e. to broaden the peaks, so that the standards no longer are valid because they were obtained using a good crystal. Appropriately degraded standards can be simulated.

(d) Changes in spectrometer alignment. Although

the downhole circuitry is automatically aligned by means of channel calibrating signals, residual shifts or gain errors must be corrected for, because the standard spectra were made with a well-aligned tool.

Much of this corrective processing is now performed in real-time in the CSU. Subsequent reprocessing can be performed off-line in the CSU or computing centre, should it be necessary.

9.2.7.5. Interpretation

The relations between the measured spectral yields and physical atomic concentrations depend on neutron energy distribution and interaction cross-sections, integrated over all space and over all neutron energies, in addition to the obvious dependence on atomic concentrations. A rigorous treatment of elemental yields would produce a response equation of the form (taking elemental carbon as an example):

$$X_c = \int_E \int_r \psi(E, r)\sigma(E)V_c(r)\,\mathrm{d}E\mathrm{d}r \qquad (9\text{-}6)$$

where $\psi(E, r)$ is the flux of neutrons of energy E at distance r from the source, $\sigma(E)$ is the total interaction cross-section for neutron energy E, and $V_c(r)$ is the concentration of atoms of carbon at r. Further terms describing the transport of gamma rays from point of emission to detector would also be required.

Thus the elemental yield for carbon, for example, cannot be easily converted to its abundance in the formation. Furthermore, the conversion coefficient would not be the same as for oxygen.

Interpretation of elemental yields or yield ratios is a complex problem, therefore. WLS analysis has the advantage of extracting true spectral fractions from the measured spectrum, rather than apparent ones, making the development of a rigorous interpretation model more feasible.

Induced gamma-ray yield responses are being researched using computerized mathematical modelling and laboratory calibrations. A simplified interpretation model has been used on field data with good results. This model assumes that the elemental yields depend in a linear manner on the concentration of elements averaged over the borehole and formation, and the average interaction cross-sections. This approach has been particularly successful in the interpretation of the C and O yields for hydrocarbon evaluation.

9.2.7.5.1. Hydrocarbon saturations from the C and O yields

The ratio spectral carbon yield/spectral oxygen yield (COR) can be approximated by:

$$\mathrm{COR} = \frac{\sigma_c}{\sigma_o}\left[\frac{\alpha(1-\phi) + \beta\phi(1-S_w) + B_c}{\gamma(1-\phi) + \delta\phi\,S_w + B_o}\right] \qquad (9\text{-}7)$$

where:

σ_c, σ_o are the effective fast neutron interaction cross sections for C and O, averaged over all interaction types and all neutron energies.

α, β are the atomic concentrations of carbon in the rock matrix volume $(1 - \phi)$ and in the hydrocarbon volume $\phi(1 - S_w)$, respectively.

γ, δ are the atomic concentrations of oxygen in the rock matrix volume $(1 - \phi)$ and the pore water volume ϕS_w, respectively.

B_c, B_o represent carbon and oxygen background contributions from the borehole. These are important terms because they are relatively large and cause systematic shifts to the log readings. They depend on hole configuration (size, casing, cement) and hole fluid.

ϕ is the effective formation porosity, most accurately obtained from open-hole logs.

The equation can be easily modified to include shale terms. Figure 9-34 shows some laboratory data plotted to show the dependence of COR on porosity, hydrocarbon saturation and lithology. The response fans, computed with an appropriate choice of constants, in eq. 9-7 fit the data points closely.

Table 9-2 lists the theoretical α and γ matrix carbon and oxygen concentrations for a variety of lithologies, together with the β of an oil of density 0.85 g/cm^3, type $C_n H_{2n}$, and the δ of pure water. These constants are easily computed from the bulk density ρ (g/cm^3), gram molecular weight (GMW), and number of atoms per n molecule (in units of Avogadro's number).

$$\text{Atoms/cm}^3 = \rho \times n/\text{GMW}$$

For example, for pure limestone; CaCO$_3$: $\alpha = 2.71 \times 1/100 = 0.027$; and $\gamma = 2.72 \times 3/100 = 0.081$. For

TABLE 9-2

Some atomic densities in units of Avogadro's number (6.023×10^{23} atoms/mole)

	Density (g/cm^3)	Oxygen (atoms/cm^3)	carbon (atom/cm^3)
Limestone	2.71	0.081	0.027
Dolomite	2.87	0.094	0.031
Quartz sandstone	2.65	0.088	–
Anhydrite	2.96	0.087	–
Oil	0.85	–	0.061
Water	1.0	0.056	–
Feldspars			
Microcline	2.56	0.074	–
Albite	2.62	0.080	–
Anorthite	2.76	0.079	–
Micas			
Muscovite	2.83	0.085	–
Biotite	3.08	0.081	–
Glauconite			
Sample I	2.51	0.074	–
Sample II	2.65	0.075	–
Clays			
Illite	2.53	0.082	–
Kaolinite	2.42	0.084	–
Chlorite	2.77	0.082	–
Montmorillonite	2.12	0.077	–

silica, SiO$_2$: $\alpha = 0$; and $\gamma = 2.65 \times 2/60 = 0.088$. Note that the concentration of atomic oxygen is, unlike any other element, fairly constant in many minerals. This suggests that the COR measurement should be relatively insensitive to variations in non-carbon-bearing mineralogy (tuffites, clays, feldspars, etc.), all other things being equal.

The borehole terms B_o and B_c have been derived from laboratory and field data. Most importantly, in water-filled open hole or casing, $B_c = 0$. In oil-filled open-hole $B_o = 0$. In oil-filled cased-hole, both B_c (oil) and B_o (cement sheath) will be non-zero.

Equation 9-7 defines a unique response curve for a given lithology and hole configuration, resembling a fan graduated in terms of S_w. In Fig. 9.-35, four COR vs porosity crossplots are shown to demonstrate this. Figures 9-35a and 9-35b correspond to the oil-filled and water-filled sections of the example log of Fig. 9-33, which is a clean limestone reservoir. Note the change of well-bore fluid from oil to water causes a large downward shift of the COR ratio, and a reduction in the dynamic range (i.e. the spread of the fan from $S_w = 1$ to $S_w = 0$). This example in fact shows two prominent interpretational features: (a) a major oil–water transition in the formation at 4986' and (b) a breakthrough of water above this OWC, 4970–4959'.

These features can also be discerned on the crosplots. Figure 9-36 presents the CPI analysis of the COR data, compared with Σ-derived saturations.

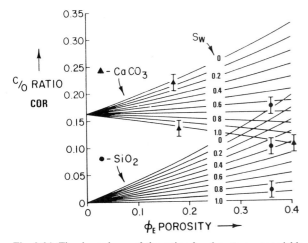

Fig. 9-34. The dependence of the ratio of carbon to oxygen yields (COR) on porosity and water saturation S_w, in limestone and sandstone lithologies, water-filled hole. Both laboratory data and postulated response curves are shown (adapted from Westaway and Hertzog, 1980).

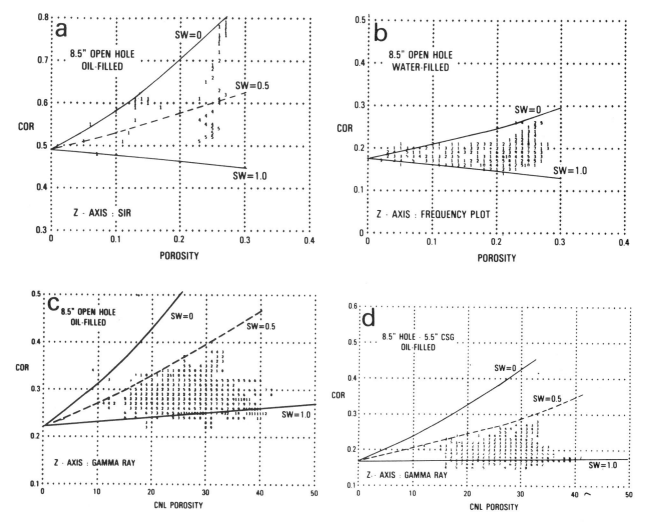

Fig. 9-35. COR versus porosity crossplots in different environments; (a) and (b) are open hole limestone responses; (c) and (d) are open and cased-hole sandstone responses (from Westaway and Hertzog, 1980).

Discrepancies can arise where formation water salinity varies, the COR being independent of salinity, unlike Σ. This is quite feasible since this saline reservoir was being subjected to a fresh water drive.

Figs. 9-35c and 9-35d demonstrate how the insertion of a casing and a cement sheath reduce the reading level and dynamic range of COR readings made across a tuffaceous sandstone reservoir in open and cased hole. Note shales and wet sands tend to lie on the same line. The four crossplots can be compared with Fig. 9-34 where hole size was 10″ with a 7″ water-filled casing. Note the line corresponding to $S_w = 1$ for the sandstone in the laboratory example is at COR = 0 because $B_c = 0$.

Figure 9-37 is a shale-sand sequence drilled in 1940. The ES resistivity log shows three oil-bearing sands A, B, C. Formation water is only 50.000 ppm NaCl salinity. Cased hole monitoring logs were run in 1978 to evaluate remaining hydrocarbons, these zones being produced from nearby wells. Owing to the freshwater, the TDT Σ curve cannot resolve oil

and water, and the log delineates only the sands and shales. The COR curve, made from stationary readings, clearly shows an oil-water transition in sand B and at the very bottom of sand A.

9.2.7.5.2. The yield ratios as qualitative indicators

The yield ratios of Table 9-1 are named according to the macroscopic reservoir parameters to which they most strongly respond.

LIR, the lithology ratio, serves as a quick-look indication of lithology changes. Si/(Si + Ca) should read 1.0 in sands and non-calcareous shales, and zero in carbonates (compare Figs. 9-33 and 9-41). In Fig. 9-38, the reading in sandstone is less than unity and in limestone greater than zero, because of the presence of a cement sheath (containing calcium and silicon).

IIR, the iron indicator ratio, responds in particular to iron-bearing shales (Fig. 9-41 where IIR correlates very well with Σ). It also responds to the presence of casing (see the top of Fig. 9-33a).

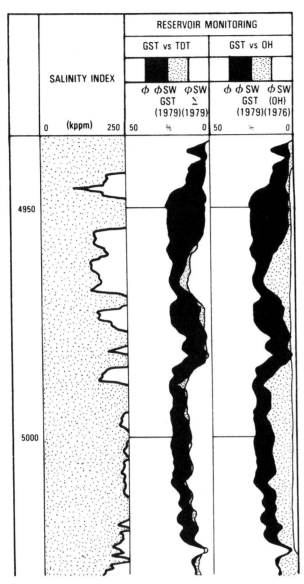

Fig. 9-36. CPI presentation based on the data of Fig. 9-33. Porosity and original open-hole saturations ($S_{w(OH)}$) were derived from conventional open-hole logs (from Westaway and Hertzog, 1980).

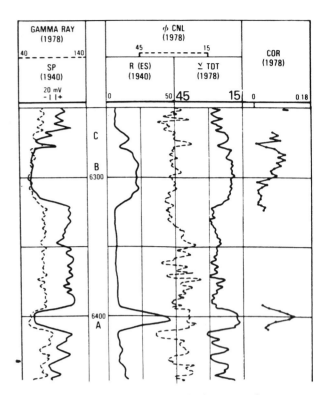

Fig. 9-37. Oil–water contact in very fresh water environment are clearly distinguished in sands A and B by the inelastic carbon to oxygen ratio (COR) of the GST (from Westaway and Hertzog, 1980).

PIR, the porosity indicator ratio, resembles a hydrogen index measurement and usually correlates with the neutron porosity. The presence of (Si + Ca) in the denomenator reduces its sensitivity to lithology changes. Figures 9-39 and 9-41 demonstrate the correlation with the CNL porosity, plotted as $\phi/(1 - \phi)$.

SIR, the salinity indicator ratio, varies with formation water salinity and hydrocarbon saturation since it is sensitive to the amount of chlorine present. It behaves rather like the capture cross-section. Since the concentration of hydrogen is fairly constant regardless of salinity or saturation, there is a fairly linear dependence between SIR and the product salinity $\times S_w$ (Fig. 9-40).

COR, already discussed as a hydrocarbon measurement, is also lithology-dependent, reading higher

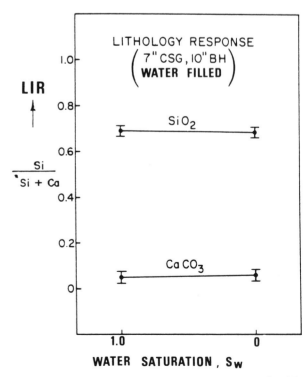

Fig. 9-38. Laboratory responses of LIR to quartz and calcite lithology in a cased hole (from Hertzog et al., 1978).

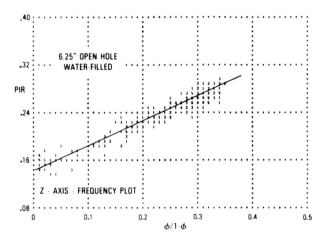

Fig. 9-39. PIR correlates with the neutron porosity, plotted here as $\phi/(1-\phi)$ (from Westaway and Hertzog, 1980).

in carbonates than shales and sands, for the same porosity and saturation.

All yields and ratios are affected strongly by the borehole, and other logs such as the CBL, caliper, and gradio-manometer (for well-bore fluid density) may be necessary to adequately describe hole conditions.

9.2.7.5.3. A summary of environmental effects
A. The borehole

(a) *Borehole fluid*: C and O yields are particularly sensitive to whether the fluid is oil or water. (Fig. 9-33). The Cl yield responds of course to the water or mud salinity (Fig. 9-33). Borehole signal will be negligible in a gas-filled hole.

(b) *Casing*: causes a general reduction in all gamma ray counts because of Compton scattering. The casing affects the Fe yield in particular, causing a large shift (Fig. 9-33). Casing size determines the degree to which the well-bore fluid and cement sheath will influence readings.

(c) *Borehole size*: determines the volume of well-

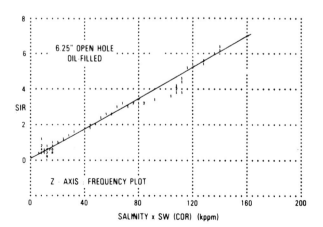

Fig. 9-40. SIR correlates with the product of saturation (S_w) and formation water salinity for a fairly constant porosity (from Westaway and Hertzog, 1980).

bore fluid (open hole) or thickness of cement sheath (cased hole).

(d) *Cement sheath*: increases the Ca and Si contributions from the borehole (Fig. 9-38). It introduces a borehole oxygen yield which will reduce the COR level in an oil-filled hole, but will have little effect in a water-filled hole. The H yield is fairly similar to that of water and oil. Cl will respond to whether the cement was made with fresh or salty water. The presence of any liquid in channels in the cement will have an effect on the readings.

B. The formation

(a) *Lithology:* All of the elemental yields are dependent to a greater or lesser extent on the minerals present. The next section discusses how they can be used to evaluate mineral fractions. An accurate description of the lithology is necessary when computing hydrocarbon saturations from eq. 9-7. The α- and γ terms can be linked to mineral composition via, for instance, matrix density from open-hole logs if available, or from a mineral analysis obtained from the GST itself. Examples of lithology responses are given in the references. Figures 9-33 and 9-41 show some interesting differences, the former being a clean carbonate, the latter a tuffaceous sand/shale sequence. Note the increasing S yield in the anhydrite cap-rock in Fig. 9-33a.

(b) *Porosity*: represents a reduction in matrix yields and an increase in pore-fluid yields (see Fig. 9-35 for instance). Porosity must be known in order to interpret COR (eq. 9-7) and is obtained from open-hole logs if available, or from PIR. Among the yield ratios, COR and PIR are the most porosity-dependent.

(c) *Saturation*: An increase in hydrocarbon saturation causes an increase in C yield and a decrease in O and Cl yields. The H yield remains fairly constant for medium to heavy oils. Oil, then, appears as an increase in COR and a decrease in SIR. Gas, having low carbon and hydrogen densities, produces a low COR and could be interpreted as water were it not for the low associated H yield which produces a low PIR and high SIR. Hydrocarbon density and type are taken into account in the β term of eq. 9-7.

(d) *Salinity*: strongly affects the Cl yield.

(e) *Shaliness*: may or may not contribute to the Fe yield (chlorite and glauconite contain iron for instance, while illite, kaolinite and montmorillonite do not). Figure 9-41 shows two iron-bearing clay types; one has a high natural potassium yield (from the NGS) and may be glauconitic, the other is low in potassium and may be chlorite. Shales usually contain freshwater, Si, O, H and possibly Ca and C (source rock and carbonaceous fragments). Non hydrocarbon-bearing shales often have COR levels similar to those of adjacent wet sands.

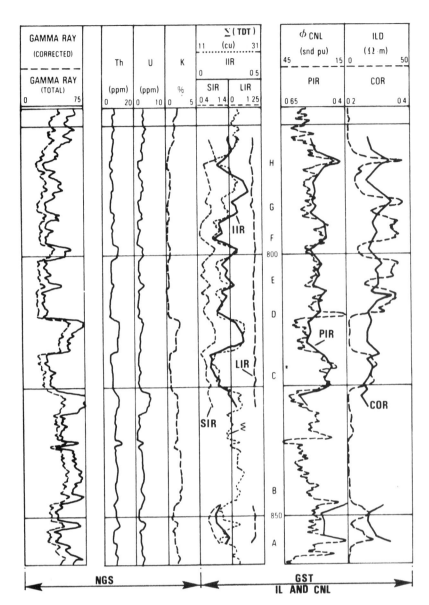

Fig. 9-41. Natural and induced gamma ray spectroscopy, showing the lithology and hydrocarbon responses in a shale-sand sequence. Note the correlation between neutron capture cross-section Σ and IIR in these iron-bearing shales (here increases to the right, the reverse of the conventional presentation). In combination with the natural gamma-ray spectrometry (NGS) log it is possible to distinguish clay types (from Westaway and Hertzog, 1980).

9.2.7.5.4. *Quantitative determination of lithology with the GST*

The elemental yields can be interpreted in terms of volume fractions of minerals by associating each yield with a mineral dominated by the corresponding element (e.g. the key element to identify quartz would be silicon, for calcite it would be calcium, and so on).

The capture gamma ray yield Y_x due to element X present in mineral f(X) is approximated by *:

$$Y_x = a\sigma_x E_x C_{f(X)} V_{f(X)} \qquad (9\text{-}8)$$

----
* This approach is analogous to the simplified model for COR interpretation, eq. 9-7.

where:

a is the neutron flux;

σ_x is the atomic thermal neutron capture cross-section of element X;

E_x is the number of gamma rays detected per neutron capture by element X;

$C_{f(X)}$ is the volume fraction of element X in mineral f(X);

$V_{f(X)}$ is the volume fraction of mineral f(X) in the formation.

Since σ_x, E_x and $C_{f(X)}$ are constant for a given element X in a mineral f(X), they can be grouped into one term $S_{f(X)}$ which can be regarded as the "sensitivity" of the mineral to gamma ray emission/detection from element X. We then have:

$$Y_x = aS_{f(X)}V_{f(X)} \qquad (9\text{-}9)$$

More generally, if several minerals contain X, the total gamma ray yield is:

$$Y_x = a\left(S_{f(X)}V_{f(x)} + S_{g(X)}V_{g(x)} + \ldots\right)$$

The neutron flux term, a, which is not a constant, can be eliminated by considering the ratio of sensitivities; for instance for minerals f(X) and g(X):

$$S_{g(X)}/S_{f(X)} = C_{g(X)}/C_{f(X)} \qquad (9\text{-}10)$$

Core and laboratory data are used to calibrate the relative sensitivities.

Consider a limestone-sandstone-clay lithology. We can write the appropriate eq. 9-9 for Ca, Si and H, which are the key elements for evaluating the calcite, quartz, clay and porosity, respectively:

$$Y_{Ca} = aS_{LST}V_{LST}$$

$$Y_H = aS_\phi\phi_t$$

$$Y_{Si} = a\left(S_{SND}V_{SND} + S_C V_C\right)$$

Also:

$$V_{LST} + V_C + V_{SND} + \phi_t = 1.0$$

Note V_C is the dry clay mineral fraction.

Core data are necessary at this stage to calibrate

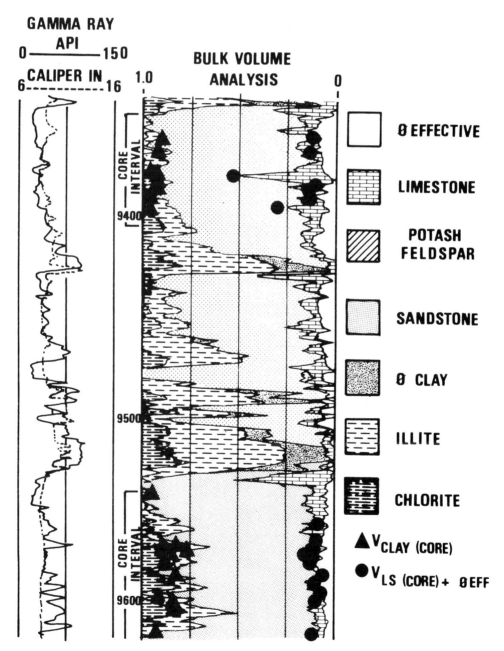

Fig. 9-42. Lithology analysis using the GST, with NGS for clay-typing (from Flaum and Pirie, 1981).

the sensitivity terms. The outcome will be the required relations between elemental yields and mineral fractions, of the form:

$$V_{LST} = A_{LST} Y_{Ca}, \text{ etc.}$$

where A_{LST} is the calibrated conversion term for limestone, and so on. This will be dealt with more fully in volume 2.

Figure 9-42 is an example of an analysis made by this method. NGS (Natural Gamma-ray Spectrometry) data has been included for clay-typing. The clay porosity is defined as the GST-derived total porosity minus the effective porosity derived from open-hole logs. Core data is included in the figure for comparison.

9.2.7.6. *High Resolution Spectroscopy (HRS)*

A germanium (Ge) crystal detector is capable of far sharper resolution than the conventional NaI. The responses of the two crystals are compared in Fig. 9-43. The Ge system preserves the sharp features of the gamma-ray spectrum. This arises from the semiconductor properties of germanium. Each incident gamma ray liberates many more electrons in the crystal; the reduction in statistics inherent in an increase of electrons in each pulse produces sharper peaks in the spectrometer circuitry. In addition, the electrons liberated in the crystal travel directly to an accelerating electrode, thus transferring the incident gamma-ray energy directly to the counting circuitry. The NaI system has the intermediate optical-photon "scintillation" stage (Fig. 7-11) in the transfer, which allows for some degradation of energy.

The Ge crystal must, however, be kept at $-196°C$ to ensure sufficiently noise-free operation. Early Ge crystals were doped with lithium (Li), (in the same way that the NaI crystals are doped with thallium). These Ge-Li crystals had to be stored at low temperature. Recently developed pured Ge crystals can be stored at room temperature, but must be operated at $-196°C$.

The excellent resolution of the germanium crystal is offset by its poor efficiency (about one fifth that of NaI) and about the same logging speeds apply to both systems. However, the fine peak definition permits identification of many more elements than is possible with the GST, by means of a large number of inelastic, neutron reaction and activation emissions not discernible with the NaI detector.

9.3 REFERENCES

Culver, R.B., Hopkinson, E.C., and Youmans, A.H., 1974. Carbon/oxygen (C/O) logging instrumentation. SPE J, pp. 463–470.

Czubek, J., 1971. Recent Russian and European developments in nuclear geophysics applied to mineral exploration and mining. The Log Analyst, 1971.

Dewan, J.T., Stone, O.L. and Morris, R.L., 1961. Chlorine logging in cased holes. J. Pet. Technol., pp. 531–537.

Edmundson, H. and Raymer, L.L., 1979. Radioactive logging parameters for common minerals. The Log Analyst, XX (5): 38.

Enge, H.A., 1966. Introduction to Nuclear Physics. Addison-Wesley, Inc. pp. 210.

Engesser, F.C. and Thompson, W.E., 1967. Gamma rays resulting from interactions of 14.7 MeV neutrons with various elements". J. Nuclear Energy, pp. 487–507.

Fertl, W.H. and Frost, E., 1980. Recompletion, workover, and cased hole exploration in clastic reservoirs utilizing the continuous carbon/oxygen (C/O) log. The CHES III Approach, SPE-9028.

Flaum, C. and Pirie, G., 1981. Determination of lithology from induced gamma-ray spectroscopy. SPWLA, 22nd Annu. Log. Symp. Trans., June 23–26, 1981, Paper H.

Fletcher, J.W. and Walter, J., 1978. A practical shale compensated chlorine log. SPWLA 19th Annu. Log. Symp. Trans., June 13–16, 1978, paper GG.

Glasgow, D.W. et al., 1976. Differential elastic and inelastic scattering of 9- to 15-MeV neutrons from carbon. Nuclear Sci. Eng., 61: 521–533.

Greenwood, R.C. and Reed, J.A., 1965. Prompt gamma rays from radioactive capture of thermal neutrons. The Div. of Isotopes Dev., U.S.A.E.C. October 1965.

Hertzog, R., 1979. Laboratory and field evaluation of an inelastic-neutron-scattering and capture gamma ray spectroscopy tool. Paper SPE 7430. Presented at the SPE 53rd Annu. F. Tech. Conf. and Exhibition, Houston (1978), J. Pet. Technol., Oct. 1980.

Hertzog, R. and Plasek, R., 1979. Neutron-excited gamma ray spectrometry for well logging. IEEE Trans. Nuclear Sci., Vol. NS-26. Feb. 1979.

Kerr, M.E., 1981. Carbon/oxygen or resistivity/pulsed neutron logging—A case study. SPE Middle East Oil Tech. Conf., Bahrain (1981), Paper SPE-9612.

Lawrence, T.D., 1981. Continuous carbon/oxygen log interpretation techniques. J. Pet. Technol., Aug. 1981.

Lawson, B.L., Cook, C.F. and Owen, J.D., 1971. A theoretical and laboratory evaluation of carbon logging: laboratory evaluation. SPE J., June, pp. 129–137.

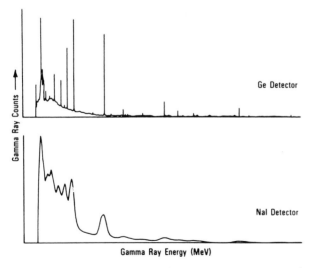

Fig. 9-43. A comparison of induced gamma-ray spectra measured with sodium iodide and germanium crystal detectors (courtesy of Schlumberger).

Lock, G.A. and Hoyer, W.A., 1974. Carbon/oxygen (C/O) log: use and interpretation. J. Pet. Technol., Sept., pp. 1044–1054.

McKinlay, P.F. and Tanner, H.L., 1975. The shale compensated chlorine log. J. Pet. Technol., Feb., pp. 164–170.

Moran, H.H. and Tittman, J., 1970. Analysis of gamma-ray energy spectrum for constituent identification. U.S. Patent 3,521,064 (July 21st 1970).

Oliver, D.W. et al., 1981. Continuous carbon/oxygen logging-instrumentation, interpretive concepts and field applications. SPWLA, 22nd Ann. Log. Symp. Trans., Paper TT.

Rabson, W.R., 1969. Chlorine detection by the spectral log. Pet. Eng., Mar. 1969, B102–111.

Ready, R., Arnold, J. and Trombka, J., 1973. Expected gamma-ray emissions spectra from the lunar surface as a function of chemical composition. Gold Space Flight Center, Publ. No. X-641-73-68.

Reike, H. et al., 1980. Successful application of carbon/oxygen logging to coal bed exploration. SPE 9464.

Schultz, W.E. and Smith, H.D., Jr., 1973. Laboratory and field evaluation of a carbon/oxygen well logging system. SPE-4638.

Smith, H.D. and Schultz, W.E., 1974. Field experience in determining oil saturations from continuous C/O and Ca/Si logs independent of salinity and shaliness. The Log Analyst, Nov-Dec. 1974, pp. 9–18.

Stroud, S.G. and Schaller, H.E., 1960. New radiation log for the determination of reservoir salinity. J. Pet. Technol., Feb., pp. 37–41.

Tittman, J. and Nelligan, W.B., 1960. Laboratory studies of a pulsed neutron-source technique in well logging. SPE AIME, Trans. 219: 375–378.

Waddell, C. and Wilson, J.C., 1960. Chlorinity and porosity determination in cased and open holes by nuclear logs. SPE of AIME, Paper 1456-G, Beaumont, Texas, Mar 8, 1960.

Westaway, P. and Hertzog, R., 1980. The gamma spectrometer tool: inelastic and capture gamma-ray spectroscopy for reservoir analysis. Paper SPE-9461.

Westaway, P., Wittmann, M. and Rochette, P., 1979. Applications of nuclear techniques to reservoir monitoring. Proc. SPE Middle East Tech. Conf., Bahrain, Paper SPE-7776.

Youngblood, W., 1980. The application of pulsed-neutron decay time logs to monitoring water floods with changing salinity. J. Pet. Technol., 32/6.

10. THERMAL DECAY TIME MEASUREMENTS

10.1 BACKGROUND THEORY

We have seen in Chapter 8 that the probability of capture of thermal neutrons by an element depends on its capture cross section and its concentration in the rock. From this we deduce that the life of a thermal neutron is shorter the greater the probability of capture, or the richer the rock is in elements of high capture cross section.

In a vacuum the average life-time of a neutron is of the order of 13 minutes, but in rock it varies between 5 microseconds for salt and 900 microseconds for quartz. A measurement of the lifetime of the thermal neutron population can therefore give information on the concentration of absorbing elements, (in particular chlorine, which will ultimately permit us to distinguish oil from saline water).

10.2. TOOL PRINCIPLE

The formation is subjected to a pulse of high-energy neutrons (14 MeV) from a neutron generator. This pulse is repeated at a fast rate. The thermal neutron population is sampled between pulses and its rate of decay computed. Either the thermal neutrons are counted directly, or the gamma rays emitted at each capture event are detected.

10.2.1. Neutron capture

10.2.1.1. *The exponential decay*

Emitted neutrons are rapidly slowed down to thermal state by collisions with the nuclei of elements present in the formation. They are eventually captured by nuclei at a rate which depends on their capture cross sections and respective abundances in the rock. At each point in the formation a certain fraction of the thermal neutrons is absorbed in unit time. This absorption rate depends on the product $v \Sigma_{abs}$, v being the mean thermal neutron velocity (a constant for a given temperature), and Σ_{abs} being the macroscopic capture cross-section of the formation at that point. If neutron capture is the only phenomenon occurring, the number of neutrons decays exponentially (Fig. 10-1). Figure 10-2 demonstrates how the decay rate increases when oil is replaced by water in the pores of a rock, because the water has a larger Σ cross-section than oil (in this and ensuing dia-

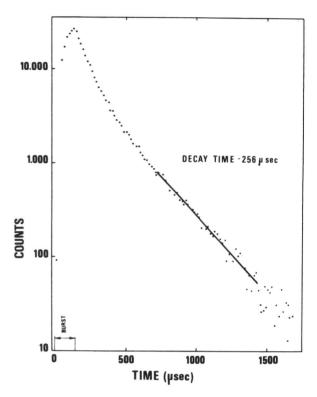

Fig. 10-1. Build-up and decay of the thermal neutron population after a burst of high-energy neutrons (courtesy of Schlumberger). The decay time is 256 μs in this example.

grams, the capture gamma ray counts are shown, rather than the neutron counts. As we discuss in 10.2.3., they are proportional).

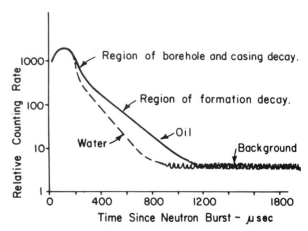

Fig. 10-2. The thermal neutron population monitored by counting the gamma rays. Decay rate in a water-bearing formation is greater than in an oil-bearing one. Note the early borehole/casing decay (courtesy of Schlumberger).

At a time t_1:

$$N_1 = N_0 e^{-v\Sigma_{abs}t_1} \qquad (10\text{-}1)$$

where: N_1 = number of thermal neutrons remaining per unit volume at time t_1; N_0 = number of thermal neutrons per unit volume at an arbitrary time $t = 0$; t_1 = the elapsed time since an arbitrary time zero. Σ_{abs} is the sum of the cross-sections of all the atomic nuclei in unit volume of formation for an average neutron velocity v of 2200 m/s (at 75°F). It is measured in cm^2/cm^3 (or cm^{-1}). If we repeat the measurement at time t_2 we will obtain:

$$N_2 = N_0 e^{-v\Sigma_{abs}t_2} \qquad (10\text{-}2)$$

where: N_2 = the number of thermal neutrons per unit volume at time t_2; and t_2 = the elapsed time since an arbitrary time zero.

The decrease in thermal neutron population density (Fig. 10-2) can be quantified by a comparison of the two measurements:

$$N_2 = N_1 e^{-v\Sigma_{abs}(t_2-t_1)} \qquad (10\text{-}3)$$

from which we can deduce

$$\Sigma_{abs} = \frac{1}{v(t_1-t_2)} \ln \frac{N_1}{N_2} \qquad (10\text{-}4a)$$

Using logarithms to the base 10 with $v = 2200$ m/s, Δt in microseconds and Σ in cm^{-1} we have:

$$\Sigma_{abs} = \frac{10.5}{\Delta t} \log_{10} \frac{N_1}{N_2} \qquad (10\text{-}4b)$$

Since the decay is exponential, we can represent it in another way by introducing an intrinsic decay time, τ_{int}, corresponding to the time necessary for the initial number of neutrons N_0 per cm of formation to decay by 63%, or to 37% of its original value (which is 1/e). We have simply:

$$N_t = N_0 e^{-t/\tau_{int}} \qquad (10\text{-}5)$$

N_t being the number of neutrons per cm^3 at time t. Comparing with eq. 10-1 we have the equivalence:

$$\tau_{int} = 1/v\Sigma_{abs} \qquad (10\text{-}6)$$

The decay-time τ is almost independent of temperature. It is also referred to as the neutron "die-away time", or "life-time". With T in microseconds, and $v = 0.22$ cm/μs, eq. 10-6 becomes:

$$\tau_{int_{(\mu s)}} = \frac{4.55}{\Sigma_{abs}(cm^2/cm^3)} \qquad (10\text{-}7)$$

Σ is more conveniently scaled in units of 10^{-3} cm^{-1} [capture units (c.u.), or sigma units (s.u.)]. Equation 10-7 now becomes:

$$\tau_{int} = \frac{4550}{\Sigma_{abs}(c.u.)} \qquad (10\text{-}7a)$$

N.B. If one measures the time L necessary for the initial number of neutrons, N_0, to decay by one half, eq. 10-7 becomes: L (μs)$= 3150/\Sigma_{abs}$; and the equivalence between L and τ_{int} is: $L = 0.693 \tau_{int}$.

10.2.1.2. Theoretical capture cross-sections of materials

The total capture cross-section of a mixture of materials is:

$$\Sigma_{abs} = \sum_{i=1}^{i=n} V_i \Sigma_i \qquad (10\text{-}8)$$

where: V_i = volume percentage of constituent i (mineral, fluid); Σ_i = capture cross section of constituent i; and n = number of constituents.

The capture cross-section Σ_i for a mineral or fluid can be computed from its chemical formula:

$$\Sigma_i = \frac{602.2}{GMW}(n_a\sigma_a + n_b\sigma_b + \ldots + n_j\sigma_j +) \qquad (10\text{-}9)$$

σ_j is the microscopic or nuclear capture cross-section for element j in mineral i. Table 10-1 lists the nuclear capture cross-sections of the commoner elements. Units are barns, i.e. 10^{-24} cm^2 per nucleus.

n_j is the number of atoms type j present per molecule.

GMW is the gram molecular weight.

Σ is referred to as the *macroscopic* capture cross-section to contrast it with σ, which is nuclear or *microscopic*.

TABLE 10-1

Microscopic capture cross-sections (σ) of some common elements (from Edmundson et al., 1979)

Element	Abbreviation	Atomic number	Atomic weight	Microscopic thermal neutron capture cross-section (barns)
Hydrogen	H	1	1.00800	0.33200
Boron	B	5	10.81000	759.00000
Carbon	C	6	12.01100	0.00340
Nitrogen	N	7	14.00700	0.00000
Oxygen	O	8	16.00000	0.00027
Fluorine	F	9	19.00000	0.00980
Sodium	Na	11	22.99000	0.53000
Magnesium	Mg	12	24.30500	0.06300
Aluminum	Al	13	26.98200	0.23000
Silicon	Si	14	28.08600	0.16000
Phosphorous	P	15	30.97400	0.01900
Sulphur	S	16	32.06000	0.52000
Chlorine	Cl	17	35.45300	33.20000
Potassium	K	19	39.10000	2.10000
Calcium	Ca	20	40.08000	0.43000
Titanium	Ti	22	47.90000	6.41000
Manganese	Mn	25	54.93800	13.30000
Iron	Fe	26	55.84700	2.55000
Copper	Cu	29	63.54600	3.82000
Zinc	Zn	30	65.38000	0.50490
Strontium	Sr	38	87.62000	0.08600
Zirconium	Zr	40	91.22000	0.29000
Barium	Ba	56	137.34000	0.06500
Lead	Pb	82	207.20000	0.18800

Table 10-3 is a list of macroscopic cross-sections of common minerals.

10.2.2. Neutron diffusion

In practice, the decline in the population of thermal neutrons in the neighbourhood of the detector is dependent not only on the capture of thermal neutrons by the surroundings but also on the phenomenon of migration or neutron diffusion.

In effect the decline is not identical at each point in the medium investigated even if it is perfectly homogeneous. While in the thermal phase, neutrons tend to migrate in a more or less random manner until they are absorbed. During this period they may undergo collisions with nuclei, without interaction. There is a net tendency for the neutrons to migrate, or diffuse, away from regions of dense population towards those less dense.

Considering a small volume of the formation, we can see what this produces. A certain fraction of the thermal neutrons present, $v\Sigma_{abs}$, will be captured per second. But during this same time and under the effect of diffusion, neutrons will also drift into the region, and others will leave.

If, for example, the flux leaving is larger than that entering, the resulting "leak" will increase the rate of depletion of the neutrons. The contribution of diffusion to the population decay is given by the equation:

$$\frac{1}{\tau_D} = -D\frac{\nabla^2 N}{N} \qquad (10\text{-}10)$$

where D is the diffusion constant of the medium and $\nabla^2 N$ is the Laplacien of the neutron density.

We could measure the intrinsic decay time by counting the *total* thermal neutron population (integrating over all space the population $N(r, t)$ at distance r from the detector at time t) and thereby eliminate the diffusion term. However, a logging tool detector samples only a small volume of space and it is not feasible to monitor the entire neutron cloud. Practical decay time measurements are therefore affected by diffusion. Equation 10-10 may be positive, negative, or zero. In the zero case, measured $\tau =$ intrinsic τ. (This rarely occurs.)

10.2.3. Measuring the neutron population

This can be done in two ways: either by the counting of thermal neutrons, or the gamma rays of capture, in discrete time-windows after the neutron burst. (The number of capture gamma rays is proportional to the number of neutrons being captured, which is in turn proportional to the number of thermal neutrons remaining. The two approaches are therefore equivalent.) The second solution is generally preferred as the time needed to wait for the

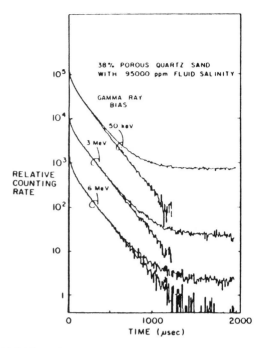

Fig. 10-3. Thermal neutron decay as measured using three different capture gamma ray energy cut-offs (from Wahl et al., 1970). The lower curve of each pair has been corrected for background.

elimination of hole effects is shorter and the depth of investigation is larger.

Dresser-Atlas count gamma rays whose energy is above 2.2 MeV in order to eliminate the influence of natural radioactivity. Schlumberger records gamma rays down to about 50 KeV energy and make a background correction. Statistical variations are smaller when a low energy threshold is used (Fig. 10-3).

10.2.4. Measurement of capture cross-section

We saw previously that the capture cross-section can be deduced from the measurement of the neutron

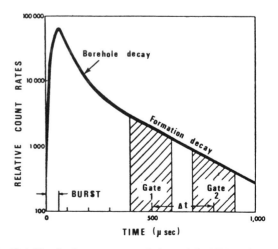

Fig. 10-4. The fixed measurement windows of the NLL tool (from Dresser Atlas).

population at two given times—applying eq. 10-4—or by determining the intrinsic decay time, τ_{int}, from a measurement of the decay rate—using eq. 10-7a for example.

10.2.4.1. *Dresser-Atlas tool (Neutron Lifetime Log)*

Dresser-Atlas uses the first method. Capture gamma rays are counted during fixed time windows at two pre-selected times after the emission of neutrons, (Fig. 10-4). We then have from eq. 10-4b:

$$\Sigma = \frac{10\,500}{\Delta t} \log_{10} \frac{N_1}{N_2} \qquad (10\text{-}11)$$

where: Σ is in 10^{-3} cm^2/cm^3 or 10^{-3} cm^{-1} or (c.u.) and Δt is equal to $(t_2 - t_1)$ in microseconds. N_1 is the number of gamma rays detected during gate 1 centred on time t_1. N_2 is the number of gamma rays detected during gate 2 centred on time t_2.

In the Dresser-Atlas N.L.L. (Neutron Lifetime Log), the interval Δt is fixed at 300 μs, and eq. 10-4 becomes:

$$\Sigma = 35 \log_{10}(N_1/N_2) \qquad (10\text{-}12)$$

10.2.4.2. *Schlumberger tools*

Schlumberger uses the second method in its TDT® (Thermal Neutron Decay Time) tools. The detection is set in phase by use of a scale factor method, which varies the opening of the measurement gates, and the gate-widths, according to the magnitude of the τ being measured.

If the decay time is short, a fixed-delay gating system may result in taking measurements to near the tail-end of the decay curve, risking both an increase in statistical variation and the influence of background radiation. On the other hand, if the decay time is long, gates opening too early can give measurements influenced by early hole effects and inelastic collisions. The operating range of a fixed gating system is therefore limited.

The TDT-K uses a self-adjusting sliding-gate technique called the "Complete Scale factor"®, based on two main detection windows with a third window for background correction (Fig. 10-5). The timing is continuously varied according to the magnitude of τ, so that the main windows are always set across the most useful part of the decay curve (Fig. 10-6).

The TDT-M samples the decay slope with 16 windows and computes τ from appropriate algorithms. The gating regime is not varied continuously as with the TDT-K, but is automatically selected from among four fixed time scalings, according to the value of τ (Fig. 10-7).

An earlier version of the TDT, the TDT-G, is now obsolete. It employed a simplified version of the

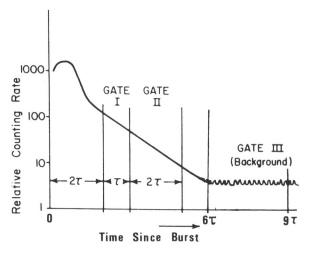

Fig. 10-5. The sliding measurement windows of the TDT-K (courtesy of Schlumberger).

complete scale factor system, and is described in the references.

10.2.4.2.1. *TDT-K*

With the complete scale factor method, the gate lengths, burst width, and delays are all functions of τ. The third gate is placed sufficiently late in the cycle to measure the background radiation (natural and activated). During each cycle the background radiation level is automatically subtracted from the counts in Gates I and II (Fig. 10-6). The initial 2τ delay after the burst allows much of the borehole signal to dissipate.

To determine τ, a negative feedback system (the tau-loop) is used. It establishes the ratio N_1/N_2 of the counts detected in two successive windows (gates) lasting times T and 2T respectively. This ratio is compared to 2 which is theoretically the value of N_1/N_2 when T is equal to τ (this can be quite simply verified mathematically).

A comparator produces an error signal indicating by how much the ratio N_1/N_2 differs from 2.

$$\frac{N_1}{N_2} = \frac{1 - e^{-T/\tau}}{e^{-T/\tau} - e^{-3T/\tau}} \qquad (10\text{-}13)$$

This signal varies the frequency of an oscillator which controls the opening and closing of the gates. Whenever the error signal appears, the frequency changes and modifies T and hence the position and duration of the gates, until they are in the correct position (i.e. the ratio N_1/N_2 is equal to 2, the error signal is nil). The oscillator stays at this frequency $f = 1/T = 1/\tau$, until τ changes.

This frequency is transmitted to the surface. It is proportional to Σ; τ is computed as $4550/\Sigma$ by a function former.

The TDT-K is a two-detector tool (near (N) and

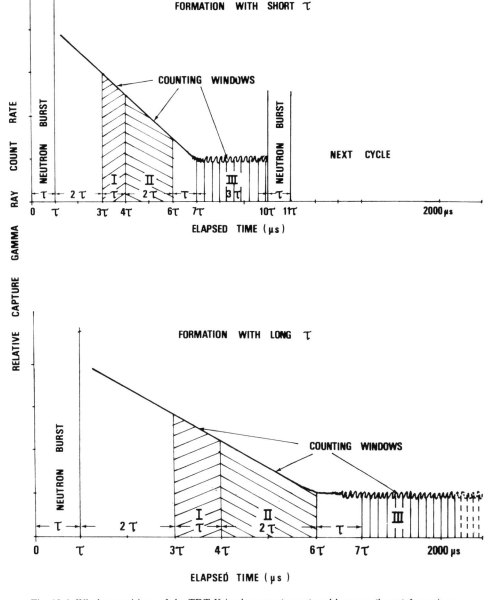

Fig. 10-6. Window positions of the TDT-K in short tau (upper) and long tau (lower) formations.

far (F) detectors). The tau-loop is operated from the near-detector counts. The resulting $T = \tau$ scale factor is imposed on both detector gating systems. Gate 1, 2 and 3 counts for both detectors (N_1, N_2, N_3 and F_1, F_2, F_3) are recorded on tape in addition to Σ and τ. The log example of Fig. 10-8 shows the conventional display of Σ, N_1, F_1, F_3. A natural gamma ray curve and CCL are also recorded.

The *Ratio*, a function of detector count-rates, is computed:

$$Ratio = \frac{f(N_1 \text{ (net)})}{f(F_1 \text{ (net)})}$$

This ratio is recorded in Track 2 of Fig. 10-8. It is similar to the CNL count-rate ratio, and behaves like

a conventional neutron log: deflections to the right indicating either a gas effect or a drop in porosity, and deflections to the left, shales or porous reservoirs.

10.2.4.2.2. TDT-M

The TDT-M is a two-detector system, identical in configuration to the TDT-K.

A more powerful neutron generator produces about four times as many neutrons, providing an improvement in statistics.

The gate timings are shown in Fig. 10-7. The basic timing unit, T, can have one of four fixed values. This allows, therefore, four gate programs corresponding to each of four scale factors F, related to T. The appropriate T is automatically selected according to the value of τ measured; each T is appropriate for a

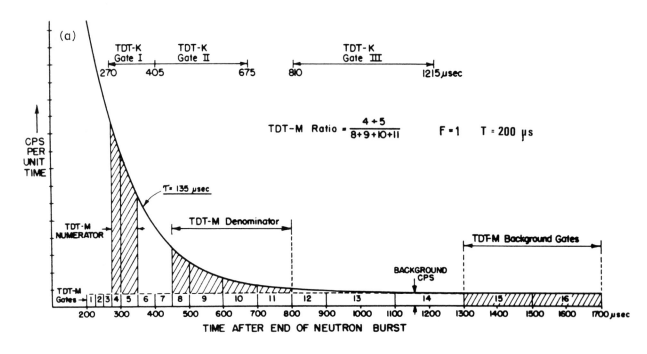

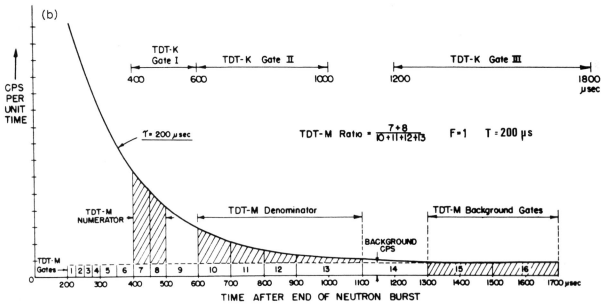

Fig. 10-7. The TDT-M gating system for two different decay times (TDT-K window positions are shown for comparison) a. $\tau = 135$ μs. b. $\tau = 200$ μs. (after Hall et al., 1980) (SPE 9462).

limited range of τ values only (Table 10-2).

τ is computed from an algorithm based on certain of the gate counts, (usually six or seven of them, never all sixteen). The gates actually considered, and the algorithm used, also vary according to the magnitude of τ, even though T might remain fixed over the range of τ involved. Thus for a τ which varies between 106.3 μs and 187.5 μs, a $T = 200$ μs would be selected by the circuitry ($F = 1$). As τ increases within this range, the gate-selections and corresponding algorithms vary as shown in Table 10-2. Note that the measuring gates are effectively being selected later as

τ increases, as for the TDT-K. Should τ exceed 187.5 μs, a new $T = 346.4$ μs ($F = \sqrt{3}$) will automatically be selected, and appropriate gates and equations used.

The same method is applied to both detectors, providing a τ_N and τ_F measurement. The scale factor and T are set using the near-detector count-rates, and imposed on the far detector system for the τ_F computation.

τ_N, τ_F data are transmitted uphole along with $N_1 \ldots N_{16}$ and $F_1 \ldots F_{16}$ count-rates. Σ_N, Σ_F and *Ratio* are computed on the surface.

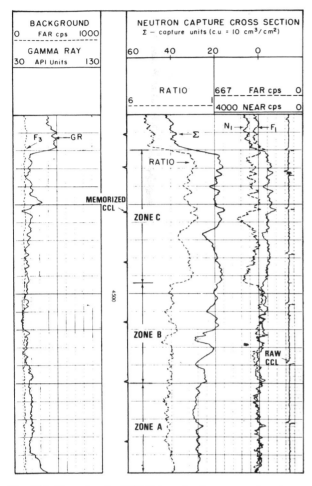

Fig. 10-8. Example of a TDT-K log, showing water, oil and gas zones (A, B and C respectively) in a shaly sand (courtesy of Schlumberger).

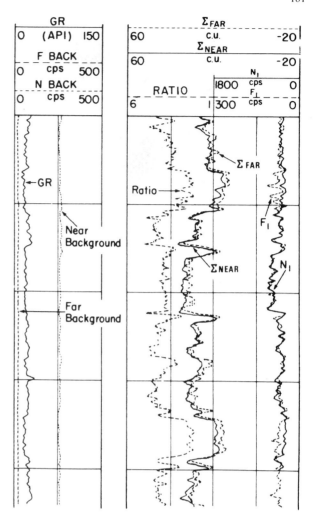

Fig. 10-9. Example of a TDT-M log, showing near and far detector curves (from Hall et al., 1980) (SPE 9462).

Figure 10-9 is a section of TDT-M log. The separation between the Σ_F and Σ_N curves is due to the reduced borehole and diffusion effects on the far-detector measurement.

10.3. NEUTRON SOURCE

The source is a neutron generator or "mini-tron"—an accelerator tube containing ions of deuterium and tritium (Fig. 10-10). A high-voltage ladder (Fig. 10-11) provides the voltage (about 100 kV) necessary for ion acceleration. Deuterium ions in

TABLE 10-2

Example of TDT-M count window selection for computation of tau for scale factor $F = 1$ ($T = 200$ μs) (from Hall et al., 1980).

τ_{NEAR} range	F	Ratio (integers are gate numbers)	Equation for τ
< 106.3 μs	1	$\dfrac{1+2}{5+6+7+8+9}$	$\tau = 38.0 + 69.9\,R^{-1}$
$106.3 - 118.8$	1	$\dfrac{2+3}{6+7+8+9+10}$	$\tau = 48.4 + 73.1\,R^{-1}$
$118.8 - 131.3$	1	$\dfrac{3+4}{7+8+9+10+11}$	$\tau = 58.2 + 76.1\,R^{-1}$
$131.3 - 143.8$	1	$\dfrac{4+5}{8+9+10+11}$	$\tau = 61.3 + 124.1\,R^{-1}$
$143.8 - 162.5$	1	$\dfrac{5+6}{9+10+11+12}$	$\tau = 67.6 + 164.2\,R^{-1}$
$162.5 - 187.5$	1	$\dfrac{6+7}{9+10+11+12+13}$	$\tau = 53.6 + 130.3\,R^{-1}$
> 187.5	1	$\dfrac{7+8}{10+11+12+13}$	$\tau = 63.2 + 164.6\,R^{-1}$

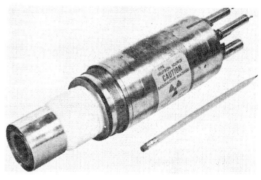

Fig. 10-10. Ion accelerator neutron generator.

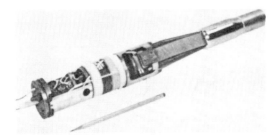

Fig. 10-11. High-voltage stage providing the accelerating voltage for the neutron generator.

collision with tritium absorbed in a carbon target produce a neutron flux with energy equal to 14 MeV following the reaction:

$$\,_1^2H + \,_1^3H \rightarrow \,_2^4He + \,_0^1n \; (14 \; MeV) \qquad (10\text{-}14)$$

Neutron emission can be shut-off from surface.

In the Dresser-Atlas tool neutron emission lasts for 30 microseconds and is repeated 1000 times per second. In the Schlumberger TDT-K tool, once the system is synchronised, the emission time is equal to τ and the time between two emissions is 10τ. This allows a gain in time and a reduction in statistics. For the TDT-M, emission time is equal to one time unit T, and repetition time is $17\ T$.

10.4. DETECTORS

These are scintillation detectors (NaI crystals) with circuitry sensitive to capture gamma rays of above a certain threshold energy. Detection circuitry is regulated to be synchronised with the emission time (Dresser-Atlas), or following the Complete Scale Factor (Schlumberger, TDT-K) or multi-gate method (TDT-M).

10.5. SPACING

In the Schlumberger TDT the source to near-detector spacing is approximately 13 inches. The Dresser-Atlas tool has about the same value. The far detectors are 12 inches further out.

10.6. UNITS

The units for the time τ are microseconds (μs). The macroscopic capture cross-section is theoretically in cm^2/cm^3 or cm^{-1}, but is most often shown in sigma units (s.u.) or capture units (c.u.), corresponding to $10^{-3}\ cm^{-1}$, for convenience.

10.7. CALIBRATION (see Appendix 5)

This is obtained by use of a crystal oscillator which generates calibration levels for $\tau\Sigma$ and the count-rate channels. A special jig containing a radium source is used to calibrate the two detector sensitivities for the *Ratio* curve.

10.8. MEASURE POINTS

The measure point is in the middle of the detector-source spacing for single detector measurements (τ_N, τ_F, Σ_N, Σ_F, count-rates) and the middle of the detector spacing for two-detector measurements (*Ratio*).

10.9. VERTICAL RESOLUTION

This is a function of the distance between source and detectors and the detectors themselves, and also to some extent on the time constant used and the logging speed. For a time constant of 3 or 4 seconds and a logging speed of 20 ft per minute it is of the order of 3 ft.

10.10. DEPTH OF INVESTIGATION

This depends on the capture cross-section of the formation surrounding the tool. Owing to the relative high energy of the neutrons the depth of investigation is slightly higher than in other nuclear tools.

Antkiw (1976) has determined the two geometrical factors J_τ and J_Σ for the Schlumberger two-detector tool. The depth of investigation is of the order of 12 to 14 inches (Figs. 10-12 and 10-13), and the near detector tends, not surprisingly, to respond more to the near-hole formation.

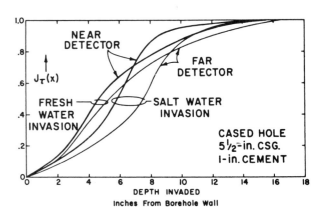

Fig. 10-12. Geometrical factor vs depth of invasion for a cased hole (from Antkiw, 1976).

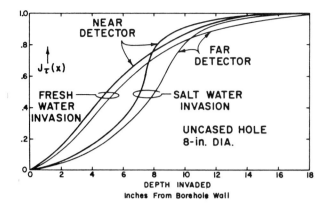

Fig. 10-13. Geometrical factor vs depth of invasion for an open hole (from Antkiw, 1976).

10.11. FACTORS INFLUENCING THE Σ MEASUREMENT

The most important factor in the capture of thermal neutrons is the capture cross section of the materials surrounding the tool, particularly the formation.

The bulk capture cross-section of the formation is equal to the sum of the capture cross-sections of the individual constituents, multiplied by their volumetric percentages, as we have seen in Section 10.2.1.2.

Table 10-3 gives the macroscopic capture cross-sections of the principal sedimentary rocks, as calculated in c.u. from eq. 10-9.

We can see that, in the absence of boron and lithium, it is chlorine that plays the major role in neutron absorption. The TDT log determines essentially the amount of chlorine and, from that, the saturation S_w.

The different materials making up the volume investigated by the tool will influence the measurement as a function of their own characteristics.

10.11.1. The matrix (Σ_{ma})

Σ is sensitive to lithology (Table 10-3). The measured capture cross-section of rocks differs considerably from that of pure minerals by virtue of impurities. Discrepancies also occur if log data are not corrected for diffusion and borehole effects.

Apparent Σ_{ma} can be determined in water-bearing porous formations by considering Σ_{log} as a function of the effective porosity (Fig. 10-14a). 100 percent wet points will lie on the "water line" which will pass through Σ_{ma} at $\phi_e = 0$, and Σ_w at $\phi_e = 100\%$. Both Σ and ϕ must be corrected for shaliness if necessary (Fig. 10-14b) (see 10.11.3).

If a sufficient range of porosity is not available, an alternative method is to crossplot Σ with $1/\sqrt{R_t}$

TABLE 10-3

Theoretical macroscopic capture cross-sections (Σ) of selected minerals (adapted from Schlumberger document, 1976)

Mineral		Σ_{ma}(c.u.)
Basic minerals		
Quartz	SiO_2	4.3
Calcite	$CaCO_3$	7.3
Dolomite	$CaCO_3 - MgCO_3$	4.8
Feldspars		
Albite	$NaAlSi_3O_8$	7.6
Anorthite	$CaAl_2Si_2O_8$	7.4
Orthoclase	$KAlSi_3O_8$	15
Evaporites		
Anhydrite	$CaSO_4$	13
Gypsum	$CaSO_4 - 2 H_2O$	19
Halite	$NaCl$	770
Sylvite	KCl	580
Carnalite	$KCl - MgCl_2 - 6 H_2O$	370
Borax	$Na_2B_4O_7 - 10 H_2O$	9000
Kermite	$Na_2B_4O_7 - 4 H_2O$	10500
Coal		
Lignite		30
Bituminous coal		35
Anthracite		22
Iron-bearing minerals		
Iron	Fe	220
Goethite	$FeO(OH)$	89
Hematite	Fe_2O_3	104
Magnetite	Fe_3O_4	107
Limonite	$FeO(OH) - 3 H_2O$	80
Pyrite	FeS_2	90
Siderite	$FeCO_3$	52
Iron-potassium-bearing minerals		
Glauconite	(green sands)	25 ± 5
Chlorite		25 ± 15
Mica (biotite)		35 ± 10
Others		
Pyrolusite	MnO_2	440
Manganite	$MnO(OH)$	400
Cinnabar	HgS	7800

Typical values of capture cross-sections derived from log data are:

35– 55	c.u.	Shale
5– 12	c.u.	Matrix
22	c.u.	Fresh water
22–120	c.u.	Formation water
0– 12	c.u.	Gas
18– 22	c.u.	Oil

from the open-hole resistivity log. This plot requires fairly constant porosity. Where reservoir conditions have not changed since open-hole logging, data points lie along a clean oil–water line as shown in Fig. 10-15. It can be shown that this line has a slope of $\sqrt{aR_w}(\Sigma_w - \Sigma_h)$ and passes through $\Sigma_{0.0}$ such that:

$$\Sigma_{ma} = \Sigma_{0.0} - \frac{\phi_{ave}}{1 - \phi_{ave}}(\Sigma_h - \Sigma_{0.0})$$

Hence Σ_{ma} and Σ_w can be derived (the latter if R_w is

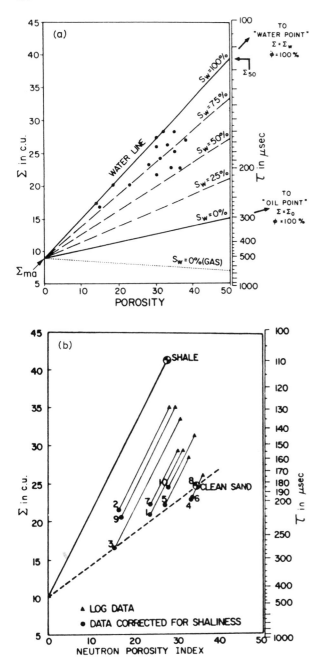

Fig. 10-14 a. Determination of Σ_{ma}, Σ_w and S_w from the Σ_{log} vs ϕ cross-plot (courtesy of Schlumberger). b. Shale corrections applied to a Σ_{log} vs ϕ_N cross-plot (triangles are raw data) (courtesy of Schlumberger).

known. If salinity is not known, it can be calculated from the slope, since Σ_w and R_w are related. This is most easily done using Fig. 10-16, where $\Sigma_{4.0}$ corresponds to the Σ value at $C_t = 4$ mho/m ($R_t = 0.25$ ohm-m)).

Figure 10-15 also shows the disposition of points corresponding to a depleted zone (Σ has increased because of water influx), and fresh filtrate that has not dissipated (Σ_{mf} less than Σ_w).

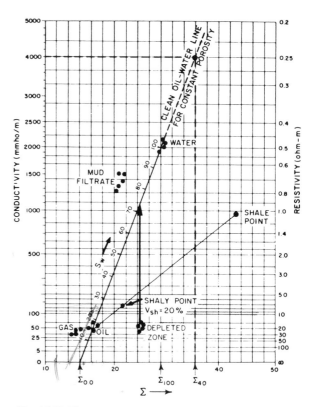

Fig. 10-15. The R_t vs Σ_{log} crossplot (courtesy of Schlumberger).

10.11.2. Porosity, fluids

In porous rocks the porosity, fluid types and saturations will affect the Σ measurement.

10.11.2.1. Formation water (Σ_w)

Pure water has a capture cross-section of 22.2 c.u. at 25°C, but formation waters contain dissolved salts. The charts shown in Fig. 10-17 give the value of Σ_w

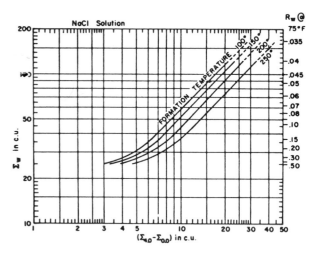

Fig. 10-16. Determination of R_w and Σ_w from the Σ vs R_t crossplot (courtesy of Schlumberger).

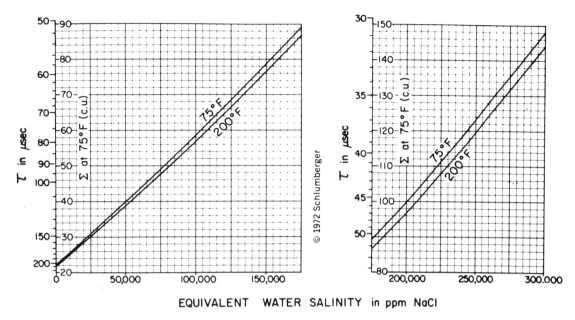

Fig. 10-17. Capture cross-section of water as a function of NaCl salinity (courtesy of Schlumberger).

and τ_w as a function of the temperature and the salinity (NaCl). The strong effect of the presence of chlorine is clearly seen. These charts are only valid for NaCl waters.

If the salinity is not known, the graphical method shown in Fig. 10-14 can be used to find Σ_w. The 100% water line is simply extrapolated to $\phi = 100\%$. This has the advantage that non-NaCl salts are automatically taken into account. (On the crossplot of Fig. 10-14, the extrapolated Σ_{50} at $\phi = 50\%$ can be used to compute Σ_w from $\Sigma_w = (2\ \Sigma_{50} - \Sigma_{ma}) = 69$ c.u.) Alternatively, the approach of Fig. 10-15 may be used. Σ_w can also be computed from a chemical analysis using eqs. 10-8 and 10-9. The most effective approach, however, is to obtain a representative sample of the fluid (DST, RFT) and measure Σ_w directly, using for instance the portable measuring cell known as the SFT-156 (Fig. 10-18).

N.B. Since the TDT is a shallow investigating measurement, it must be born in mind that the near-hole pore-fluid at the time of logging may contain connate water and/or mud-filtrate. It can often be safely assumed that filtrate will have dissipated by the time the first TDT is run and that near-hole conditions represent the reservoir at large. (It can be appreciated from Fig. 10-14 that residual fresh filtrate (Σ_{mf} less than Σ_w) will appear as oil to the TDT.)

10.11.2.2. Hydrocarbon (Σ_h)

The capture cross-section of hydrocarbon is attributed mainly to the hydrogen present. For gases, it is

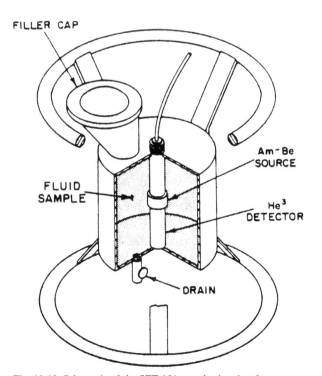

Fig. 10-18. Schematic of the SFT-156 sample chamber for measuring the Σ of water samples (courtesy of Schlumberger).

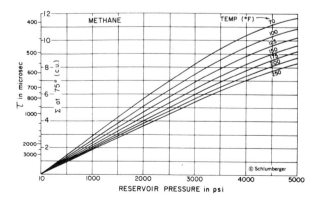

Fig. 10-19. Capture cross-section of methane gas as a function of temperature and pressure (from Clavier et al., 1971).

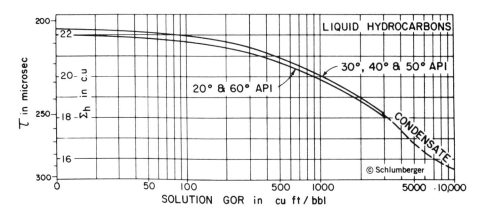

Fig. 10-20. Capture cross-section of a typical oil as a function of gravity and GOR (from Clavier et al., 1971).

sensitive to temperature and pressure. The charts of Figs. 10-19 and 10-20 are for the determination of Σ_h, knowing the composition, API, temperature and the pressure. For gases other than methane the capture cross section can be estimated by the following:

$$\Sigma_g = \Sigma_{methane} \times (0.23 + 1.4 \, \gamma_g) \qquad (10\text{-}17)$$

where γ_g = specific gravity of the gas relative to air. Note that oils with a GOR less than about 1000 have a Σ_h of 20–22 c.u., and it is usually safe to assume a value of 21 c.u. for oil.

10.11.3. Shales

From chemical analysis it appears that clays frequently contain significant traces of boron. Besides this they often contain iron and a lot of water. All this implies a high capture cross-section—typically 20–50 c.u. If, then, we make a TDT interpretation to determine water saturation in shaly formations we need to correct the reading for the effect of shale using the following relationship:

$$\Sigma_{cor} = \Sigma_{log} - V_{sh}(\Sigma_{sh} - \Sigma_{ma}) \qquad (10\text{-}18)$$

where Σ_{sh} is taken from the reading in adjacent shales and is assumed to represent the shale in the porous zone.

Σ is an excellent clay indicator and can be used as such in open-hole log interpretation, particularly where natural gamma radioactivity cannot be correlated to clay content.

10.11.4. Acidization

Since the TDT is most commonly run in producing wells, any well stimulation such as fracturing or acidization will affect the log readings. Acidization in carbonate reservoirs increases the value of Σ because of increased porosity and the presence of soluble $CaCl_2$, (a by-product of the reaction between hydrochloric acid and $CaCO_3$) which probably remains in

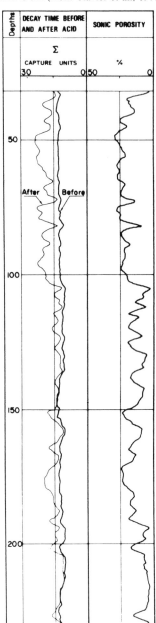

Fig. 10-21. A comparison of thermal neutron capture cross-section (Σ) logged before and after acidization of a carbonate reservoir (from Saif et al., 1975). Note how the "acid effect" is strongest where the sonic log (pre-acid) shows good porosity.

the irreducible formation water fraction. A similar increase occurs after treating sandstone reservoirs with "mud-acid" (containing hydrofluoric acid).

Evidence suggests that this "acid effect" is permanent and unchanging, at least while only oil is being produced. The magnitude of the acid shift on Σ can be estimated by comparing a post-acid TDT with a pre-acid survey, or with a "synthetic sigma" recomputed from the open-hole data. This shift must be allowed for in subsequent time-lapse monitoring. An example is shown in Fig. 10-21.

10.12. ENVIRONMENTAL EFFECTS

10.12.1. Borehole signal and diffusion

10.12.1.1. *Borehole signal*

The borehole affects the Σ-measurement in two ways. Firstly, it has its own capture cross-section (made up of contributions from the mud, casing, cement sheath, even the tool housing). The measured decay slope is in fact the sum of the formation decay and the borehole decay. In Fig. 10-1 the borehole capture events predominate just after the burst, but die away rapidly until it is essentially only the formation capture events that are being detected. The 2τ gating delay preceding Gate I (Fig. 10-5) is intended to skip most of the unwanted borehole signal. This is indeed the case for a highly captive borehole; for instance, saline well-bore fluid, casing, perhaps tubing—all having large Σ's. Residual borehole signal during the detector gates is negligible.

Were the casing fluid fresh water or oil, however, the early borehole decay slope of Fig. 10-1 would be flatter, and significant counts would still be picked up in the measuring gates despite the 2τ delay. The apparent Σ computed from this composite decay slope will differ from the formation Σ, according to the relative captive properties of the well-bore and formation. Σ_{BH} may still be larger than $\Sigma_{formation}$, in which case the decay slope is steepened slightly and Σ_{meas} is too large (τ too short). Should Σ_{BH} be less than $\Sigma_{formation}$ (shales, saline porous zones) the decay slope is rendered too shallow and Σ_{meas} is too small.

Although gas has a very low Σ, it also has a low HI, and most neutrons pass straight across a gas-filled well-bore into the formation. There are therefore almost no borehole capture events to be counted.

10.12.1.2. *Diffusion*

Because the borehole and formation have different Σ's, the concentrations of uncaptured thermal neutrons will differ at any time in these two regions. Usually there is a lower neutron density in the well-

bore, except when $\Sigma_{formation}$ is very large (section 10.2.2). The neutron cloud diffuses towards the low density regions—usually outwards into the reservoir, and inwards towards the borehole. This appears to the TDT as a more rapid decay of thermal neutrons in the formation, and the measured Σ is too large in this case.

The combined effects of borehole signal and diffusion usually result in Σ_{meas} being too large by some 20–30% for Σ_{int} less than about 30 c.u.

Combined borehole and diffusion correction charts are available to correct the measured Σ to Σ_{int} for a variety of hole configurations. Figure 10-22 is an example of the near-spaced TDT detector. Since the corrections are very nearly linear, at least for Σ less than about 30 c.u., a quite reasonable interpretation of Σ can be made using uncorrected data and apparent values of Σ_{ma}, Σ_w, etc. derived from crossplots. Environmental corrections can be applied in the computing centre prior to interpretation. Figure 10-25 shows both raw and corrected Σ.

10.12.1.3. *Borehole configuration*

10.12.1.3.1. *Fluids*

Fluids were discussed in 10.12.1.1. To summarize, high salinity or gas results in negligible borehole signal in the decay slope. Freshwater or oil have a measureable effect. However, diffusion effects are present in all cases. Charts like Fig. 10-22 correct for borehole and diffusion.

10.12.1.3.2. *Casing*

Steel has a large Σ, and casing has a reducing effect on the borehole signal that appears in the measuring gates. In addition, it displaces well-bore fluid. The cement sheath behind casing also has a large Σ. The TDT is able to read through several concentric casings and tubing.

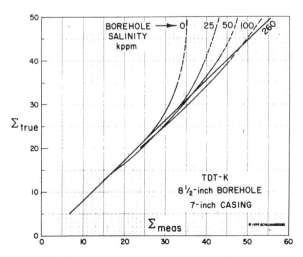

Fig. 10-22. Example of a correction chart for TDT near detector borehole and diffusion effects (courtesy of Schlumberger).

EFFECT OF CENTERING
on 1 11/16" SONDES

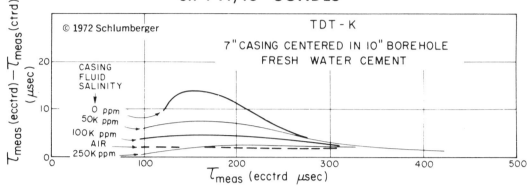

Fig. 10-23. Effect of eccentering of the TDT-K sonde in casing (courtesy of Schlumberger).

10.12.1.3.3. Hole-size

Increased hole size compounds the effects discussed in 10.12.1.1. and 10.12.1.2.

10.12.2. **Tool eccentralization**

The effect of tool eccentering are shown in Fig. 10-23. It introduces a small shift in measured τ. Eccentralization is usually constant because the tool follows the low side of the casing. It is otherwise a factor difficult to quantify.

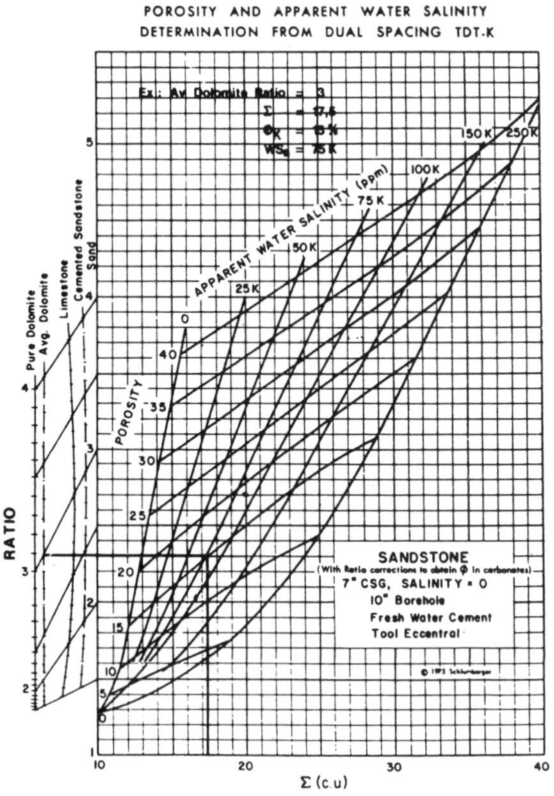

10.12.3. **Invasion**

The TDT can be run in open or cased hole.

In open hole, the depth of investigation is such that most of the measurement is made in the flushed zone. Most of the formation water and some hydrocarbon are replaced by mud filtrate, and the TDT measures S_{xo}. The appropriate value of Σ_{mf} must be used instead of Σ_w.

In cased hole, the mud filtrate may dissipate after a few days or weeks. If the log is run while the well is producing, the original drilling mud invasion will have dispersed in the perforated zones, and probably elsewhere provided sufficient time has elapsed. Σ_w should therefore be used in interpretation, but keeping in mind the possibility of residual filtrate.

If the TDT is run with the well shut-in, the porous, permeable zones can be deeply invaded by the standing fluid column. In this case, the log interpretation is no longer meaningful, at least for saturation values in the reservoir at large.

10.12.4. **Time constant, logging speed, bed thickness and vertical resolution**

It is estimated that bed thickness should be three to four feet for measurements to be correct. In any case, by adjusting the time constant (or depth smoothing) and logging speed the vertical resolution can be improved. The minimum thickness is given by:

$$h = 2 + \frac{(\text{speed} \times \text{TC})}{40}$$

where TC = time constant (typically 4 sec)
h = thickness in feet

Fig. 10-24. One of the charts used to compute a neutron porosity ϕ_k from the TDT Σ and *Ratio*. An apparent salinity (WS_a) can also be computed (courtesy of Schlumberger).

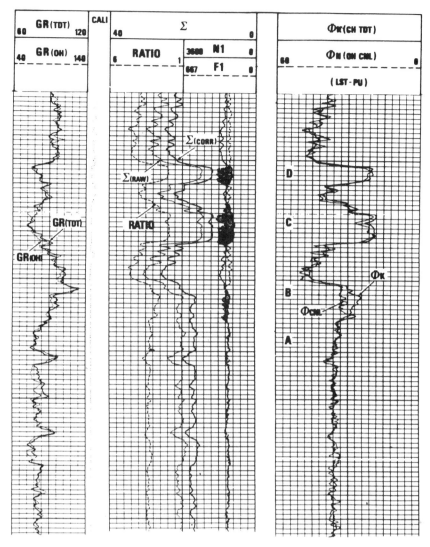

Fig. 10-25. Gas indication on the TDT porosity ϕ_k.

speed = logging speed in feet/minute (typically 20 fpm)

10.13. GEOLOGICAL FACTORS AFFECTING THE Σ MEASUREMENT

From the points already discussed it seems that these are essentially:

10.13.1. Composition of the rock

(a) The minerals making up the matrix contribute according to the capture cross-sections of the constituent elements and the volume percentages of the minerals present. Σ is particularly sensitive to the presence of clay minerals and halite.

(b) The fluids contained in the rock affect the log as a function of their own capture cross-sections, and of their relative volumes in the rock (porosity) and in the percentages (saturation) in the volume investigated by the tool.

10.13.2. Rock texture

Rock texture has a direct effect as it affects the porosity, the permeability and the invasion. Except that it has generally no direct influence as the formation appears homogeneous to the tool as long as any heterogeneities are smaller than about one inch.

10.13.3. Temperature

Its effect is mainly seen in gas (Fig. 10-19). Σ_w has a very slight temperature dependence (Fig. 10-17). TDT operating limit is 325°F.

10.13.4. Pressure

Σ_g is sensitive to pressure (Fig. 10-19).
The diameter of invasion is a function of the

difference between the formation pressure and the mud pressure. Hence an effect will be seen where logs are run in open hole or where a static column of production fluid fills the casing.

10.14. POROSITY AND GAS INDICATION

10.14.1. Porosity

A neutron porosity (ϕ_k) is derived from Σ and *Ratio* using charts such as the one in Fig. 10-24. Charts exist for different hole and casing configurations, and casing fluids. ϕ_k is valid for liquid-filled porosity, but reads low in the presence of gas because both Σ and *Ratio* are decreased. In Fig. 10-25, a cased-hole TDT ϕ_k is compared with an open-hole CNL porosity.

ϕ_k, suitably corrected for matrix effects (see the left-hand edge of Fig. 10-24) can be used as a neutron porosity. Gas may be indicated when ϕ_k reads low (Fig. 10-25, zones B, C and D).

An apparent water salinity, WS_a, is also obtained from the chart. WS_a is equal to $P \times S_w$ where P is the true water salinity, or the value of WS_a in a wet zone), hence a quick-look estimation of S_w in liquid-filled porosity can be obtained from:

$$S_w = \frac{WS_a \, (\text{oil zone})}{WS_a \, (\text{wet zone})}$$

In gas zones, very approximately:

$$S_w = \frac{\phi_k \, (\text{gas zone})}{\phi_k \, (\text{wet zone of same true porosity})}$$

10.14.2. Gas indication from the count-rates

If the N_1 and F_1 count-rates are scaled so as to overlay in a wet zone, (zone A in Fig. 10-8), they provide a distinctive separation in gas-bearing sections with F_1 moving strongly to the left, and N_1 usually decreasing slightly to the right. In Fig. 10-8, zone C is a gas-sand. Note that in shale, the reverse separation occurs. Shaliness therefore tends to reduce the gas separation, but this quick-look aspect of the TDT is usually effective in moderately shaly formations. The Σ and *Ratio* curves are also strongly affected by the gas. Zone B is oil-bearing. There is a slight separation on N_1, F_1, and a reduction in Σ relative to zone A.

10.15. APPLICATIONS

The most widely exploited application of the measurement of thermal neutron decay time is the de-

termination of water saturation in hydrocarbon reservoirs. Best results are obtained when Σ_w is at least 10–20 c.u. greater than Σ_h. This usually implies a salinity of at least 30 kppm (NaCl equiv.). This is usually a cased-hole application. Initial hydrocarbons having been evaluated with open-hole logs, the TDT is used to monitor the depletion profile behind casing at any time during the producing life of the well. This information gives:

(a) remaining (possibly residual) hydrocarbon reserves;

(b) location of oil–water, gas–water, gas–oil contacts;

(c) warning of impending gas or water breakthrough at producing zones.

10.15.1. Response equations

10.15.1.1. General

In the case of a shaly porous reservoir the generalized response equation can be written as follows:

$$\Sigma_{\log} = \phi_e S_w \Sigma_w + \phi_e (1 - S_w) \Sigma_h$$
$$+ V_{sh} \Sigma_{sh} + (1 - \phi_e - V_{sh}) \Sigma_{ma} \qquad (10\text{-}19)$$

and so, knowing ϕ_e, (effective porosity), V_{sh}, Σ_w, Σ_h, Σ_{sh}, Σ_{ma} we can obtain S_w from:

$$S_w = \frac{(\Sigma_{\log} - \Sigma_{ma}) - \phi_e(\Sigma_h - \Sigma_{ma}) - V_{sh}(\Sigma_{sh} - \Sigma_{ma})}{\phi_e(\Sigma_w - \Sigma_h)}$$

$$(10\text{-}20)$$

10.15.1.2. Dual-water model

Alternatively, using the "dual-water" model for shaly formations, we can write:

$$\Sigma_{\log} = \phi_t (S_{wt} - S_{wb}) \Sigma_{wf} + \phi_t S_{wb} \Sigma_{wb}$$
$$+ \phi_t (1 - S_{wt}) \Sigma_h + (1 - \phi_t) \Sigma_{ma} \qquad (10\text{-}21)$$

where:

ϕ_t = total porosity (including shale bound-water);

S_{wt} = total water saturation (including shale bound-water);

S_{wb} = shale bound-water fraction; which can usually be taken as equal to V_{sh};

Σ_{wf} = "free" water capture cross-section equivalent to Σ_w in eq. 10-19;

Σ_{wb} = shale bound-water capture cross-section;

Σ_{ma} = capture cross-section of all the dry solids (matrix, silt, dry clay colloids).

$$S_{wt} = \frac{(\Sigma_{\log} - \Sigma_{ma}) - \phi_t(\Sigma_h - \Sigma_{ma}) - \phi_t S_{wb}(\Sigma_{wb} - \Sigma_{wf})}{\phi_t(\Sigma_{wf} - \Sigma_h)}$$

$$(10\text{-}22)$$

This is easily converted to the conventional S_w (free water saturation) since:

$$S_w = \frac{S_{wt} - S_{wb}}{1 - S_{wb}}$$

Note also that $\phi_e = \phi_t(1 - S_{wb})$

This is discussed in detail in the references.

10.15.1.3. *Time lapse equation*

This technique eliminates the need for accurate evaluation of shale (or bound-water) fractions, Σ_{ma} and Σ_{sh} (or Σ_{wb}), these being the greatest sources of possible error in the previous two approaches.

The TDT is used to follow changes in S_w during the life of the well. It is reasonable to assume that any change in Σ between successive TDT surveys (say six months apart) is caused by a change in hydrocarbon saturation (provided formation water salinity and hydrocarbon nature do not change, and no well-stimulation such as acidizing has occurred between surveys, of course). A *Base-log* is run shortly after well-completion, when it is assumed that mud-filtrate has dissipated, and no depletion has yet occurred. This base Σ therefore reflects the initial reservoir conditions.

If we compare any subsequent TDT survey, Σ_n, with the base log Σ_1, changes (usually increases) in Σ are caused by changes in S_w.

Writing Eq. 10-19 or 10-21 for the two logs, in terms of S_{w1} and S_{wn}, and subtracting, we obtain:

$$\Sigma_n - \Sigma_1 = (S_{wn} - S_{w1})\phi_e(\Sigma_w - \Sigma_h)$$

i.e. $\Delta S_w = \Delta\Sigma/\phi_e(\Sigma_w - \Sigma_h)$ (10-23)

We can obtain the actual saturation S_{wn} simply by adding ΔS_w to S_{w1}. Ideally, a good estimate of S_{w1} has already been made from the *open-hole* logs, i.e. $S_{w1} = S_{wOH}$. So:

$$S_{wn} = S_{wOH} + \Delta S_w \qquad (10\text{-}24)$$

Equations 10-23 and 10-24 are the basic equations of the time-lapse technique. Note in eq. 10-23, shale and matrix parameters do not appear; the problems of formation evaluation have been taken care of in the open-hole CPI.

This approach permits greater precision in the evaluation of the depletion profile.

Qualitatively, a time-lapse comparison such as Fig. 10-26 can indicate fluid contact movement. Figure 10-27 shows a series of four TDT monitoring runs presented in full CPI format (i.e. S_w has been computed from Σ each time). Bulk volume remaining hydrocarbons at the time of each survey are shown in black, moved hydrocarbons in grey.

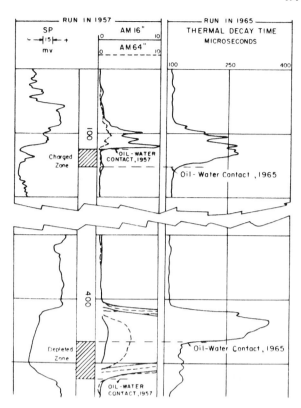

Fig. 10-26. Comparison of a cased hole TDT survey with the original open hole resistivity survey run 8 years earlier. Movements of the oil-water contact are clearly seen in this saline formation water-bearing reservoir (courtesy of Schlumberger).

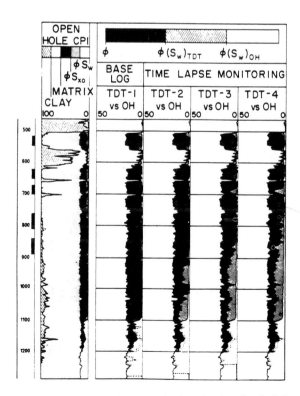

Fig. 10-27. CPI of time-lapse monitoring, showing the depletion profiles at the times of four different TDT monitoring surveys (from Westaway et al., 1979).

10.15.2. **Residual oil saturation**

Residual Oil Saturation can be estimated in cased-hole or observation wells. Various specialized techniques such as "Log-inject-log" (see References) have been developed to enhance the accuracy of the ROS evaluation.

10.15.3. **Formation fluid**

Identification of the formation fluid—gas, oil, water contacts—with the aid of count-rate data, and Σ.

10.15.4. **Old wells**

Evaluation of old wells for possible recompletion, or assessment of initial hydrocarbon reserves. The TDT log provides information on porosity, clay fraction, hydrocarbon saturation and type. In an old well where perhaps only an early electrical survey has been run, an evaluation can be made of porosity and initial hydrocarbon reserves with the aid of a TDT log. Gas zones can be identified. Where formation water salinity permits, present-day hydrocarbon saturations can also be estimated.

10.15.5. **Supplementary uses**

Where bad open-hole conditions prevent a comprehensive logging suite from being made, the TDT can be run through a drill-pipe, or after completion, to supplement the logging data (lithology, porosity, hydrocarbons).

Although rarely run in open hole, the TDT can be used alongside other tools to give a better definition of the lithology, particularly the clay fraction. An obvious application here would be in shaly carbonate reef deposits, where the clay fraction does not generally correlate with the GR (unless the NGS is run).

10.16. **REFERENCES**

Antkiw, S., 1976. Depth of investigation of the dual-spacing thermal neutron decay time logging tool. SPWLA 17th Annu. Log. Symp. Trans., Paper CC.

Boyeldieu, C. and Horvath, S., 1980. A contribution to the evaluation of residual oil from well logs for tertiary recovery. SPWLA 21st Annu. Log. Symp., Trans.

Clavier, C., Hoyle, W.R. and Meunier, D., 1969. Quantitative interpretation of TDT Logs, paper SPE 2658 presented at the SPE 44th Annu. Meet., Denver, Sept. 28–Oct. 1, 1969. J. Pet. Technol., June 1971.

Dewan, J.T., Johnstone, C.W., Jacobson, L.A., Wall, W.B. and Alger, R.P., 1973. Thermal neutron decay time logging using dual detection. Log Analyst 14(5): 13.

Edmundson, H. and Dadrian, C., 1973. Thermal neutron decay time logging applications in the Eastern Hemisphere. 2nd Annu. Symp. SAID, Paris, Oct.

Edmundson, H. and Raymer, L., 1979. Radioactive logging parameters for common minerals. SPWLA 20th Annu. Log. Symp. Trans.

Fertl, W.H., 1972. Occurrence of the neutron-absorbing tracing element boron. Part II-boron in oil-field water. The Log Analyst, Vol. XIII, Jan-Feb, 1972.

Golder, D., 1978. Thermal decay time logging at Ekofisk, case study of a multiple use tool. European Offshore Petroleum Conference and Exhibition, Paper No. EUR 91.

Hall, J., Tittman, J. and Edmundson, H., 1981. A system for wellsite measurement of Σ fluid. SPWLA. 21st Annu. Log. Symp. Trans., Paper W.

Hall, J., Johnstone, C., Baldwin, J. and Jacobson, L., 1980. A new thermal neutron decay logging system–TDT-M. SPE of AIME, Paper No. SPE 9462.

Jameson, J.B., McGhee, B.F., Blackburn, J.S. and Leach, B.C., 1977. Dual spacing TDT applications in marginal conditions. J. Pet. Technol., Sept.

Kidwell, C. and Guillory, A., 1980. A recipe for residual oil saturation determination J. Pet. Technol. Nov.

Locke, S., and Smith, R., 1975. Computed departure curves for the thermal neutron decay time log. SPWLA, 16th Annu. Log. Symp. Trans.

McGhee, B.F., McGuire, J.A. and Vacca, H.L., 1974. Examples of dual spacing thermal neutron decay time logs in Texas Coast oil and gas reservoirs. SPWLA 15th Annu. Log. Symposium, Trans., Houston, Paper R.

Nelligan, W.B. and Antkiw, S., 1976. Accurate thermal neutron decay time measurements with the dual-spacing TDT—a laboratory study. 51st Annu. Fall Techn. Conf. SPE/AIME, Paper SPE 5156.

Nelligan, W.B., Wahl, J.S., Frentrop, A.H., Johnstone, C.W. and Schwartz, R.J., 1970. The thermal neutron decay time log. Soc. Pet. Eng., J. Dec: 365–380.

Preeg, W.E. and Scott, H.D., 1981. Computing thermal neutron decay time (TDT) environmental effects using Monte Carlo techniques. Paper SPE 10293 presented at the SPE 56th Annual Technical Conference and Exhibition, San Antonio, Oct. 4–7, 1981.

Rinehart, C.E. and Weber, H.J., 1975. Measuring thermal neutron absorption cross sections of formation water. SPWLA 16th Annu. Log. Symp. Trans. June, 1975.

Robinson, J.D., 1974. Neutron decay time in the subsurface: theory, experiment, and an application to residual oil determination. Soc. Pet. Eng. of AIME 49th Annu. Fall Meet., Houston, October 6–9, 1974, Paper SPE 5119.

Saif, A.S., Cochrane, J.E., Edmundson, H.N. and Youngblood, W.E., 1975. Analysis of pulsed-neutron decay-time logs in acidized carbonate formations. SPE J., Dec. Paper No. SPE 5443.

Schlumberger, 1974. Log Interpretation; Vol. II-Applications.

Schlumberger, Mar., 1976. The Essentials of Thermal Decay Time Logging.

Schlumberger, Mar., 1980. Dual-Water Model Cased Reservoir Analysis.

Serpas, C.J., Wickmann, P.A., Fertl, W.H., Devries, M.R. and Randall, R.R., 1977. The dual detector neutron lifetime log—theory and practical applications. SPWLA 18th Annu. Log. Symp. Trans., Houston Paper CC.

Smith, R.L. and Patterson, W.F., 1976. Production management of reservoirs through log evaluation. SPE of AIME, Paper No. SPE 5507.

SPWLA, 1976. Pulsed Neutron Logging, Reprint Volume.

Tittman, J., 1956. Radiation Logging. Fundamentals of Logging, University of Kansas.

Tixier, M.P. and Vesperman, F.A. 1976. The porosity transform-A shale-compensated interpretation using TDT and neutron Logs, SPWLA 17th Annu. Log. Symp., Trans., Paper GG.

Westaway, P., Wittman, M. and Rochette, P., 1979. Application of Nuclear Techniques to Reservoir Monitoring. SPE of AIME, Paper No. SPE 7776.

Wahl, J.S., Nelligan, W.B., Frentrop, A.H., Johnstone, C.W. and

Schwartz, R.J., 1970. The thermal neutron decay time log. SPE J., Dec.

Youmans, A.H., Hopkinson, E.C. and Wichmann, P.A., 1966. Neutron lifetime logging in theory and practice SPWLA Seventh Annual Logging Symposium, May 8–11, Tulsa, Eq. 5.

Youngblood, W.E., 1980. The application of pulsed neutron decay time logs to monitor water floods with changing salinity. SPE of AIME, Paper No. 7777.

11. FORMATION DENSITY MEASUREMENTS
(the gamma-gamma log or density log)

11.1. PRINCIPLE

The formation is subjected to gamma rays emitted by a special source (^{60}Co or ^{137}Cs). Gamma rays are particles having no mass and moving at the speed of light. These gamma rays or photons collide with matter in three different types of interaction, depending on their incident energy.

11.1.1. Pair production

When the photon energy is above 1.02 MeV, the interaction of photon and matter leads to *pair production*, it means the production of a *negatron* (or negative electron) and a *positron* (or positive electron), each with an energy of 0.51 MeV (Fig. 11-1).

These two previously non-existent electron masses appear as a result of the disappearance of the photon energy. All of the photon energy is given up to the two electrons, with the exception of a very small amount going into the recoiling nucleus. The reaction is:

$$h\nu_0 \rightarrow e^+ + e^- + 2T \qquad (11\text{-}1)$$

This phenomenon can only occur when the energy of the photon is higher than $2mc^2$ ($= 1.02$ MeV). Requirement of conservation of energy and momentum of the system allow this process to occur only in the electric field of a nucleus.

After the pair has been created, positron and electron lose energy by ionization as they move off from the point of origin. When the positron energy becomes low it annihilates by combination with an electron in a process that is the reverse of that by which it was created:

$$e^+ + e^- \rightarrow 2h\nu \qquad (11\text{-}2)$$

It means that the rest energy of the two mutually annihilating particles is transformed into radiant energy. The two photons, or gamma rays, in eq. 11-2 are called *annihilation radiation*. The two photons

have an energy of 0.51 MeV and are emitted in almost exactly opposite direction (Fig. 11-2).

The pair production cross-section increases with increasing energy (Fig. 11-5). It is directly proportional to Z^2 (Z being the atomic number).

The influence of pair production is negligeable for the types of source used in density tools.

11.1.2. Compton scattering

When the incident photon collides with an electron (Fig. 11-3) its energy $h\nu_0$ is divided between the kinetic energy $E = mv^2$ given to the electron ejected from its atom with initial velocity v and a photon "scattered" in a direction making an angle θ with the original incident direction.

This elastic collision between the gamma ray and an individual electron simply allows the gamma ray to proceed, deflected by its encounter, but reduced in energy. This process occurs mainly with outer electrons of the atoms.

This type of reaction is called *Compton scattering* and it is the reaction figuring mainly in density measurements. The scattering effect is sensitive to the electron density of the formation (number of electrons per unit volume). This is developed in Paragraph 11-2.

An important feature of Compton scattering is that if the incident energy of the gamma ray, E_γ, is known, then the scattering angle θ and the scattered gamma ray energy, E_γ, are quite simply related (Fig. 11-3).

11.1.3. Photo-electric effect

In the course of a collision with an electron a photon can transfer all its energy to the electron in the form of kinetic energy. The electron is ejected from its atom and the photon disappears. The gamma ray is absorbed (Fig. 11-4).

The microscopic cross-section of this reaction, τ, has been found to be related to the atomic number of the target atom, Z, and the energy of the incident gamma ray, E_γ, by the following relation:

$$\tau = 12.1 \frac{Z^{4.6}}{(E_\gamma)^{3.15}} \qquad \text{barns/atom} \qquad (11\text{-}3)$$

The photo-electric effect is highest when the gamma-

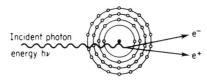

Fig. 11-1. Schematic of pair production at a nucleus.

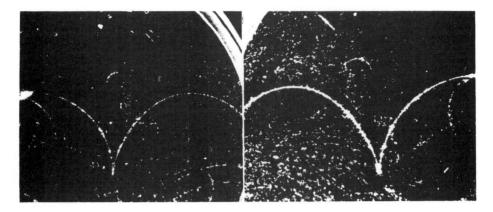

Fig. 11-2. Cloud chamber photograph of pair production with track curvatures produced by a field of 1500 gauss (in Lapp and Andrews, 1972).

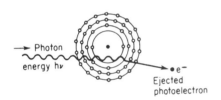

Fig. 11-3. Schematic of the Compton process. Geometrical relations in the Compton process.

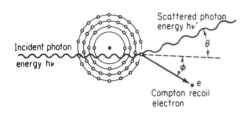

Fig. 11-4. Schematic of the photoelectric process.

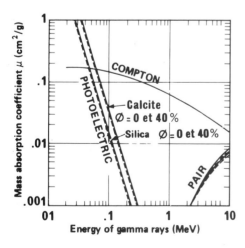

Fig. 11-5. Gamma ray mass absorption coefficient over the energy range of interest in Formation Density Logging (from Tittman and Wahl, 1965).

ray energy is small and the atomic number of the element high.

The relative importance of the three major types of interactions between photon and matter for different photon energies and for different atomic numbers is represented by the schematic diagram of Fig. 11-5. The lines indicate the values of absorber atomic number, Z, and photon energy at which neighboring effects are equally probable.

If we compare the mass absorbtion factor μ with the energy of the gamma rays (Fig. 11-5) we can see that for an energy band of 0.2–2.0 MeV absorption is due essentially to Compton scattering.

The density tool measures the intensity of scattered gamma rays at a fixed distance from the source.

This intensity is smaller the larger the number of collisions experienced by a photon, hence when the electron density is higher. To a first approximation the electron density is proportional to the density of the formation. Hence, the number of gamma rays detected is smaller when the formation density is higher.

So, in dense formations few gamma rays are detected since the number of collisions is high and gamma rays are absorbed (photo-electric effect) having lost a proportion of their energy in each collision. In low-density formations few collisions are made and so less energy is lost. More gamma rays are detected.

11.2. ABSORPTION EQUATION

If L is large enough the intensity of the gamma rays is an exponential function of the electron density of the formation and is given by the equation:

$$I = I_0 e^{-\mu \rho_e L} \qquad (11\text{-}4)$$

TABLE 11-1

Values of $C = 2 Z/A$ for the most common elements

Element	Z	A	$2 Z/A$
H	1	1.0079	1.9843
C	6	12.0111	0.9991
N	7	14.0067	0.9995
O	8	15.9994	1.000
Na	11	22.9898	0.9569
Mg	12	24.312	0.9872
Al	13	26.9815	0.9636
Si	14	28.086	0.9969
P	15	30.9738	0.9686
S	16	32.064	0.998
Cl	17	35.453	0.959
K	19	39.102	0.9718
Ca	20	40.08	0.998
Fe	26	55.847	0.9311
Ba	56	137.34	0.8155

where:

I = intensity of gamma rays measured at the detector.

I_0 = intensity of gamma rays at the source.

ρ_e = electronic density of the formation in the interval L (number of electrons per unit volume).

L = detector-source spacing.

μ is, to a first approximation, a constant depending on the tool geometry, the energy of the gamma rays emitted by the source and the detector characteristics.

Taking natural logarithms of eq. 11-4 gives:

$$L_n I = L_n I_0 - \mu \rho_e L \qquad (11-5)$$

which says in effect that the electron density is a linear function of logarithm of the intensity of gamma rays detected.

11.3. THE RELATION BETWEEN THE ELECTRONIC DENSITY AND THE BULK DENSITY

The bulk density (ρ_b) of the formation is what we seek to determine. This is related to the electron density (ρ_e) by the following:

$$\rho_e = \rho_b (Z/A) N \qquad (11-6)$$

where:

Z = atomic number

A = atomic mass

N = Avogadro's number (6.02×10^{23})

For the majority of elements and constituents of rocks (Z/A) is very close to 0.5 except for hydrogen for which it is almost 1 (Tables 11-1 and 11-2).

We define an electron density index by:

$$(\rho_e)_i = 2\rho_e/N \qquad (11-7)$$

As the tool is calibrated in a freshwater-saturated limestone the apparent global density ρ_a is linked to $(\rho_e)_i$ by the equation:

$$\rho_a = 1.07(\rho_e)_i - 0.1883 \qquad (11-8)$$

For sands, limestones and dolomites (liquid saturated) ρ_a is practically equal to ρ_b. For some substances or for gas-filled formations corrections have to be made (Fig. 11-6).

11.4. GAMMA-RAY SOURCES

The most widely used are:

[60]Cobalt which emits photons at energies of 1.17 and 1.33 MeV.

[137]Cesium that emits photons of energy 0.66 MeV.

TABLE 11-2

Values of C, electronic and bulk densities of the most common minerals found in rocks

Compound	Formula	Actual density ρ_b	$\frac{2\Sigma Z's}{mol.wt.}$	Electronic density ρ_e	Apparent density (as seen by tool) ρ_a
Quartz	SiO_2	2.654	0.9985	2.650	2.648
Calcite	$CaCO_3$	2.710	0.9991	2.708	2.710
Dolomite	$CaCO_3 MgCO_3$	2.870	0.9977	2.863	2.876
Anhydrite	$CaSO_4$	2.960	0.9990	2.957	2.977
Sylvite	KCl	1.984	0.9657	1.916	1.863
Halite	$NaCl$	2.165	0.9581	2.074	2.032
Gypsum	$CaSO_4 2 H_2O$	2.320	1.0222	2.372	2.351
Anthracite coal		$\begin{cases} 1.400 \\ 1.800 \end{cases}$	1.030	$\begin{cases} 1.442 \\ 1.852 \end{cases}$	$\begin{cases} 1.355 \\ 1.796 \end{cases}$
Bituminous coal		$\begin{cases} 1.200 \\ 1.500 \end{cases}$	1.060	$\begin{cases} 1.272 \\ 1.590 \end{cases}$	$\begin{cases} 1.173 \\ 1.514 \end{cases}$
Fresh water	H_2O	1.000	1.1101	1.110	1.00
Salt water	200.000 ppm	1.146	1.0797	1.237	1.135
"Oil"	$n(CH_2)$	0.850	1.1407	0.970	0.850
Methane	CH_4	ρ_{meth}	1.247	1.247 ρ_{meth}	1.335 ρ_{meth} −0.188
"Gas"	$C_{1.1}H_{4.2}$	ρ_g	1.238	1.238 ρ_g	1.325 ρ_g −0.188

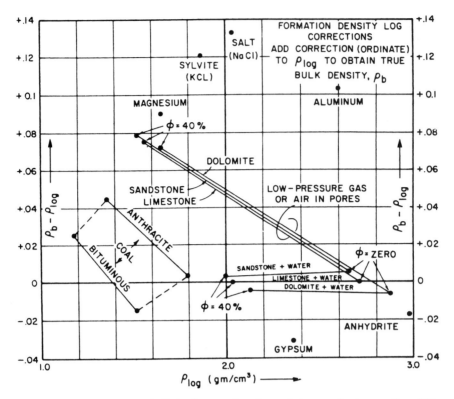

Fig. 11-6. Corrections to be applied to apparent bulk density, ρ_{log}, in order to derive true density, ρ_b. (from Tittman and Wahl, 1965).

11.5. DETECTORS

These are scintillation detectors set to detect gamma rays above a certain energy level.

11.6. CALIBRATION UNITS

The initial laboratory calibrations are made in pure limestone saturated with freshwater, where the density is known exactly.

Secondary calibrations are made in blocks of aluminium and sulphur or magnesium. Finally, at the wellsite a calibration jig is used that gives a radioactivity level of known intensity designed to test the detection system.

Originally the units of measurement were the standard counts per sec (Schlumberger) or the standard unit of density (PGAC). These were then transformed to density by charts given by the service companies. However, since the introduction of two detector tools the transformation is made inside the surface instrumentation and the log is given directly in grams per cubic centimeter: g/cm^3.

11.7. THE TOOLS

The first tools used only one detector (Fig. 11-7). Although pushed against the borehole wall by a spring, the measurement suffered from the effects of mud-cake, its type, thickness and density.

To eliminate mud-cake effects service companies now offer two-detector systems in so-called compensated tools (Fig. 11-8). The corrections are made within the surface equipment as a function of the readings from the two detectors. Figure 11-9, known as the *spine- and ribs plot*, shows that for a given value of ρ_b, the readings from the two detectors fall on an average curve, whatever the density or thickness of the mud-cake. By using the set of curves it is

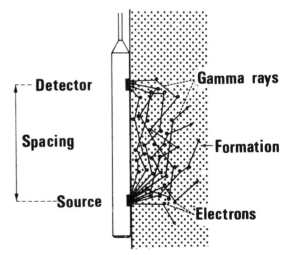

Fig. 11-7. Principle and schematic of the one detector formation density tool.

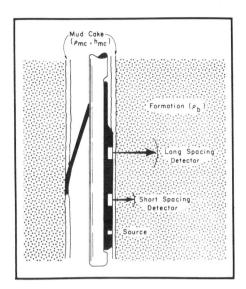

Fig. 11-8. Schematic of the Dual-spacing formation density log (FDC) (courtesy of Schlumberger).

possible to determine the correction $\Delta\rho_b$. The log presents both ρ_b and the correction applied $\Delta\rho_b$.

11.8. DEPTH OF INVESTIGATION

This is lower the higher is the density of the rock. Figure 11-10 gives an idea of the investigated zone. We can see that it is small and does not go above 6 inches. In porous and permeable formations the density tool investigates essentially the invaded zone.

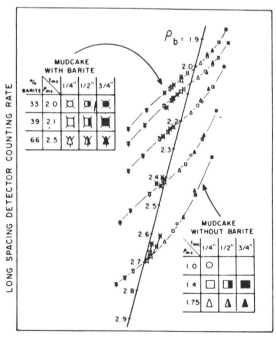

Fig. 11-9. "Spine-and-ribs" plot, showing response of FDC counting rates to mud cake (from Wahl et al., 1964) (courtesy of SPE of AIME).

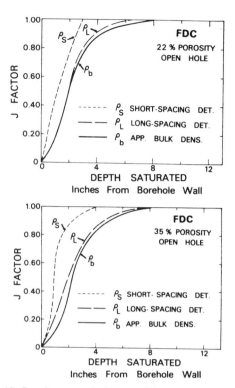

Fig. 11-10. Pseudo-geometrical factor (J-factor) curves for FDC density tool, for 22% and 35% porosity (from Sherman and Lock, 1975, courtesy of SPE of AIME).

11.9. VERTICAL RESOLUTION

For single-detector tools it corresponds to the source-detector spacing, around 16 inches for the FDL. For two-detector tools the distance between the detectors (span) gives the resolution, about 10 inches for the FDC.

11.10. MEASURE POINT

This is either at the midpoint of the source-detector spacing (FDL one-detector system) or the midpoint of the two detector span (FDC or two-detector system).

11.11. FUNDAMENTAL FACTORS INFLUENCING THE MEASUREMENT

We have seen that for a given source and spacing the main factors are the bulk density and the ratio Z/A.

As a first approximation Z/A is considered a constant. The global density in the region investigated by the tool depends then on: (a) the density and percentage of different mineralogical constituents in the rock; and (b) in the case of porous rocks, the density of the different fluids and their percentage in the reservoir (porosity) and in the pore space (saturation). However, a correct interpretation implies that we take account of the influence Z/A as this differs from 1 for oil, gas and water (see Table 11-2).

We should consider in particular the influence on the measurement of three important parameters.

11.11.1. Shales

If we try to interpret the density in terms of porosity the influence of shale for density logs is much less than for neutron logs. The density of "dry" clays is somewhere around that of quartz and so they have approximately the same effect within the matrix.

In any case when an interpretation is made for lithology or porosity it is preferable to take into account the effect of shales, above all if their density is very different from the other minerals composing the rock. We have the following equation:

$$\rho_{bc} = \rho_b + V_{sh}(\rho_{ma} - \rho_{sh}) \qquad (11-9)$$

where:

ρ_{bc} = shale corrected bulk density;
ρ_b = log reading of bulk density;
ρ_{ma} = matrix bulk density;
ρ_{sh} = shale bulk density;
V_{sh} = shale percentage.

11.11.2. Water

The fluid in the zone investigated by the tool is mainly mud filtrate, if the rock is porous and invaded. As the density of this can vary with temperature and pressure as a function of its salinity, we need to make corrections in the interpretation (Fig. 11-11).

11.11.3. Hydrocarbon

Hydrocarbon densities, especially gas, are less than that of water (Fig. 11-12), which means that the same formation full of gas appears much lighter and therefore more porous than if water-saturated. There is a case for making a hydrocarbon correction especially when determining porosity. If we write $\Delta\rho_{bh}$ for the variation in density caused by hydrocarbons we can write:

$$\rho_b = \rho_{bc} + \Delta\rho_{bh} \qquad (11-10)$$

where:

ρ_{bc} = bulk density corrected for hydrocarbon (i.e. where the pores are saturated with mud filtrate ρ_{mf});
ρ_b = bulk density log reading.

It can be shown that $\Delta\rho_{bh}$ is given by:

$$\Delta\rho_{bh} = -1.07 \, \phi S_{hr}(C_{mf}\rho_{mf} - C_h\rho_h) \qquad (11-11)$$

and hence

$$\rho_{bc} = \rho_b + 1.07 \, \phi S_{hr}(C_{mf}\rho_{mf} - C_h\rho_h) \qquad (11-12)$$

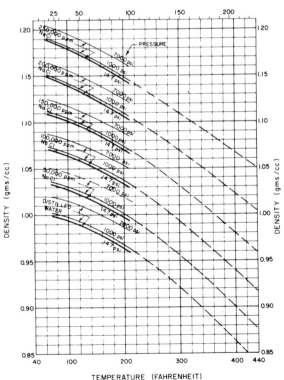

Fig. 11-11. Densities of water and NaCl solution at varying temperatures and pressures (courtesy of Schlumberger).

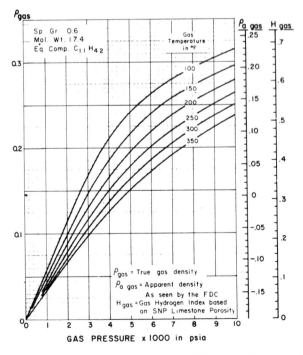

Fig. 11-12. Gas density and hydrogen index as a function of pressure and temperature for a gas mixture slightly heavier than methane ($C_{1.1}H_{4.2}$) (courtesy of Schlumberger).

with:

ϕ = porosity;

S_{hr} = hydrocarbon residual saturation in invaded zone;

C_{mf} = electronic density coefficient for mud-filtrate;

C_h = electronic density coefficient for hydrocarbon;

ρ_h = hydrocarbon density;

ρ_{mf} = mud filtrate density.

11.12. INTERPRETATION

The bulk density is exactly equal to the sum of the densities of the component parts of the formation multiplied by their respective percentage volume. For a porous formation, this implies:

$$\rho_b = \phi\rho_f + (1-\phi)\rho_{ma} \qquad (11\text{-}13)$$

with

$$\rho_f = S_{hr}\rho_h + S_{xo}\rho_{mf} \qquad (11\text{-}14)$$

$$\rho_{ma} = \sum_1^n V_n\rho_{ma_n} \qquad (11\text{-}15)$$

$$S_{hr} + S_{xo} = 1 \qquad (11\text{-}16)$$

$$\phi + \sum_1^n V_n = 1 \qquad (11\text{-}17)$$

Equation 11-13 can be rewritten in terms of porosity as

$$\phi_D = (\rho_{ma} - \rho_b)/(\rho_{ma} - \rho_{mf}) \qquad (11\text{-}18)$$

with

ϕ_D = porosity;

ρ_{ma_n} = bulk density of mineral n;

S_{xo} = water saturation in the invaded zone;

V_n = volumic percentage of mineral n.

The chart in Fig. 11-13 shows this relation.

In general for most fluids, with the exception of gas or light hydrocarbons, and most minerals making up a rock ρ_b is obtained directly from the apparent ρ_a

given by the tool (the coefficient C being nearly equal to 1).

In the case of a gas or light hydrocarbon formation where the coefficient C is no longer near 1, this has to be taken into account when ϕ is calculated. It can be shown that eq. 11-3 may be rewritten as:

$$\rho_{log} = -1.0704\,\phi[C_{ma}\rho_{ma} - S_{hr}C_h\rho_h - S_{xo}C_{mf}\rho_{mf}] + \rho_{ma} \qquad (11\text{-}19)$$

and that eq. 11-18 becomes:

$$\phi_D = \phi\frac{C_{ma}\rho_{ma} - S_{hr}C_h\rho_h - S_{xo}C_{mf}\rho_{mf}}{C_{ma}\rho_{ma} - C_{mf}\rho_{mf}} \qquad (11\text{-}20)$$

11.13. ENVIRONMENTAL EFFECTS

11.13.1. Time constant, recording speed, dead time, bed thickness

These factors have been dealt with in the general discussion of nuclear logs (Chapter 5).

11.13.2. The borehole

The effect of the borehole is easily seen for uncompensated density logs but less so for the compensated (Fig. 11-14). Several factors are important.

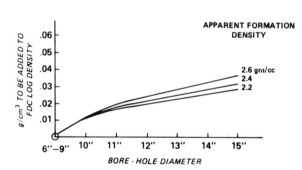

MUD – FILLED HOLES

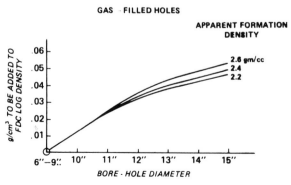

GAS – FILLED HOLES

Fig. 11-14. Corrections for hole size (from Wahl et al., 1964; courtesy of SPE of AIME).

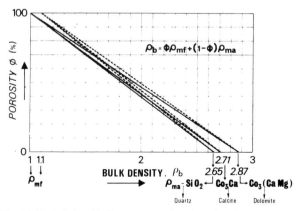

Fig. 11-13. Relationship between ϕ and ρ_b knowing the density of the mineral (ρ_{ma}).

11.13.2.1. *Hole diameter*

Its influence on the FDC is more evident in holes above 10 inches in diameter.

11.13.2.2. *The drilling fluid*

The corrections needed are more important for air-drilled than for mud-drilled wells, as air, being less dense, stops fewer gamma rays than mud.

11.13.2.3. *Hole rugosity*

If the borehole well is not smooth the FDC pad is not correctly applied to the formation and isolates zones full of mud which strongly affect the measurement. In this case, the FDC correction curve shows a different influence on the two detectors (Fig. 11-15).

11.13.3. **Mud-cake**

The effect of mud-cake depends on its thickness and type. The FDC has a sharp-edged pad which is pushed firmly against the formation. This tends to cut through part of the mud-cake. The two-detector system allows compensation for the remaining mud-cake layer up to a certain thickness corresponding to a correction of about 0.15 g/cm³.

11.13.4. **Casing**

Iron is an absorber of gamma rays. Hence, if a casing exists between the tool and the formation, the number of gamma rays reaching the detector is severely reduced. For this reason, it is not recommended to run the FDC in cased hole. Besides, the

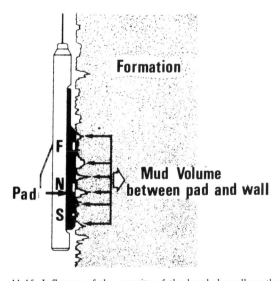

Fig. 11-15. Influence of the rugosity of the borehole wall on the FDC log response.

volume of mud and/or cement between the casing and the formation is generally not known. It follows that revised density measurements are difficult since the corrections needed are not known. In some cases calibration of cased hole density measurements is possible using core data of neighbouring wells or repeat sections run previously in open hole.

N.B. When making repeat sections there may appear differences with the main log above those due to statistical variations. This is often due to heterogeneous formations having for example fractures, vugs or fissures more on one side of the hole than the other. If the FDC pad does not trace the same path on the two runs there will be a resulting difference.

11.13.5. **Invasion**

We have seen that invasion depends on the pressure difference ΔP between the formation pressure and the mud column, the porosity and the permeability. Invasion changes the nature of the fluids in the zone of investigation of the tool (replacement of none, part or all of R_w by R_{mf}, saturations ranging from S_w to S_{xo}). In order to determine the effect of the invaded zone on the tool measurement we need to know the nature of the fluid within its zone of investigation. By the use of one or more resistivity tools the resistivity of the water in the pores of the invaded zone and the saturations there can be found.

Among the different environmental effects the most important are hole rugosity and invasion. The first, hole rugosity, can only be corrected empirically, giving a density obtained from other sections of hole where pad application is good and is low, or by attributing theoretical values (the case of salt, for example) or values defined statistically using nearby wells.

The second effect, universal where we have porosity and permeability, is only important where gas or light hydrocarbons are in the reservoir. A correction has to be made to obtain the density of the virgin formation ie. for geophysical purposes. For this, a general equation is used:

$$(\rho_b)_{cor} = \phi S_w \rho_w + \phi(1 - S_w)\rho_h$$
$$+ V_{sh}\rho_{sh} + (1 - \phi - V_{sh})\rho_{ma} \qquad (11-21)$$

ϕ, S_w, V_{sh} and possibly ρ_{ma} being defined from a quantitative interpretation of well logs. This interpretation and correction can be made automatically using an appropriate program.

11.14. **GEOLOGICAL FACTORS**

As for previous logs the geological factors of importance are as follows:

11.14.1. Rock composition

The minerals making up the rock (grains, matrix, cement) are seen through their respective electron densities and volume percentage in the formation.

Any fluids contained in the rock come into play via their type, volume percentage in the formation (porosity) and volume percentage in the pore space (saturations) seen by the tool.

11.14.2. Rock texture

If we except the porosity which depends on grain size and shape, sorting, packing, percentage of matrix and cement, on textural parameters, the texture has only an indirect effect via the permeability and invasion, thus on the nature of the fluids in the zone of investigation of the tool.

11.14.3. Sedimentary structure

If the thickness of the beds is lower than the vertical resolution of the tool, this will affect the log response. High apparent dips will also have an influence on the response, the investigated zone, near a boundary including the bed behind. Consequently the bed boundary will not be so well defined (Fig. 11-16).

11.14.4. Temperature

This has little effect except that it changes fluid densities, especially gas (Fig. 11-12).

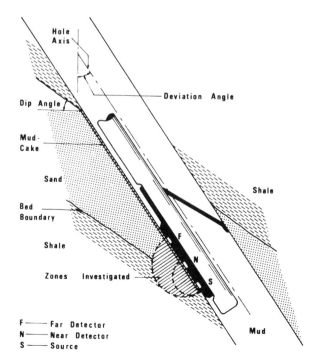

Fig. 11-16. Influence of the apparent dip on the FDC log response.

11.14.5. Pressure

This affects both the density of the formation fluids, particularly gas, and the depth of invasion.

11.14.6. Depositional environment-sequential evolution

These factors control bed thickness and the evolution of the lithology which in turn are reflected in the amplitude and shape of the density curve.

11.15. APPLICATIONS

(a) The measurement of density is of interest itself in geophysical studies. It can help in interpretation of gravity measurements and by association with the speed of sound in formations of different density it permits an interpretation of seismic profiles passing by the wellbore.

(b) The porosity can be calculated directly if the density of the mineral component (matrix) and fluid are known, or, if not, by combination with the neutron log.

(c) The density provides a base log for the determination of mineral component either on its own in the case of non-porous formations or in combination with other logs, (LDT, CNL, BHC, NGS) for porous reservoirs.

(d) The study of the vertical evolution of shale or sand densities with depth is used in the study of compaction.

(e) Comparison of density, neutron and resistivity logs allows a fast and accurate identification of reservoir fluids and of gas-oil, gas-water, oil-water contacts.

(f) The density is used in the definition of electrofacies.

(g) Finally the density curve can be used for correlation of facies.

11.16. REFERENCES

Baker, P.E., 1957. Density logging with gamma-rays. Petrol. Trans. AIME, 210.

Dresser-Atlas, 1973. Density log. Techn. Bull.

Evans, R.D., 1955. The Atomic Nucleus. McGraw-Hill, New York.

Lapp, R.E. and Andrews, H.L. 1972. Nuclear Radiation Physics, 4th ed. Prentice-Hall, Inc., Englewood Cliffs, N.J.

Newton, G.R., 1954. Subsurface formation density logging. Geophysics, 19 (3).

Pickell, J.J. and Heacock, J.G., 1960. Density logging. Geophysics, 25 (4).

Schlumberger, 1972. Log interpretation. Vol. 1: Principles.

Sherman, H. and Locke, S., 1975. Effect of porosity on depth of investigation of neutron and density sondes. SPE of AIME, paper SPE 5510.

204

Tittman, J. and Wahl, J.S., 1965. The physical foundation of formation density logging (Gamma-Gamma). Geophysics, 30 (2).

Wahl, J.S., Tittman, J., Johnstone, C.W. and Alger. R.P. (1964). The dual spacing formation density log. J. Pet. Technol., 16 (12).

12. MEASUREMENT OF THE MEAN ATOMIC NUMBER

(Litho-density tool)

This measurement has been proposed by Schlumberger since 1977. This is a new generation of density tool which provides additional lithology information, through the measurement of the photo-electric cross-section, and an improved density measurement.

12.1. PHYSICAL PRINCIPLE OF THE TOOL

As previously seen (see chapter 11) when the formation is submitted to a gamma-ray flux, gamma photons interact with matter in different ways. Only two of them are of practical interest since the energy of the gamma rays given by the ^{137}Cesium source is 662 keV. These interactions are:

(a) Compton scattering of gamma rays by electrons.

(b) Photo-electric absorption of gamma rays by electrons.

The probability that a gamma ray is involved in one of these two reactions is measured by the characteristic cross-section of this reaction. The cross-section can be compared with the area of a target under the fire of a person shooting at random: the larger the target, the more often it is hit.

We have previously studied the first interaction (see Chapter 11). So we will focus our attention on the second one.

12.1.1. Photo-electric interaction

When a gamma ray of low energy (below 100 keV) interacts with an electron (Fig. 12-1), it is totally absorbed. About all its energy is given to the electron which is freed from the atom to which it was attached. The energy of the electron, E_e, is given by the following relation:

$$E_e = h\nu - B_e \qquad (12-1)$$

where B_e is the binding energy of the ejected electron, and $h\nu$ is the photon energy. The most tightly bound electrons have the greatest ability to absorb gamma rays. Photo-electric effect is highly selective: it has been proved that 80% of the process concerns the electrons of the innermost shell, the K shell, and that the peripheral electrons are seldom involved (Fig. 12-2).

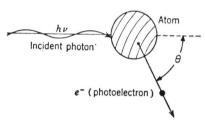

Fig. 12-1. Schematic representation of the photoelectric effect. The primary gamma photon is completely absorbed and a photo-electron is ejected at an angle θ with energy $E_e = h\nu - B_e$.

The absolute probability of a photo-electric interaction is described by the atomic cross-section σ_e. It has been found that σ_e is related to the atomic number of the target atom, Z, and the energy of the incident gamma ray, E_γ, by the following relationship:

$$\sigma_e = K \frac{Z^n}{(E_\gamma)^m} \qquad \text{barns/atom} \qquad (12-2)$$

where K is a constant. The exponents n and m are functions of the incident gamma ray energy, $E_\gamma = h\nu$. n is found to increase from about 4.0 to 4.6 as E_γ increases from 0.1 MeV to 3 MeV, as shown in Fig. 12-3.

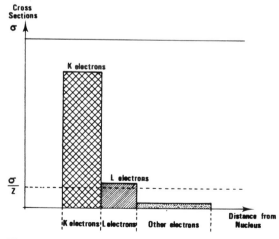

Fig. 12-2. Photo-electric absorption cross-section. σ_e is the photo-electric absorption cross-section per atom. The cross-sections for K electrons, L electrons and further electrons are different (courtesy of Schlumberger).

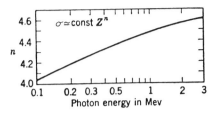

Fig. 12-3. Approximate variation of the photo-electric cross section σ_e (cm²/atom) with Z^n, for various values of $h\nu$. (from Rasmussen).

For a given atom—so, for a given value of Z^n—Fig. 12-4 shows that m decreases when E_γ increases. It can be seen from this figure that: (a) σ_e changes continuously as E_γ increases; and (b) in a given energy range the variation of σ_e with E_γ is different for different Z, the lower Z, the higher is the exponent m of E_γ.

Considering the mean value of Z encountered in the most common sedimentary rocks and the range of energy in which this atomic cross-section, σ_e, is mea-

sured, the following relationship has been found:

$$\sigma_e = 12.1 \frac{Z^{4.6}}{E_\gamma^{3.15}} \qquad \text{barns/atom} \qquad (12.2\text{b})$$

12.1.2. Definition of the photo-electric absorption index

Though the photo-electric absorption cannot be easily described at the level of the electron, it has been found convenient, by analogy with the Compton effect, to define a parameter P_e, the photo-electric absorption index, proportional to the "average cross-section by electron", σ_e/Z:

$$P_e = \frac{1}{K} \frac{\sigma_e}{Z} \qquad (12\text{-}3)$$

K is a constant coefficient characteristic of the energy E_γ where the photoelectric absorption is observed.

$$K \sim \frac{48 \times 10^3}{E_\gamma^{3.15}}$$

Practically, the detection is made over a range of energies and:

$$\sigma_e = \Sigma(\sigma_e)_i \qquad (\sigma_e)_i \text{ cross-section at energy } E_i$$
$$K = \Sigma K_i \qquad K_i \text{ coefficient at energy } E_i$$

The energy dependencies of σ_e and K cancel out and P_e is a parameter which does not depend on the energy. For the range of energies and minerals currently encountered P_e is well approximated by:

$$P_e = \left(\frac{Z}{10}\right)^{3.6} \qquad (12\text{-}4)$$

is expressed in barns/atom *. P_e is expressed in barns/electron. The value of P_e for a few common minerals are listed in Table 12-1.

12.1.3. P_e of a composite material

A binary mixture of atoms type 1 (Z_1 electrons per atom, n_1 atoms/cm³, electronic density ρ_{e1}) and type 2 (Z_2 electrons per atom, n_2 atoms/cm³, electronic density ρ_{e2}) is considered. $(\sigma_e)_1$ and $(\sigma_e)_2$ are the respective photo-electric absorption cross-sections per atom, P_{e1} and P_{e2} are the respective photo-electron indices. Volumetric fractions of atoms 1 and 2 are V_1 and V_2 with:

$$V_1 + V_2 = 1$$

σ_e, ρ_e, P_e, n, characterize the mixture.

$$n\sigma_e = n_1 V_1 (\sigma_e)_1 + n_2 V_2 (\sigma_e)_2 \qquad (12\text{-}5)$$

Replacing the cross-sections as a function of P_e (12-3)

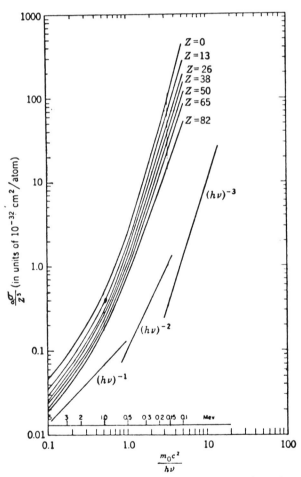

Fig. 12-4. Approximate "best" theoretical values of the photo-electric cross-section σ_e (cm²/atom), obtained by blending data from Sauter-Stobbe (below 0.35MeV), Hulme, McDougall, Buckingham and Fowler results (between 0.35 and 2 MeV), Hall's high-energy approximation (above 2 MeV).

---

* 1 barn $= 10^{-24}$ cm².

TABLE 12-1

The photo-electric cross-section per electron of the most common rocks and minerals (courtesy of Schlumberger)

Name	Formula	Atomic or molecular weight	Photo-electric absorption index P_e	Atomic number or Z_{eq}	Density e	Electronic density ρ_e	Bulk density, ρ_b	C	U
A. Elements									
Hydrogen	H	1.000	0.00025	1				1.984	
Carbon	C	12.011	0.15898	6				0.9991	
Oxygen	O	16.000	0.44784	8				1.0000	
Sodium	Na	22.991	1.4093	11				0.9566	
Magnesium	Mg	24.32	1.9277	12				0.9868	
Aluminum	Al	26.98	2.5715	13	2.700	2.602	2.596	0.9637	
Silicium	Si	28.09	3.3579	14				0.9968	
Sulfur	S	32.066	5.4304	16	2.070	2.066	2.022	0.9979	
Chloride	Cl	35.457	6.7549	17				0.9589	
Potassium	K	39.100	10.081	19				0.9719	
Calcium	Ca	40.08	12.126	20				0.9980	
Titanium	Ti	47.90	17.089	22				0.9186	
Iron	Fe	55.85	31.181	26				0.9311	
Copper	Cu	63.54	43.5	29					
Strontium	Sr	87.63	122.24	38				0.8673	
Zirconium	Zr	91.22	147.03	40				0.8770	
Barium	Ba	137.36	493.72	56				0.8154	
Thorium	Th	232.05	2724.4	90					
Uranium	U	238.07	2948.74	92					
B. Minerals									
Anhydrite	CaSO$_4$	136.146	5.055	15.69	2.960	2.957	2.977	0.9989	14.9
Barite	BaSO$_4$	233.366	266.8	47.2	4.500	4.011		0.8913	1065.0
Calcite	CaCO$_3$	100.09	5.084	15.71	2.710	2.708	2.710	0.9991	13.8
Carnallite	KCl-MgCl$_2$-6 H$_2$O	277.88	4.089	14.79	1.61	1.645	1.573	1.0220	
Celestite	SrSO$_4$	183.696	55.13	30.4	3.960	3.708		0.9363	
Corundum	Al$_2$O$_3$	101.96	1.552	11.30	3.970	3.894		0.9808	
Dolomite	CaCO$_3$-MgCO$_3$	184.42	3.142	13.74	2.870	2.864	2.877	0.9977	9.00
Gypsum	CaSO$_4$-2 H$_2$O	172.18	3.420	14.07	2.320	2.372	2.350	1.0222	
Halite	NaCl	58.45	4.169	15.30	2.165	2.074	2.031	0.9580	9.68
Hematite	Fe$_2$O$_3$	159.70	21.48	23.45	5.240	4.987		0.9518	
Illmenite	FeO-TiO$_2$	151.75	16.63	21.87	4.70	4.46		0.9489	
Magnesite	MgCO$_3$	84.33	0.829	9.49	3.037	3.025	3.049	0.9961	
Magnetite	Fe$_3$O$_4$	231.55	22.08	23.65	5.180	4.922		0.9501	
Marcasite	FeS$_2$	119.98	16.97	21.96	4.870	4.708		0.9668	
Pyrite	FeS$_2$	119.98	16.97	21.96	5.000	4.834		0.9668	82.1
Quartz	SiO$_2$	60.09	1.806	11.78	2.654	2.650	2.648	0.9985	4.78
Rutile	TiO$_2$	79.90	10.08	19.02	4.260	4.052		0.9512	
Siderite	FeCO$_3$	115.86	14.69	21.09	3.94		3.89		55.9
Sylvite	KCl	74.557	8.510	18.13	1.984	1.916	1.862	0.9657	
Zircon	ZrSiO$_4$	183.31	69.10	32.45	4.560	4.279		0.9383	
C. Liquids									
Water	H$_2$O	18.016	0.358	7.52	1.000	1.110	1.000	1.1101	0.398
Salt water	(120,000 ppm)		0.807	9.42	1.086	1.185	1.080	1.0918	0.850
Oil	CH$_1$-6		0.119	5.53	0.850	0.948	0.826	1.1157	0.136 ρ_{oil}
	CH$_2$		0.125	5.61	0.850	0.970	0.849	1.1407	
D. Miscellaneous									
Berea Sandstone			1.745	11.67	2.308	2.330	2.305	0.9993	
Pecos Sandstone			2.70	13.18	2.394	2.414	2.395	1.0000	
Average Shale			3.42	14.07	2.650	2.645	2.642	0.998	
Anthracite coal	C:H:O = 93:3:4		0.161	6.02	1.700	1.749	1.683	1.0287	
Bituminous coal	C:H:O = 82:5:13		0.180	6.21	1.400	1.468	1.383	1.0485	

and n, number of atoms by n_e/Z:

$$\frac{n_e}{Z} P_e ZK = \frac{n_{e1}V_1}{Z_1} P_{e1} Z_1 K + \frac{n_{e2}V_2}{Z_2} P_{e2} Z_2 K$$

it becomes:

$$n_e P_e = n_{e1}V_1 P_{e1} + n_{e2}V_2 P_{e2}$$

Hence:

$$P_e \rho_e = V_1 P_{e1} \rho_{e1} + V_2 P_{e2} \rho_{e2}$$

or introducing $U = P_e \rho_e$, the volumetric photo-electric absorption index:

$$U = V_1 U_1 + `V_2 U_2 \qquad (12\text{-}6)$$

These equations can be generalized for a molecule with several atoms or for a mineral made of different components, i being the subscript of each individual atom or component:

$$P_e \rho_e = \sum_i V_i P_{ei} \rho_{ei}$$

$$U = \sum_i V_i U_i \qquad (12\text{-}7)$$

U is expressed in barns/cm³.

Table 12-1 gives the values of U for the most common minerals.

12.2. The Schlumberger Litho-density tool (LDT *)

The LDT tool is composed of:

(a) 1.5 curie ^{137}Cesium source which emits a collimated constant flux of gamma rays with an energy of 661 keV.

(b) A near detector (short-spacing detector) which is a scintillation counter connected to a photomultiplier.

(c) A far detector (long-spacing detector) which is also a scintillation counter connected to a photomultiplier with a beryllium window which allows the low-energy gamma rays to reach the counter.

The two counters function in the proportional mode. It means that the output pulse height is proportional to the incoming gamma-ray energy. Two small reference sources are attached to the counter. A feedback loop interacting with the high voltage level operates in such a way that the counts in the energy windows are maintained equal. In this manner temperature effects are continuously compensated even though the counters are not in Dewar flasks. The diameter of the tool is equal to 4″. Its ratings are 350°F and 20,000 psi. The maximum hole diameter is 22″. It is combinable with the standard gamma ray tool, and the compensated neutron tools (CNL *). With the Cable Communication System (CCS *) version of the tool additional combinations with Natural

---

Gamma-ray Spectrometry (NGS *) tool or Electromagnetic Propagation Tool (EPT *) are available.

12.3. Principle of measurement

When the gamma rays, emitted by the ^{137}Cesium source arrive in the formation, they interact with the electron of the atoms. Collision after collision they lose their energy (Compton scattering) until reaching such a low level of energy that they will be absorbed (photo-electric effect). As a result of this interaction, at a point placed at a certain distance from the energy source, the gamma rays have an energy which can vary between 661 keV, for a gamma ray which has never collided with an electron, and practically zero for a gamma ray after several collisions.

Figure 12-5 shows the gamma-ray energy spectrum seen by a detector located at a given distance from the source. The area H represents the number of gamma rays which reach the counter with relatively high energy. It corresponds to the region of Compton scattering. The quantity of gamma rays which finally reach the counter in this area H is a function of the electronic density ρ_e.

On the other hand, the area S represents the quantity of gamma rays of low energy that reach the counter. This quantity depends on the photo-electric cross-section U, as well as on the electronic density, ρ_e, of the formation. Therefore, H and S areas are defined as follows:

$$H = f_1(\rho_e) \qquad (12\text{-}8)$$

$$S = f_2(U, \rho_e) \qquad (12\text{-}9)$$

Figure 12-5 shows the effect of U on the spectrum

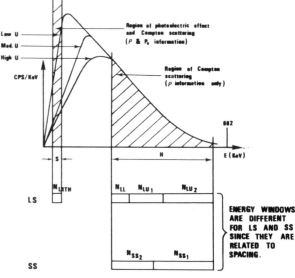

Fig. 12-5. Gamma-ray detection in long-spacing and short-spacing energy windows S and H. Influence of U values on the gamma-ray spectrum (courtesy of Schlumberger).

---

* Mark of Schlumberger.

shape and finally how it affects S.

The long-spacing detector (which has several energy windows) reads in H and S areas, while the short-spacing detector (which has two energy windows) reads in the H area.

The energy windows used, the source-detector spacings, the detector characteristics and the geometrical configuration of the tool are such that the ratio S/H is an unequivocal function of P_e (or U/ρ_e) under the conditions usually found in the formation.

12.4. RADIUS OF INVESTIGATION

It is very similar to that of the compensated formation density tool.

12.5. VERTICAL RESOLUTION

The LDT long-spacing being smaller than the one of the FDC, the vertical resolution is better.

12.6. MEASURING POINT

The measuring point of the density depends on the formation density. In any case it is closer to the detector than to the source. The measuring point of P_e is at the long-spacing detector.

12.7. STATISTICAL VARIATIONS

Due to the smaller long spacing and to more efficient detectors and electronics, the statistical variations have been significantly reduced as shown by Fig. 12-6.

This advantage is particularly noticeable at high density, where the LDT long-spacing count rates practically double the FDC ones.

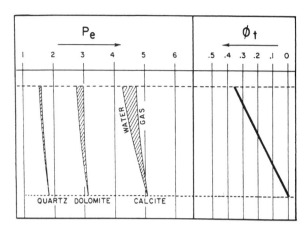

Fig. 12-7. Variation of P_e with porosity and fluid content (courtesy of Schlumberger).

12.8. GEOLOGICAL FACTORS WHICH AFFECT THE MEASUREMENTS

For the density measurement these factors have been previously discussed (see Chapter 11).

For the photo-electric absorption index, the fundamental factor is the mean atomic number of the bulk formation and through it:

(a) Essentially the minerals with which the formation is composed. Of course minerals composed with elements having a high atomic number, will have a big influence on the measurement. Table 12-1 lists the P_e values for the most common elements and minerals. It appears clearly that the common iron-bearing minerals (hematite, siderite, pyrite, chlorite, glauconite,...) will be easily detected. It is obvious also that formations rich in uranium or thorium will show relatively high P_e values:

(b) Secondarily the fluids. Their influence will depend on their nature, their volume (porosity, saturation) in the rock. But due to the fact that most of them are composed of low atomic number elements ($H = 1$; $C = 6$; $O = 8$) their influence is very

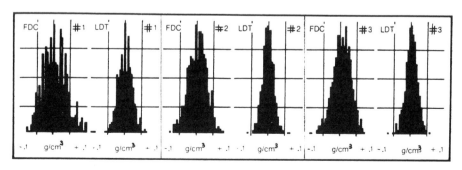

Fig. 12-6. Comparison of ρ_b histograms for FDC and LDT run in the same wells. As observed the dispersion is smaller for LDT (courtesy of Schlumberger).

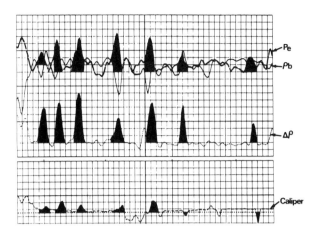

Fig. 12-8. Barite effect on log response. This can be used as fracture detection (courtesy of Schlumberger).

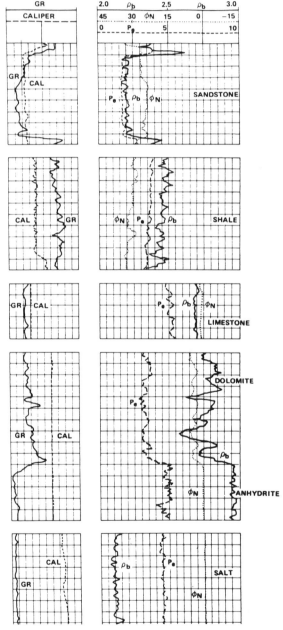

Fig. 12-9. Example of log responses in different sedimentary rocks (from Felder and Boyeldieu, 1979).

small, except if it is salty water.

Figure 12-7 shows the variations of P_e with porosity and fluid content for sandstone, limestone and dolomite.

It is for that reason that the P_e measurement is a very good lithology indicator, especially in gas-bearing reservoirs, in front of which the density and neutron are much more affected.

12.9. ENVIRONMENTAL EFFECTS ON THE MEASUREMENTS

For the density measurement the borehole influence has previously been discussed. But we must add that, in the LDT, the barite effect correction on the density measurement can be done because the gamma-ray spectrum is known. This is not the case with FDC.

For the photo-electric absorption index, the borehole has generally a very small influence due to the fact that the tool is on a pad and that the P_e value for the standard muds is very low compared to the formation ones. But if the mud contains barite, due to the high P_e value of barite, its influence will be high and will depend on either the mud-cake thickness in front of porous and permeable levels, or the

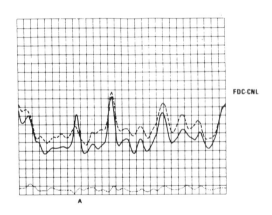

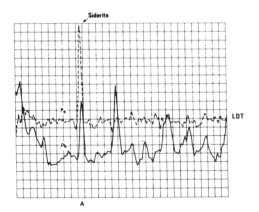

Fig. 12-10. Example of heavy mineral detection.

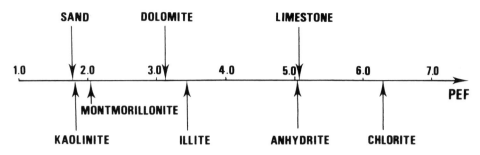

Fig. 12-11. P_e values of some common minerals (courtesy of Schlumberger).

mud-volume between the pad and the formation in rugosed intervals, or the filling of the open fractures by barite mud (Fig. 12-8).

12.10. APPLICATIONS

12.10.1. Mineralogical composition of the formation

This is the main application of the tool, especially in complex lithology and/or in gas bearing formations. The P_e information, or its derived U_{LDT} parameter combined with ρ_b, ϕ_N or $(\rho_{ma})_a$, as well as with NGS data, allows a better mineralogy definition and clay mineralogy recognition (Figs. 12-9 to 12-14).

12.10.2. Fracture detection

As previously discussed, if the drilling mud is barite bearing, the open fractures will appear as peaks on the P_e curve (Fig. 12-8).

12.10.3. Sedimentological studies

By a more accurate mineralogy determination allowed with this tool, the electrofacies analysis using

this information will be improved and consequently the environment of deposition better defined.

All these applications will be developed in volume 2.

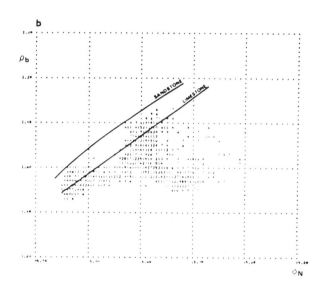

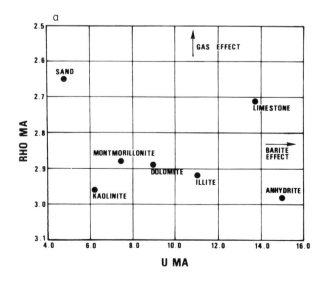

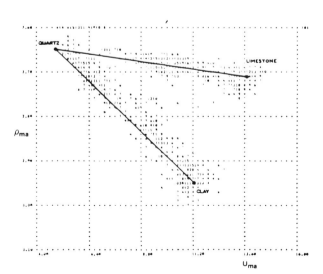

Fig. 12-12. a. ρ_{ma} vs U_{ma} cross-plot for mineral identification and computation of their percentage in the rock (courtesy of Schlumberger). b. Example of ρ_b vs ϕ_N and $(\rho_{ma})_a$ vs $(U_{ma})_a$ for mineralogical identification.

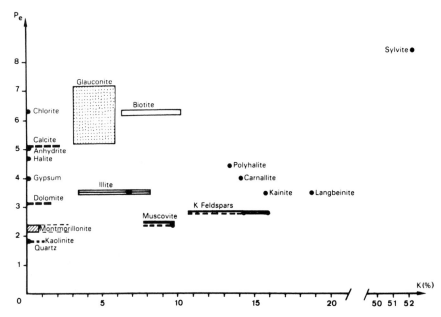

Fig. 12-13. P_e vs potassium (K) cross-plot for mineral identification (courtesy of Schlumberger).

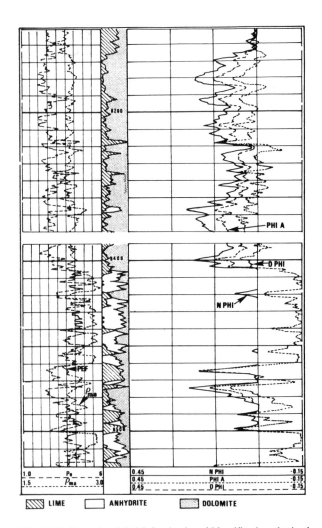

12.11. REFERENCES

Edmundson, H. and Raymer, L.L., 1979. Radioactive logging parameters for common minerals. Log Analyst, 20 (5).

Evans, R.D., 1955. The Atomic Nucleus. McGraw-Hill, New York, N.Y.

Felder, B. and Boyeldieu, C., 1979. The lithodensity Log. 6th European Symposium Transactions, SPWLA London Chapter, Mar. 26–27, 1979, paper O.

Gardner, J.S. and Dumanoir, J.L., 1980. Lithodensity log interpretation.

Fig. 12-14. Example of Quick-Look mineral identification obtained at the well site with a CSU (courtesy of Schlumberger).

13. ACOUSTIC LOG GENERALITIES—FUNDAMENTALS

We can group together as acoustic logs those that involve recording a parameter linked with the transmission of sound waves in the formation.

These parameters are mainly:

(a) The propagation speed of a wave in the formation calculated from the time taken to travel through a certain thickness of formation. This is the *sonic log*.

(b) The amplitude at the receiver of the first or second wave in the signal, either on arrival of the compressional wave or the shear wave. This is the *sonic amplitude log* which has an important application as the Cement Bond Log (CBL).

(c) The amplitude and position of the positive sections of the received signal. This is the *variable density log* (VDL).

13.1. ACOUSTIC SIGNALS

An acoustic signal is the sound wave resulting from the release of acoustic energy (Fig. 13-1).

13.1.1. Period, T (Fig. 13-2)

The period of the wave is defined as the duration of one cycle, and is generally measured in microseconds. It corresponds to the time separating two successive positive wave peaks (or negative peaks), measuring the same amplitude value in the same direction each time.

13.1.2. Frequency, f (Fig. 13-2)

This corresponds to the number of complete cycles per second and is measured in Hertz (Hz). 1 Hertz = 1 cycle/second. Frequency is the inverse of the period, hence

$$f = 1/T \tag{13-1}$$

13.1.3. Wavelength, λ

This is the distance travelled in one cycle by a wave front. It is equal to the ratio of the propagation speed (v) and the frequency (f):

$$\lambda = v/f \tag{13-2}$$

13.2. ACOUSTIC WAVES

There are several types of sound waves, each one characterized by the particular kind of particle movement.

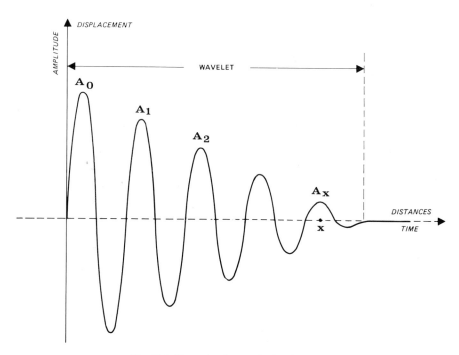

Fig. 13-1. Example of acoustic signal.

214

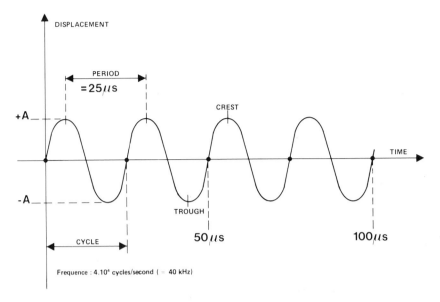

Fig. 13-2. Definition of the terms used in acoustics.

13.2.1. Compressional or longitudinal waves (or P wave)

In this wave the particles move in a direction parallel to the direction of propagation (Fig. 13-3). The speed of propagation is largest for this kind of wave compared to others and so it arrives first. It is the only wave propagated in liquids.

13.2.2. Transverse or shear waves (or S wave)

Particle movement is in a direction perpendicular to the wave direction (Fig. 13-4). As mentioned, the speed of propagation is less than the P-wave with a ratio of about 1.6 to 2. No shear waves are transmitted in liquids.

In the formation sound energy is transmitted by both compressional and shear wave. In the mud, energy is transmitted solely by compressional waves.

The energy transmitted by the slower shear wave is much higher than that of the compressional wave which is first to arrive. In the wave pattern received we can identify the shear wave by this feature. The ratio of amplitudes is of the order of between 15 and 20 to 1 (Fig. 13-5). As shales are less rigid structurally they do not transmit transverse waves very well.

N.B. Transverse waves may be polarized and so can be subdivided into horizontal and vertical components

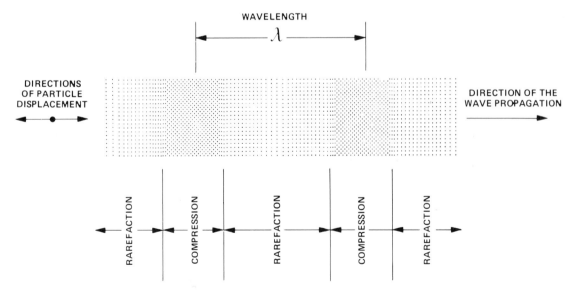

Fig. 13-3. Compressional wave.

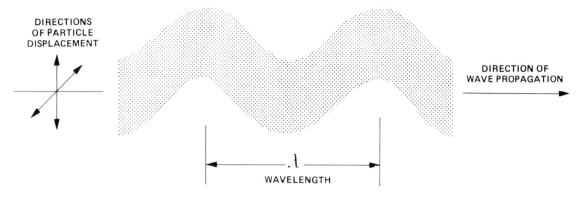

DIRECTIONS
OF PARTICLE
DISPLACEMENT

DIRECTION OF
WAVE PROPAGATION

λ
WAVELENGTH

Fig. 13-4. Shear wave.

13.2.3. Surface waves

These are waves transmitted on the surface within a layer whose thickness is about equal to the wavelength. They are divided as follows:

(a) *Rayleigh waves* in which the *particle motion is elliptical*, and retrograde with respect to the direction of propagation. These waves are not transmitted in liquids and their velocity is around 90% of that of transverse waves.

(b) *Love waves* in which particle *motion is transverse to the direction of propagation* but without any vertical movement. They are faster than Rayleigh waves.

(c) *Coupled waves* in which the *movement is diagonal*. These are the fastest surface wave.

(d) *Hydrodynamic waves* in which *movement is elliptical* but symmetrical to Rayleigh waves.

(e) *Stoneley waves* that are boundary acoustic waves at a liquid–solid interface resulting from the interaction of the compressional wave in the liquid and the shear wave in the solid. By definition, the Stoneley wave must have a wavelength smaller than the borehole diameter. Particle motion in the solid will be elliptical and retrograde similar to a Rayleigh wave. The velocity of the Stoneley wave will be less than that of the compressional wave in the fluid or the shear wave in the solid.

N.B. Sound waves transfer energy step by step by the movement of particles under elastic forces.

13.3. ELASTIC PROPERTIES OF ROCKS

These are the properties that define the ability of a body or rock to resist permanent deformation when

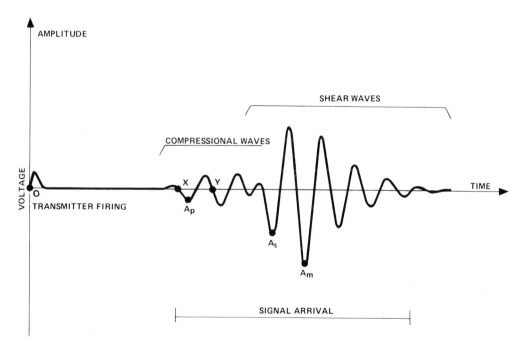

Fig. 13-5. Separation of compressional and shear waves from their transit travel time and amplitude.

deformed slightly. All solids, including rocks, follow Hooke's law which gives the proportional relation between the tension and the constraint (force).

(a) The ratio of constraint to tension in a simple linear compression or dilation is known as *Young's modulus*.

$$E = \frac{F/S}{dL/L} \qquad (13\text{-}3)$$

where (F/S) is the constraint or force applied per surface area and dL/L is the stretch or compression per unit length under the effect of the force.

(b) The ratio of force to tension under hydrostatic compression or dilatation corresponds to the *elastic bulk modulus*.

$$k = \frac{F/S}{dV/V} \qquad (13\text{-}4)$$

in which (dV/V) is the change per unit volume under the effect of the force.

(c) The ratio of force to tension under a shear force or one that is applied tangential to the displaced surface is known as the *shear modulus*, μ:

$$\mu = \frac{F/S}{dL/L} \qquad (13\text{-}5)$$

in which (F/S) is the shear force and dL/L is the shear tension or the deformation without a change in total volume.

To these elastic moduli we should add:

(d) The *compressibility*, c, (or β) which is the inverse of the elastic modulus k.

(e) *Poisson's ratio*, σ, which is a measure of the change in shape, or ratio of the lateral contraction to the longitudinal dilation.

$$\sigma = \frac{dl/l}{dL/L} \qquad (13\text{-}6)$$

(dl/l) being the transverse or lateral change.

N.B. We also talk about the *space modulus, M*, which is given by:

$$M = k + \tfrac{4}{3}\mu \qquad (13\text{-}7)$$

It is a measurement of the resistance to deformation from compressional and shear force in an elastic medium.

(f) The relationship between the various elasticity coefficient: the above coefficients can all be expressed in terms of any two. If for example we use μ and k we have:

$$E = \frac{9k\mu}{3k + \mu} \qquad (13\text{-}8)$$

and

$$\sigma = \frac{3k - 2\mu}{6k + 2\mu} \qquad (13\text{-}9)$$

13.4. SOUND WAVE VELOCITIES *

The velocity of sound in elastic media can be expressed using the elastic moduli. The longitudinal wave velocity, v_L is given by:

$$v_L = \left(\frac{k + \frac{4}{3}\mu}{\rho_b} \right)^{1/2} = \left(\frac{E}{\rho_b} \frac{1 - \sigma}{(1 + \sigma)(1 - 2\sigma)} \right)^{1/2}$$

$$(13\text{-}10)$$

and the transverse velocity, v_τ, by

$$v_\tau = \left(\frac{\mu}{\rho_b} \right)^{1/2} = \left(\frac{E}{\rho_b} \frac{1}{2(1 + \sigma)} \right)^{1/2} \qquad (13\text{-}11)$$

13.5. SOUND WAVE PROPAGATION, REFLECTION AND REFRACTION

Huyghens' principle states that each point reached by a wave oscillation acts as a new source of oscillation radiating spherical waves (Fig. 13-6).

The surface described by the in-phase oscillation at any given time is known as the *wave front*. The surface of separation between the set of points not in motion and those that are (or were) makes up a particular wave front called the *wave surface*.

Sound waves follow Descartes' law. Suppose we have two homogeneous media, isotropic and infinite with velocities v_{L_1}, v_{T_1} and v_{L_2}, v_{T_2}, separated by a plane surface.

Every wave L, even those purely longitudinal or transverse, incident at a point I on the separating surface at an angle i_1, gives rise to four new waves. Two of these are reflected waves, one compressional L_r at an angle r_1, one transverse T_r at an angle r_2, and the other two are refracted waves, one longitudinal L_R at an angle R_1 and the other transverse T_R at angle R_2 (Fig. 13-7).

Other sound waves may appear due to diffraction or dispersion.

If we only consider longitudinal waves the reflection law is:

$$i_1 = r_1$$

and the refraction law is:

$$\sin i_1 / v_{L_1} = \sin R_1 / v_{L_2} \qquad (13\text{-}12)$$

If $v_{L_2} > v_{L_1}$ the angle of total refraction or the critical angle of incidence, l, $(R_1 = 90°)$ is given by the equation:

$$\sin l = v_{L_1} / v_{L_2} \qquad (13\text{-}13)$$

---

* In liquids the shear modulus being nil, we have: $v_L = (k/\rho_b)^{1/2}$, and $v_T = 0$

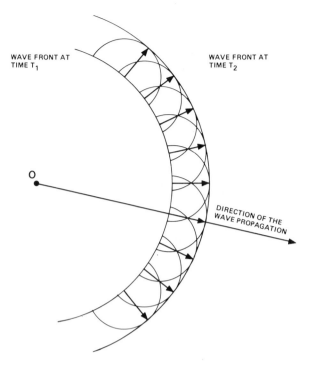

WAVE FRONT AT TIME T_1

WAVE FRONT AT TIME T_2

O

DIRECTION OF THE WAVE PROPAGATION

Fig. 13-6. Wave propagation.

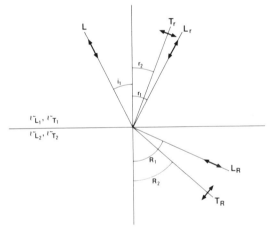

Fig. 13-7. Reflection and refraction of an acoustic wave.

We can define two critical angles of incidence, one for longitudinal and the other for transverse waves.

If we consider the reflected and refracted waves given by a transverse wave, we have:

$$\frac{\sin i_1}{v_{L_1}} = \frac{\sin r_2}{v_{T_1}} \qquad (13\text{-}14)$$

and

$$\frac{\sin i_1}{v_{L_1}} = \frac{\sin R_2}{v_{T_2}} \qquad (13\text{-}15)$$

N.B. In the case of a wellbore where medium 1 is a fluid the reflected transverse wave does not exist.

13.6. ACOUSTIC IMPEDANCE

This is given by the product of the density of a medium by the velocity of sound in the medium:

$$r = v_1 \rho_1 \qquad (13\text{-}16)$$

13.7. REFLECTION COEFFICIENT

In the case of an incident wave normal to a surface this is the ratio of reflected energy to incident energy:

$$R_{1\text{-}2} = \frac{v_2 \rho_2 - v_1 \rho_1}{v_2 \rho_2 + v_1 \rho_1} \qquad (13\text{-}17)$$

When the angle of incidence varies, the ratio changes from $R_{1\text{-}2}$ and depends on i_1, v_{T_1} and v_{T_2}.

13.8. WAVE INTERFERENCE

Interference occurs when waves of the same frequency arrive at the same point. At this time particles are subjected to two different forces which can either reinforce or tend to neutralise depending on the phase difference (Fig. 13-8).

In the case of the sonic, emitted waves are re-

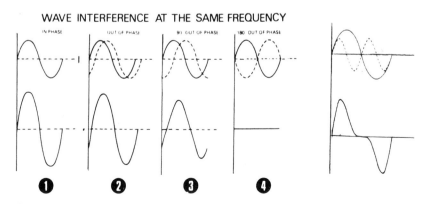

Fig. 13-8. Interference phenomenon of acoustic waves of the same frequency 1, 2, 3 and 4 are of the same wavelength.

218

flected or refracted at the borehole wall. However, as this is not perfectly cylindrical and the tool axis is rarely at the hole axis, wave interference may occur.

13.9. REFERENCES

See end of chapter 14.

14. MEASUREMENT OF THE SPEED OF SOUND (Sonic Log)

14.1. PRINCIPLE

A magnetostrictive transducer, excited from the surface by a signal, emits a sound wave (Fig. 14-1) whose average frequency is of the order of 20 to 40 kHz. The duration of the emission is short but it is repeated several times per second (10 to 60 times depending on the tool). The wave spreads in all directions from the transmitter, so producing spherical wavefronts. The wavefront passing through the mud is incident upon the borehole wall with increasing time and increasing angle of incidence as the distance from the transmitter increases (Fig. 14-3).

We can consider several cases:

(a) If the angle of incidence is less than the critical angle each incident longitudinal wave * gives rise to: (1) two longitudinal waves, one reflected, one refracted; and (2) one refracted transverse wave (the reflected transverse wave cannot propagate in the mud).

(b) If the angle of incidence is larger than the critical angle the incident longitudinal wave produces a single reflected longitudinal wave.

The incident or reflected longitudinal waves travelling in the mud are slower than the refracted compressional waves propagated in the formation, since the speed of sound in the ground is greater than that in mud.

Among the refracted longitudinal waves we are particularly interested in those waves refracted at the critical angle (Fig. 14-2), since they propagate along the borehole wall at a speed v_{L_2}.

Each point reached by this wave acts as a new source transmitting waves, so creating effectively cones of waves in the mud travelling at a speed v_{L_1}.

If we place two receivers, R_1 and R_2 at certain distances from the transmitter and along the axes of the tool and of the hole (supposed the same) they are reached by the sound at times T_{R_1} and T_{R_2}, respectively given by:

$$T_{R_1} = \frac{\overline{EA}}{v_{L_1}} + \frac{\overline{AB}}{v_{L_2}} + \frac{\overline{BR_1}}{v_{L_1}}$$

$$T_{R_2} = \frac{\overline{EA}}{v_{L_1}} + \frac{\overline{AB}}{v_{L_2}} + \frac{\overline{BC}}{v_{L_2}} + \frac{\overline{CR_2}}{v_{L_1}} \qquad (14\text{-}1)$$

We can either:

---

* There are no incident shear waves as these do not propagate in the mud.

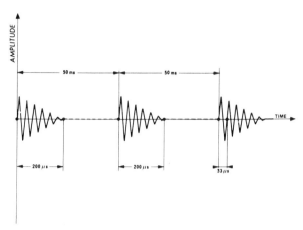

Fig. 14-1. Schematic representation of the signal emitted by the transducer.

(a) measure the time T_{R_1} (or T_{R_2}) taken for the sound wave to reach R_1 (or R_2). This is the method known as the single-receiver time. However, the total time has to be corrected for the time spent crossing and recrossing the mud. This gives:

$$\frac{\overline{AB}}{v_{L_2}} = T_{R_1} - \frac{\overline{EA}}{v_{L_1}} - \frac{\overline{BR_1}}{v_{L_1}} \qquad (14\text{-}3)$$

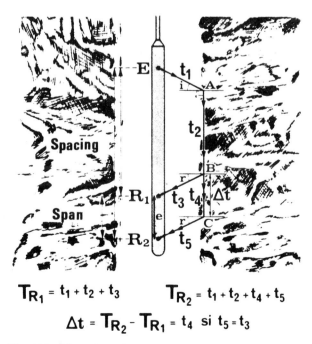

$$T_{R_1} = t_1 + t_2 + t_3 \qquad T_{R_2} = t_1 + t_2 + t_4 + t_5$$

$$\Delta t = T_{R_2} - T_{R_1} = t_4 \text{ si } t_5 = t_3$$

Fig. 14-2. Schematic of the principle for measuring the interval transit time (tool with two receivers).

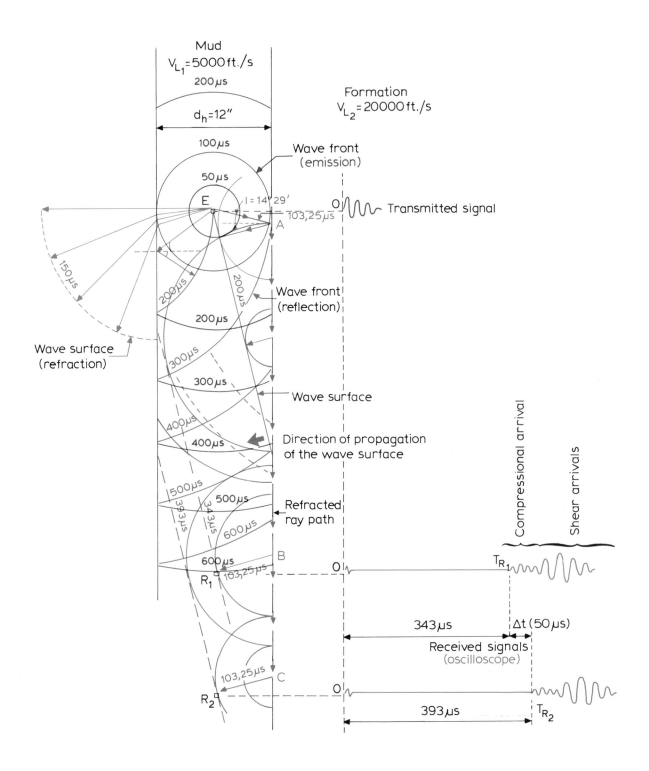

Fig. 14-3. Propagation of acoustic waves in a well. Principle for measuring the interval transit time (tool with two receivers).

Knowing $\overline{AB}$ (which is not equal to $\overline{ER_2}$) we can deduce v_{L_2}. However, we need to know the hole diameter, possibly from a caliper log.

or

(b) measure the time Δt that elapsed between the wave arrival at R_1 and R_2. This is the two-receiver method. The time Δt known as the transit time is directly proportional to the speed of sound in the formation and the distance between the two receivers R_1 and R_2.

If the tool is at the centre of the hole and the hole is of uniform diameter we have, in effect:

$$\Delta t = T_{R_2} - T_{R_1} = \frac{\overline{BC}}{v_{L_2}} \qquad (14\text{-}4)$$

as $\overline{BR_1}/v_{L_1} = \overline{CR_2}/v_{L_1}$ and $\overline{BC} = \overline{R_1R_2}$

N.B. If the distance $\overline{R_1R_2}$ is one foot the measurement gives transit time for one foot. From this time we can derive velocity using:

$$\Delta t_{(\mu s/ft)} = 10^6/v_{(ft/s)} \qquad (14\text{-}5)$$

The measurement of the first wave arrival relates only to those waves refracted at the critical angle as these are the fastest. In fact: (a) other longitudinal waves refracted into the formation travel at the same speed as the first arrival but due to their path length generally arrive later (see the waves shown in green in Fig. 14-3); and (b) transverse waves refracted into the formation travel more slowly than the longitudinal wave and so give rise to later waves at the receiver. However, as their energy is higher they are easily seen on the oscilloscope and can be detected (see further).

N.B. The compressional interval transit time, derived from the difference between the first compressional arrival times (found by threshold detection) at the two receivers, is reasonably accurate since the compressional arrival is easily detected because it arrives first and because at the spacings generally used it stands out against the background noise preceding it.

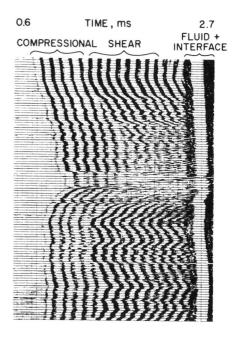

Fig. 14-5. Use of variable density log for identification of arrivals (from Aron et al., 1978).

In any case the arrival of the wave front corresponding to the refracted shear wave could be detected by raising the wave detection threshold to an amplitude level higher than that used to detect compressional arrivals.

Indeed, the shear arrivals generally have higher energy and so can be separated from the compressional. This allows us to measure the shear interval transit time (Fig. 14-4).

This measurement can be obtained from a variable density log (see further for explanation). On Fig. 14-5 one can recognize the compressional arrivals which form the first set of bands and which also show practically uniform variations in time delay with changing depth. The shear arrivals can be picked out

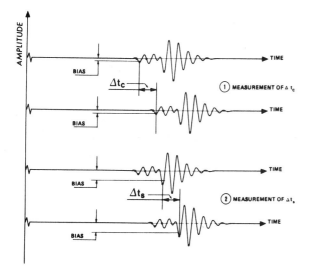

Fig. 14-4. Method for measuring the interval transit time of the shear wave.

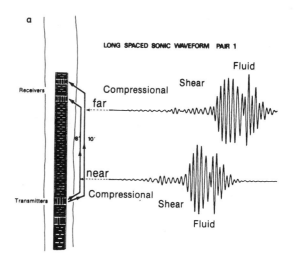

Fig. 14-6a. (see following page for legend).

222

as a later set of bands, which are generally of higher energy (higher amplitude and therefore a darker trace. See paragraph 5, chapter 15). They again show variations in time delay with depth, but they are different in shape from those of the compressional bands, due to their lower velocity (the angle is higher). The fluid arrivals can be identified in a similar manner. Also handy for identification, the use of the variable den-

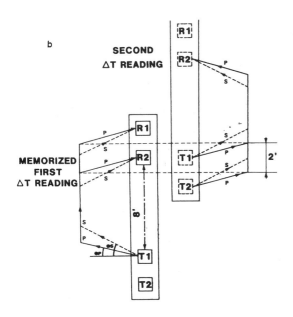

C and S indicate onset of compressional and shear
F indicates arrival time for DT= 189 μsec / ft.

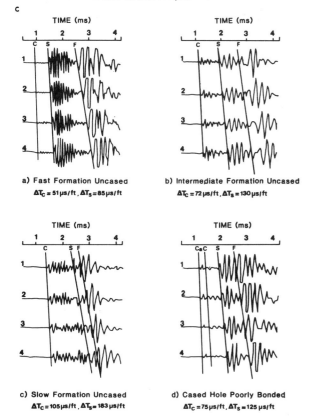

a) Fast Formation Uncased
ΔT_C=51μs/ft, ΔT_S=85μs/ft

b) Intermediate Formation Uncased
ΔT_C=72μs/ft, ΔT_S=130μs/ft

c) Slow Formation Uncased
ΔT_C=105μs/ft, ΔT_S=183μs/ft

d) Cased Hole Poorly Bonded
ΔT_C=75μs/ft, ΔT_S=125μs/ft

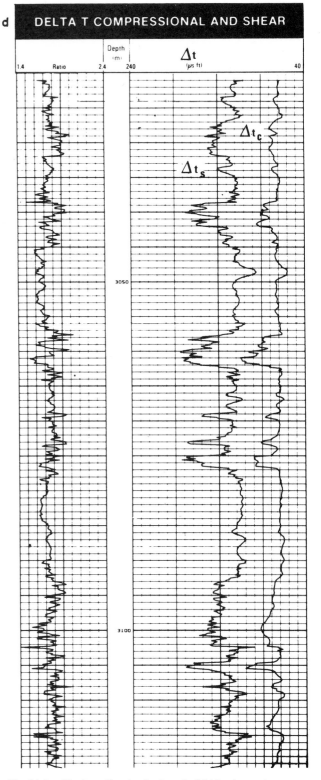

Fig. 14-6. a. The Long Spacing Sonic tool of Schlumberger (courtesy of Schlumberger). b. Principle of measurement. c. Feature of the complete signal as recorded by the Long Spacing Sonic tool (courtesy of Schlumberger). d. Example of Δ*t* compressional and shear (courtesy of Schlumberger).

sity log for shear interval transit time estimation is not reliable because of the lack of time resolution in the display.

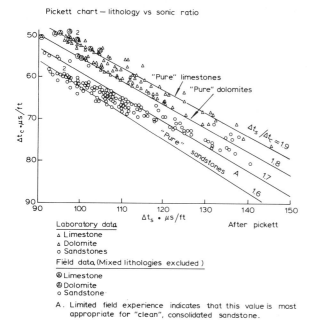

Pickett chart — lithology vs sonic ratio

"Pure" limestones
"Pure" dolomites
$\Delta t_s / \Delta t_c = 1.9$
"Pure" sandstones A
1.8
1.7
1.6

Laboratory data
△ Limestone
▲ Dolomite
○ Sandstones

Field data (Mixed lithologies excluded)

Ⓐ Limestone
Ⓐ Dolomite
○ Sandstone

A. Limited field experience indicates that this value is most appropriate for "clean", consolidated sandstone.

Fig. 14-7. Definition of the lithology from the v_L/v_T ratio. (from Pickett, 1963).

The shear interval transit time can automatically be obtained by using the Schlumberger Long Spacing Sonic tool (Fig. 14-6). The advantages of this new tool are the following:

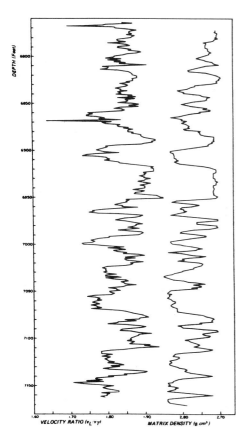

Fig. 14-8. Example of v_L/v_T ratio log and comparison with the matrix density measured from cores (from Pickett, 1963).

(a) It digitally records the entire received waveform (not the first arrival) from which compressional, shear and fluid arrivals can be separated and studied.

(b) The spacing can be varied; an increase of the spacing between transmitter and receivers allows for adequate time separation between the various arrivals, a good signal-to-noise ratio and a minimum signal distortion.

(c) The frequency used is lower (= 11 kHz instead of 20 kHz).
This allows a lower attenuation of the signal.

The interest of the shear interval transit time measurement has been shown by Pickett (1963). He has demonstrated that the ratio of the compressional to the shear velocities (v_L/v_T) can be used as a lithology indicator (Figs. 14-7 and 14-8). Further investigations seem to confirm this but to indicate also the textural influence on the measurement.

14.2. BOREHOLE COMPENSATED SONIC

If there are caves in the borehole wall or, where the tool axis is inclined to the hole axis, for some reason, the transit time is in error since the mud travel time is not the same for both receivers.

A way to counteract this is to use a tool that has two transmitters and four receivers arranged in pairs, two to each transmitter. Figures 14-9 and 14-10 show how the cave or tool inclination effect is inverted for the second transmitter-receiver pair. The average of two measurements, one for each receiver, should eliminate the effect. The tool first transmits from E_1 using receivers R_1 and R'_1 and then from E_2 using R_2 and R'_2. The average is taken of the two measurements. It is this average that is recorded (Fig. 14-11).

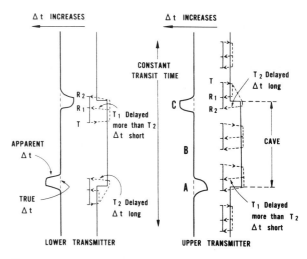

Fig. 14-9. Influence of a cave on the Δt measurement. a. Transmitter above receivers. b. Transmitter below receivers (from Kokesh et al., 1969).

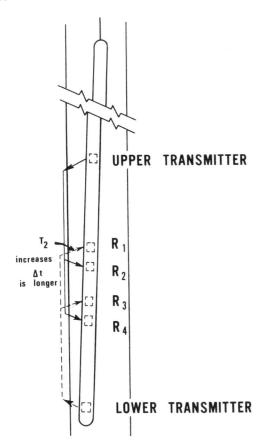

UPPER TRANSMITTER

T_2
increases
Δt
is longer

R_1
R_2
R_3
R_4

LOWER TRANSMITTER

Fig. 14-10. Influence of the tool inclination on the Δt measurement (from Kokesh et al., 1964).

14.3. MEASURE POINT

In the BHC configuration it corresponds to the middle of the interval between the two outside receivers.

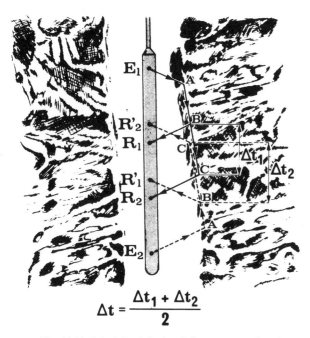

E_1
R'_2
R_1
R'_1
R_2
E_2

A
B
C
C
B
A
Δt_1
Δt_2

$$\Delta t = \frac{\Delta t_1 + \Delta t_2}{2}$$

Fig. 14-11. Principle of the borehole-compensated sonic.

14.4. DEPTH OF INVESTIGATION

As we have just seen, in the BHC configuration, the first arrival is detected, that is the fastest and that is linked with the longitudinal waves refracted at the critical angle.

Intuitively, in these conditions, the depth of investigation should be of the order of a few centimeters. However, we have to take into account the wavelength λ. Laboratory experiments show that a thickness of at least 3λ is needed to propagate a pressure wave through several feet of formation *.

Knowing the frequency f and that v ranges between 5000 and 25,000 ft/s, λ varies between 8 and 40 cm (for f equal to 20 kHz) and from 4 to 20 cm (for f equal to 40 kHz).

Hence, the depth of investigation varies between 12 cm and 100 cm. It should be a function of the formation velocity. In fact, when the invaded zone is deep and full of a fluid whose sonic velocity is less than the fluid in the virgin zone, a short spacing will give the speed of sound in the invaded zone and a long spacing that in the virgin formation.

This is also the case in formations micro-fractured by drilling (in some shales), hence the interest of long spacing sonic tools in massive shales.

Remark. One must take into account that an increase of the spacing implies a decrease of the signal amplitude received at the detector. If the attenuation is too strong the amplitude of the first arrival can be too low and consequently the first arrival not detected giving a wrong interval transit time measurement. So the long spacing will improve the Δt measurement only if the attenuation of the signal is not too important.

By contrast, if the invaded zone is of a depth of 50 cm and has a speed higher than the virgin formation (for example gas reservoirs) it seems that Δt will always be that of the invaded zone.

14.5. VERTICAL RESOLUTION

This is about equal to the distance between the receiver pairs, generally 2 feet, but sometimes 1, 3 or 6 feet.

---

* There is no contradiction between this and the detection of waves transmitted in casing, where the casing thickness is small. In the case of tubing, the surface wave is detected where its speed is given by:

$$v_S = \left(\frac{E}{\varrho(1 - \sigma^2)} \right)^{1/2} \tag{14-6}$$

14.6. UNITS OF MEASUREMENT

The petroleum industry uses microseconds per foot as units for transit time. It is related to the velocity in ft/s by eq. 14-5. To convert $\mu s/ft$ to $\mu s/m$, multiply by 3.28084; to convert ft/s to m/s divide by 3.28084.

14.7. FACTORS INFLUENCING THE MEASUREMENT

14.7.1. The matrix

The speed of sound in the formation depends on the kind of minerals making up the rock. The effect of the minerals is determined by their densities and their parameters of elasticity.

These parameters are not always well known. However, the transit time has been measured for a few of the common minerals (Table 14-1).

In the case of complex lithologies the individual mineral effect is determined by their volume fraction and their individual speed of sound. The way the minerals are distributed is also important: (a) laminations: in this case the rock transit time is given by:

$$\Delta t_{ma} = V_1 \Delta t_{ma_1} + V_2 \Delta t_{ma_2} + \ldots + V_n \Delta t_{ma_n} \quad (14-7)$$

(b) dispersed: in this case we need to bring in the concept of a continuous phase (cf. 14.7.4.).

TABLE 14-1

Interval transit time and speed of compressional wave of the most common minerals and rocks

	$\Delta t\,(\mu s/ft)$		$v_L\,(ft/s \cdot 10^3)$		Bulk modulus k
	Mean value	Extreme values	Mean value	Extreme values	
Hematite		42.9		23.295	
Dolomite	44.0	(40.0– 45.0)	22.797	(22.222–25.000)	82
Calcite	46.5	(45.5– 47.5)	21.505	(21.053–22.000)	67
Aluminium		48.7		20.539	
Anhydrite		50.0		20.000	54
Granite	50.8	(46.8– 53.5)	19.685	(18.691–21.367)	
Steel		50.8		19.686	
Tight limestone	52.0	(47.7– 53.0)	19.231	(18.750–21.000)	
Langbeinite		52.0		19.231	
Iron		51.1		19.199	
Gypsum	53.0	(52.5– 53.0)	19.047	(18.868–19.047)	40
Quarzite	55.0	(52.5– 57.5)	18.182	(17.390–19.030)	
Quartz	55.1	(54.7– 55.5)	18.149	(18.000–18.275)	38
Sandstone	57.0	53.8–100.0)	17.544	(10.000–19.500)	
Casing (steel)		57.1		17.500	
Basalt		57.5		17.391	
Polyhalite		57.5		17.391	
Shale		(60.0–170.0)		(5.882–16.667)	
Aluminium tubing		60.9		16.400	
Trona		65.0		15.400	
Halite		66.7		15.000	23
Sylvite		74.0		13.500	
Carnalite		83.3		12.000	
Concrete	95.0	(83.3– 95.1)	10.526	(10.526–12.000)	
Anthracite	105.0	(90.0–120.0)	9.524	(8.333–11.111)	
Bituminous coal	120.0	(100.0–140.0)	8.333	(5.906–10.000)	
Sulphur		122.0		8.200	
Lignite	160.0	(140.0–180.0)	6.250	(3.281–7.143)	
Lead		141.1		7.087	
Water, 200.000 ppm NaCl, 15 psi		180.5		5.540	
Water, 150.000 ppm NaCl, 15 psi		186.0		5.375	
Water, 100.000 ppm NaCl, 15 psi, 25°C		192.3		5.200	2.752
Water, pure (25°C)		207.0		4.830	2.239
Ice		87.1		11.480	
Neoprene		190.5		5.248	
Kerosene, 15 psi		214.5		4.659	
Oil		238.0		4.200	
Methane, 15 psi		626.0		1.600	
Air, 15 psi		910.00		1.100	

226

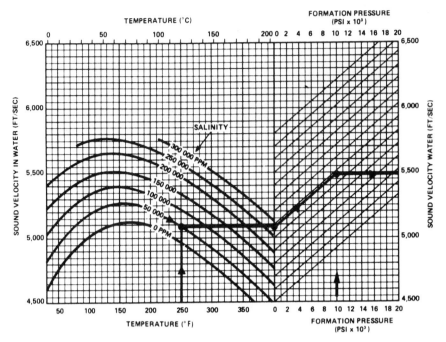

Fig. 14-12. Speed of sound in water as a function of its salinity, temperature and pressure.

14.7.2. Porosity and fluids

The speed of sound also depends on the porosity and the pore space fluids:

(a) Everything else being equal the higher the porosity the lower the speed.

(b) Generally if for a constant porosity and matrix

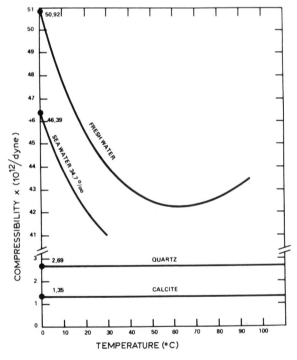

Fig. 14-13. Variation of the compressibility with the temperature (from Shumway, 1958).

we replace water by oil or oil by gas the speed goes down, at least down to a certain depth.

(c) The speed of sound in water depends on the salinity: the higher the salinity the higher the speed. The chart in Fig. 14-12 gives the speed of sound in water at various salinities and also as a function of pressure and temperature. The variation in velocity seems to depend above all on the variation in compressibility, as Fig. 14-13 demonstrates.

N.B. For a water of constant salinity above 150°F the change in velocity due to an increase in temperature is compensated by the pressure effect. The speed goes up as the salinity increases. This is really in some way the effect of density changes, as the increase in salinity implies a density increase.

14.7.3. Temperature and pressure

As we have seen these have an effect in saline water. This is equally true in gas or oil and also in the matrix itself, as experiments at the French Institute of Petroleum (I.F.P.) seem to show (see Figs. 14-14 and 14-15). Examination of Fig. 14-15 shows that the speed tends towards a limit, known as the terminal velocity, as the pressure increases.

On the other hand, at constant external pressure, the speed of sound is a monotonic increasing function of the pressure difference ΔP (Fig. 14-16) between the internal and external pressures.

N.B. From the various experiments we can come to no conclusion on the influence of pressure and temperature on the speed of sound in the matrix as the variations in speed observed may be explained by the influences of these factors on the fluid alone. In any case it seems that the matrix speed is affected by pressure and temperature if we refer to the modifications necessary to Δt_{ma} for

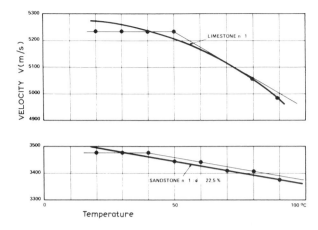

Fig. 14-14. Speed of sound of rocks as a function of temperature (courtesy of Institut Française du Petrole).

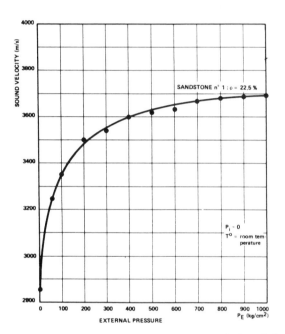

Fig. 14-15. Speed of sound of a sandstone as a function of the external pressure (courtesy of Institut Française du Petrole).

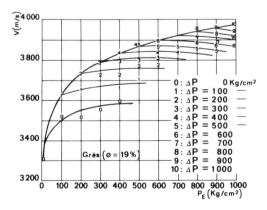

Fig. 14-16. Influence of ΔP on the speed of sound in a sandstone (courtesy of Institute Française du Petrole).

a sand, limestone and dolomite to get sonic porosity values compatible with core data and minor variations in compressibility.

14.7.4. Texture

The way in which the grains and the porosity of a formation are arranged both have an influence on the speed of sound.

Sarmiento (1961) has shown that the type, size and distribution of pores (intergranular, vugs and fractures) all have an effect on the speed: "Below a certain size the pores are probably included in the elastic character of rocks, but for large pores, or vugs, it is probable that the sound vibration follows the shortest path in the matrix, thus around the pores rather than across them. The critical size of the pores is directly related to the wavelength".

We can also see that for the same porosity the speed will depend on the kind of intergranular contact. These may be of a point (the case of an arrangement of spheres), line (spheroids) or surface type (polyhedra: cubes, dodecahedra, or flakes).

This leads to the idea of anisotropy in the speed of sound, that is, it is not the same if it is measured parallel to or perpendicular to the grains. From this we see the influence of bed dip on measured speed and it also brings us to the concept of a continuous phase.

(a) In formations with low porosity, which means with pores more or less isolated and randomly distributed, the matrix constitutes the continuous phase and it seems logical that the first arrival wave, and so the fastest, travels in this phase and avoids pores. Consequently, until porosity reaches a certain value (5–10%) the transit time does not vary significantly from the Δt_{ma}. This is why the sonic log is considered as not "seeing" secondary porosity of a vug type.

(b) Conversely, if a grain is in suspension in the fluid, as in the case of low compaction shale series and surface sands with high porosity (higher than 48–50%), the continuous phase is the fluid, and what is measured is the interval transit time of the sound in the fluid. This was confirmed by measurements made on shallow shales subjected to the permafrost.

In this case it is the transit time of sound in ice that is measured. This means that transit time in fluid is reached as soon as the porosity is higher than 50%. Consequently in this case it is impossible to measure the interval transit time in the formation, the first arrival being this travelling in the mud. This analysis seems confirmed by investigations made by Raymer et al. 1980 (see paragraph 14-8).

Besides, the existence of microfissures, either natural or caused by drilling, will equally reduce the speed of sound by the production of microporosity in the form of planes which the waves have to cross. This is why in some kinds of formation, usually shales and

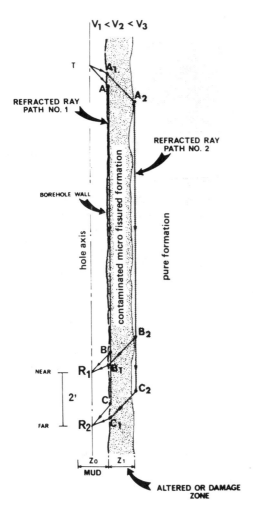

Fig. 14-17. Two refracted paths for a step profile of alteration. The use of a long-spacing sonic allows us to measure the interval transit time of the undisturbed formation.

carbonates but rarely sands, there is a discrepancy between seismic and sonic log velocities. It is recommended to measure interval transit times by using long-spacing sonic tools that avoid the area of cracked, altered formation (Fig. 14-17).

14.8. INTERPRETATION

From the factors influencing the measurement, as in the previous sections, we can see that the sonic log can be used as an indicator of lithology and of porosity, both intergranular and intercrystalline.

For any given lithology, with the zone of investigation of the tool mainly in the invaded zone containing mud filtrate, the speed of sound (or the interval transit time Δt) is a function of porosity.

In fact, for rocks that are sufficiently compacted we can, to a first approximation, accept that the variation of the speed of sound with the depth (temperature and pressure) of the fluid and matrix are

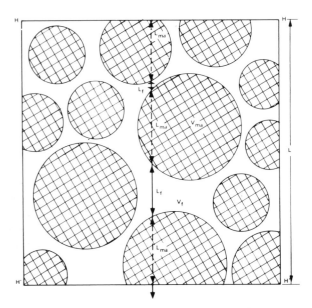

Fig. 14-18. Trajectory of the compressional wave in a water-saturated sand (from Wyllie et al., 1956).

negligeable and that a terminal velocity is reached.

Wyllie et al. (1956) have proposed an empirical equation based on numerous laboratory experiments on clean formations. This links interval transit time directly with porosity by taking the total interval transit time as equal to the sum of the interval transit times in the grains of the matrix and in the pores (Fig. 14-18):

$$\Delta t = \frac{t}{L} = \frac{\Sigma(L_f/L)}{v_f} + \frac{\Sigma(L_{ma}/L)}{v_{ma}} \qquad (14\text{-}8)$$

which can be written as:

$$\Delta t = \phi \Delta t_f + (1 - \phi)\Delta t_{ma} \qquad (14\text{-}9)$$

where we assume a relationship between L_f and ϕ.

From eq. 14-9 it follows that:

$$\phi_s = \frac{\Delta t - \Delta t_{ma}}{\Delta t_f - \Delta t_{ma}} \qquad (14\text{-}10)$$

N.B. The Wyllie equation establishes a linear relationship between Δt and ϕ which is not really in agreement with previous remarks on rock texture (cf. 14.7.4.). In any case it is approximately correct in

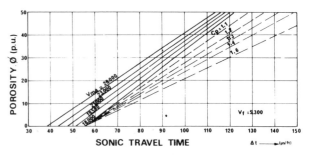

Fig. 14-19. Relationship between interval transit time, Δt, and porosity, ϕ, from the Wyllie's eq. (14-10).

the range of usual porosities encountered, that is from 5 to 25%, and in the case of an arrangement of almost spherical grains. There remains the difficulty of chosing Δt_{ma} and Δt_f for the matrix and the fluid.

Equation 14-10 is represented by the chart in Fig. 14-19.

In uncompacted formations the Wyllie equation gives porosities that are too high. It is therefore not directly applicable. A correction factor is needed, to take into account the effects of temperature and pressure, or in other words that the terminal speed of sound is not reached. Equation 14-10 is then written:

$$(\phi_s)_c = \frac{\Delta t - \Delta t_{ma}}{\Delta t_f - \Delta t_{ma}} \cdot \frac{100}{c \cdot \Delta t_{sh}} \qquad (14\text{-}11)$$

and is given by the chart in Fig. 14-19. The best way to compute $c\Delta t_{sh}$ is to compare computed sonic porosities with the true porosity from another source.

Where this is possible several approaches can be tried:

(a) $\rho_b - \Delta t$ cross-plot method (Fig. 14-20). ρ_b and Δt are plotted on linear grids for water-bearing clean formations close to the zone of interest. From this a clean formation line is established that can be scaled in porosity units using the density log. Similarly, a theoretical porosity line using eq. 14-10 can be drawn. For any value of porosity a corresponding value of Δt can be found. Using the actual value of Δt and this new value in the chart of Fig. 14-19 $c\Delta t_{sh}$ can be determined.

(b) The neutron method (SNP or CNL). The porosity is obtained from the neutron for water-bearing sands. This value should be about equal to the actual porosity. Hence we have:

$$c\Delta t_{sh} = 100\phi_S/\phi_N \qquad (14\text{-}12)$$

N.B. We can also plot the values of ϕ and Δt (Fig. 14-21) to make a statistical evaluation of $c\Delta t_{sh}$. This can also give the value of Δt_{ma}.

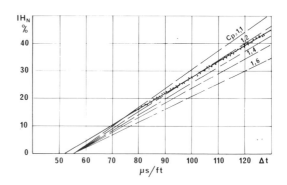

Fig. 14-21. ϕ_N vs Δt cross-plot to determine $c\Delta t_{sh}$.

(c) The R_0 method. In clean-water-bearing sands we can estimate ϕ from R_{IL}, knowing R_w:

$$F_R = \frac{R_{IL}}{R_w} = \frac{a}{\phi_R^m} \qquad (14\text{-}13)$$

then giving $c\Delta t_{sh}$ using ϕ_R:

$$c\Delta t_{sh} = \frac{100\phi_S}{\phi_R} \qquad (14\text{-}14)$$

From studies, Geerstma (1961) proposed the following equation:

$$v_L = \left[\left(M + \frac{(1-\beta)^2}{(1-\phi-\beta)c_{ma} + \phi c_f} \right) \frac{1}{\rho_b} \right]^2 \qquad (14\text{-}15)$$

where:

M = elastic modulus (or space modulus)

$$M = k + \frac{4}{3}\mu = \frac{3}{c_b}\frac{(1-\sigma)}{(1+\sigma)} \qquad (14\text{-}16)$$

$$M = \frac{\beta}{c_{ma}} + \frac{4}{3}\mu \qquad (14\text{-}17)$$

σ = Poisson's coefficient (ratio);
β = c_{ma}/c_b;
ϕ = porosity;
c_{ma} = the compressibility of the matrix ($c_{quartz} = 25 \times 10^{-13}$ baryes^{-1});
c_b = compressibility of the empty matrix;
c_f = compressibility of the fluids

$$c_f = c_w S_w + (1 - S_w)c_h \qquad (14\text{-}18)$$

c_h = compressibility of hydrocarbons;
c_w = compressibility of water ($c_w = 42 \times 10^{-12}$ baryes^{-1});
ρ_b = bulk density

$$\rho_b = \phi\rho_w S_w + \phi\rho_h(1 - S_w)$$
$$+ (1 - \phi)\rho_{ma} \qquad (14\text{-}19)$$

S_w = water saturation;
ρ_w = water density;
ρ_h = hydrocarbon density;
ρ_{ma} = matrix density.

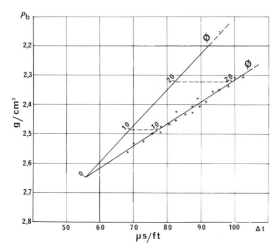

Fig. 14-20. ρ_b vs Δt cross-plot to determine $c\Delta t_{sh}$.

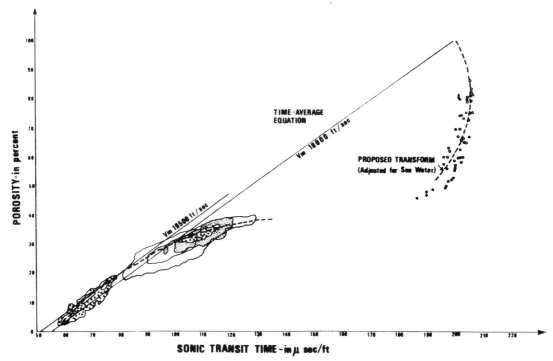

Fig. 14-22. Comparison of sonic transit time to core porosity from published data (from Raymer et al., 1980).

Equation 14-15 which corresponds to an infinite medium of fluid and matrix, may seem complicated. It has the advantage of including all the different factors influencing the speed of sound according to classical theories. Besides, when most of the parameters are known it should be possible to get porosity knowing the saturation, or vice versa.

More recently, Raymer et al. (1980), proposed another transit time-to-porosity transform. This seems more in agreement with observations made. It is illustrated by Figs. 14-22 and 14-23 and provides superior transit-time-porosity correlation over the entire porosity range. It suggests more consistent matrix velocities for given rock lithology and permits the

determination of porosity in unconsolidated low velocity sands without the need to determine a "lack of composition", or similar correction factor.

14.9. ENVIRONMENTAL AND OTHER EFFECTS

14.9.1. Transit time stretching

As the sound arriving at the second receiver has a longer path, the signal is generally weaker. As the

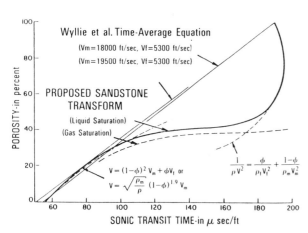

Fig. 14-23. The proposed sonic transit time to porosity transform, showing comparison to Wyllie time average equation and to suggested algorithms (from Raymer et al., 1980).

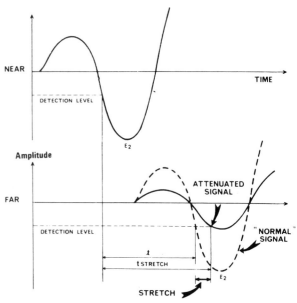

Fig. 14-24. Schematic explaining transit time stretching.

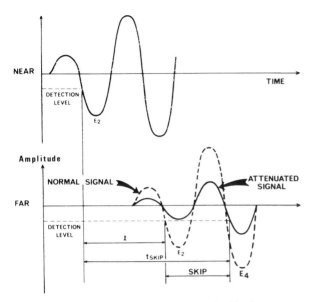

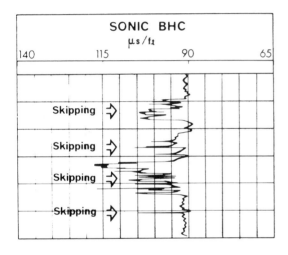

Fig. 14-25. Schematic explaining cycle skipping.

Fig. 14-26. Example of cycle skipping.

detector threshold is the same for both receivers, the detection may occur later on the further receiver. This gives a Δt that is too large (Fig. 14-24).

14.9.2. Cycle skipping

In some cases the signal arriving at the second receiver is too low to trigger the detection on the first arrival. The detection then occurs at the second or third arriving cycle (Fig. 14-25). We therefore have missed or skipped cycles. This shows up as sudden and abrupt increases in the interval transit time (Fig. 14-26).

If cycle skipping appears on only one of the far detectors, the increases in Δt is between 10 and 12.5 μs/ft for the second cycle and 20 to 25 μs/ft for the third. If cycle skipping occurs on both far receivers the error on Δt is between 20 and 25 μs/ft for one cycle missed and 30 to 37.5 μs/ft for two.

This sudden jump in Δt is often linked to the presence of gas and sometimes oil. It can also happen in fractured zones. This is due to a strong attenuation of the signal.

14.9.3. Kicks to smaller Δt

This happens when the signal to the first receiver is weaker than that arriving at the second or, where Δt is suddenly diminished by the detection jumping forward from the usual sound arrival to detect on noise appearing before the actual sonic signal (Fig. 14-27).

14.9.4. The borehole

Hole size. This comes into effect only when the sum of the transit times from emitter to borehole wall, and the wall to the receiver is greater than the distance from the transmitter to the receiver directly. In this case the first arrival is straight through the mud. This only happens in holes larger than about 24 inches for common rocks (Fig. 14-28). To eliminate

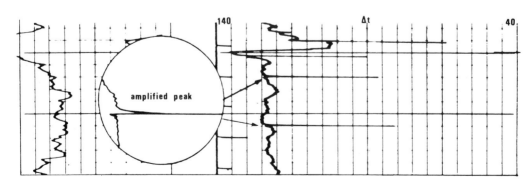

Fig. 14-27. Example of kicks of smaller Δt.

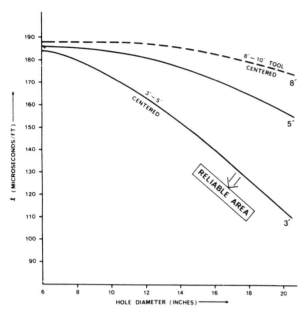

Fig. 14-28. Effect of large hole size. For a given centered sonic tool, the maximum Δ*t* detectable is read at a given hole diameter from the curve identifying the transmitter–near receiver spacing (from Goetz et al., 1979).

this effect the sonic tool is run excentralized.

For the borehole, compensated sonic (BHC) caving has little effect except where the caving is very large.

The drilling mud. If the borehole is air-filled or if the mud is gas-cut, the attenuation of the sonic signal is too high to allow detection on the first arrival. This problem may arise in front of zones that are producing gas into the mud (Fig. 14-26).

14.9.5. Invasion

There is little invasion effect in water-bearing zones. However, in gas or oil zones, even with high water saturations, the interval transit time in the invaded zone may be very different to that in the virgin zone. We can look at several cases.

(a) Deep invasion: the sonic reads only the flushed

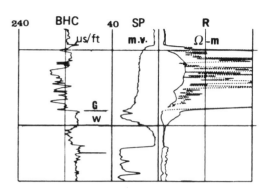

Fig. 14-30. Example showing the sonic reading in virgin zone (no invasion). Observe the attenuation of the SP deflection in front of the gas bearing zone.

zone and does not see the gas or oil at all (Fig. 14-29). There is no problem in calculating porosity in this case. The reading may need correction using the Geerstma equation before comparison or use in seismic work.

(b) Little or no invasion: the gas or light oil affects the measurement. If the interval transit time is less in the hydrocarbon formation than through the mud, the reading is representative of the virgin formation (Fig. 14-30).

However this transit time needs careful interpretation in terms of porosity. If the formation transit time is longer than that in the mud (for example very porous, shallow depths) then of course the sonic just gives the mud transit time (Fig. 14-31). The virgin formation transit time cannot be less. To determine it, we have to use the Geerstma equation but no porosity determination is possible.

(c) No invasion and slight production: in this case bubbling occurs in the mud and cycle skipping occurs due to the strong attennuation (Fig. 14-26).

It is possible sometimes to see very low sonic velocities or high transit times in formations that are apparently water bearing from other logs (Fig. 14-32). This is generally due to a very small percentage of gas in the form of micro-bubbles in the water, leading to

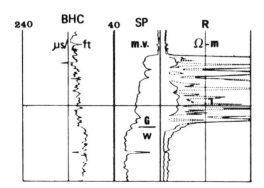

Fig. 14-29. Example showing the sonic reading in the flushed zone.

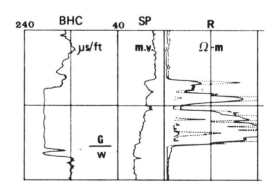

Fig. 14-31. Example showing the sonic reading the mud interval transit time.

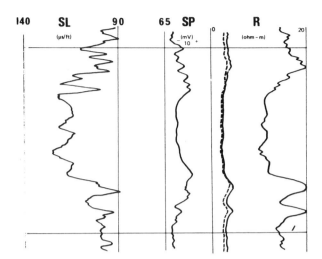

Fig. 14-32. Example of the effect of low gas saturation on the interval transit time. The resistivity log shows a water-bearing sand.

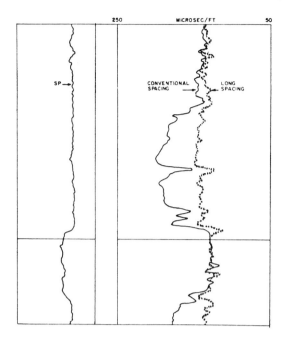

Fig. 14-34. Example showing the effect of radial cracking (alteration) on the short-spacing sonic measurement of the interval transit time in the non-altered formation.

a strong attenuation of the sonic wave. As shown by Domenico (1976), a gas saturation of 15% reduces drastically the velocity with respect to a water-saturated formation (Fig. 14-33).

14.9.6. Radial cracking effects

We have already seen that microfractures in the rock linked with radial cracking caused by drilling leads to an increase in the interval transit time.

Rocks liable to this phenomenon (shales, shaly limestone, etc.)—which are also liable to caving caused by particles dropping down from the fracturing or breaking—will show a sonic Δt too large, or a velocity that is too low. To obtain a measurement in the non-fractured formation a long spacing sonic is needed that reads beyond the damaged zone (Fig. 14-34).

14.10 TRAVEL TIME INTEGRATION

In order to use the sonic in seismic work a travel time integration is provided in the logging equipment. The integrated time is shown by a series of pips, usually to the left of the sonic track (Fig. 14-35). The small pips indicate an increase of integrated time of one millisecond, whereas the large pips are for ten milliseconds. The average travel time between two depths can , therefore, be found by simply counting the pips.

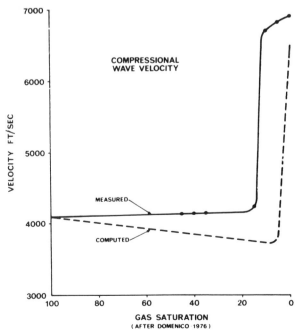

Fig. 14-33. Effect of gas saturation on velocity in a shallow sandstone (from Domenico, 1976).

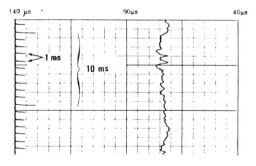

Fig. 14-35. Example of interval transit time measurement showing on the left the integration pips.

Of course the integration is valid so long as there are no cycle skips or jumps on the log. In another way it can be used to verify the calibration, for example in an homogeneous zone, by counting the number of peaks in a zone and comparing the product of the sonic (Δt) and the length of the interval (h):

$$\Delta t(\mu s/ft) \times h(ft) = t(\mu s)$$

14.11 SONIC LOG RESCALING

The sonic log is the basis for calibration of surface seismic data and in favourable cases for detailed seismic interpretation.

For reliable detailed studies, the adjustment of the sonic log is important. The various previous comments on the factors affecting the sonic measurement show that the sonic data must be carefully checked and, if necessary, the integrated sonic transit-time adjusted on seismic time derived from check shot times. The check shots are known as Well Velocity Survey.

The basic principle of this technique is to measure the time needed for a pressure pulse created at surface to reach a receiver anchored at a selected depth in the borehole (Fig. 14-36).

The equipment consists of: (a) a seismic source creating a pressure pulse (generally an air gun); (b) a receiver in the downhole tool, anchored at selected depths; and (c) surface equipment which records a surface signal from a detector near the source, and a downhole signal from the receiver. Both signals are recorded on a time base provided by a quartz clock.

The check shot time corresponds to the elapsed time between the arrivals of the surface and the downhole signals. This time must be corrected to a vertical time and referred to the Seismic Reference Datum plane, which is the time origin for surface seismic data.

This assumption is that check shot times can be considered as measured on vertical straight paths. This is basically true if altogether the wells are vertical, the offset of the source to the well is small, and the formation does not show significant dip. In that case the check shot time corresponds to the seismic time.

The difference between the seismic time and the Transit Time Integration of sonic TTI is known as the *drift*.

drift = seismic time − TTI

At each check shot depth, a drift can be computed, and the successive values plot versus depth make a drift plot (Fig. 14-37).

Between two levels the difference between the drift of the deeper and the shallower levels is the amount of time correction to be applied to the sonic time. This conducts to an adjustment by interval. This practice is not recommended because, as explained by Goetz et al. (1979), it can have an adverse effect. It is better to use the drift plot to draw drift curves.

The procedure consists of selecting zones in which the character of the sonic log is about constant. In each zone, drift points are fitted by segment of a

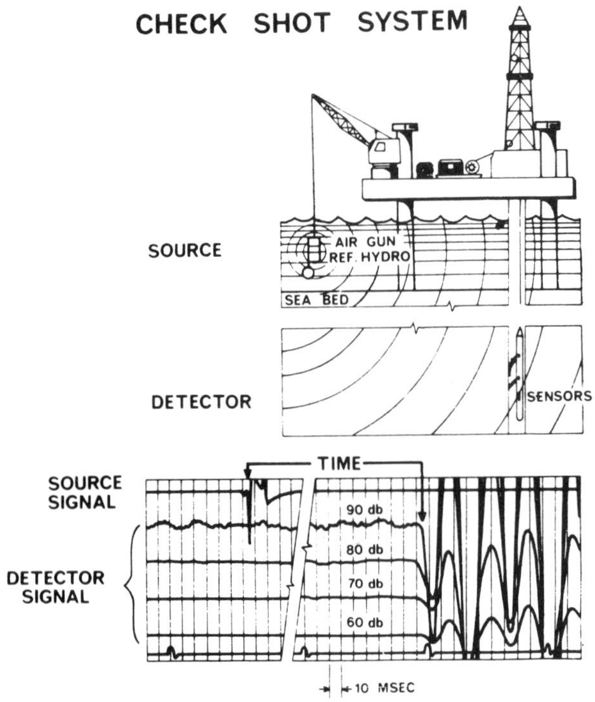

Fig. 14-36. Check shot system (from Goetz et al., 1979).

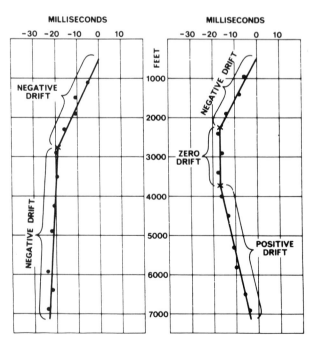

Fig. 14-37. Example of drift plot.

straight line. From one zone to the next, these segments are joined at knees, that form the common boundaries between zones. The slope of these segments of straight line joining two consecutive knees is the gradient of drift. This gradient is the average correction to be applied to the sonic transit times between the two knees.

When the slope is negative we say that we have a "negative" drift. It means that the sonic time is longer than the seismic times between the knees considered.

When the slope is zero, in spite of whether the plotted points fall within the negative or positive region of the drift plot, the sonic time and the seismic time are equal.

When the slope is positive, we say that we have a "positive" drift. In that case the sonic time is shorter than the seismic time.

Very often we observe a negative drift. This can be easily explained and related to one of the phenomena previously analysed: stretch, cycle skipping, formation alteration, large hole conditions, noise...; but positive drifts also occur as explained by Goetz et al., (1979). They can be related to one of the following cases:

(a) Noise, which can trigger the far receiver before the real signal. It should appear only as spikes on the log (Fig. 14-27).

(b) Negative stretch and cycle skipping when the signal to the near receiver is weaker than that arriving at the far one.

(c) Formation with a longer transit-time than mud. This occurs in formations such as gas-bearing shales or in shallow gas reservoirs, or in some formations close to the surface.

(d) Velocity invasion near the borehole which occurs when the invaded or damaged zone close to the borehole wall is faster than the undisturbed formation and both are faster than the mud. A characteristic example is given by gas-bearing reservoirs invaded by mud-filtrate.

(e) High dips relative to the wellbore. In this case the sonic signal may travel along refracted paths more than along the borehole, leading to a shorter transit-time.

(f) Frequency-dependent velocities. Acoustic velocities are dependent on the frequency of the signal. So, sonic velocities must be faster than check shot velocities. The frequency used in sonic tool is equal to 20 kHz compared to the roughly 50 Hz of the air gun or vibrosis.

This is related to wave dispersion in rocks (see studies made by Futterman (1962), Wuenschel (1965), Kalinin (1967), Strick (1971), White (1975), Anstey (1977) and a review of these studies made by Peyret and Mons (1980).

14.12. APPLICATIONS

Sonic transit-time is mainly measured to determine the porosity in a reservoir. We have to recognize that, partly through the difficulty of interpretation and factors affecting the measurement and partly because of the introduction of new nuclear devices for porosity measurement, this is no longer as important as before. It is used though both as a safeguard in porosity determination, especially as the measurement is not very sensitive to borehole size, and to compute secondary porosity in carbonate reservoirs.

The sonic is also an aid in lithology determination, especially with the hydrogen index (neutron log) or density (FDC). The M and N plot or Mid-plot methods can be used for this purpose. As the measurement is practically not affected by hole size variations, it can be used to study compaction in sand shale sequences or through the computation of the shear to compressional transit-time ratio.

Since the transit-time is directly linked to the speed of sound in the formation it can be used in combination with the density to establish an acoustic impedance log ($r = v\rho$) and to calculate a reflection

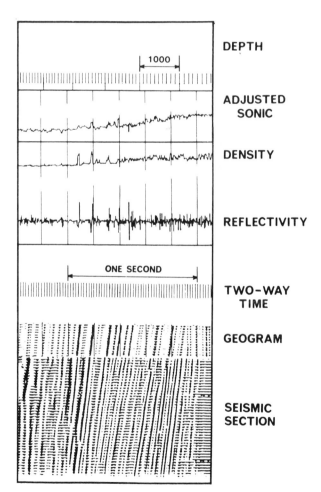

Fig. 14-38. Example of a Geogram from a reflectivity log (courtesy of Schlumberger).

236

coefficient;

$$R_{1-2} = \frac{v_2 \rho_2 - v_1 \rho_1}{v_2 \rho_2 + v_1 \rho_1}$$

which leads to the realization of an impulse log and of a synthetic curve which help in the interpretation of vertical seismic profiles in terms of seismofacies, porosity fluid determination (Fig. 14-38).

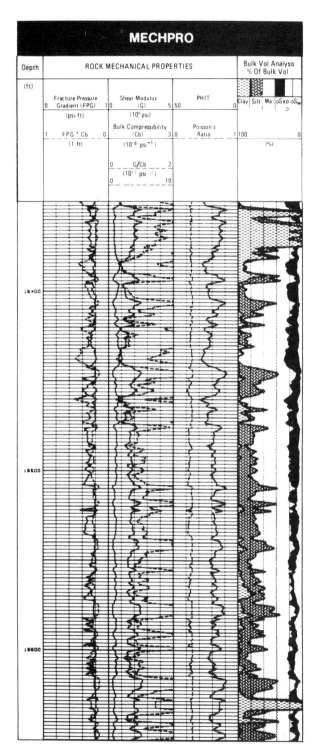

Fig. 14-39. Example of mechanical properties log (courtesy of Schlumberger).

The formation fluid can be identified by comparing sonic values with other logs (neutron, density, resistivity).

As with other logs, interval-transit-time is used in correlations and in sedimentology studies (electrofacies definition).

With a relatively good vertical resolution (about 60 cm) the sonic log can be used to determine bed thickness.

The knowledge of the compressional and shear velocities allows us to determine the mechanical properties of the rocks (Fig. 14-39).

14.13. DETERMINATION OF ELASTICITY PARAMETERS USING LOGS

If we can record, systematically, the density and the speeds of both the longitudinal and the transverse waves, the elastic moduli of the rock can be determined using the following equations. Youngs modulus, E:

$$E = \frac{9k \rho v_T^2}{3k + \rho v_T^2} = \left[\frac{\rho}{\Delta t_T^2} \right] \left[\frac{3\Delta t_T^2 - 4\Delta t_L^2}{\Delta t_T^2 - \Delta t_L^2} \right]$$
$$\times 1.34 \times 10^{10} \text{ psi}$$

Bulk modulus, k:

$$k = \rho \left[v_L^2 - \frac{4}{3} v_T^2 \right] = \rho \left[\frac{3\Delta t_T^2 - 4 \Delta t_L^2}{3 \Delta t_T^2 - \Delta t_L^2} \right]$$
$$\times 1.34 \times 10^{10} \text{ psi}$$

Shear modulus, μ:

$$\mu = \rho v_T^2 = \left[\frac{\rho}{\Delta t_T^2} \right] \times 1.34 \times 10^{10} \text{ psi}$$

Poisson's ratio, σ:

$$\sigma = \frac{1}{2} \frac{\left[\dfrac{v_L^2}{v_T^2} \right] - 2}{\left[\dfrac{v_L^2}{v_T^2} \right] - 1} = \frac{1}{2} \left[\frac{\Delta t_T^2 - 2 \Delta t_L^2}{\Delta t_T^2 - \Delta t_L^2} \right]$$

14.14. REFERENCES

Anderson, W.L. and Riddle, G.A., 1964. Acoustic amplitude ratio logging. J. Pet. Technol., 16 (11).

Anstey, N.A., 1977. Seismic interpretation: the physical aspects. pp. 2.78A–288B.

Aron, J., Murray, J. and Seeman, B., 1978. Formation compressional and shear interval transit time logging by means of long spacings and digital techniques. SPE of AIME, paper No. SPE 7446.

Domenico, S.N., 1974. Effect of water saturation on seismic reflectivity of sand reservoirs encased in shale. Geophysics, 39 (6).

Domenico, S.N., 1976. Effect of brine gas mixture on velocity in an unconsolidated sand reservoir. Geophysics, 41 (5).

Dupal, L., Gartner, J. and Vivet, B., 1977. Seismic applications of well logs. 5th European Logging Symposium.

Futterman, W.I., 1962. Dispersive body waves: J. Geophys. Res., 67 (13), pp. 5269–5291.

Geerstma, K., 1961. Velocity log interpretation: the effect of rock-bulk compressibility. J. Soc. Pet. Eng., 1.

Goetz, J.F., Dupal, L. and Bowler, J., 1979. An investigation into discrepancies between sonic log and seismic check shot velocities.

Guyod, H. and Shane, L.E., 1969. Geophysical Well Logging, Vol. 1. Guyod, Houston.

Kalinin, A.V., 1967. Estimate of the phase velocity in absorbing media: Izv. Earth Physics (English translation) No. 4, pp. 249–251.

Kokesh, F.P., Schwartz, R.J., Wall, W.B. and Morris, R.L., 1960. A new approach to sonic logging and other acoustic measurements. Soc. Petrol. Eng., Paper No. SPE 991. J. Pet. Technol., 12 (3).

Kowalski, J., 1975. Formation strength parameters from well logs. SPWLA, 16th ann. Log. Symp. Trans., Paper N.

Ladefroux, J., 1961. Mesure en laboratoire de la vitesse du son dans les roches sédimentaires consolidées. Rev. Inst. Franç. Pétrol., 16 (4).

Morlier, P. and Sarda, J.P., 1971. Atténuation des ondes élastiques dans les roches poreuses saturées. Rev. Inst. Franç. Pétrol., 26 (9).

Morris, R.L., Grine, D.R. and Arkfeld, T.E., 1963. The use of compressional and shear acoustic amplitudes for location of fractures. Soc. Petrol. Eng., Paper No. SPE 723.

Nations, J.F., 1974. Lithology and porosity from acoustic shear and compressional wave transit time relationships. SPWLA, 15th Annu. Log. Symp. Trans., Paper Q.

Peyret, O. and Mons, F., 1980. Sonic versus seismic velocities, positive drift study, recording frequency effect.

Pickett, G.R., 1963. Acoustic character logs and their applications in formation evaluation. J. Pet. Technol., 15 (6).

Raymer, L.L., Hunt, E.R. and Gardner, J.S., 1980. An improved sonic transit time to porosity transform. SPWLA, 21st Annu. Log. Symp. Trans., Paper P.

Sarmiento, R., 1961. Geological factors influencing porosity estimates from velocity logs. Bul. Am. Assoc. Pet. Geol., 45 (5).

Shumway, G., 1958. Sound velocity vs temperature in water-saturated sediments. Geophysics, 23 (3).

Strick, E., 1971. An explanation of observed time discrepancies between continuous and conventional well velocity survey. Geophysics, 36(2); 285–295.

Summers, G.C. and Broding, R.A., 1952. Continuous velocity logging. Geophysics, 17 (3).

Tixier, M.P., Alger, R.P. and Doh, C.A., 1959. Sonic logging. J. Pet. Technol., 11 (5).

White, J.E., 1975. Computed seismic speeds and attenuation in rocks with partial gas saturation. Geophysics, 40 (2): 224–232.

Wuenschel, P.C., 1965. Dispersive body waves—an experimental study. Geophysics, 3G. 539–551.

Wyllie, M.R.J., Gregory, A.R. and Gardner, L.W., 1956. Elastic wave velocities in heterogeneous and porous media. Geophysics, 21 (1).

15. MEASUREMENT OF SONIC ATTENUATION AND AMPLITUDE

The *amplitude* of an acoustic wave decreases as it propagates through a medium. This decrease is known as *attenuation* (Fig. 15-1).

The attenuation as the wave moves through the formation depends on several factors, mainly:

(a) The wavelength of the wave and its type (longitudinal or transversal).

(b) The texture of the rock (pore and grain size, type of grain contact, sorting), as well as the porosity, permeability and the specific surface of the rock pores.

(c) The type of fluid in the pores and in particular its viscosity.

(d) Rock fractures or fissures.

This means that the measurement of attenuation can be of real use in the analysis of formations.

In cased wells the attenuation depends mainly on the quality of the cement around the casing. This can be indirectly measured by recording the sonic amplitude. This application is known as the *Cement Bond Log* (or CBL).

15.1. THEORETICAL CAUSES OF ATTENUATION

These are fundamentally of two types.

15.1.1. Loss of energy through heat loss

This loss of energy can have several causes.

15.1.1.1. *Solid-to-solid friction*

The vibration caused by the passage of a sonic wave causes the grains or crystals of the rock to move minutely one against another. This fractional movement generates heat and so a loss of energy. This phenomenon occurs mainly inside the formation.

15.1.1.2. *Solid-to-fluid friction*

As the forces acting on the solid grains and on the fluids cause different amounts of movement, frictional forces are generated at fluid to solid boundaries, with energy loss in the form of heat. This occurs in porous formations and also in muds that contain solid particles.

15.1.1.3. *Fluid-to-fluid friction*

When a formation contains two different non-miscible fluids the wave forces act to create fluid to fluid friction leading to acoustic energy loss. This occurs in porous formations containing water and hydrocarbons.

15.1.2. Redistribution of energy

This may occur in several ways.

15.1.2.1. *Transfers along the media limits*

Consider the cycle of a plane primary longitudinal wave moving in a solid M_1 (Fig. 15-2) which presents a vertical boundary with a liquid M_2, with the speed v_2 in M_2 less than that v_1 in M_1. The cycle is bounded by the wave fronts FF' and BB'. The direction of propagation is given by the arrow H. The region under compression is C and that of rarefaction is R. These two regions are separated by a plan NN' where M_1 is neither compressed nor dilated.

The compression in region C causes medium M_1 to bulge out into medium M_2, as the liquid is more compressible than the solid. Likewise the rarefaction R allows M_2 to expand slightly. This undulation of the boundary, shown very exaggerated in Fig. 15-2, moves towards the bottom with the primary wave and generates in the medium M_2 a compressional wave whose forward and rear fronts are shown by F'F'' and B'BB''. This secondary wave propagates in the direction P, forming with the original direction H,

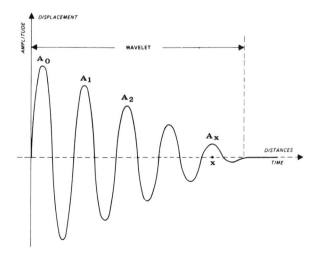

Fig. 15-1. Amplitude attenuation of acoustic wave with distance.

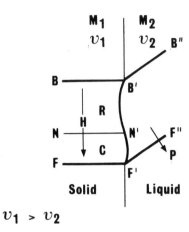

$$v_1 > v_2$$

Fig. 15-2. Mecanism of transfer of acoustic energy by radiation along a boundary.

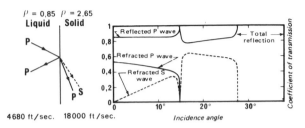

Fig. 15-3. Variation of the coefficients of reflection and transmission as a function of the angle of incidence (from Gregory, 1965).

an angle equal to the critical angle of incidence.

The energy of the secondary wave comes from the primary wave and so a fraction of the original energy is transferred to the medium M_2.

A transverse wave moving in the medium M_1 also transfers part of its energy to M_2 in the form of a compressional wave.

All this occurs in open hole along the borehole wall where the mud-formation boundary occurs and also in cased hole where the casing is not well cemented.

15.1.2.2. *Transfers across media boundaries*

When a wave crosses the boundary between two media M_1 and M_2 of different acoustic impedance we either have, depending on the angle of incidence, total reflection of the wave or part of the wave refracted into the medium M_2 and part reflected back into M_1.

In the second case, there is attenuation of the wave. The ratio of the amplitudes of the incident and transmitted waves is called the coefficient of transmission of amplitude T_c.

$$T_c = A_0/A_x \qquad (15-1)$$

For a normal wave incidence we have for P and S waves:

$$T_c = \frac{2}{(r_1/r_2)^{1/2} + (r_2/r_1)^{1/2}} \qquad (15-2)$$

with r_1 and r_2 equal to the acoustic impedance of the media M_1 and M_2 ($r = v\rho$ with v = velocity and ρ = density) and the reflection coefficients of amplitude are:

$$R'_c = \frac{1 - r_1/r_2}{1 + r_1/r_2} \qquad \text{for P waves} \qquad (15-3)$$

$$R''_c = \frac{1 - r_2/r_1}{1 + r_2/r_1} \qquad \text{for S waves.} \qquad (15-4)$$

When the angle of incidence is no longer normal the calculation of the reflection and transmission coefficients is more complex. Figure 15-3 gives in graphical form the variation of the coefficients as a function of the angle of incidence.

This phenomenon is produced either at the boundary of formation and mud, or between layers of different lithologies or at fracture planes when the fractures are full of fluid or cemented.

In a cased hole it occurs at the boundaries of casing-cement-formation when the cement is good.

15.1.2.3. *Dispersion*

When the sonic wave encounters particles, whose dimensions are less than the wavelength, the sonic energy is dispersed in all directions, whatever is the shape of the reflection surface.

15.2. CAUSES OF ATTENUATION IN THE BOREHOLE

We must distinguish two cases:

15.2.1. **Open-hole**

From what we have said the principal causes of attenuation are:

15.2.1.1. *Attenuation in the mud*

This is due to acoustic losses by frictional losses, solid to fluid, and to dispersion losses at particles in suspension in the mud.

In a pure liquid this attenuation follows an exponential law at least for one unique frequency:

$$\delta_m = e^{mx} \qquad (15-5)$$

in which m is the attenuation factor in the liquid, proportional to the source of the frequency, and x is the distance over which the attenuation is measured.

For fresh water and at standard conditions of temperature and pressure, for a frequency of 20 kHz the attenuation factor is of the order of 3×10^{-5}

db/ft. It is higher for salt water and oil. It decreases as the temperature and pressure increase.

For normal drilling muds which contain solid particles we have to add the effect of dispersion. It is estimated that the total dispersion is of the order of 0.03 db/ft for a frequency of 20 kHz.

In gas cut muds the attenuation caused by dispersion is very large, so making all sonic measurements impossible.

15.2.1.2. *Attenuation by transmission of energy across the mud-formation boundary for waves arriving at an angle of incidence less than critical*

The coefficient of transmission depends on the relative impedances of the rock and the mud. As the impedance of the mud is about constant and there is little variation in rock density, the ratio of impedances is effectively proportional to the speed of sound in the rock.

15.2.1.3. *Attenuation in the rock*

Several factors are important:
(a) Frictional energy loss

In non-fractured rocks the attenuation of longitudinal and transverse waves is an exponential function of the form:

$$\delta_F = e^{al} \qquad (15-6)$$

in which a is the total attenuation factor due to different kinds of friction: solid to solid (a'), fluid to solid (a'') and fluid to fluid (a'''):

$$a = a' + a'' + a''' \qquad (15-7)$$

and l is the distance travelled by the wave. It is given by the equation:

$$l = L - (d_h - d_{tool})\mathrm{tg}i_c \qquad (15-8)$$

where L is the spacing, d_h and d_{tool} are the diameters of the hole and the tool, i_c is the critical angle of incidence, which goes down as the speed in the formation increases.

When the rock is not porous, the factors a'' and a''' are zero. When the rock is water saturated, $a''' = 0$.

In porous rocks, the attenuation factor a'' depends on the square of the frequency, whereas the factors a' and a''' are proportional to the frequency.

The factor a'' depends equally on porosity and permeability. It increases as the porosity and permeability increase.

The attenuation factors a' and a'' decrease as the differential pressure ΔP (geostatic pressure − internal pressure of the interstitial fluids) increases. Figure 15-4 from Gardner et al., (1964), gives the relationship for dry rock—the energy losses are then due to solid to solid friction (a')—and for a water-saturated

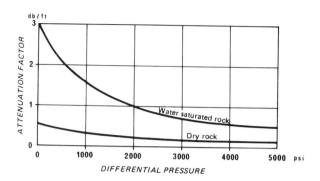

Fig. 15-4. Influence of the differential pressure on attenuation (from Gardner et al., 1964).

rock—the difference in attenuation (gap between the curves) is due to fluid to solid friction (a'').

When the rock contains hydrocarbons a greater attenuation of the longitudinal wave is observed in the case of gas than for oil (the factor a''' non zero). From this we can deduce that the viscosity of the fluid has an effect on the attenuation factor a'''.

Hence, if we resume all the different parameters acting on the attenuation, we can write that for a given tool:

$$a = f(f, v, \phi, k, S, \mu, \Delta P, \rho) \qquad (15-9)$$

with:
f = frequency;
v = velocity of the sound;
ϕ = porosity;
k = permeability;
S = saturation;
μ = viscosity of the fluids;
ΔP = differential pressure;
ρ = density of the formation.

(b) Loss of energy through dispersion and diffraction: this appears mainly in vuggy rocks.

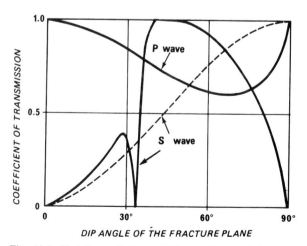

Fig. 15-5. Variations of the coefficients of transmission as a function of the apparent dip angle of a fracture plan with regard to the propagation direction (solid lines from Knopoff et al., 1957; dash lines from Morris et al., 1964).

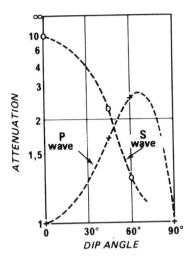

Fig. 15-6. Attenuation across a fracture as a function of the apparent dip angle of a fracture plan with regard to the propagation direction (from Morris et al., 1964).

(c) Transmission across the boundaries of a medium.

When a formation is made up of laminations of thin beds of different lithology at each boundary some or all of the energy will be reflected according to the angle of incidence. This angle is dependent on the apparent dip of the beds relative to the direction of the sonic waves.

In the case of fractured rocks the same kind of effect occurs with the coefficient of transmission as a function of the dip angle of the fracture with regard to the propagation direction. The chart given in Fig. 15-5 is only applicable in the case of thin fractures that are open. It does not include the acoustic losses due to friction in the fracture.

Figure 15-6 is derived from Fig. 15-5, but applies to experimental data.

From these two figures we can draw the following conclusions: (a) the P waves are only slightly attenuated when they cross a horizontal or vertical fracture. The attenuation is large when the angle of the fracture plane is between 25 and 85 degrees and (b) The S waves are strongly attenuated when they cross a fracture at a slight angle. The attenuation decreases as the dip increases.

15.2.1.4. *Transfer of energy along the borehole wall*

This phenomenon, described previously, leads to an attenuation of the signal at the receiver. We can generally conclude that the attenuation linked to this is a function of the tool transmitter to receiver spacing, the diameter of the hole, and the frequency and speed of the P and S waves.

15.2.2. **Cased hole**

The attenuation is affected by the casing, the quality of the cement and the mud. If the casing is free and surrounded by mud, it can vibrate freely. In this case, the transfer factor of energy to the formation is low and the signal at the receiver is high.
Remarks: In some cases, even when the casing is free we can see the formation arrivals (on the VDL). This can happen if the distance between the casing and the formation is small (nearer than one or two wavelengths), or when the casing is pushed against one side of the well but free on the other. Transmission to the formation is helped by the use of directional transmitters and receivers of wide frequency response.

If the casing is inside a cement sheath that is sufficiently regular and thick (one inch at least) and the cement is well bonded to the formation the casing is no longer free to vibrate. The amplitude of the casing vibrations is much smaller than when the casing is free and the transfer factor to the formation is much higher. Just how much energy is transferred to the formation depends on the thickness of the cement and the casing. As energy is transferred into the formation the receiver signal is, of course, smaller. Between the two extremes (well bonded casing, free pipe) the amount of energy transferred and hence the receiver signal will vary.

15.3. **MEASUREMENT OF ATTENUATION**

This is not possible directly so an indirect measurement of amplitude is used.

15.3.1. **Cement Bond Log**

In the case of the Cement Bond Log (CBL), the general method is to measure the amplitude of the first arrival in the compressional wave at the receiver(s) (Fig. 15-7).

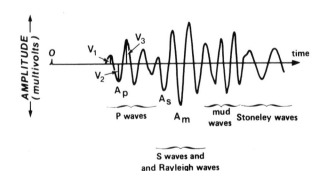

Fig. 15-7. Complete theoretical acoustical signal received from the formation showing the arches usually used for the amplitude measurement. A_P: amplitude of compressional wave (P), A_S: amplitude of shear wave (S). A_m: maximum amplitude.

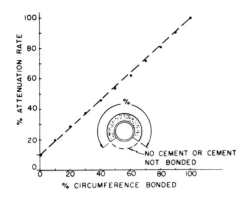

Fig. 15-8. Per cent attenuation vs per cent circumference bonded (courtesy of Schlumberger).

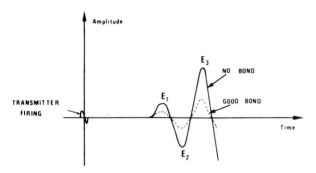

Fig. 15-9. Schematic receiver output signal with unbonded casing and with bonded casing (courtesy of Schlumberger).

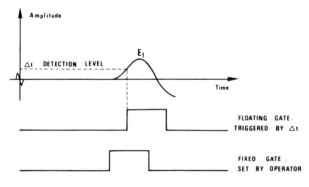

Fig. 15-10. Gating systems (courtesy of Schlumberger).

These arrivals have a frequency between 20 and 25 kHz. The amplitude of the first arrival is a function partly of the type of tool (particularly the tool spacing) and of the quality of the cementation: the nature of the cement and the percentage of the circumference of the tubing correctly bound to the formation (Fig. 15-8).

As we have seen the amplitude is a minimum, and hence the attenuation a maximum, when the tool is in a zone where the casing is held in a sufficiently thick annulus of cement (one inch at least). The amplitude is largest when the casing is free (Fig. 15-9).

The amplitude is measured using an electronic gate (or window) that opens for a short time and measures the maximum value obtained during that time.

In the Schlumberger CBL (Fig. 15-10) there is a choice of two systems for opening the gate:

(a) Floating gate: the gate opens at the same point in the wave as the Δt detection occurs and remains open for a time set by the operator, normally sufficient to cover the first half cycle. The maximum amplitude during the open time is taken as the received amplitude measurement.

(b) Fixed gate: the time at which the gate opens is chosen by the operator and the amplitude is measured as the maximum signal during the gate period. The fixed gate measurement is therefore independent of Δt.

In the case of the fixed gate care must be taken to follow the position of the E_1 arrival if for any reason true Δt varies, as, for example, when the fluid inside the casing changes. Normally however, when Δt is properly detected at E_1 the two systems give the same result. If E_1 is too small then Δt detection will cycle skip to E_3 (the case where the casing is very well cemented). The two systems then give: (a) fixed gate: E_1 is still measured and is small; and (b) floating gate: E_1 is measured and is usually large (Fig. 15-11).

The measurement and recording of transit time at the same time as amplitude allows cycle skipping to be detected (Fig. 15-12).

Excentralization of the tool may cause a drop in the transit time (Figs. 15-13 and 15-14): the wave that has the shorter path through the mud arrives before the theoretical wave coming from a centred sonde, and triggers the measurement of Δt, even if its amplitude is attenuated (Fig. 15-13b).

The transformation to attenuation from the amplitude measured in a CBL tool in millivolts depends mainly on the transmitter receiver spacing (Fig. 15-15). We can establish that smaller spacings (3 feet) always give better resolution than a large spacing. Figure 15-16 is an example of a CBL log.

The interpretation of the CBL consists of the determination of the *bond index* which is defined as the ratio of the attenuation in the zones of interest to the maximum attenuation in a well cemented zone. A bond index of 1 therefore, indicates a perfect bond of casing to cement to formation.

Where the bond index is less than 1, this indicates a less than perfect cementation of the casing. However, the bonding may still be sufficient to isolate zones from one another and so still be acceptable. Generally some lower limit is set on the bond index, above which the cementation is considered acceptable (Fig. 15-17). The interpretation of the bond index is helped by the use of the Variable Density Log or VDL.

Attenuation can be calculated from the amplitude by using the chart shown in Fig. 15-18, which also

244

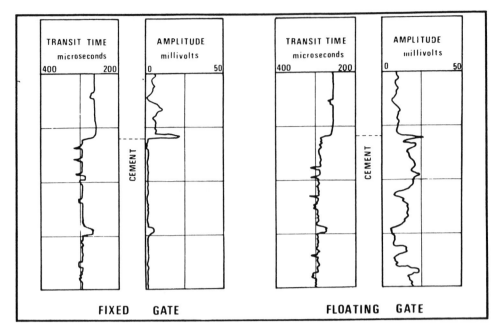

Fig. 15-11. Example of logs recorded with fixed gate and floating gate showing the influence of the gate system on the measurement (courtesy of Schlumberger).

allows determination of the compressional strength of the cement. Some companies offer this computation directly from their well-site equipment (e.g. the Schlumberger CSU).

15.3.2. Attenuation index

In its use in open hole Lebreton et al. (1977) proposed a calculation of an index I_c defined by the relation:

$$I_c = (V_2 + V_3)/V_1 \qquad (15-11)$$

where V_1, V_2 and V_3 are the amplitude of the three first half-cycles of the compressional wave. According

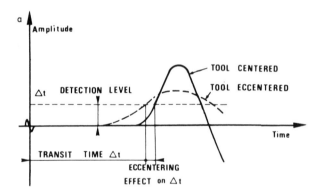

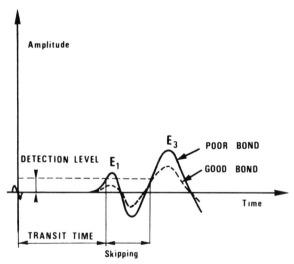

Fig. 15-12. Cycle skipping with floating gate system recording in the case of good bond (courtesy of Schlumberger).

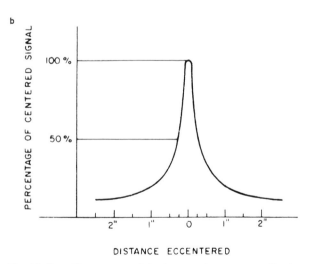

Fig. 15-13. a. Phenomenon of Δt decrease due to eccentralization of the tool (courtesy of Schlumberger). b. Compressional arrival amplitude vs the distance by which a tool is eccentered in open hole (from Morris et al., 1963).

245

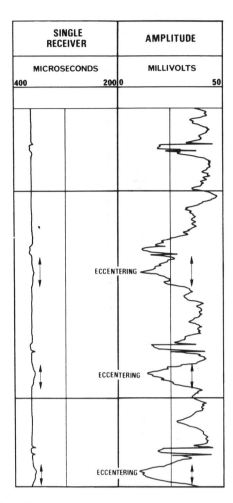

Fig. 15-14. Example of eccentering effects on Δt and amplitude recorded with the CBL (courtesy of Schlumberger).

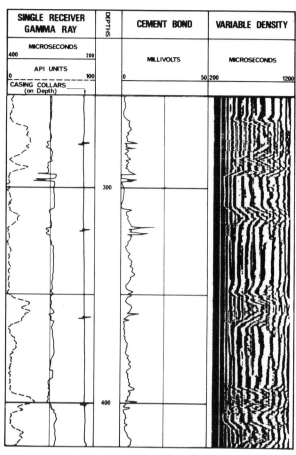

Fig. 15-16. Example of a CBL recorded with a VDL. Case of a well bonded casing (courtesy of Schlumberger).

to these authors this index should be a function of the permeability:

$$I_c = \alpha \log(k_v/\mu) + \beta \tag{15-12}$$

where:

k_v = permeability measured along the axis of the core;

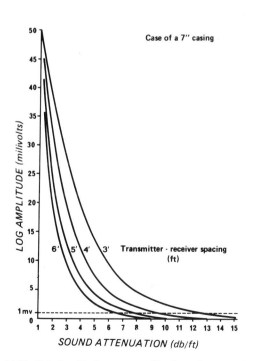

Fig. 15-15. Relationship between amplitude and attenuation for different spacings (from Brown et al., 1971).

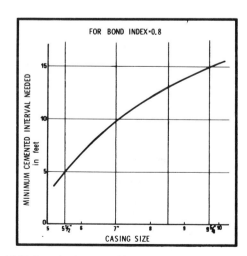

Fig. 15-17. Length of cemented interval required for zone isolation (for Bond Index = 0.8).

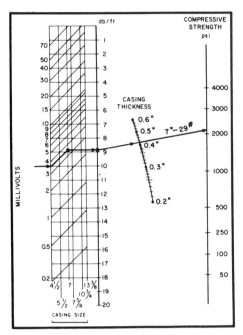

Fig. 15-18. CBL interpretation chart for centered tool only and 3 foot-spacing (courtesy of Schlumberger).

μ = viscosity of the wetting fluids in the rock;
α and β are constants for a given tool and well. However, we have to take into account the fact that it is not possible to record the whole wave using the

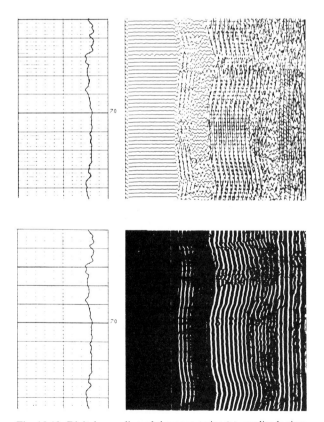

Fig. 15-19. Digital recording of the wave train: (a) amplitude-time mode or wiggle-trace; (b) intensity modulated-time mode (VDL).

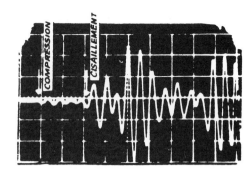

Fig. 15-20. Photograph of the signal received on the oscilloscope.

CBL, except if we use the long-spacing sonic. In that case, we can record the entire signal as shown by Fig. 15-19 and of course process it. In the other cases, it is necessary to use a photographic system on the screen of an oscilloscope (Fig. 15-20) or to use a digitization of the wave train that can later be played back on film.

We can equally, as proposed by Welex, measure the amplitude of the same half-wave of the signal at two different receivers, and calculate the ratio, which should be only a function of the acoustic transmission factor of the formation. The ratio is converted to attenuation by using a chart (Fig. 15-21).

We have to remember, however, that the transmission coefficients in open hole are very sensitive to the hole size and rugosity of the wall which in effect implies that attenuation as measured in open hole cannot normally be used.

15.4. AN EXPRESSION FOR THE LAW OF ATTENUATION IN OPEN HOLE

By using experimental laboratory measurements Morlier and Sarda (1971) proposed the following

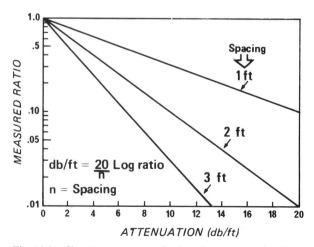

Fig. 15-21. Chart to convert amplitude ratio to attenuation (from Anderson and Riddle, 1963).

equations for the attenuation of the longitudinal and transverse waves in a saturated porous rock:

$$S_p = 1.2 \times 10^{-3}\, \frac{S}{\phi} \left(2 \frac{Mk}{\mu} f\rho_f\right)^{1/3} \qquad (15\text{-}13)$$

$$S_s = 2.3\, S_p \qquad (15\text{-}14)$$

where:

S = specific surface (surface area of the pores per unit volume);
ϕ = porosity;
k = permeability;
ρ_f = fluid density;
μ = fluid viscosity;
f = signal frequency.

15.5. VARIABLE DENSITY LOG (VDL)

The principle is shown in Figs. 15-22 and 15-23. A record is made of the signal transmitted along the logging cable during a 1000-μs period using a special camera. We can then either reproduce the trace (Fig. 15-19a) by using an amplitude-time mode in which the wave train is shown as a wiggle trace or translate it into a variable surface by darkening the area depending on the height of the positive half-waves of the sonic signal (Fig. 15-19b). This last method is known as the intensity modulated-time mode.

The different arrivals can be identified on the VDL. Casing arrivals appear as regular bands whereas the formation arrivals are usually irregular. It is sometimes possible to distinguish amongst the arrivals between those linked with compressional waves and those with shear waves, by the fact that the latter arrive later and that they are at a sharper angle (Fig.

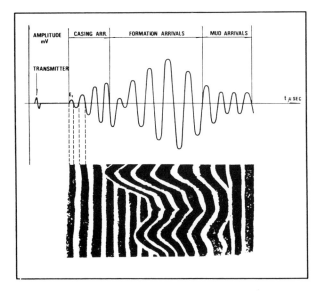

Fig. 15-22. Principle of operation of the Variable Density Log (courtesy of Schlumberger).

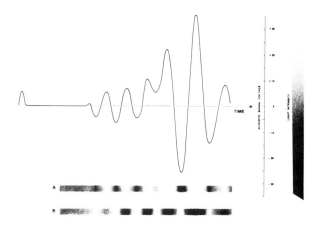

Fig. 15-23. Schematic explaining the conversion of amplitude in white and black (from Guyod and Shane, 1969).

15-24). They are often of higher energy (higher amplitude and therefore a darker trace).

In the case of the Schlumberger VDL the five foot receiver is used in order to improve the separation between waves.

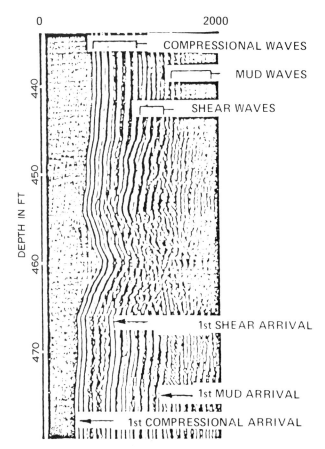

Fig. 15-24. VDL recording showing how the P-waves can be separated from the S-waves by using time and angle (courtesy of Schlumberger).

248

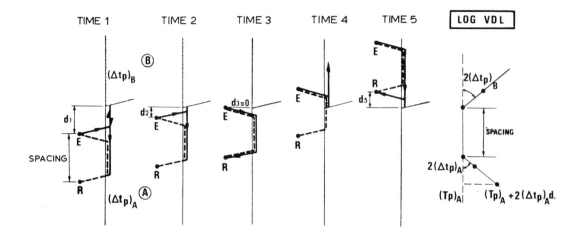

Fig. 15-25. Schematic explaining the formation of chevrons (courtesy of Schlumberger).

VDL-recording often has distinguishable chevron patterns. These are related to secondary arrivals caused by reflections and conversion of the primary waves at the boundaries of media with different acoustic characteristics, perhaps corresponding to: (a) bed boundaries; (b) fractures; (c) hole size variations; and (d) casing joints. Chevrons appear on longitudinal as well as transverse waves.

The appearance of this phenomenon is explained by the schemes in Fig. 15-25. At time 1, when the wave leaves the transmitter E, part of the wave is refracted downwards as far as the receiver R. Its travel time is T_p. Another part of the wave is re-fracted upwards, is reflected at a bed boundary and goes down to the receiver. Its transit time is:

$$T_1 = \left(T_p\right)_A + 2d_1\left(\Delta t_p\right)_A$$

At time 2, the tool has moved up, d decreases (d_2) and T_2 as well.

At time 3, d_3 is zero and $T_3 = (T_p)_A$.

As soon as the transmitter is above the bed boundary (time 4), the wave is reflected upwards and so does not arrive at the receiver. This carries on happening as long as the transmitter and the receiver are on different sides of the boundary.

When the receiver itself passes to the other side of

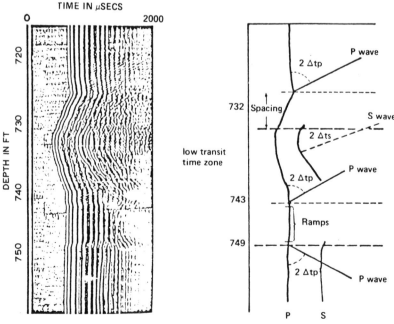

Fig. 15-26. Example of VDL showing chevron patterns on P- and S-waves (courtesy of Schlumberger).

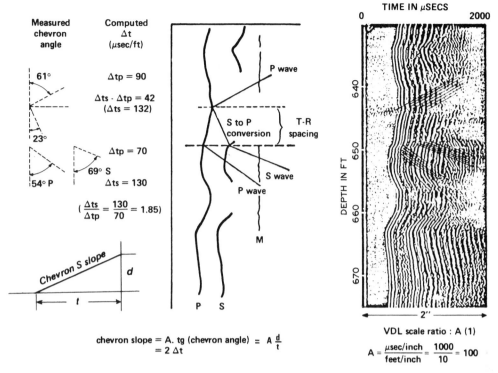

Fig. 15-27. Computation of Δt_s from the slope of the chevron pattern.

the boundary (time 5) it detects once again an arrival reflected from the boundary and at a transit time equal to:

$$T_5 = (T_p)_B + 2d_5(t_p)_B$$

An example of chevron patterns is given in Fig. 15-26.

The main applications of a study of chevron patterns are:

(A) *Fracture detection*

Depending on the angle that the fracture planes make with the hole we have to consider three different cases:

(a) Fractures whose inclination is less than 35°. The amplitude of the compression wave is hardly reduced. We can expect only a small amount of reflection. The VDL will have the following characteristics: (1) Strong amplitude of the compressional wave (E_1 or E_2); (2) weak or no P-chevron pattern; (3) low amplitude of the shear waves; and (4) well defined S-chevron pattern.

(b) Fractures with an inclination between 35° and 85°. The amplitude of the P-wave is reduced. The amplitude of the S-wave goes up and the VDL has the following characteristics: (1) low-amplitude P wave (E_1 and E_2); (2) little or no S-chevron patterns; and (3) some P-chevron patterns.

(c) Fractures with an inclination of over 85°. These are very difficult to detect by acoustic methods.

(B) *The calculation of Δt_s*

This can be done if there are S-chevrons. In this case, Δt_s is given by the gradient (Fig. 15-27). We have in fact:

$$\Delta t_s = \tfrac{1}{2}d/t$$

We can also calculate it by a method analyzing the whole wave train. The changes in Δt_s are larger than those of Δt_p, which explains why the S-wave arrivals are not parallel to the P-wave arrivals. The difference in time between P- and S-wave arrivals can be approximated by the equation:

$$T_s - T_p = (\text{spacing}) \, (\Delta t_s - \Delta t_p) \text{ from which we solve}$$
for Δt_s:

$$\Delta t_s = \Delta t_p + \frac{T_s - T_p}{\text{spacing}}$$

Thus, in the example of Fig. 15-23 between 4277 and 4281 the VDL, whose spacing is five feet, gives:

$$T_s - T_p = 200 \, \mu s \qquad (\text{approx.})$$

The Δt_p read by the uncompensated tool is around 45 μs/ft from which we have:

$$\Delta t_s = 40 + 45 = 85 \, \mu\text{s/ft}$$

Remarks: Another method to find Δt_s is to record two VDL's, one with a 3-ft, the other a 5-ft spacing, and to determine the S-wave arrivals on each one. Then, the respective arrival times are measured and the difference divided by 2 ft.

15.6. **REFERENCES**

1. **CBL**

Brown, H.D., Grijalva, V.E. and Raymer, L.L., 1971. New developments in sonic wave train display and analysis in cased holes. Log Analyst, 12 (1).

Grosmangin, M. et al., 1960. The Cement Bond Log. Soc. Pet, Eng. AIME, Pasadena Meeting, Paper no. 1512 G.

Grosmanging, M., Kokesh, F.P. and Majani, P., 1961. A sonic method for analyzing the quality of cementation of borehole casings. J. Pet. Technol., 165–171; Trans., AIME, 222.

Muir, D.M. and Latson, B.F., 1962. Oscillographs of acoustic energy and their application to log interpretation. Document P.G.A.C., Houston, 1962.

Pardue, G.H., Morris, R.L., Gollwitzer, L.H., and Moran, J.H., 1963. Cement Bond Log. A study of cement and casing variables. Soc. Pet. Eng. AIME, Los Angeles Meeting, Paper No. 453.

Poupon, A., 1964. Interpretation of Cement Bond Logs. Document Schlumberger, Paris.

Putman, L., 1964. A progress Report on Cement Bond Logging. J. Petr. Technol.

Riddle, G.A., 1962. Acoustic Propagation in Bonded and Unbonded Oil Well Casing. Soc. Pet. Eng. AIME, Los Angeles Meeting, Paper no. 454.

Thurber, C.H. and Latson, B.F., 1960. SATA Log Checks Casing Cement Jobs. Petrol. Eng., Dec. 1960.

Schlumberger, 1976. The essentials of cement evaluation.

2. **Attenuation and VDL, VSP**

Anderson, W.L. and Riddle, G.A., 1963. Acoustic amplitude ratio logging. Soc. Petr. Eng. AIME, 38th Annu. Fall Meeting, New Orleans, Paper 722.

Anderson, W.L. and Walker, T., 1961. Application of open hole acoustic Amplitude Measurements. Soc. Petr. Eng. AIME, 36th Annu. Fall Meeting, Dallas, Paper 122.

Fons, L.H., 1963. Use of acoustic signal parameters to locate hydrocarbons and fractured zones. P.G.A.C. Service report, Houston.

Fons, L.H., 1963. Acoustic scope pictures. Soc. Pet. Eng. AIME, 38th Annu. Fall Meeting, New Orleans, Paper 724.

Fons, L.H., 1964. Utilisation des paramètres du signal acoustique pour la localisation directe des hydrocarbures et la détermination des zones fracturées. Bull. Assoc. Franç. Techn. Pét., 167 (in English: P.G.A.C., no. 64.5).

Gardner, G.H.F., Wyllie, M.R.J. and Droschak, D.M., 1964. Effect of pressure and fluid saturation on the attenuation of the elastic waves in sands. J. Pet. Technol., 16 (2).

Gregory, A.R., 1965. Ultrasonic pulsed-beam transmission and reflection methods for measuring rock properties. Some theoretical and experimental results. 6th Annu. Log. Symp. Trans. SPWLA.

Guyod, H. and Shane, L.E., 1969. Geophysical Well Logging, Vol. 1. Guyod, Houston.

Knopoff, L. et al., 1957. 2nd Annual Report, Seismic Scattering Project. Inst. Geophysics, UCLA.

Kokesh, F.P. et al., 1964. A new Approach to sonic Logging and other acoustic measurements. Soc. Pet. Eng. AIME, Paper S.P.E., Houston Meeting, Paper 991.

Lebreton, F., Sarda, J.P., Trocqueme, F. and Morlier, P., 1977. Essais par diagraphie dans des milieux poreux pour évaluer l'influence de leur perméabilité sur des impulsions acoustiques. 5th Europ. Log. Symp. SAID (oct. 1977, Paris).

Morlet and Schwaetzer, 1961. Mesures d'amplitude des ondes soniques dans les sondages. Geophys. Prospect., 9, 4.

Morlet and Schwaetzer, 1962. Le log d'atténuation. 22ᵉ réunion EAEG (oct. 1977, Paris).

Morlier, P. and Sarda, J.P., 1971. Atténuation des ondes élastiques dans les roches poreuses saturées.

Morris, R.L., Grine, D.R., Arkfeld, T.E., 1963. The use of compressional and shear acoustic amplitudes for the location of fractures. Soc. Pet. Eng. AIME, 38th Annu. Fall Meeting, New Orleans, Paper 723.

Morris, R.L. et al., 1964. Using compressional and shear acoustic amplitudes for the location of fractures. J. Pet. Technol., 16 (6).

Muir, D.M. and Fons, L.H., 1964. Case histories: Hydrocarbon location from acoustic parameter logs. Trans. S.P.W.L.A. 5th Annu. Log. Symp., 13–15 may 1964.

Muir, D.M. and Fons, L.H., 1964. The new acoustic parameter log: what it is...how it operates? World Oil, May, 1964.

Pickett, G.R., 1963. Acoustic character logs and their Applications in Formation Evaluation. J. Petrol. Technol., 15, 6.

Schlumberger, 1964. Brevet français no 1401258 demandé le 2 avr. 1964.

Thurber, C.H., 1961. Acoustic log interpretation. Document P.G.A.C., Houston.

Tixier, M.P., Alger, R. and Doh, C.A., 1959. Sonic Logging. J. Pet. Technol., 11 (5).

Tixier, M.P., Loveless, G.W. and Anderson, R.A., 1973. Estimation of formation strength from the mechanical properties log. SPE of AIME, paper no SPE 4532.

Walker, T., 1962. Fracture Zones Vary Acoustic Signal Amplitudes. World Oil, May, 154, 6.

Walker, T., 1962. Progress Report on Acoustic Amplitude Logging for Formation Evaluation. Soc. Pet. Eng. AIME, 37th Annu. Fall Meeting, Los Angeles, Paper 451.

Walker, T. and Riddle, G., 1963. Field investigation of full acoustic wave recording. 4th Annu. Log. Symp. SPWLA.

Wyllie, M.R.J., Gardner, G.H.F. and Gregory, A.R., 1962. Studies of elastic wave attenuation in porous media. Geophysics, 27 (5).

16. MEASUREMENT OF THE PROPAGATION TIME AND ATTENUATION RATE OF AN ELECTROMAGNETIC WAVE
(Electromagnetic Propagation Tool, EPT *)

16.1. GENERALITIES

Three parameters uniquely characterize a rock electrically. They are:

(a) magnetic permeability, μ, (in Henry/metre);

(b) electrical conductivity, C, (in Siemens/metre or mho/m);

(c) dielectric permittivity, ϵ, (in Farad/metre).

Because in most cases, rocks are composed of non-magnetic minerals, their magnetic permeability is the same as that of free space, μ_0 ($= 4\pi \times 10^{-7}$ Henry/m). The variation of this parameter is generally too small to be of much interest.

As previously seen (Chapter 3), the electrical conductivity or its inverse, the resistivity, is of most interest in evaluating the water saturation in porous hydrocarbon-bearing media. However, when the formation water is of low or varying salinity, the detection of hydrocarbons has proved to be a difficult task.

The dielectric permittivity, ϵ, measured at very high frequency (GHz), is primarily a function of the water-filled porosity, and practically independent of the salinity. This allows the measurement of the water saturation in the flushed zone, and an evaluation of moved hydrocarbon by comparison of the water-filled porosity with other (total) porosities as seen by other tools (FDC, CNL...). From this comparison the presence of hydrocarbons can be detected.

16.2. BASIC CONCEPTS

The dielectric permittivity, ϵ, is one of the main factors which affect electromagnetic propagation. The dielectric permittivity of any medium is proportional to the electric dipole moment per unit volume. Several effects contribute to this electric dipole moment: electronic, ionic, interfacial and dipolar. The electronic contribution is due to the displacement of electron clouds and is the only one that operates at optical frequencies.

The ionic and interfacial contributions arise from displacement and movement of ions, and hence are confined to low frequencies. The dipolar contribu-

---

* Trade Mark of Schlumberger

TABLE 16-1

Dielectric constants and propagation times for several minerals and fluids (courtesy of Schlumberger)

Mineral	$\epsilon'_r = \epsilon'/\epsilon_0$	t_{pl} (nanos/m)
Sandstone	4.65	7.2
Dolomite	6.8	8.7
Limestone	7.5–9.2	9.1–10.2
Anhydrite	6.35	8.4
Halite	5.6–6.35	7.9– 8.4
Gypsum	4.16	6.8
Muscovite *		8.3– 9.4
Biotite *		7.1– 8.2
Talc *		7.1– 8.2
K-Feldspar *		7 – 8.2
Siderite *		8.8– 9.1
Limonite *		10.5–11.0
Sylvite *		7.2– 7.3
Apatite *		9.1–10.8
Sphalerite *		9.3– 9.5
Rutile *		31.8–43.5
Petroleum	2.0–2.4	4.7– 5.2
Fresh water (@ 25°C)	78.3	29.5

* Values estimated from published literature, not verified by large measurements.

tion, which is the dominant effect at the frequency used in the EPT (gigahertz), is due to the presence of permanent dielectric dipoles which orient themselves in the direction of an applied electric field.

With the exception of water, there are very few materials commonly found in nature which have permanent electric dipoles.

The dipolar nature of water is inherent to its molecular structure, and not to the content of dissolved salts. Thus a dielectric permittivity logging measurement in the gigahertz frequency region should lead to a measurement of water content which does not have a major dependence on salinity.

Table 16-1 summarizes typical values of relative dielectric constants and propagation times for several minerals and fluids. The values of propagation time correspond to propagation without dissipation of energy. They will be longer when energy losses occur.

16.3. THEORY OF THE MEASUREMENT

The EPT measures the travel time and attenuation of an electromagnetic wave. These parameters are related to the dielectric permittivity, ϵ, as explained below.

Electromagnetic waves are not plane waves, owing partly to the closeness of the transmitter to the receiver (8 cm). However, this fact affects only the amplitude of the signal; one can correct this effect by using a spreading loss term (see later). So for practical purposes, we can assume that the propagating fields are planar. For a plane wave travelling in the Z-direction in an homogeneous isotropic medium, the magnitude of the electric field at the second receiver is given by:

$$E = E_0 e^{-\gamma^* z + j\omega t} \qquad (16\text{-}1)$$

where:

E_0 = magnitude of the electric field at the first receiver;
z = distance between the two receivers;
ω = angular frequency = $2\pi f$;
t = time taken by the wave to travel the distance z;
j = vectorial operator = $\sqrt{-1}$;
f = frequency (Hz).

γ^* is the complex propagation constant given by:

$$\gamma^* = \alpha + j\beta \qquad (16\text{-}2)$$

where: α = attenuation factor in nepers/metre; and β = phase shift in radians/metre

For a "loss-less" formation (see later for definition or losses) the attenuation is nil.

The phase velocity, v_{p0}, is given by:

$$v_{p0} = \frac{dz}{dt} = \frac{\omega}{\gamma_0} = \frac{1}{t_{p0}} \qquad (16\text{-}3)$$

(The subscript "0" indicates lossless conditions). From Maxwell's equations, it can be shown that:

$$\gamma_0 = j\omega\sqrt{\mu_0\epsilon} = j\omega t_{p0} \qquad (16\text{-}4)$$

where: μ = magnetic permeability (Henry/metre) = μ_0; and ϵ = dielectric permittivity (Farad/metre).

When the formation is "lossy" γ and ϵ are complex (γ^* and ϵ^*).

$$\gamma^* = j\omega\sqrt{\mu_0\epsilon^*} = j\omega\sqrt{\mu_0(\epsilon' - j\epsilon'')} = j\omega\sqrt{\mu_0\left(\epsilon' - j\frac{C}{\omega}\right)} \qquad (16\text{-}4a)$$

Squaring eqs. 16-2 and 16-4a and equating the real and imaginary parts, we have:

$$\omega^2\mu_0\epsilon' = \beta^2 - \alpha^2 \qquad (16\text{-}5)$$

and

$$\omega\mu_0 C = 2\alpha\beta \qquad (16\text{-}6)$$

where C is the equivalent conductivity (in mho/m). Dividing eq. 16-5 by ω^2 we have:

$$\mu_0\epsilon' = \frac{\beta^2}{\omega^2} - \frac{\alpha^2}{\omega^2} \qquad (16\text{-}7)$$

Since from eq. 16-4 $\mu_0\epsilon' = t_{p0}^2$ and $\beta/\omega\ (= t_{p1})$ is the travel time in the lossy medium, eq. 16-7 can be written as follows:

$$t_{p0}^2 = t_{p1}^2 - \frac{\alpha^2}{\omega^2} \qquad (16\text{-}8)$$

16.4. MEASUREMENT TECHNIQUE

The Electromagnetic Propagation Tool (EPT) transmits a 1.1 GHz electromagnetic wave into the formation and measures its propagation time (t_{p1} in nanosecond/metre), attenuation rate (eatt in decibels/metre) between a pair of receivers 4 cm apart (Fig. 16-1).

The EPT is a mandrel type tool. The antennae are mounted in a pad which is rigidly fixed to the tool body, and contact with the borehole wall is achieved by means of a powered mechanical back-up arm.

The antenna pad consists of two transmitters and two receivers. The two transmitters are alternately energized to allow a Bore Hole Compensation (BHC) mode of operation, similar to that employed in the sonic tool (Fig. 16-2). This technique reduces the effects of tool-tilt, varying mud cake thickness and slight instrumentation imbalance. The antennae are cavity-basked slot radiators designed for efficient radiation at 1.1 GHz.

The tool contains a 1.1 GHz microwave transceiver, a frequency down-converter and low-

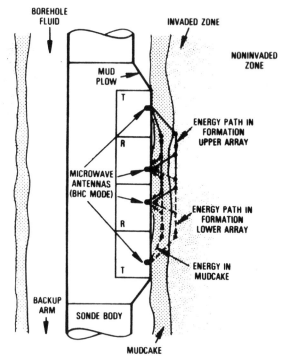

Fig. 16-1. EPT antenna configuration and signal paths (from Wharton et al., 1980) (courtesy of SPE of AIME).

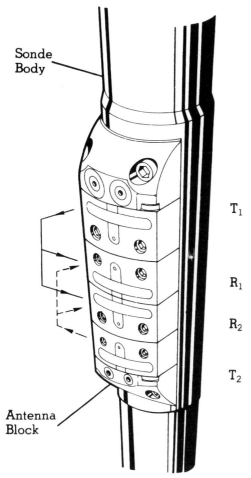

Fig. 16-2. The antenna assembly (courtesy of Schlumberger).

frequency processing circuitry.

The high-frequency solid-state transmitters, are capable of generating more than 2 watts of output power.

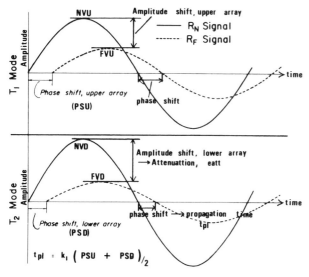

Fig. 16-3. Single cycle representation of the borehole compensated measurement of phase and amplitude shifts (from Calvert et al., 1977; courtesy of SPE of AIME).

The receivers are basically single channel, superheterodyne type with automatic gain control. They can process a typical 0.3 picowatt signal.

In order to measure the average propagation time, t_{pl}, an average phase-shift is computed (Fig. 16-3) which is given by the following relation:

$$\frac{PSU + PSD}{2}$$

where PSU and PSD are the phase-shifts measured in the upward and downward directions respectively.

The propagation time is equal to:

$$t_{pl} \, (ns/m) = K_1 \left(\frac{PSU + PSD}{2} \right) \qquad (16-9)$$

where K_1 is a fixed constant.

In the same manner, to measure the average attenuation, eatt (or A) the average amplitude shift is computed (Fig. 16-3). This is achieved by measuring the differences in the two receiver signal levels, in the upward and downward directions. The two receiver signals are input to the receiver alternately and the corresponding voltage change, which is proportional to the difference in db between the received signal levels, is measured. So the average attenuation is given by the following relation:

$$eatt \, (db/m) = K_2 \frac{\dfrac{(NVU + NVD)}{2} - \dfrac{(FVU + FVD)}{2}}{NVR - FVR}$$

$$(16-10)$$

where K_2 is a fixed constant, NVU and NVD are the near-voltage measured on the way up and the way down respectively, and FVU and FVD are the far-voltages measured on the way up and the way down. NVR and FVR are near-voltage and far-voltage references.

All measurements of voltage and phase are made and stored digitally. A complete BHC mode measurement is made once every 1/60th of a second and transmitted to the surface equipment via a digital telemetry system.

On the surface, measurements are accumulated and averaged over either 2 inch (5 cm) or 6 inch (15.2 cm) depth intervals.

The number of measurements averaged per depth interval depends on the logging speed. It can be computed as follows:

$$N = \frac{18,000 \, D}{LS} \qquad (16-11)$$

where: N = number of measurements averaged per interval; D = averaging interval in inches; and LS = logging speed in feet/hour.

A microlog pad mounted on a large caliper arm of the EPT tool body (Fig. 16-4) provides standard microinverse and micronormal signals. A short-arm caliper close to the pad measures hole rugosity and

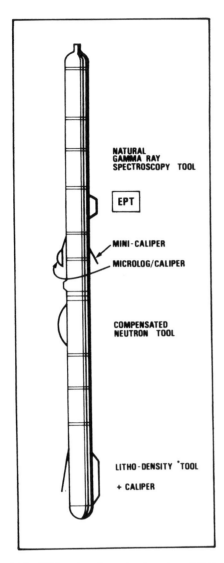

Fig. 16-4. The EPT tool configuration (courtesy of Schlumberger).

indicates where pad contact may be poor. The EPT tool can be combined with a number of telemetry compatible tools. Figure 16-4 gives an example of such a combination.

The present EPT tool has the specifications listed in Table 16-2.

TABLE 16-2

Specifications for the EPT tool (courtesy of Schlumberger)

Temperature rating	350°F or 175°C
Pressure rating	20.000 psi or 1400 bars
Tool diameter	
Housing	$3^{3/8}''$ or 8.5 cm
Antenna	$5^{5/8}''$ or 14 cm
With ML Pad	$6^{7/8}''$ or 17.5 cm
Minimum hole size	
with ML	$8^{1/4}''$ or 21 cm
without ML	$6^{1/2}''$ or 16.5 cm
Maximum hole size	22'' or 56 cm
Length of combination	96' or 29.3 m
Weight of combination	1862 lbs or 845 kg

16.5. DEPTH OF INVESTIGATION

The exact depth of investigation of the sensor is rather complicated to compute, owing to the geometry of the antennae. It may be approximated by the "skin depth", δ, which is inversely proportional to the attenuation;

$$\delta = \frac{8.68}{A_c} \, (m) \qquad (16\text{-}12)$$

where A_c is the corrected attenuation (see later).

The attenuation is a function of the conductivity (Fig. 16-5). So the depth of investigation is about 1 to 6 inches (2.54 to 15.2 cm) depending on the conductivity, and the measurements are normally related to the flushed zone, where the water is primarily mud filtrate and bound water.

16.6. VERTICAL RESOLUTION

The distance between the two receivers being short (4 cm), the EPT has a nominal vertical resolution of 2 inches (5 cm). But the actual resolution obviously depends on the data acquisition averaging mentioned previously. A vertical resolution of 2 or 6 inches may result. Consequently, a very good bed-boundary definition can be obtained with this tool.

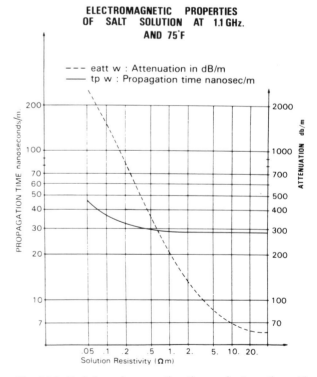

Fig. 16-5. Variation of propagation time and attenuation with salinity at 25°C (courtesy of Schlumberger).

16.7. ENVIRONMENTAL FACTORS INFLUENCING THE RESPONSE

16.7.1. Hole-size and shape

The EPT being a mandrel type tool, the bore-hole size has no significant effect on the measurement as long as good pad contact is maintained.

However, the rugosity of the hole-wall will cause erratic measurements because of the intermittent presence of mud between the pad and the formation.

16.7.2. Fluid

When the drilling fluid is air or oil-base mud it seems that even if the layer of fluid between pad and formation is very thin, the tool responds only to the fluid, not to the formation. This is due to the short travel time of these fluids compared to those of the common formations. Conductive muds, where the filtrate salinity exceeds about 30×10^3 ppm, generally lead to excessive attenuation of the signal. The EPT is not applicable in such conditions at moderate porosities.

16.7.3. Mud-cake

It has no effect on the measurement until its thickness exceeds about 3/8 in (0.9 cm), when the

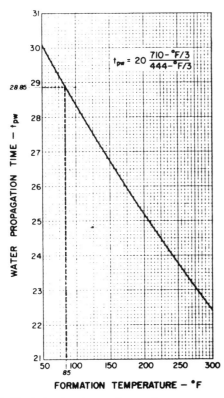

Fig. 16-6. Chart showing temperature and pressure corrections, assuming average temperature and pressure gradients (from Calvert et al., 1977).

measurement is strongly affected and may even respond entirely to the mud-cake.

16.7.4. Temperature

The propagation time of water decreases with increasing temperature and varies slightly with increasing pressure (Fig. 16-6).

The travel time of lossless (nonconductive) water can be obtained from the following relation:

$$t_{pw0} = 20\left(\frac{710 - T/3}{444 + T/3}\right) \quad (16\text{-}13)$$

T is the temperature expressed in degrees Fahrenheit.

16.8. INTERPRETATION

The data must be corrected for "losses" before interpretation.

16.8.1. Energy losses

An electromagnetic wave propagating through a formation undergoes energy losses and is attenuated. These losses are categorized into three types:

(a) *Spreading losses* always occur and are virtually independent of the propagating media. The energy loss is small and mainly dictated by the system geometry.

(b) *Conductivity losses* increase with the conductivity (Fig. 16-5), and very high attenuations are encountered under high salinity conditions.

(c) *Dielectric losses* are associated mainly with shales because of their high bound-water content.

16.8.2. Interpretation in lossless formations

We can consider lossless formations which are free of shale and have a low-salinity formation water. In these formations, an empirical formula, similar to Wyllie's formula for acoustic propagation, is used to relate the lossless propagation time, t_{p0}, to the propagation times of the formation constituents.

$$t_{p0} = t_{pw0}\phi S_{x0} + t_{ph}\phi(1 - S_{x0}) + (1 - \phi)t_{pma} \quad (16\text{-}14)$$

where t_{pw}, t_{ph} and t_{pma} are respectively the propagation time of water (at formation temperature), hydrocarbon, and matrix; ϕ is the porosity and S_{x0} the water saturation in the flushed zone.

When there is no hydrocarbon, the EPT porosity, computed from eq. 16-14 is similar to that of other total porosities given by other porosity-tools (CNL, FDC):

$$\phi_{EPT} = (t_{p0} - t_{pma})/(t_{pw0} - t_{pma}) \quad (16\text{-}15)$$

In hydrocarbon zones, since t_{ph} is close to t_{pma}, the

EPT tool "sees" only the water filled-porosity and we can to a first approximation derive an estimation of the water saturation in the flushed zone:

$$(S_{x0})_{EPT} = \phi_{EPT}/\phi_{ND} \qquad (16\text{-}16)$$

This approximation appears valid also in moderately shaley formations, where the porosities now include bound-water.

A more accurate value of S_{x0} derived from eq. 16-14, can be computed if we know the type of hydrocarbon and consequently its t_{ph}:

$$(S_{x0})_{EPT} = \frac{(t_{p0} - t_{pma}) + \phi_{ND}(t_{pma} - t_{ph})}{\phi_{ND}(t_{pw0} + t_{ph})} \qquad (16\text{-}17)$$

16.8.3. Interpretation in lossy formations

In formations with conductive water or with shale, the energy losses are not negligible. Their effect is to lengthen t_{pl} (see Fig. 16-5). Attenuation values from the log are used to correct t_{pl} to give a propagation time which is equivalent to that of a loss-free formation.

The apparent lossless formation travel time is given by the following relation:

$$t_{p0} = \left(t_{pl}^2 - \frac{A_c^2}{3604} \right)^{1/2} \qquad (16\text{-}18)$$

where A_c is the corrected attenuation computed as follows:

$$A_c = A_{LOG} - SL \qquad (16\text{-}19)$$

SL is the geometrical spreading loss. It is not a fixed number but is somewhat porosity dependent. SL has been determined from experimental data as follows:

$$SL = 44.6 + 1.32\, t_{pl} + 0.18\, t_{pl}^2 \dots \qquad (16\text{-}20)$$

A simple overlay of EPT porosity and true total porosity (eg from the neutron-density combination) provides an effective "quick-look" indication of hydrocarbons in the flushed zone. This can be performed at the wellsite with the CSU.

Attenuation spreading losses are approximated as 50 db (eq. 16-20). Log t_{pl} is corrected as in eq. 16-18. t_{pw0} is determined according to the temperature and ϕ_{EPT} computed for a chosen t_{pma} (eq. 16-15). A more accurate computation can be made in clayey sediment by applying the following response equations:

$$t_{pl} = \phi S_{x0} t_{pf} + \phi(1 - S_{xo})t_{ph} + V_{cl}t_{pcl}$$
$$+ (1 - \phi - V_{cl})t_{pma} \qquad (16\text{-}21)$$

$$eatt = \phi S_{x0} A_{cf} + V_{cl} A_{ccl} \qquad (16\text{-}22)$$

t_{pf} and A_{cf} are given by the following relations:

$$t_{pf} = \frac{1}{0.3} \text{Real } \sqrt{\epsilon_f^*} \qquad (16\text{-}23)$$

$$A_{cf} = 200 \text{ Imag } \sqrt{\epsilon_f^*} \qquad (16\text{-}24)$$

16.9. GEOLOGICAL PARAMETERS AFFECTING THE MEASUREMENT

16.9.1. Mineralogical composition

As seen in Table 16-1, the dielectric constants and propagation times of the most common minerals (except clays) vary over a small range. This shows that the mineralogical composition will have a minor influence except if some minerals with high dielectric constants (such as rutile) are present, or if the rock is clayey.

In this last case, the clay minerals, being both highly surface active and plate-like, will increase the dielectric constant (from Sen, 1980). The presence of bound-water causes a further increase.

16.9.2. Texture

Its influence is much more important. It is essentially expressed through the porosity, especially if it is water-filled.

From Sen (1980), the shape of the grains has a high influence when the electric field is applied perpendicular to the plates. So sandstones rich in mica might show a higher dielectric constant than would be expected from the respective values of the minerals.

16.9.3. Structure

Owing to its very good vertical resolution the EPT can theoretically detect thin beds. In fact, due to the averaging we lose this advantage. Consequently, the response corresponds to the sum of the elementary transit times in each bed or lamination constituting the averaged interval.

The shape of the grains and their orientation affecting the response, we might expect oblique laminations, cross-bedding and dipping beds to have an influence on the recorded value.

16.9.4. Fluids

As previously seen, the fluids have an influence dependent on their type (gas, oil or water) and their quantity (porosity and saturation). Their influence is higher if the fluid is water. The salinity of the water has a small effect on the transit time but a stronger one on the attenuation (Fig. 16-5).

16.10. APPLICATIONS

The EPT requires fresh mud filtrate (less than 30×10^3 ppm at moderate to high porosities) to operate effectively. Under correct conditions, the measurement can be used for:

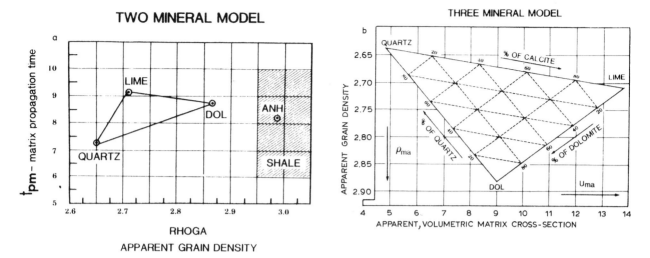

Fig. 16-7. a. Two-mineral matrix identification plot. b. Three-mineral matrix identification plot.

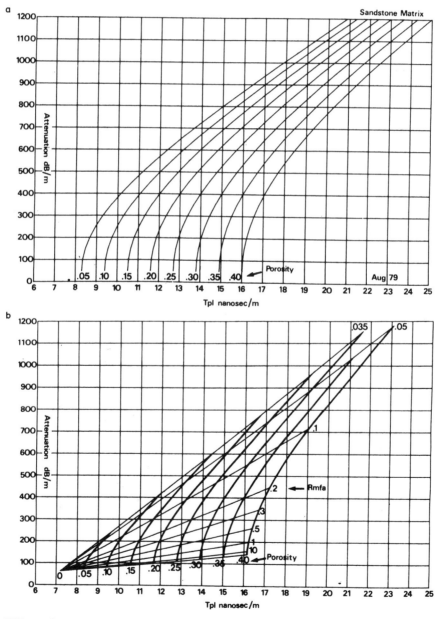

Fig. 16-8. a. EPT porosity t_{p0} method (courtesy of Schlumberger). b. EPT porosity CRIM method (courtesy of Schlumberger).

258

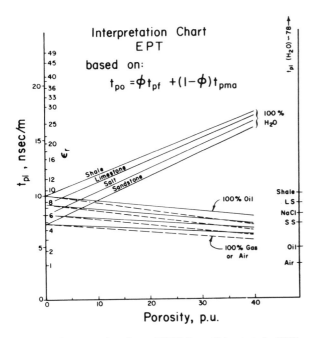

Fig. 16-9. Interpretation chart of EPT (from Calvert et al., 1977).

(a) Estimation of the water-filled porosity (Fig. 16-8) and from this, computation of the saturation in the flushed zone if we know the total porosity (by a neutron-density combination for instance). Detection of hydrocarbon is possible even if the formation water (Fig. 16-9) is of low or of varying salinity.

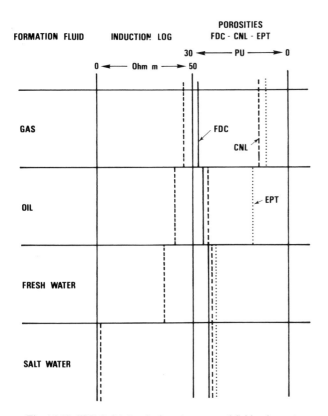

Fig. 16-10. EPT Quick-Look chart (courtesy of Schlumberger).

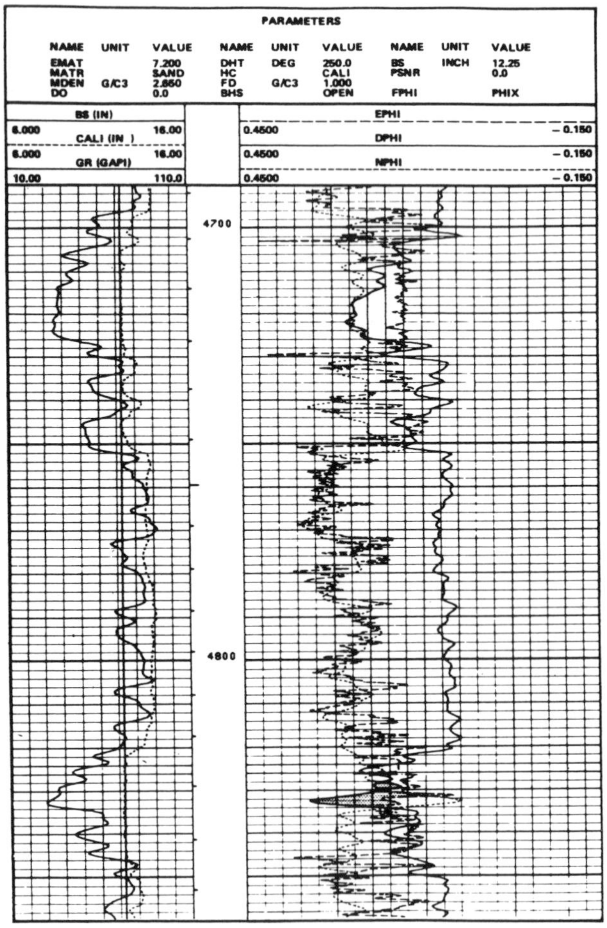

Fig. 16-11. Example of EPT Quick-Look: porosity comparison.

These two applications can be obtained by processing of the data at the wellsite as a CSU quick-look. (Figs. 16-10 to 16-12). The EPT is run in combination with the density-neutron, and such a quick-look is possible in a single run;

(b) Determination of the mineralogical composition of the rocks by combination with the apparent matrix density $(\rho_{ma})_a$ from the FDC-CNL, or with other logs such as LDT, NGS, (Fig. 16-13);

(c) Estimation of the shaliness from the attenuation. Identification of thin shale laminations;

(d) Contribution to a better definition of the electrofacies;

(e) Identification of hydrocarbons in thinly laminated sand-shale sequences.

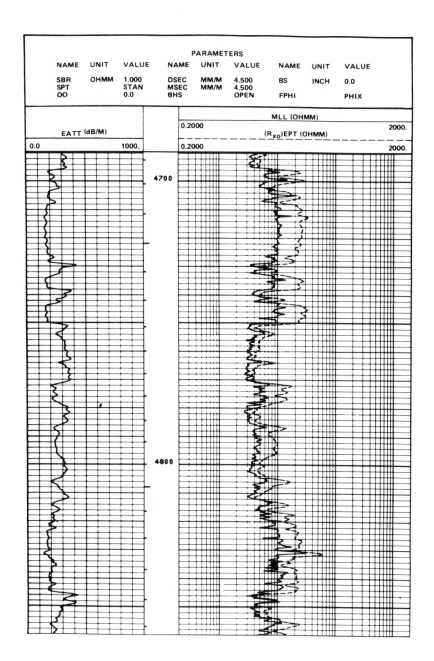

Fig. 16-12. Example of EPT Quick-Look: $(R_{xo})_{EPT}$ vs R_{MLL} for hydrocarbon detection.

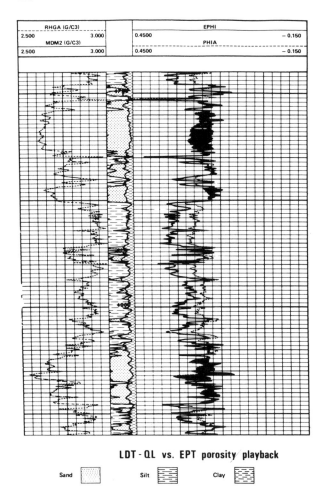

RHGA (G/C3)			EPHI		
2.500	3.000		0.4500		− 0.150
MDM2 (G/C3)			PHIA		
2.500	3.000		0.4500		− 0.150

LDT - QL vs. EPT porosity playback

Sand Silt Clay

Fig. 16-13. Example of EPT-LDT combination for Quick-Look interpretation (lithology, porosity, fluid content) (courtesy of Schlumberger).

16.11. REFERENCES

Calvert, T., Rau, R. and Wells, L., 1977. Electromagnetic propagation: a new dimension in logging. SPE Paper No. 6542.

Rau, R. and Wharton, R., 1980. Measurement of core electrical parameters at UHF and microwave frequencies. SPE Paper No. 9380.

Schlumberger. The Electromagnetic Propagation Tool (Brochure).

Schlumberger, WEC Venezuela (1980) Section III-d. WEC South East Asia (1981) Chapter 7c. Electromagnetic Propagation.

Sen, P., 1980. Dielectric and conductivity response of sedimentary rocks. SPE Paper No. 9379.

Taylor, L.S., 1965. Dielectric properties of mixtures. IEEE Trans. Antennae and Propagation, vol. AP-13 (6).

Wharton, R., Hazen, G., Rau, R. and Best, D., 1980a. Electromagnetic propagation logging: Advances in technique and interpretation. SPE Paper No. 9267.

Wharton, R., Hazen, G., Rau, R. and Best, D., 1980b. Advancements in electromagnetic propagation logging. SPE Rocky Mountain Regional Meeting. SPE Paper No. 9041.

Wobschall, D., 1977. A theory of the complex dielectric permittivity of soil containing water: the semidisperse model. IEEE Trans. Geoscience Electronics, GE-15 (1).

17. BOREHOLE CALIPER MEASUREMENTS

17.1. PRINCIPLE

The measurement of the diameter of the borehole is made using two arms, symmetrically placed on each side of a logging tool. The arms are linked to the cursor of a potentiometer (Fig. 17-1).

Variations in hole diameter cause the arms to close or open and the movement is reflected in resistance changes in the potentiometer.

A simple calibration allows the changes in resistance to be scaled to changes in diameter.

17.2. TOOLS

The majority of tool combinations include a caliper tool, giving a log of the hole diameter. This may be a separate tool or included in another tool (for example the FDC).

As the caliper arms are spring loaded the tool preferentially opens to the maximum hole diameter, in an ellipse along the major axis. It may, therefore, be useful to use a tool that has four arms, in order to get a more precise idea of the hole shape and of the hole volume.

Schlumberger offers the Borehole Geometry Tool (or BGT, see Fig. 17-2) that includes:

(a) Four arms coupled in pairs, each opening to a

maximum hole diameter of 30 inches. If required, arm extensions are added to increase this to 40 inches. The BGT gives a measure of the hole diameter in two vertical perpendicular planes.

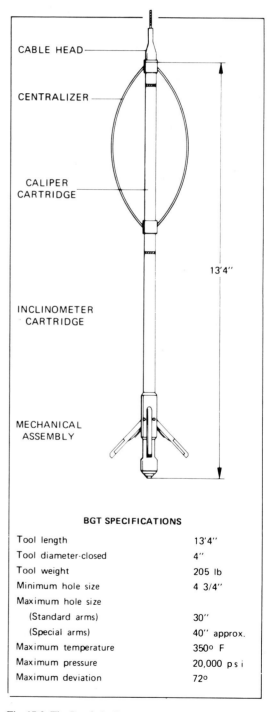

BGT SPECIFICATIONS

Tool length	13'4''
Tool diameter-closed	4''
Tool weight	205 lb
Minimum hole size	4 3/4''
Maximum hole size	
(Standard arms)	30''
(Special arms)	40'' approx.
Maximum temperature	350° F
Maximum pressure	20,000 p s i
Maximum deviation	72°

Fig. 17-2. The Borehole Geometry Tool (courtesy of Schlumberger).

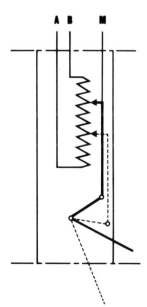

Fig. 17-1. Schematic explaining the measurement of the diameter. The cursor of a potentiometer is linked to the arm.

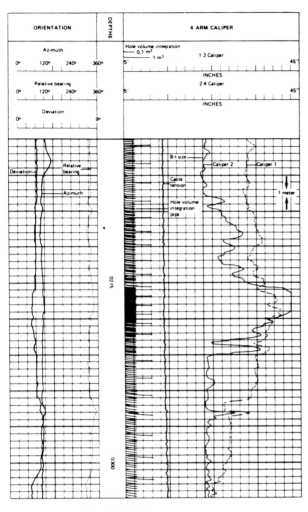

Fig. 17-3. Example of a BGT log.

(b) A hole volume integrator.

(c) An inclinometry cartridge allowing a continuous measurement of the angle and azimuth of the hole deviation and the orientation of the tool with regard to magnetic north. As the sonde is powered the tool may be opened and closed when required and while downhole.

The data from a BGT may be recorded on film or on magnetic tape. The tool is combinable with a gamma ray.

An example of a log is given in Fig. 17-3.

17.3. GEOLOGICAL FACTORS INFLUENCING THE HOLE DIAMETER

The diameter of the hole depends mainly on:

(a) The lithology, since certain rocks may:
  (1) be soluble in the drilling mud for example salt;
  (2) disintegrate and cave in (for example sands, gravel, shales), in which case hole caves will appear;
  (3) flow, as in the case of swelling shales or

low compaction shales, when the hole will close in;
  (4) be consolidated, in which case the hole will be in gauge.

(b) The texture and structure of the rock. These influence the porosity and permeability of the rock and hence determine whether a mud cake will develop and its thickness, leading of course to a reduced hole diameter. Note here that while some caliper tools cut through the mud cake and hence measure to the borehole wall, others ride on the mud cake. The texture and structure of the formation will also determine the bedding, the shale distribution, and also the possibility of microfractures brought about by drilling and radial cracking away from the borehole (as with consolidated shales and carbonate laminae). Fracturing linked with tectonic changes may also be mentioned here.

17.4. APPLICATIONS

Hole diameter measurements are used for:

(a) The detection of porous and permeable zones (mud cake presence) and the determination of mud cake thickness (Fig. 17-4): $h_{mc} = (d_{bit} - d_h)/2$.

(b) The measurement of hole volume in order to obtain an estimation of cement volume.

hole volume: $V_h \simeq \dfrac{d_h^2}{2} + 1.2\%$ (in l/m)

volume of cement: $V_{cement} \simeq \frac{1}{2}\left(d_h^2 - d_{casing}^2\right)$
$+ 1\%$ (in l/m)

d_h and d_{casing} being measured in inches.

(c) Detection of consolidated and in gauge sections for the scaling of packers for well testing;

(d) The correction of several log types for the effect of the borehole and of mud cake in order to obtain a more precise interpretation;

(e) A guide to lithology.

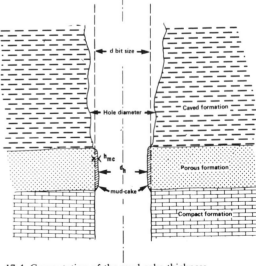

Fig. 17-4. Computation of the mud-cake thickness, $h_{mc} = (d_{bit} - d_h)/2$.

18. TEMPERATURE MEASUREMENTS (TEMPERATURE LOGS)

Generally temperature increases with depth and in undisturbed conditions it has a rate of increase with depth known as the geothermal gradient (Fig. 18-1).

This gradient varies according to the geographical location and the thermal conductivity of the formation. The gradient (Figs. 18-2 and 18-3) is generally low in formations of high thermal conductivity (salt or anhydrite for example) and high in the opposite case (e.g. shales) (see Table 18-1).

The overall temperature in a borehole will depend on the geothermal gradient except that temperature changes will occur due to the circulation of drilling muds that cool the formation as it is drilled. The transfer of heat at the mud–rock contact is by convection whereas in the formation itself it is by conduction.

The cooling effect of the mud at the bottom of the hole and the heating up effect higher up the hole will change the temperature profile and hence the thermal gradient. This is shown in Fig. 18-4. The temperature

at the bottom of the hole approaches the initial ground temperature, though there is still a difference

TABLE 18-1

Thermal conductivity of the most common minerals and rocks

Quartz	$6-30 \times 10^3$ CGS
Sand	$3-5 \times 10^3$ CGS
Calcite	10×10^3 CGS
Chalk	$2-3 \times 10^3$ CGS
Limestone (porous)	$3-5 \times 10^3$ CGS
Limestone (tight)	$5-8 \times 10^3$ CGS
Shale	$2-4 \times 10^3$ CGS
Coal	$0.5-1 \times 10^3$ CGS
Halite	$8-15 \times 10^3$ CGS
Granite	$5-8 \times 10^3$ CGS
Basalt	$5-7 \times 10^3$ CGS
Air	0.05×10^3 CGS
Natural gas	0.1×10^3 CGS
Oil	0.3×10^3 CGS
Water	1.4×10^3 CGS

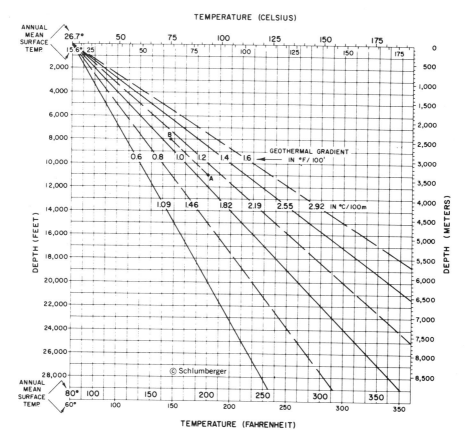

Fig. 18-1. Evolution of the temperature with depth and geothermal gradient (courtesy of Schlumberger).

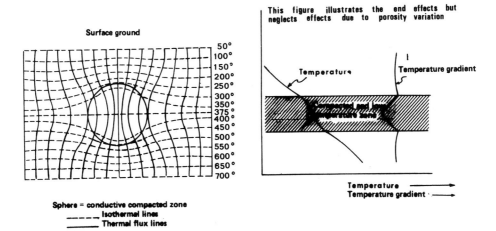

Sphere = conductive compacted zone
----- Isothermal lines
———— Thermal flux lines

Fig. 18-2. Repartition of the thermal flux and temperature profile in the case of a high thermal conductivity rock (from Lewis and Rose, 1970).

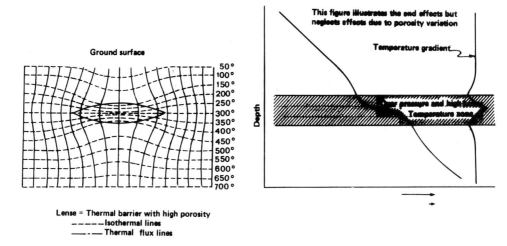

Lense = Thermal barrier with high porosity
----- Isothermal lines
——·—— Thermal flux lines

Fig. 18-3. Repartition of the thermal flux and temperature profile in the case of a low thermal conductivity rock (from Lewis and Rose, 1970).

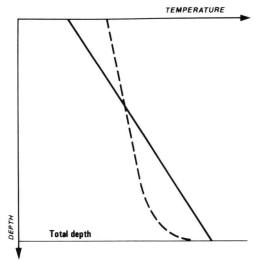

Fig. 18-4. Temperature profile in a well.

of between 10 and 20°C (circulation has stopped in this case).

Once circulation has stopped heat exchange from rock to rock means that there will be a gradual warming up of the rock to its original stable temperature. This will depend on the thermal conductivity of the rock but will generally take some time.

18.1. METHODS OF TEMPERATURE MEASUREMENT

18.1.1. Point measurements

Generally each logging tool run in the hole has attached a maximum reading thermometer, that gives a reading of the temperature at the bottom of the hole. If for each tool run the temperature is measured an increase in temperature is observed with time. This is the tendency to restore equilibrium and for the ground to return to its original temperature. From these measurements it is possible to extrapolate the initial temperature of the ground at the bottom of the hole, using a Horner plot of temperature.

18.1.1.1. *Principle*

In 1959, Lachenbruch and Brewer proposed a formula that allowed the temperature evolution with time to be extrapolated to the stabilised temperature (T_∞), identified as the initial ground temperature.

The data used in this formula are:

(a) the cooling time at the bottom of the hole (drilling + circulation time): t_k;

(b) the "warming up" time t_1 corresponding to the time at which the first logging tool reaches bottom and makes a temperature measurement. t_1 is measured from the end of circulation;

(c) the temperature measured at time t_1 is T_1;

(d) and in general, the temperatures are T_i measured at times t_i for each successive tool run. The formula was:

$$T = K \log\left(1 + \frac{t_k}{\Delta t}\right) + T(\infty)$$

which is a linear function of $\log[(\Delta t + t_k)/\Delta t]$ that is of the type:

$$y = ax + b$$

with

$$y = T \qquad x = \log\left(\frac{\Delta t + t_k}{\Delta t}\right);$$

$$a = K \qquad b = T(\infty)$$

Knowing two pairs of points (x, y) written as:

$$\left[T_1, \log\left(\frac{\Delta t_1 + t_k}{\Delta t_1}\right)\right] \text{ and } \left[T_2, \log\left(\frac{\Delta t_2 + t_k}{\Delta t_2}\right)\right]$$

we can construct the correct straight line. $T(\infty)$ is the ordinate of the intersection of this line with the y-axis where:

$$\log\left(\frac{\Delta t + t_k}{\Delta t}\right) = 0$$

(that is $x = 0$) or where:

$$\frac{\Delta t + t_k}{\Delta t} = 1.$$

This implies a t_k negligeable in comparison with Δt.

18.1.1.2. *Theoretical and experimental critique*

This technique of the estimation of the initial temperature of the ground has been taken up and codified by Timko and Fertl in 1972. They have

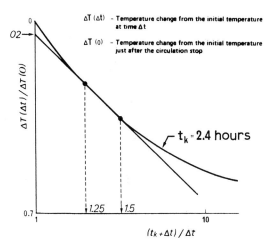

Fig. 18-5. Horner plot of temperatures for a well with a circulation time of 2.4 hours.

recommended the use of a Horner plot as shown in Fig. 18-5.

Evans and Coleman (1974) have used this method to interpret data given by BP, UK; CONOCO, UK; AMOCO, UK for the southern North Sea and have obtained a map of temperature gradients that are apparently very consistent.

More recently, in 1975, Dowdle and Cobb have published a theoretical and experimental critique of the method. They have shown that the Horner plot leads to an excellent temperature estimation when the cooling time t_k is no greater than a few hours (say less than 5 hours) (Fig. 18-5). However, the estimation becomes very poor when t_k is large (greater than, say, 10 hours) (Fig. 18-6).

Whatever the case, a Horner plot of temperature measurements made when the logs are run is the only way to get at the initial temperature, and in most practical cases this estimation will be good enough so long as the temperatures and time of measurement are correctly recorded.

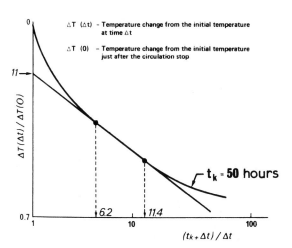

Fig. 18-6. Horner plot of temperatures for a well with a circulation time of 50 hours.

18.1.1.3. *Recommendations*

From what has been said it appears that we can estimate the initial formation temperature in the case where we have at least two pairs of temperature-time measurements taken in favourable conditions. This implies: (a) the recording of the maximum temperature reading each time a logging tool is run in the hole; (b) the least circulation time possible compatible with safety and being able to go down with logging tools. This is so as not to cool the formation too much before the temperature is measured (thus to minimize t_k); and (c) collecting enough data to get a good definition of Δt and t_k.

18.1.1.4. *Use of the Horner method*

This supposes:
(A) The collection of the following data
   (1) before recording logs;
      – bottom hole depth;
      – time at which drilling ceased (day, hour, minute);
      – time circulation stopped (day, hour, minute);
      – rate of penetration of the bit during the last ten metres drilled;
      – time taken to drill the last metre (t_{k1}) in minutes;
      – circulation time (t_{k2}) in minutes,

$$t_k = t_{k_1} + t_{k_2}$$

   (2) when each log is run:
      – logging tool type,
      – maximum depth reached,
      – time at which log started coming up (day, hour, minute),
      – time between logging started and circulation stopped, Δt,
      – maximum temperature recorded.

Remarks: although two temperature-time points are sufficient to draw the line:

$$T = k \log \frac{\Delta t + t_k}{\Delta t} + T(\infty)$$

it is preferable to have at least three or four points. This is why a maximum thermometer is used with each tool.

(B) The Horner plot itself.
This presents no particular difficulty. A semi-logarithmic grid is used.

In abscissa (*x*-axis) $(\Delta t + t_k)/\Delta t$ is plotted on logarithmic scale.

In ordinate (*y*-axis) the temperature is plotted on a linear scale.

The plotted points (Fig. 18-7) allow a line to be drawn. T, or the initial formation temperature, is read as the intersection of the line with the *y*-axis, for

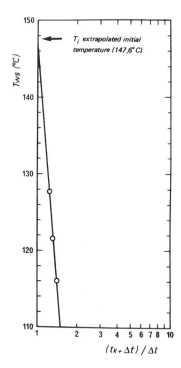

Fig. 18-7. Example of Horner plot in a well of Gulf Coast.

$x = 1$, that is for $(\Delta t + t_k)/\Delta t = 1$. In fact, in this case, the equation:

$$T = K \log\left(\frac{\Delta t + t_k}{\Delta t} \right) + T(\infty)$$

reduces to

$$T = T(\infty)$$

18.1.1.5. *Establishing the temperature profile of a well*

The temperature calculated by the Horner method is plotted against depth following each set of logs

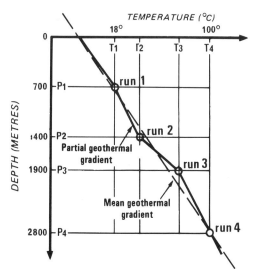

Fig. 18-8. Construction of the geothermal profile of a well.

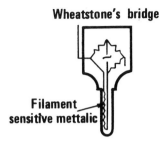

Fig. 18-9. Principle of the high-resolution thermometer (courtesy of Schlumberger).

(Fig. 18-8). We can then determine a geothermal profile for the well and from that an average geothermal gradient (in degrees per hundred metres or feet).

18.1.2. Continuous temperature measurements

These are made using a thermometer whose element consists of a temperature sensitive metal resistor, whose resistivity changes with temperature. The metal is a corrosion proof alloy that has a linear response in the normal temperature range (0 to 350°F), a low time constant and a low Joule effect (or heating effect) due to the measurement current.

The element makes up the fourth arm of a Wheatstone bridge (Fig. 18-9) that controls the frequency of an oscillator located in the tool cartridge.

The recording of temperature is usually made while going down in the hole in order not to disturb the thermal equilibrium by stirring up the mud with the movement of the sonde and cable (Fig. 18-10).

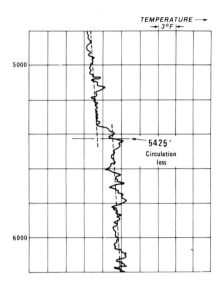

Fig. 18-11. Example of localization of a lost circulation zone from the temperature survey (courtesy of Schlumberger).

18.2. APPLICATIONS

18.2.1. Open hole

Temperature measurements allow the definition of changes in the geothermal balance (geothermal energy, thermal flux, maturing organic matter, etc.). We can then define the average geothermal activity of a well or a zone.

The thermal equilibrium destroyed by drilling re-establishes itself more or less quickly according to the thermal conductivity of the rock. Variations in temperature can then give some indication of lithology. This is one way of detecting low compacted shales

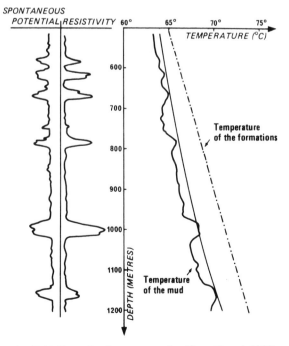

Fig. 18-10. Example of temperature log (from Guyod, 1946).

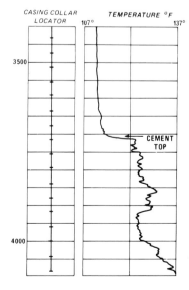

Fig. 18-12. Detection of the cement height behind casing from a temperature survey (courtesy of Schlumberger).

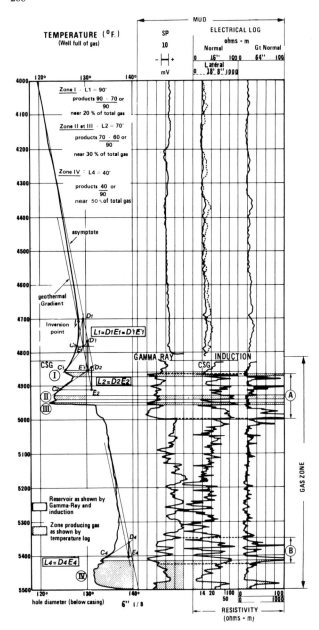

Fig. 18-13. Example of detection of gas producing zones from a temperature survey (from Kunz and Tixier, 1955).

(the temperature goes up more quickly, the geothermal gradient rises).

We can locate lost circulation zones (Fig. 18-11) or, on the other hand, fluid flow into the well, and in particular, gas, which is detected by the cooling effect brought about by gas expansion.

18.2.2. Cased hole

The main applications of temperature logs are in cased holes and in particular in production logging:

(a) detection of the cement height behind casing (Fig. 18-12) and channelling zones (fluid circulation behind pipe);

(b) detection of producing zones (Fig. 18-13);

(c) determination of the depth of the bubble point;

(d) detection of zones of fluid injection entry.

18.3. REFERENCES

Dowdle, W.L. and Cobb, W.M., 1975. Static formation temperature from well logs. An empirical method. J. Pet. Technol., 27 (11).

Evans, T.R. and Coleman, N.C., 1974. North Sea geothermal gradients. Nature, 247 (January 4, 1974).

Guyod, H., 1946. Temperature well logging. Oil Weekly, Oct. 21 + 28, Nov. 4 + 11, Dec. 2, 9 + 16.

Hutchins, J.S. and Kading, H., 1969. How to interpret temperature surveys. Oil Gas J., August 1969.

Kunz, K.S. and Tixier, M.P., 1955. Temperature surveys in gas producing wells. J. Pet. Technol.

Lachenbruch, A.H. and Brewer, M.C., 1959. U.S. Geol. Surv. Bull., 1083 C, 78.

Loeb, J. and Poupon, A., 1965. Temperature logs in production and injection wells. 27th Meet. Europ. Assoc. Explor. Geophys. (Madrid, 5-7 May 1965).

Riley, E.A., 1967. New temperature log pinpoint water loss in injection wells. World Oil, January 1967.

Schoeppel, R.J. and Gilarranz, S., 1966. Use of well log temperatures to evaluate regional geothermal gradients. Soc. Petrol. Eng of AIME, Paper 1297.

Timko, D.J. and Fertl, W.H., 1972. How Downhole Temperature Pressures Affect Drilling. World Oil, October 1972.

Witterholt, E.J. and Tixier, M.P., 1972. Temperature logging in injection wells. Presented at Fall Meeting of SPE of AIME, San Antonio, Paper No. SPE 4022.

19. DIP MEASUREMENTS (DIPMETER LOGS)

19.1. OBJECTIVE

The aim of this log is to determine the angle to the horizontal and the azimuth referenced to magnetic north and geographical north of the dip of the planes cut by the well. These planes can be
–bed boundaries;
–an open or closed fracture;
–an erosional surface;
–a stylolitic joint...
The planes can be planar, or can correspond to a convex or concave surface intersecting the well.

19.2. PRINCIPLE

A plane is defined by a minimum of three points not lying in a straight line. It should be sufficient, therefore, to know the coordinates (X, Y, Z) of three points in space to define the plane. These three points will be the intersection of three generatrices of the borehole wall with the plane (Fig. 19-1).

19.3. THE MEASUREMENT PROCESS (Fig. 19-2)

The tool consists of at least three electrodes[1] mounted on pads in a plane perpendicular to the axis

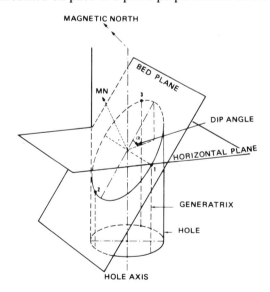

Fig. 19-1. A plane is defined by a minimum of three points not lying on a straight line. The three points are the intersection of three generatrices of the borehole wall with the plane (i.e. bed boundary).

(1) Nuclear or acoustic dipmeters do not exist until now.

of the tool and situated at angles of 120 degrees (3-pad tool) or 90 degrees (4-pad tool) to each other.

The three electrodes each make a resistivity measurement of the borehole wall. Because of the size of the electrode and the current focusing that occurs on each pad the resistivity measurement is assumed to be a point measurement at the electrode. When the tool crosses the boundary between two formations the corresponding response change is recorded for each pad at different depths according to the apparent dip (that is to say in relation to the borehole axis). The relative depth differences for the curves give the necessary information to evaluate the dip and the azimuth if we also know:

(a) the orientation of the sonde defined by the azimuth and one of the pads (pad number 1). This azimuth is the angle formed by the horizontal projection of the lines perpendicular to the sonde axis and passing through pad number 1 and magnetic north;

(b) the borehole deviation and its azimuth;

(c) the hole diameter (or more exactly the distance between the sonde and each pad).

The azimuth of pad 1 with respect to magnetic north is measured using a compass to which is fixed the cursor of a potentiometer. The movement of the compass pointer is reflected in changes in the potentiometer resistance. The borehole deviation is measured using a pendulum linked to a potentiometer whose resistance varies as a function of the deviation.

The azimuth of the deviation is measured by means of another pendulum that is continuously aligned in a vertical plane passing through the low side of the borehole. This pendulum is linked to a circular potentiometer whose resistance is a function of the angle formed by the azimuth of the deviation and of the azimuth of pad number 1 as reference (called the relative bearing).

Finally the hole diameter is measured using **potentiometers linked to the sideways movement of the pads.**

19.4. DIPMETER TOOLS *

19.4.1. Discontinuous dipmeters

19.4.1.1. *Anisotropy dipmeter*

We can mention as a reference the "anisotropy dipmeter", used by Schlumberger around 1936 and

* The toools described are those of Schlumberger.

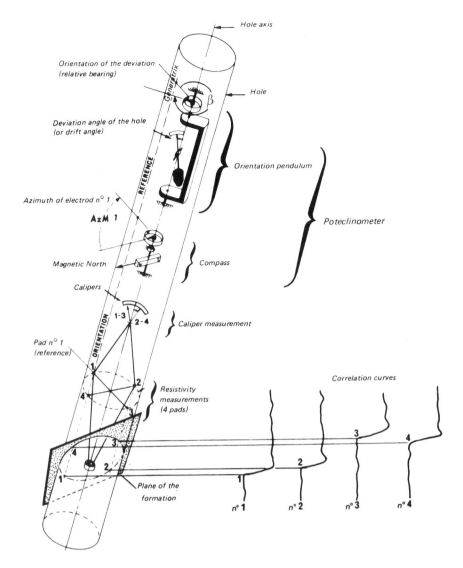

Fig. 19-2. Principle of the dipmeter showing the different measurements made by the tool (from Dresser-Atlas document).

which measured the difference in potential between two electrodes situated on axes at 90°. The orientation of the sonde and the borehole deviation were given by an induction compass. This method used the anisotropy property of shales.

19.4.1.2. *SP dipmeter*

Towards 1942, Schlumberger introduced to the market a three pad dipmeter, recording SP curves. A photoclinometer associated with the tool gave the orientation of electrode number 1 referenced to magnetic north as well as the deviation and azimuth of the well, in the form of a negative taken with the sonde stopped.

19.4.1.3. *Lateral dipmeter*

In 1945 the recording of SP curves was abandoned in favour of small "lateral" systems since the SP often had little character in salt muds or in front of very resistive formations.

19.4.2. **Continuous dipmeters**

19.4.2.1. *CDM (Continuous dipmeter)*

In 1952, Schlumberger put into service a continuous dipmeter (Fig. 19-3) that allowed the continuous recording of three parameters: the deviation and azimuth of the well and the orientation of electrode number 1, thanks to a system that transmitted to the surface the information recorded by each potentiometer.

At the same time there were changes in the pad curves. First pads of a microlog and then, later, of a focussed micro-device were used, with the pad applied to the formation by means of articulated arms (1956). According to the CDM type, hole diameters of between 4 and 19 inches could be logged.

The calculation of dips consisted first in establishing correlations between the three curves, noting the displacement by superposition of the curves, and then in defining the angle and azimuth of the dip knowing the angle and azimuth of the hole deviation.

Fig. 19-3. Three-pad dipmeter (CDM) (courtesy of Schlumberger).

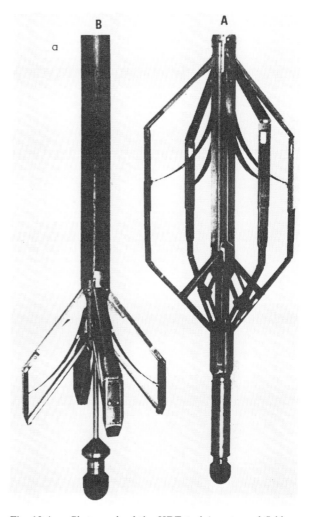

Fig. 19-4. a. Photograph of the HDT tool (courtesy of Schlumberger).

Although well developed, this method presented some problems:

(a) In holes that are caved or ovalized one of the pads may not touch the wall of the hole (floating pad) and the recorded resistivity would be about equal to the mud resistivity so giving a characterless curve. The calculation of dip becomes impossible since only two points are known. In some cases to get over this problem the log may be recorded twice in the hope that the sonde will turn and the pads take a different path.

(b) In very deviated wells, the tool presses against one side of the hole under its own weight, leading to a possible partial or complete closing of the sonde. The uppermost pad is likely to be free of the wall and so gives a characterless curve.

(c) In the case of very saline muds the resistivity curves will also be lacking in character and hence the correlations difficult or imprecise.

(d) In the case of oil base muds a film of oil comes between the pad and the formation and no current passes. Special pads called "scratcher pads" are avail-

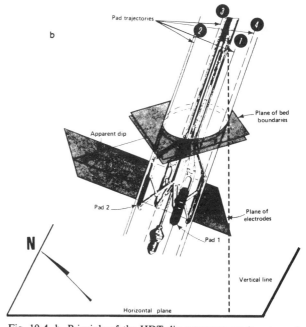

Fig. 19-4. b. Principle of the HDT dip measurement (courtesy of Schlumberger).

able to get over this. The results are often erratic as the pad's contact is very irregular as the tool moves in the hole.

(e) As already described above the correlations had to be established manually. The interpretation, itself a manual process for a long time, is now done on a computer.

Figure 19-8 shows a field example of a dipmeter log recorded with a CDM type tool.

Remark: The Dresser Atlas CDM had the advantage of having moveable pads that were free to rotate by +15° about a vertical axis at the arm end, so improving the pad contact with the wall.

19.4.2.2. *HDT (High resolution dipmeter tool: Fig. 19-4)*

Since 1967 a four-pad dipmeter has almost totally replaced the CDM (Fig. 19-4 a and b).

(a) The sonde is powered and can be opened and closed at will from the surface.

(b) It has four pads at 90° to each other. This means that if one pad does not touch the borehole wall because the hole is ovalized or the hole is deviated (floating pad) then three pads can still give good logs that allow the determination of dip planes.

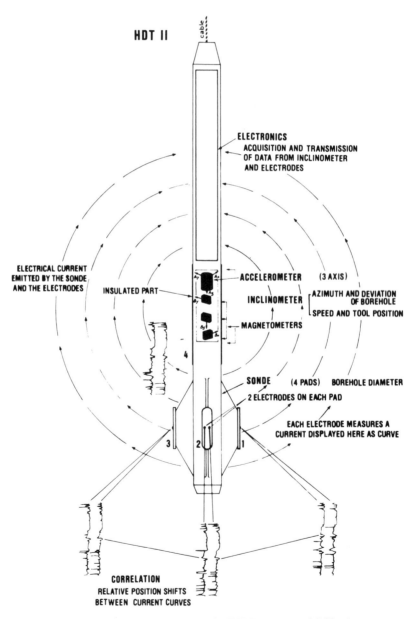

Fig. 19-5. The different measurements of the SHDT (courtesy of Schlumberger).

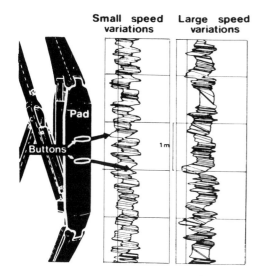

Small speed variations **Large speed variations**

Fig. 19-6. Influence of the speed variations of the tool on the dip measurement (courtesy of Schlumberger).

does not affect the second (orthogonal) pair. Generally one pair of arms will be opened along the major axis of the hole and the second pair along the minor axis (thinking of the hole cross section as an ellipse).

The minimum and maximum diameters with the available tools are $4\frac{1}{2}$ inches and 18 inches (for a given sonde type).

A new sonde type, known as a lightweight sonde (HDT-E) opens to 21 inches. In this type of tool the inclinometry section is in the sonde and not in a separate cartridge.

HDT tools have the following characteristics:

(a) The vertical definition is about 5 mm or 0.2 inches.

(b) The quantity of data collected and transmitted to the surface necessitates the use of an elaborate telemetry system: Pulse Amplitude Modulation-Frequency Modulated (PAM-FM) giving the following advantages: (1) no crosstalk (interference) between the different curves (signals); (2) transmission of signals at different frequencies; (3) improvement of the signal/noise ratio; and (4) easier conversion to digital recording.

(c) The recording is made simultaneously on film and magnetic tape. If for some reason the tape is destroyed or lost, the film can still be used to give optical correlations. The recording speed is around 3600 ft per hour.

(c) The pressure of the pads against the formation can be regulated in order to obtain better pad contact. (This is not true in all tools.)

(d) Since the arms are arranged in pairs with independant springing the pads follow the shape of the well much better. The collapse of one arm pair

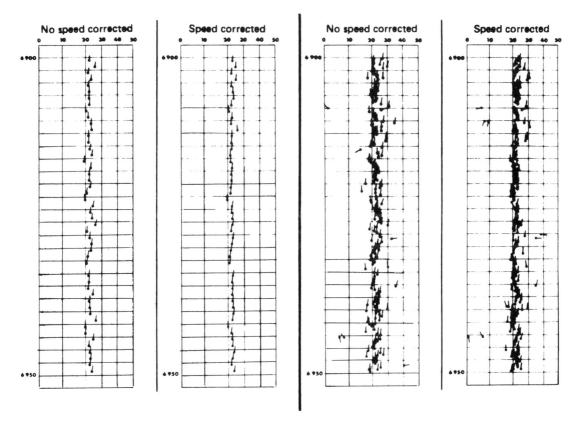

Fig. 19-7. Effect of the speed variation correction on the dip measurement (courtesy of Schlumberger).

274

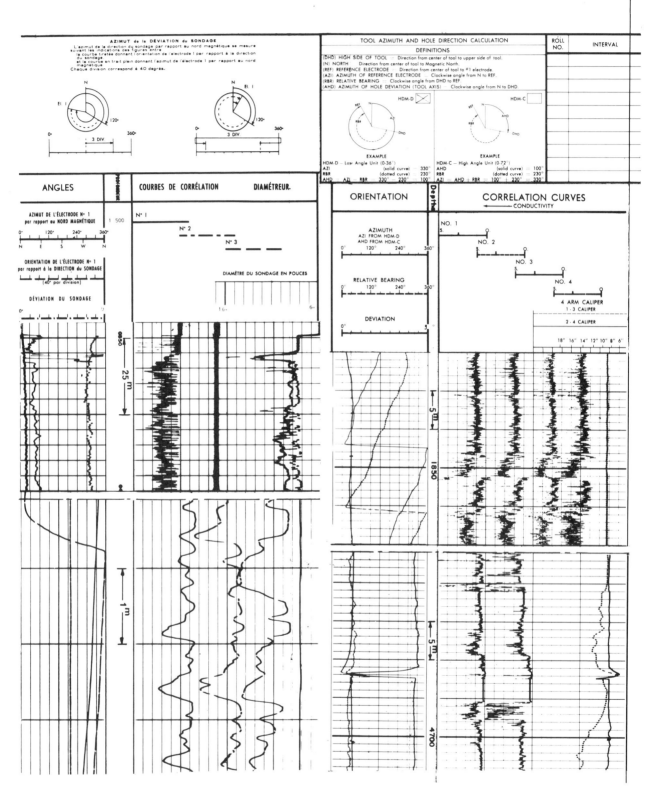

Fig. 19-8. Examples of dipmeter log (raw data). a. CDM. b. HDT.

(d) The correction for variations in the sonde speed in the hole is made automatically by computer. For this purpose a fifth resistivity curve is recorded by a second electrode situated on the pad number 1: curve 0 (Fig. 19-6). If the separation stays constant between the 0 and 1 curves and so give parallel correlations this means the sonde and the cable have the same speed. If the parallelism is bad this implies that there are differences between the speeds of the cable and the sonde (tool sticking ...). The speed correction curve is recorded on magnetic tape and not on film. The ratio of the actual distance between the 0 and 1 electrodes to the calculated separation allows correction of the displacement between the different electrodes (h_{1-2}, h_{2-3}, h_{3-4}, h_{4-1}). The effect of the speed variation correction on the dip measurements is shown in Fig. 19-7.

(e) In order to allow good depth matching between the HDT and the other logs, a resistivity curve is recorded on tape and then played back on the dipmeter processed data (arrow plot) alongside the "arrows". This curve is presented on a floating scale (that is no scale). This is related to the fact that the geometrical factor of the sonde is not well known.

Figure 19-8 shows a field example of a dipmeter log.

19.4.2.3. *PDT*

In some areas (South America) this tool is used in place of the HDT.

The PDT is a three-pad tool but with a powered sonde. The focusing of the electrodes is the same as the HDT. However, only one caliper measurement is made.

An analog system of transmission is used but the data is digitized on the surface, allowing both types of recording (film and tape).

Evidently being a three-pad tool there is always the possibility of a floating pad and again, in this case no interpretation is possible.

These two tools (PDT and HDT) are a great improvement over the CDM but still have certain limitations:

(a) Very saline muds: as mentioned the curves are often characterless, making correlation difficult.

(b) Highly deviated wells: the effect is less marked for the HDT than for the CDM or PDT. Further improvement has been brought about by Schlumberger by the introduction and use of a lighter sonde made of titanium (HDT Sonde type F) already mentioned.

(c) Turbine drilled holes: in general this drilling method causes a spiral to be cut into the wall of the well. These sections that project into the well appear more resistant and may simulate beds or layers and give dipmeter values that have no real significance.

(d) If the tool rotates rapidly in the hole and the dip is large there is a variation of the azimuth of electrode 1 between the moment when the first electrode passes across the bed boundary and when the last crosses. Since this effect is difficult to detect it is not usually corrected for. There is a case therefore, if the tool appears to rotate rapidly, to refuse the log and to repeat it, asking the engineer to reduce the tension on the cable by moving the tool in the hole.

19.4.2.4. *SHDT* (Fig. 19-5, p. 272)

Schlumberger is commercializing a new tool, the SHDT (Sedimentary Dipmeter Tool) with four independant arms, each carrying a pad with two small circular electrodes side by side and 3 cm apart. The sampling is made every 1/10 inch (2.5 mm). This system acquires more information on sedimentary structures: lithological or textural changes over very short distances can be detected. In addition the tool deviation is more precisely measured by a new magnetometer. The tool also has an accelerometer which allows better speed corrections, resulting in a more accurate dip measurement. The Emex current is automatically adjusted to always have the highest contrast in resistivity.

19.5. USE OF DIPMETER RESISTIVITY LOGS

19.5.1. Description of raw log

Where a three-pad continuous dipmeter is used, seven curves are recorded (Fig. 19-8a). These are:

(A) In the left-hand track *: (a) the azimuth of electrode number 1 referenced to magnetic north; (b) the orientation of electrode number 1 referenced to the hole direction (the relative bearing); and (c) the hole deviation.

(B) In the right-hand track: (a) the three resistivity curves, recorded by the electrodes on pads 1, 2 and 3 which are used as the correlations curves; and (b) the hole diameter.

In the case of the four-arm (pad) dipmeter (Fig. 19-8b) several presentations are possible: (a) The left-hand track has the same three curves as for the CDM (with the same differences for 36 and 72 degree cartridges); and (b) the right-hand track may have the four resistivity curves and both calipers or possibly on the 1/1000, 1/500 or 1/200 scale, to improve the readability, the third and/or the fourth resistivity curves may be left off. In particular, this allows the caliper to be more easily read in badly caved holes (Fig. 19-8b lower section). Marks may be recorded on certain curves (deviation and 2-4 caliper) and correspond to depth reference points between film and magnetic tape **.

We should note that the normal gradual progression from 0 to 360 degrees for the azimuth and relative bearing curves end with a rapid return to 0 as the tool continues to rotate in the hole.

---

* This is only true when a 36° inclinometry system is used. In the case of 72° system, high angle cartridge, a recording is made of: (a) the direction of the well directly referenced to magnetic north; and (b) the orientation of pad No. 1 referenced to the hole direction.
** Only in the case of the HDT-C where the recording is made as a function of time (DDR). For the HDT-D/E the recording is made as a function of depth (TTR) and so no depth reference marks are made on film.

19.5.2. Dip computation: manual method

19.5.2.1. Establishing correlations—measurement of the displacement between curves

A transparent film is needed to overlay another copy of the log in order to note the correlations. The displacement is determined in millimeters or fractions of an inch, expressed at the log scale, between the curves 1 and 2, then 1 and 3 * noting by the signs " + " all upward displacements of curves 2 and 3 from 1 and " − " all downward displacements. This method is very time consuming and hardly practical in view of the length of log to be processed (generally recorded at a scale of 1 : 20). Measurements of depth differences can be made by an optical correlator. In this piece of equipment the overlay of two of the correlation curves is made by moveable mirrors and the depth difference is read on a dial (Fig. 19-9).

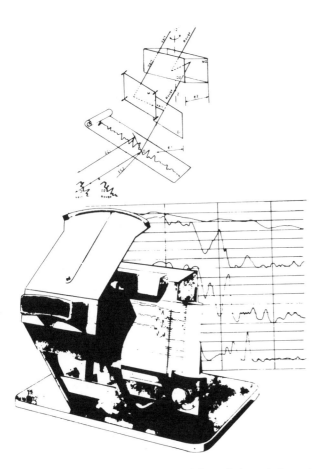

Fig. 19-9. Schematic of the principle of the optical correlation and photograph of the optical correlator (courtesy of Schlumberger and Seiscor).

---

* In the case of 3 pad dipmeters. In the case of four-pad tools we generally use the diagonal curves 1, 3 and 2, 4.

19.5.2.2. Use of the azimuth and deviation curves (left-hand track of the field log, Fig. 19-8a)

The values of the angle of deviation δ of the borehole are read directly from the log (generally to the left of the log with usual scales of 0–9°/ 18°/36°/72°).

In the case of a 36° inclinometry system, the curve giving the orientation of pad 1 in relation to hole deviation is a dashed curve, from which β = relative bearing.

The curve giving the azimuth of pad 1 (AZM 1) relation to magnetic north is a continuous curve.

These two curves allow the calculation of the azimuth of the deviation: magnetic azimuth of the deviation = magnetic azimuth of pad 1 − orientation of pad 1 relative to the well direction ($+ 360°$ if $\beta < 0$)

$$AZMd = AZM\ 1 - \beta$$

19.5.2.3. Computation principles

The majority of the methods are in three stages in the following order:

(a) Determination of the angle and the azimuth of the apparent dip in relation to the well direction (θ, F).

(b) Determination of the angle and the magnetic azimuth after correction for the deviation of the well (α, AZM).

(c) Determination of the geographic azimuth of the dip by correcting for the magnetic declination (AZG).

Several calculation methods have been proposed (see Appendix 3 for the application examples of the different methods) *.

Here we shall detail the method described by Schlumberger in their booklet "Fundamentals of Dipmeter Interpretation". It requires the use of a chart, whose mathematical justification is given in Appendix 2, and the use of a stereographic projection.

We shall only describe the general use of these charts and methods without going into the fundamentals of the theory.

(d) Determination of the apparent dip angle and the azimuth in relation to Pad 1. We enter chart 1 on axis No. 2 the depth displacement between curves 1 and 2 (up or down depending on the position of 2 in relation to 1) and as axis No. 3 the displacement between curves 3 and 1, expressed in millimeters at the log scale (1/20). The point so found allows us to note the combined displacement (see Appendix 2) expressed in millimeters (read on an ellipse or interpolated between ellipses) and the apparent dip

---

* These are only valid for three-pad dipmeters.

azimuth (AZMa) relative to electrode 1 read on the lines radiating from the centre.

The angle of apparent dip θ is calculated as a function of the hole diameter (see Appendix 2 and chart 2 in Fig. 19-10). We enter the combined displacement in millimeters and we read on a diagonal, at the intersection with the hole diameter, the apparent dip angle θ.

The apparent dip azimuth F in relation to the hole deviation is obtained by the addition of the apparent dip azimuth, AZMa, relative to pad 1 and the orientation of this same pad in relation to the hole direc-

tion (relative bearing read from the log left-hand track).

$$F = \text{AZMa} + \beta$$

(e) Determination of the angle α and the azimuth G of the dip with respect to the magnetic deviation of the hole. This requires the use of a stereographic net (Fig. 19-11).

The apparent dip azimuth F with respect to the hole and the apparent dip angle θ are entered in the net. From the point found we go down one of the small circles (see Appendix 3) by an amount equal to

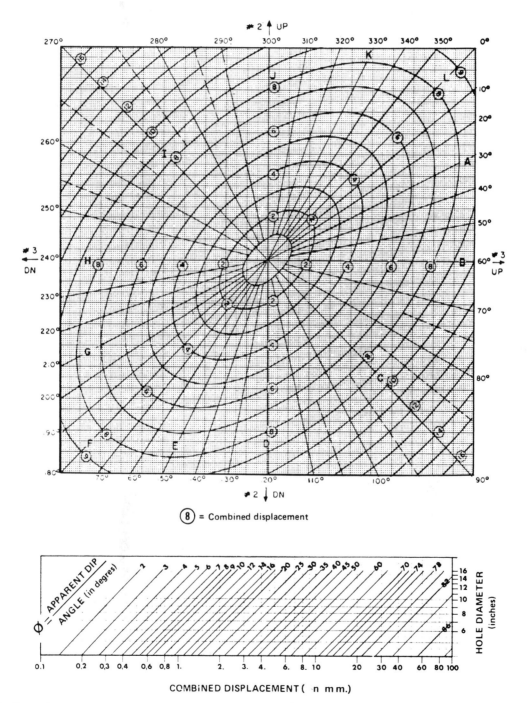

$\textcircled{8}$ = Combined displacement

COMBINED DISPLACEMENT (in mm.)

Fig. 19-10. Charts for the determination of the combined displacement and for the computation of the apparent dip (courtesy of Schlumberger).

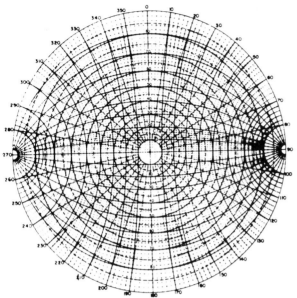

Fig. 19-11. Wulff diagram and stereonet for computation of the real magnitude and azimuth of the dip (courtesy of Schlumberger).

the deviation. From this new point, we read the actual dip angle α and the actual dip azimuth G.

(f) Determination of the magnetic azimuth of the actual dip, AZM. It is found by adding the actual dip azimuth G with respect to the deviation of the hole

and the magnetic azimuth of the deviation (AZMd).

(g) Determination of the geographic azimuth of the actual dip (AZG). The value of the actual dip azimuth in relation to geographic north is obtained by adding (or subtracting) the magnetic declination east (or west) to the magnetic azimuth value.

The second method also requires the use of a stereographic net. The displacements between curves 1-2, 1-3 and 2-3 should be measured and brought back to the actual depth scale (X20 for a film scale of 1/20).

These data allow determination of the apparent dip angle after use of the nomogram given in Fig. 19-12 also introducing the hole diameter. The value of apparent dip angle is read on the lower half circle at its intersection with the line coming from the diameter value passing through the displacement between the electrodes. The value of the dip is + or − depending on the displacement between the curves (up or down).

A series of operations made using a stereonet allows the determination of the geographic angle of the actual dip. The order of operations is given in Appendix 3.

Method Number 3 published in 1955 by Schlumberger uses a simplified graphical construction method. The necessary data are the values of the apparent dip azimuth and angle relative to the hole deviation as well as those from the left hand track of

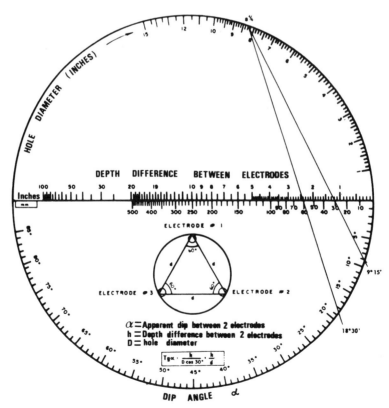

Fig. 19-12. Nomogram for determination of the apparent-dip angle.

the dipmeter log. The first can be computed using Method 1. The hole deviation correction allowing calculation of the values of the angle and magnetic azimuth of the actual dip is a graphical construction explained in the first method.

Looking at the tables describing the three methods (see appendix 3) it appears that the first method is the quickest and the most precise. The number of steps is reduced to a minimum.

This does not alter the fact that the manual method of dip computation is extremely time consuming and tedious in all its phases (determination of correlations, measurement of the displacement between curves, reading the values of deviation, azimuth and relative bearing on the log and the various steps in the calculation of dips). It can really only be made point by point in zones chosen beforehand. It has been abandoned for semi-automatic methods still used for the CDM and for optical methods for the HDT. This method consists of using an optical correlation to determine the displacement between curves and of reading the angle measurements in the left hand track. These values are assembled in a table that is then processed by computer. Obviously a continuous and fully automatic computer method would replace all of the manual or semi-automatic ones. This is the case as we shall see later.

19.5.3. Field magnetic tapes—HDT

(A) *Truck Tape Recorder (or TTR) and DDR Tapes*

This is a half-inch seven-track tape, one of which is the parity, with a bit density of 200 bpi. It contains the following information:

(a) Five resistivity curves that are sampled every $1/5$ of an inch (5 mm) * with a recording speed of 3600 ft/hr.

(b) The azimuth of the hole deviation or relative bearing.

(c) The deviation angle of the well.

(d) The orientation or azimuth of electrode (pad) 1.

(e) Two caliper curves (at 90° to one another).
These last four data are transmitted at a sampling rate of one sample per 3 or 4 inches (about $1/3$ of a second) at the usual recording speed (3600 ft/hr). We should note that only in the HDT-C is the sampling tied to time. In more recent tools (HDT-D) the sampling is strictly a function of depth.

According to the HDT type some other data are recorded concerning the cartridge temperature, the depth (with a sample every inch), the hydraulic pressure of the pads, the level of current emitted, and marks tying together depths on film and tape.

* The HDT-B samples every $1/10$ of an inch. Very few tools were made of this type.

These are three kinds of HDT tool:

(a) The HDT-C records on a DDR the data transmitted from the well. This tool is no longer in use.

(b) The HDT-D records on TTR. It transforms the data into analog form then redigitises for TTR recording. The recording is a function of depth (cable movement).

(c) The HDT-E is identical to the HDT-D except that the titanium sonde is lighter. The inclinometry cartridge is in the sonde. It also opens to 21″ instead of the nominal 16″ for other sondes.

(B) *Computer Unit Tapes*

The modern computer units used at the wellsite (for example the Schlumberger CSU) record HDT data in a new format that is similar to the edited tapes produced from Field Log Interpretation Centres (FLIC). The format used is a modified form of LIS (Log Information Standard) that records the data in discrete frames with each frame corresponding to one depth sample. While the angle and caliper channels are recorded in more or less standard LIS the resistivity or "fast" channels are recorded in blocks of raw data in the frame. In order to decode these blocks it is essential to know the prescribed order of data. However, this method of recording is vastly superior to previous methods since the depth is written onto the tape in each block of data and the general data format can be read by any computer programmed to record the data records.

Both HDT-D and HDT-E can be used with the CSU.

19.5.4. Dip computation: Automatic cross-correlation method

19.5.4.1. *Principle of automatic correlation*

This consists of a method of building a correlogram between two curves, hence determining the optimal correlation and so the depth displacement used in the dip calculation.

19.5.4.2. *Method of operation: building a correlogram*

In order to make the correlation curves comparable they are normalized to the same scale ranging from -10 to $+10$.

A section, or correlation interval Z, of one resistivity curve is compared to a similar length on a second curve. This second interval is displaced by a distance l where l varies from 0 to H, either up or down. H itself is determined by the search angle α (Fig. 19-13a and c).

Further:

(a) The interval Z on curve 1 is positioned on curve 2 but shifted by $+H$. At each point in the

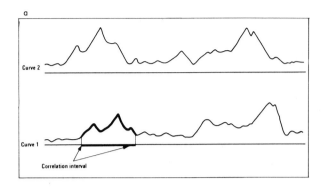

Fig. 19-13a. Correlation interval defined on Curve 1.

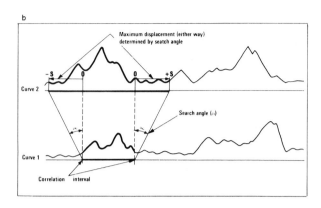

Fig. 19-13b. Displacement of the correlation interval on Curve 2 as a function of the search angle.

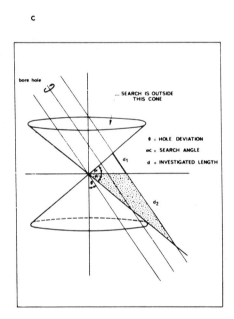

A : Schematic drawing of the normal search option in a deviated hole

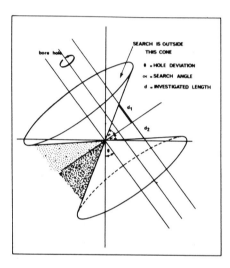

B : The "California option", low search angle

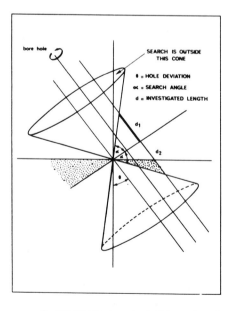

C : The "California option", high search angle

Fig. 19-13c. Determination of the displacement by the search angle. A. Schematic drawing of the normal search option in a deviated hole. B. The "California option" low seach angle. C. The "California option" high search angle.

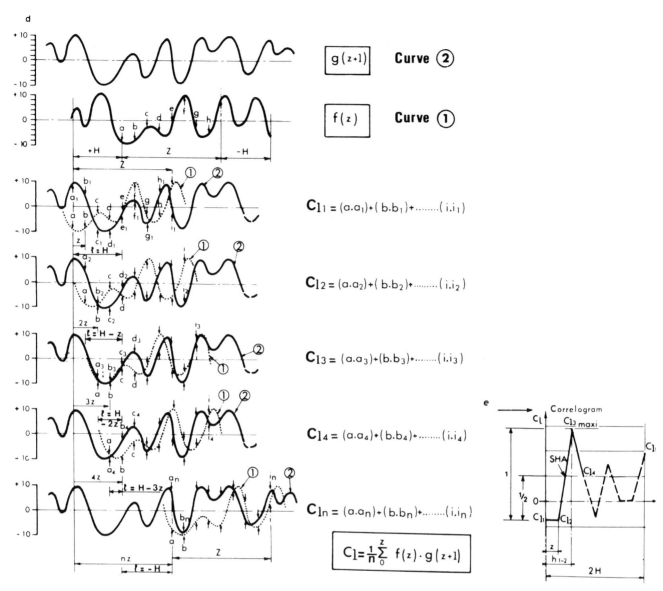

Fig. 19-13d. Principle of the computation of the correlogram function.

Fig. 19-13e. Construction of the correlogram which will be used to define the displacement of Curve 1 with respect to Curve 2.

interval Z we determine a value of the correlogram defined by the relation:

$$C_1 = \frac{1}{n} \sum_0^z f(z)g(z+l)$$

where n = number of samples in the interval Z.

(b) Then we displace curve 1 by distance z (separation between two samples) downwards on curve 2 and recompute C_1. This is repeated until the whole interval $2H$ on curve 2 has been covered (Fig. 19-13b, d).

We record C_1 as a function of the displacement l. The curve so produced defines the correlogram (Fig. 19-13e). The correlogram will have a maximum for C_1 for a certain value of l_1 if a correlation exists between curves 1 and 2. This value of L will be the displacement between correlatable points on the two

curves, and hence will be used for dip computation as was the optically found correlation and displacements used before.

In order to get to the next level we take a new correlation interval Z on curve 1, displaced from the first by the step distance D, which can be less than, equal to, or greater than Z.

19.5.4.3. *The importance of the correlation parameters and their effect on results (Fig. 19-14)*

(a) *Correlation interval*

This can vary between 0.5 and 10 m. If we want to look at fine detail we take the lower value. If we want to give less importance to detail then a large value is taken.

We can in exceptional cases go down to an inter-

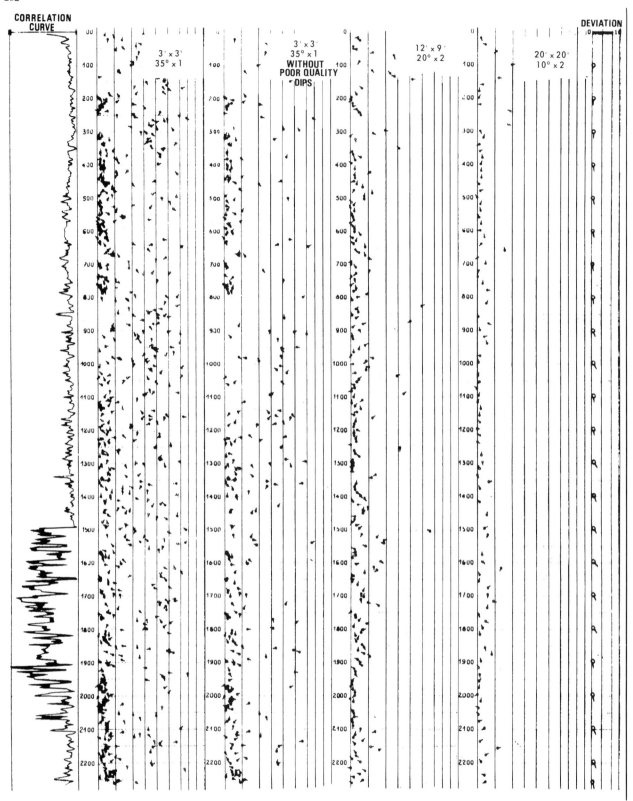

Fig. 19-14. Effect of the choice of the correlation parameters (correlation interval, search angle and step distance) on the dip measurement (courtesy of Schlumberger).

val 0.10 m but this brings into account the effects of hole rugosity, or noise, and we risk making correlations on small peaks, that may not be real dips or have any valid correlation.

For some time an interval of 1 m (3 ft) has been used, which allows us to have a maximum number of points, followed by a 3 or 5-m interval to look at more general features.

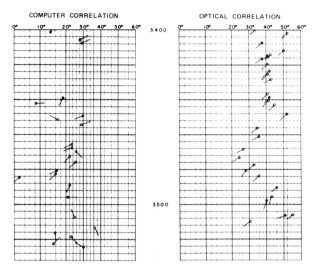

Fig. 19-15. Effect of the choice of the correlation parameters on the computation. Influence of a fixed search distance on the dip computation (on the left) compared with the dip computation made with a fixed search angle in a caved zone (from McRobertson, 1972).

(b) Search angle

The search angle may vary between 5° and 85°.

In the present programs the search angle is given

in dip degrees with the program defining for itself the search length as a function of this angle and of the diameter of the hole; in some programs a fixed search length may be specified. So, if we want to make an interpretation of the arrow plots made with these programs, it may be of interest to look at the caliper and possibly remake the processing if the hole is too caved.

This length defined for an average hole diameter leads in the case of very badly caved holes to erroneous or insufficient results, as is shown in Fig. 19-15. This can be explained by turning the (fixed) search length into a search angle using the varying hole diameter. For a very caved hole, the search angle so found, may be too small to find the correct correlation.

The search angle may be single (35° × 1) or double (35° × 2) (where 35° is just an example). It is double when we wish to emphasize low dips since in this case a search is first made in the angle 0 to 35° for a correlation and only if this is unsuccessful is a further search made in the angle 35° to 70°. Such a method can be dangerous in that a correlation found in the 0 to 35° range may be a lower quality correlation than in the 35° to 70° range, but the program

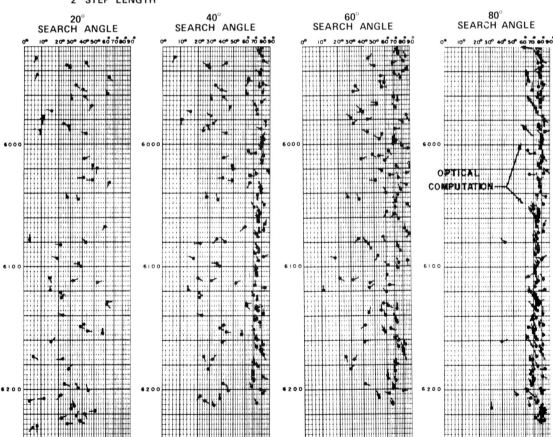

8' CORRELATION LENGTH
2' STEP LENGTH

Fig. 19-16. Effect of the choice of correlation parameters (search angle) on the dip computation (from McRobertson, 1972).

stops with the first correlation in the 0–35° range.

In fact, as the example in Fig. 19-16 shows, if the dip angle is high and the search angle too small (for example 20° × 2 or 60° × 1) no correlation can be found, but with a search angle of 40° × 2 a large number of dips appear in the 70° to 80° range and with an azimuth NNW that is almost constant. This is confirmed by an 80° × 1 pass and optical correlation.

In general we start with a search angle of 35° × 1 except where high dips are expected, in which case we should choose a search angle which when doubled allows a large interval to be searched (35° × 2 or 40° × 2). As a function of the first results we eventually choose a correlation interval of 3 m and if the dips are low a search angle of 10° × 1 or 10° × 2 for example. To increase detail the search angle has to be raised while reducing the correlation interval or vice-versa.

(c) Step distance

If this is smaller than the correlation interval good dips may be repeated. An example: 3m × 1m; in this case we risk finding the same correlation three times with a distance of 1 m between each repetition. In general, the step distance and correlation interval are chosen equal for the first pass. To choose D, the step distance, greater than Z excludes any possibility of dip repetition but means losing information.

19.5.4.4. Correlation quality coefficients

One of the dip computation programs, that is the one used until 1st November 1975 and called MARK IV, calculated and retained the following coefficients.

(a) Likeness coefficient (Likeness = MAX)

This is the maximum value of the correlogram C_h and varies from -100 to $+100$. The one that is kept and printed is the worst correlogram found between the four electrodes (the correlation between electrodes 0 and 1 on pad 1 is not taken into account).

If this is negative it is considered that there is no correlation. If it is zero or slightly negative we keep the correlation and print MAX = 0. Hence MAX is tabulated in a range of 0 to 100.

(b) Planarity (PLA)

This is defined as follows

$$PLA = 10 \frac{|h_{1\text{-}2}| + |h_{3\text{-}4}|}{|h_{1\text{-}2} + h_{3\text{-}4}|}$$

It checks that the four measure points lie in the same plane. PLA varies between 10 and infinity but is practically limited to a maximum of 100. It is taken as 0 when it cannot be calculated.

(c) Sharpness (SHA)

This is the amplitude of the largest peak of the correlogram obtained by correlation between electrodes 0 and 1, taken as half of the maximum from 0.

SHA is not tabulated but is used in the calculation of coherence.

(d) Closure (CLO)

This is defined by the equation

$$CLO = 10 \frac{\Sigma |h_{i-j}|}{|\Sigma h_{i-j}|}$$

It indicates an error in the displacements and varies from 0 to ∞. Everything that is greater than 100 is taken as equal to 100. A zero value is calculated when it is not possible to compute CLO. In the three pad tool case this happens when there is no correlation between pads 3 and 1.

(e) Coherence coefficient (COH)

This is defined by:

$$COH = K \frac{d_h}{SHA} \sqrt{\frac{100}{100 - MAX}}$$

K = coefficient generally taken as equal to 7
d_h = diameter (the sum of both calipers/2)
COH varies from 0 to ∞, but in practice is limited to 100. This coefficient serves to define dip quality but is not listed.

(f) Quality

This is represented by the letters A, B, C and D or an asterisk* and depends on the value of COH.

$$(\infty) > A > 75 > B > 50 > C > 25 > D$$

The asterisk (*) denotes a dip quality that cannot be defined either because of the absence of sonde speed correction or a closure less than 50. It corresponds to "tadpoles" on the arrow plot with white circle (Ο).

19.5.4.5. Dip sorting

In order to obtain outputs in which there are no doubtful results, Schlumberger will make, on request, a purge of points with an indetermined quality (* on the listing and white circles (Ο) on the arrow plots).

The criteria for dip sorting can be used on the coefficients CLO, PLA, MAX and SHA.

Dip sorting is established following a study of statistical analysis of calculation results (Table 19-1) made by Schlumberger on request of the user.

19.5.4.6. Simplified flow-chart

A simplified flow chart of the processing and calculations made in automatic interpretation is shown in Table 19-2.

TABLE 19-1

Listing of the quality coefficients vs. depth for a statistical analysis of the calculation results (courtesy of Schlumberger)

Scale		Percentage of planeity	Percentage of quality		Percent of 4P log in each L bracket	
100	8	8.4	9	9.0	0	.0
98	8	.0	9	.2	0	.1
96	8	.1	9	.3	1	.7
94	8	.0	10	.6	1	.7
92	8	.0	10	.1	3	1.2
90	8	.0	11	.5	4	1.1
88	8	.0	11	.5	5	1.5
86	8	.0	12	.5	6	1.0
84	9	.1	12	.3	7	.5
82	9	.1	12	.3	8	1.1
80	9	.1	13	.5	9	1.4
78	9	.2	13	.4	11	1.5
76	9	.2	14	.7	12	1.2
74	9	.0	15	.9	12	.8
72	9	.0	16	.9	14	1.3
70	9	.0	16	.3	15	1.1
68	9	.1	17	1.3	15	.3
66	9	.0	18	.8	16	.4
64	9	.1	19	.8	16	.6
62	9	.1	19	.6	17	.8
60	10	.3	20	.5	17	.4
58	10	.0	21	.9	18	.3
56	10	.0	21	.9	18	.1
54	10	.1	23	1.3	18	.2
52	10	.1	24	.9	18	.0
50	10	.2	25	1.2	18	.4
48	10	.0	26	1.5	19	.4
46	10	.3	28	1.7	19	.1
44	11	.4	30	1.7	19	.0
42	11	.4	30	.9	19	.4
40	12	.3	31	.9	20	.4
38	12	.2	33	2.1	20	.1
36	12	.3	34	1.1	20	.2
34	12	.3	36	1.3	20	.1
32	13	.3	37	1.0	20	.2
30	13	.5	38	1.5	21	.4
28	14	.7	40	1.5	21	.0
26	14	.3	41	1.3	21	.2
24	14	.2	42	.8	21	.3
22	15	.3	44	2.0	21	.2
20	15	.4	45	1.3	21	.1
18	16	.9	46	1.0	21	.2
16	16	.5	46	.5	21	.0
14	17	1.2	47	.6	21	.0
12	20	2.1	47	.5	22	.1
10	32	12.4	48	.3	22	.1
8	32	.0	48	.5	22	.0
6	32	.0	49	.4	22	.1
4	32	.0	49	.0	22	.1
2	32	.0	49	.2	22	.1
0	100	68.1	100	51.2	22	.0

19.5.4.7. *Mirror images*

By this term we mean those manual calculation results that are exactly opposite in azimuth to the azimuth of the deviation and of the same value (the image of the deviation in a mirror).

This occurs whenever the displacement between the three curves is zero. The correction for the hole deviation makes them appear as if opposite in azimuth but equal in dip.

This phenomenon can originate in two ways: (a) current jump (i.e. a technical problem); and (b) displacements that are too small to be calculated.

Automatic correlation using intervals has practically got rid of these parasitic results.

19.5.5. **Pattern recognition method of dip computation**

The automatic method described in Paragraph 19.5.4. does have a few problems.

(a) It does not allow all of the available information on the field tape to be presented since the number of dips theoretically possible is a function of the correlation interval used (one dip calculated per metre or by ten metres if the correlation interval is one or ten metres for example). From this we can estimate that at least 90% of the information is lost in certain cases.

(b) It can introduce false information since, as we have seen the results may vary depending on the parameters chosen (search angle, correlation interval, step distance) and may be very different from reality (Figs. 19-15 and 19-16).

(c) It gives information which does not correspond to any precise geological phenomena and the listing of dips may not be at the precise depth of the geological event that led to the dip.

Recognizing these problems, Schlumberger has put into use a new method of dip computation based on the mathematical recognition of patterns and hence the program is called "Pattern Recognition" (P.R.). Its commercial name is GEODIP.

This program seeks to resolve three types of problems:

(a) Better correction for the effect of speed changes on the curve shape.

(b) An increase in the density of calculated dips in order to solve sedimentary or fracture problems.

(c) Better calculation of dip in logs recorded in bad holes (caved, rugose, etc) or in deviated wells. (It is true to say however, that the P.R. technique works best where the quality of data is good. When the traditional correlogram methods kill there is little

286

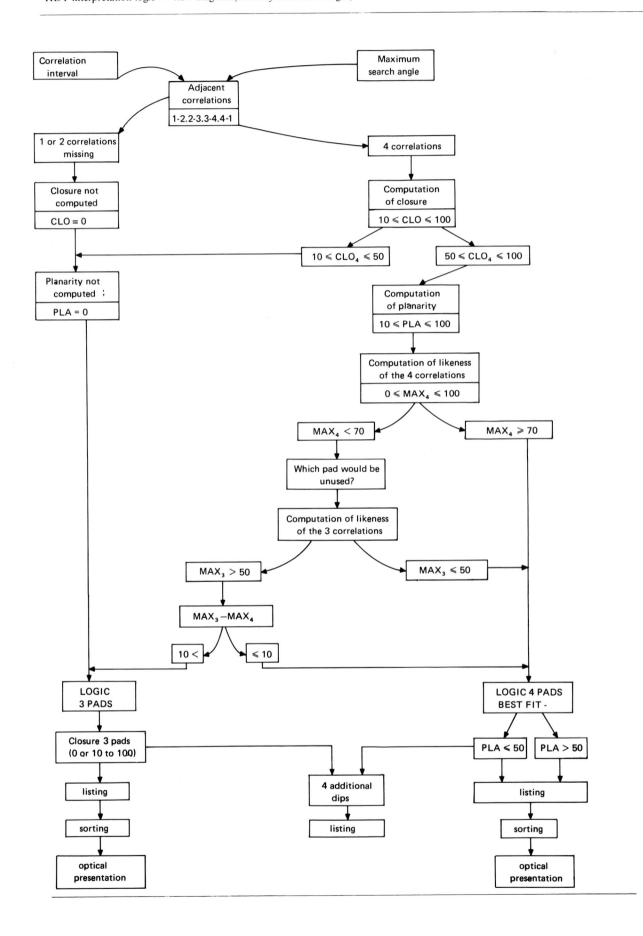

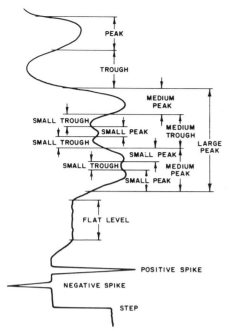

Fig. 19-17. Examples of various curve-element types (from Vincent et al., 1977).

point in attempting a P.R. technique.)
The program works in three phases.

19.5.5.1. *Phase 1: Feature extraction*

The curves are first analyzed with a reference to a catalogue of standard patterns such as: peak, trough, spike, flat level, steps, ramp (peaks, troughs and steps are the only ones actually used, Fig. 19-17). Each curve is described as a succession of these elements.

This analysis is made at three scales: large, medium and small. The attribution of each element or shape in each of the scales depends on a study of the variation of the derivative with a definition of cut-offs (a large peak corresponds to a large variation in the

derivative). The curve is filtered before calculating the derivative.

At the end of the feature extraction phase, the curves are replaced by their description in terms of elements. Each element is associated with one or two boundaries which give the position of the element on the initial curve.

Also associated with each element is a *pattern vector* which in the case of a peak is defined by use of the following nine parameters (Fig. 19-18):

(1) average ($P1$).
(2) maximum ($P2$).
(3) position of maximum, x_M, relative to boundaries, B_1 and B_2, as given by $P3 = (x_M - x_{B1})/(x_{B2} - x_{B1})$.
(4) maximum minus average ($P4$).
(5) balance left/right inflection-point smoothed-derivative values (d_1 and d_2), given by $P5 = -(d_1/d_2)/[1 + |d_1/d_2|]$;
(6) left jump ($P6$) *.
(7) right jump ($P7$) *.
(8) balance left/right jump, given by $P8 = -(P6/P7)/[1 + |P6/P7|]$.
(9) width of peak ($P9$).

Each of these parameters can be weighted or even left out if it is proved that it adds nothing to the determination of good correlations. Similarly for other types of elements, pattern vectors are defined with appropriate parameters.

19.5.5.2. *Phase 2: Correlation between patterns*

At the outset, we consider successively each large peak (or trough) and calculate the correlation between this element and the elements of the same type on another curve within a given search angle (in the standard interpretation this angle is taken as 50°).

To find this correlation a coefficient is computed. It measures the *likeness* between any two elements of the same type.

The likeness coefficient, K_L, is a positive number associated with a pair of elements of the same type or compatible types (e.g. large peak and medium peak), and computed from their pattern vectors. By definition, the smaller the likeness coefficient, the more alike the elements. (Two identical pattern vectors would give a zero coefficient value.) The likeness distance is used as a likeness coefficient

$$K_L(e,e') = \sum_{i=1}^{9} [V(e)_i - V(e')_i]^2$$

where: e and e' are two similar elements (e.g., peaks) on curves c and c'; and $V(e)_i$ and $V(e')_i$ are the ith parameters of elements e and e'.

* Note that the jumps, *P6* and *P7*, depend on minima in adjacent troughs; i.e. they are not completely contained in the peak itself.

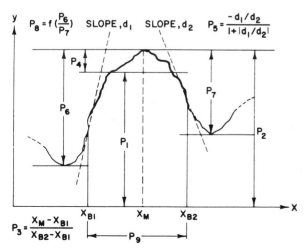

Fig. 19-18. Parameters for a peak (from Vincent et al., 1977).

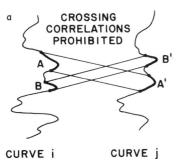

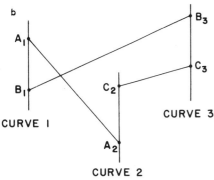

Fig. 19-19a. Crossing correlations are prohibited (from Vincent et al., 1977).

Fig. 19-19b. Combination of three curves illustrating three correlations which contradict the non-crossing-correlation rule (from Vincent et al., 1977).

The best correlation coefficient (the smallest difference in vectors) is retained as giving the correlation if on one hand it is smaller than the imposed likeness threshold (thus than the retained maximum difference) and on the other hand the displacement between the best coefficient and the adjacent in quality is greater than a fixed discernability threshold.

Each time we retain a correlation between two elements we deduce two correlations, one between the top boundary (top point of inflexion) and one between the lower boundary of the element. We always have therefore pairs of correlations, except for steps.

Rule of non-crossing correlation: once a correlation between boundaries has been retained no new correlation can cross the boundary lines (Fig. 19-19a, b).

We repeat this operation with medium and then small elements, progressively lowering the correlation thresholds. The progression is always then from the most to the least sure.

At the start we correct the curve shape as a function of the sonde speed, continuously determined using the same correlation methods on the two curves coming from the electrodes on pad 1. This speed can vary enormously, ranging sometimes from 0 and 3 times the theoretical logging speed (i.e. the cable speed at the logging unit).

19.5.5.3. *Phase 3: dip computation*

Having established the correlations the depth displacements between curves for each one are determined. The dip computation can then be made. If the planarity is poor, the average dip is calculated and the four dips using triangular correlations (three-pad). At such a depth we will then have five smaller arrows on the arrow plot.

19.6. RESULTS PRESENTATION

The results of computer processing of data recorded in a well are presented in various forms, mainly using a line printer or a plotter.

19.6.1. Listing presentations

The Service Company gives dip computation results in the form of listings.

In the case of MARK IV the listings given by Schlumberger give the following data in 13 columns: Depth; Dip angle; Dip azimuth; Deviation angle; Caliper 1-3 diameter; Caliper 2-4 diameter; LOGI corresponding to the program logic (Table 19-3).

Two asterisks are printed when the average dip is determined using four curves; the column is left blank when only three resistivity curves are used (CDM or when in the four-pad tool case one electrode has given poor quality results).

—Q = Quality of the results, is the coherence factor value (cf 18.5.4.4.) in general from 0 to 100. Letters A, B, C, D are given to different ranges of values and give the quality of the point.

D < 25 < C < 50 < B < 75 < A < 100

TABLE 19-3
Normal listing of MARK IV (courtesy of Schlumberger)

DEPTH	DIP	DIP AZM	DEV	DEV AZM	DIAM 1-3	DIAM 2-4	LO GI	Q	PLA	CLO	MAX	SPD COR
952.4	15.0	94	1.2	287	12.5	12.3		B	11	100	48	
953.3	14.3	108	1.4	293	12.5	12.4	**	C	85	53	46	**
954.3	10.2	158	1.3	299	12.4	12.4		A	0	100	35	
955.3	36.7	221	1.4	303	12.6	12.5		*	0	0	48	
956.3	10.6	352	1.4	301	12.6	12.6		D	0	100	42	
957.3	24.3	288	1.4	301	13.6	13.4		D	0	51	45	
958.3	23.7	102	1.5	296	14.2	13.4		D	0	100	82	
959.3	37.2	181	1.6	291	13.3	12.9		C	0	100	78	
960.3	26.0	170	1.6	291	13.2	13.0		D	0	95	86	
961.3	19.9	149	1.6	292	13.3	13.0		*	0	16	25	
962.3	13.9	144	1.5	289	12.5	12.4		C	0	100	81	
963.3	41.2	268	1.4	292	12.3	12.4		*	0	0	39	
964.2	4.7	154	1.3	295	12.5	12.5		A	0	100	46	**
965.2	20.6	64	1.4	296	12.6	12.5		*	0	31	41	
966.2	35.3	207	1.4	293	12.4	12.4		A	0	100	40	
967.2	37.3	70	1.5	290	12.6	12.4		B	0	84	24	
968.2	10.7	220	1.7	290	12.6	12.7	**	A	10	100	43	
968.2			4 DIPS	17.7	163	4.3	29	18.2	272	24.2	218	
969.2	21.5	331	1.8	292	12.7	12.6		*	0	21	32	
970.2	21.4	128	2.0	292	12.4	12.3	**	A	11	100	35	
970.2			4 DIPS	5.6	134	26.5	164	34.6	127	25.8	90	
971.2	18.6	74	1.9	295	12.6	12.4		B	0	68	41	
972.2	26.9	108	1.9	298	12.5	12.4		B	0	56	38	
973.2	38.9	9	1.9	302	12.5	12.7		B	0	100	26	
974.1	NO CORR		1.8	301	12.5	12.6						
975.1	36.4	239	1.7	296	12.7	12.6		*	0	35	23	
976.1	18.8	273	1.7	294	12.6	12.4	**	A	27	100	40	
976.1			4 DIPS	26.5	286	25.9	255	13.0	240	14.7	307	
977.1	17.6	174	1.7	296	12.5	12.8		*	0	10	67	
978.1	5.0	162	1.7	294	12.5	12.3		A	0	100	47	
379.1	22.6	75	1.9	292	12.9	12.6		*	0	21	42	
980.1	37.0	81	1.9	3U4	12.9	12.5		*	0	0	50	
981.1	33.6	80	2.0	306	13.0	12.6		*	0	0	29	
982.1	19.0	268	2.1	306	12.5	12.5		C	0	60	24	
983.1	NO CORR		2.2	305	12.5	12.3						
984.1	14.0	168	2.3	299	12.4	12.4		*	0	11	36	
985.0	NO CORR		2.3	297	12.8	12.5						
986.0	37.2	233	2.2	297	12.7	12.4		A	0	100	24	
987.0	NO CORR		2.2	296	13.1	12.9						
988.0	20.6	89	2.2	300	13.4	13.0		C	0	100	57	

TABLE 19-4
Listing of a Geodip processing (courtesy of Schlumberger)

BOUNDARIES BETWEEN SEDIMENTARY LAYERS ARE NOT ALWAYS PLANAR.
THE PLANARITY COEFFICIENT DEPENDS LINEARLY ON THE DISTANCE BETWEEN
THE TWO DIAGONAL CORRELATIONS.
    PLANARITY=100 WHEN DISTANCE =0 (PERFECTLY PLANE)
    PLANARITY=0 WHEN DISTANCE =5 INCHES

IF THE PLANARITY IS HIGHER THAN 80,THE AVERAGE DIP PLANE IS COMPUTED.

FOUR PAD CORRELATIONS WITH PLANARITY LOWER THAN 80 ARE RECOMPUTED WITH
THE 4-DIP 'OPTION. FOR EACH SUB 3-PAD CORRELATION ,THE DIP AND AZIMUTH
OF THE TRIANGLE PLANE ARE COMPUTED.
TR1 MEANS THE THREE PAD CORRELATION AFTER PAD 1 DELETION,ETC..

IN COLUMN B/T B MEANS:LOWER BOUNDARY OF AN ELEMENT
              T MEANS:UPPER BOUNDARY OF AN ELEMENT
              S MEANS:BOUNDARY OF A STEP

UNUSED PAD IS INDICATED BY A 0 IN THE CURVE COLUMN.

WHEN TWO(OR MORE)ELEMENTS OF DIFFERENT TYPES HAVE A COMMON BOUNDARY,
THE DIPS DERIVED FROM THE CORRELATIONS(EITHER 4 OR 3 PADS) ARE
SUCCESSIVELY LISTED WITH THE SAME DEPTH,EVEN IF THE RESULTS REPEAT.

CODING OF TYPES OF CORRELATED ELEMENTS

```
1 : SMALL PEAK
2 : SMALL TROUGH
3 : MEDIUM PEAK
4 : MEDIUM TROUGH      RESISTIVITY
5 : LARGE PEAK         INCREASES
6 : LARGE TROUGH
7 : STEP
```

```
* DEPTH     DIP AZI PLA DEV AZDEV B/T     TYPE OF ELEMENT         *
*                                      CUR1  CUR2  CUR3  CUR4     *
*****************************************************************
* 2076.30  12.8 193  0  0.4 295   T     0    6    4    2
* 2076.35  11.9 195  0  0.4 295   T     0    6    4    2
* 2076.35  11.9 195  0  0.4 295   T     0    5    3    3
* 2076.47  12.6 215  0  0.4 295   B     0    5    3    3
* 2076.47  12.6 215  0 ;0.4 295   T     0    6    6    6
* 2076.69   4.4  97  0  0.4 295   T     0    6    6    6
* 2076.80  22.6 235  0  0.0 297   T     0    6    6    4
* 2076.89  18.7 239  0  0.0 295   H     0    6    6    4
* 4 DIP COMPUTATIONS TH1 18 239 TR2 .12 182 TH3 8 85 TR4 7 297
* 2076.89   5.7 219 76  0.0 295   T     5    5    5    5
* 2077.09   8.2 214 87  0.4 295   H     5    5    3    5
* 2077.21  10.9 122  0  0.4 297   T     5    5    0    5
* 2077.30   5.6 155  0  0.4 297   B     5    5    0    3
* 2077.30   7.7 174  0  0.4 297   T     6    6    0    6
* 2077.38   4.8 218  0  0.0 297   H     6    6    0    6
* 2077.77   4.9 112 68  0.4 295   T     6    6    3    6
* 2077.91  14.5  88 80  0.4 295   B     5    5    3    5
* 2078.52  23.6 122  0  0.4 297   T     3    5    0    5
* 2078.67  22.8 135  0  0.4 297   T     3    0    0    4
* 2078.67  22.8 135  0  0.4 297   T     3    0    0    4
* 2078.82  16.5 127  0  0.4 295   H     4    0    4    4
* 2079.40  20.6  68  0  0.4 297   H     0    3    5    5
* 2079.56   7.0 102  0  0.4 295   T     0    3    5    0
* 2079.56   6.4 139  0  0.4 297   H     3    7    7    0
* 2079.54   4.3 128 93  0.4 297   T     6    6    6    6
* 2079.66  13.2 103 99  0.4 297   H     6    6    6    6
* 2079.66  13.4 102  0  0.4 297   H     0    1    1    3
* 2079.72   4.8 114  0  0.4 295   H     0    1    1    3
* 2079.93   8.8 186  0  0.4 298   T     5    5    0    5
* 2079.93   8.8 186  0  0.4 298   H     1    3    0    3
* 2079.99   9.5 133  0  0.4 298   T     1    3    0    0
* 2080.05   5.1  36  0  0.4 298   T     1    3    3    0
* 2080.11   4.1 103  0  0.4 298   H     1    3    3    5
* 2080.11  10.3 101  0  0.4 298   B     5    5    0    5
* 2080.12   6.7 127 90  0.4 298   T     6    6    6    6
* 2080.21   9.8 114 88  0.4 298   H     6    5    6    6
* 2080.21   9.8 114 88  0.4 298   T     5    5    3    5
* 2080.30  10.3 124 94  0.4 298   H     5    5    3    5
*****************************************************************
```

```
                S T A T I S T I C S   O N   R E S U L T S
THIS ZONE CONTAINS   2537   RESULTS

   1209   RESULTS ARE 4-PAD CORRELATIONS

    309   RESULTS ARE 3-PAD CORRELATIONS WITH PAD 1 DELETION

    438   RESULTS ARE 3-PAD CORRELATIONS WITH PAD 2 DELETION

    277   RESULTS ARE 3-PAD CORRELATIONS WITH PAD 3 DELETION

    304   RESULTS ARE 3-PAD CORRELATIONS WITH PAD 4 DELETION


STATISTICS ON PLANARITY( PLA ) OF 4-PAD RESULTS

P:0 5 10 15 20 25 30 35 40 45 50 55 60 65 70 75 80 85 90 95 100
*: 0 0  0  0  0  0  0  0  0  1  0  0  1  2  2  7 12 28 47


   6% OF 4-PAD RESULTS HAVE BEEN COMPUTED WITH THE
4-DIP OPTION(LOW PLANARITY)


STATISTICS ON TYPES OF CORRELATED ELEMENTS

TYPES:       1    2    3    4    5    6    7

CURVE 1:(%)  2    1   10    9   34   35    4

CURVE 2:(%)  2    1   11   11   33   33    4

CURVE 3:(%)  2    2   13   10   33   33    4

CURVE 4:(%)  3    2   12   10   33   33    4
```

Corresponding points on the arrow plot are black circles with an arrow (⬤).

The asterisk (*) indicates that calculation was impossible for closure or speed correction. Corresponding points on the arrow plot are white circles (◯).

—PLA gives the planarity, from 10 to 100. 0 indi-cates that the index cannot be calculated but not that the planarity is bad.

If the index is bad and if all four curves have been used, four values (angle and azimuth) for the dip are given as well as the average (correlations using three pads at a time).

—CLO (closure) of the three or four correlations taken into account in the calculation from 10 to 100.

As for PLA, 0 indicates that no calculation is possible (for example when no diagonal correlation is present in the case where three out of four curves are used) but not that the closure is necessarily bad.

—MAX is the value of the maximum of the worst correlogram that has been kept. It varies from 0 to 100, with 100 representing a very good correlation.

It is important in dipmeter interpretation to refer to listings in order to evaluate the calculated points. We should remember, for example, that a point whose quality is (*) and whose closure is 0, is not to be rejected if the value of MAX is 90. The 0 value noted in PLA and CLO and the * in the quality only show that no calculation of these indices was possible.

In the Geodip case the listing is as shown in Table 19-4: (a) it gives the shape of the element correlated by a numerical code (shape 4 on curve 1, shape 4 on 2, shape 2 on 4), and (b) statistics on results.

19.6.2. Graphical presentation

Dip results may be presented graphically in many ways.

19.6.2.1. *Arrow plots*

This is the most widely known presentation (Fig. 19-20) and can be produced at depth scales of 1/1000, 1/500 or 1/200.

Each calculated point is plotted:

(a) In abscissa according to the dip angle.

(b) In ordinate with depth.

The point is extended by a short straight line in the dip direction, with geographical north towards the top of the plot (or log). Also shown are: (a) the angle and azimuth of the borehole deviation (intervals of 50 m or 100 feet, usually); and (b) a scaleless resistivity curve allowing depth correlation with other logs. This curve is drawn with low resistivities to the left.

Different symbols are used to represent points:

(a) A black circle indicates a good quality result.

(b) An open or white circle indicates a point for which the quality is undefined (cf. 19.5.4.4. and 19.6.1.).

(c) A cross shows that the four-pad logic has been used, however this symbol is not used in the standard presentation.

290

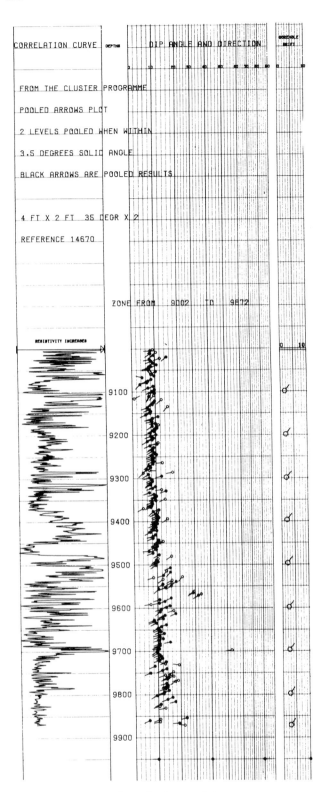

Fig. 19-20. Example of arrow-plots.

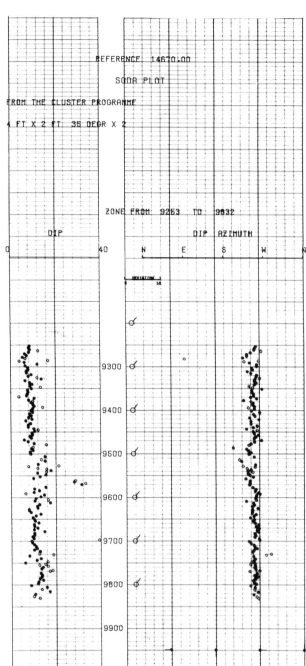

Fig. 19-21. Example of SODA plot.

19.6.2.2. *Soda * plot*

This plot presents the same information as an arrow plot but here dip angle and azimuth are drawn

* SODA means: Separation Of Dip and Azimuth.

in two different tracks (Fig. 19-21). In addition the azimuth is presented linearly which allows one to notice a preferential azimuth and its value.

Added to these log-type presentations are presentations of statistical results.

19.6.2.3. *Azimuth frequency plots (Fig. 19-22b)*

Calculated azimuth values in a depth interval are plotted in the form of circular histogram. The value of the azimuth is read clockwise from 0 to 360° and the frequency of occurrence is shown as a radial line whose length is proportional to the number of points

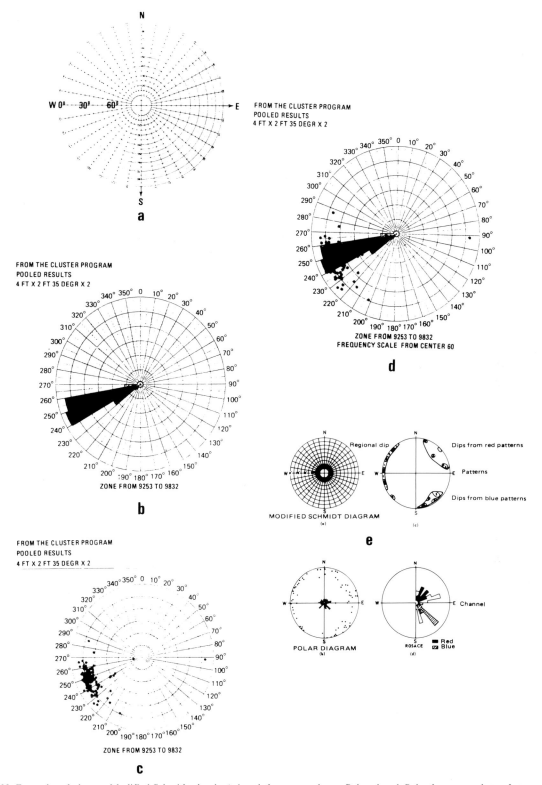

Fig. 19-22. Examples of plots. a. Modified Schmidt plot. b. Azimuth frequency plot. c. Polar plot. d. Polar frequency plot. e. Interpretation of polar plot (from Campbell, 1968).

whose azimuth falls in the range. Zero values are at the centre.

The intervals on which those plots are made should be fixed by the user and should only include one stratigraphic unit or lithological section.

The plots allow the determination of azimuth trends, the azimuth or regional dip, sedimentary dip, etc....

Structural dip removal (subtraction) is possible using this presentation (in fact in all the presentations, see 19.7.3. after this section).

19.6.2.4. *Polar plot*

This is made using the same type of diagrams as above but with concentric circles added, graduating in dip angle value from 0° on the outside to 90° at the centre. This is a modified Schmidt plot (Fig. 19-22a). The scale can be adjusted to suit the case under study (Fig. 19-22c).

Each point is plotted according to its dip angle and azimuth. Having 0° at the outside allows better localization of low dips.

We can combine in one diagram azimuth frequency and polar plots. In this case we limit ourselves to low dip angle values for the polar presentation (this combination is called a POLAR-F plot, Fig. 19-22d).

In a further stage we can draw on the polar plot the envelope curve of the plotted points. The points corresponding to structural dip (in the case of dip no greater than 2 or 3°) will appear as a thin and very long section, covering a wide range of azimuth values (influence of local variations).

By contrast, points corresponding to tectonic events, stratigraphic features, etc., are grouped in triangular forms less dispersed in azimuth but more stretched out in dip values (Fig. 19-22e).

19.6.2.5. *FAST * plot*

This is a presentation of the intersection of dip planes with the borehole considered as a cylinder in space (Fig. 19-23).

This diagram when rolled up and placed inside a transparent cylinder allows a visualization of the dips as they would appear on an orientated core.

Its major interest is in displaying important tectonic phenomena or major sedimentary discontinuities, since the depth scale usually used does not allow to go down to a detailed study.

19.6.2.6. *Stick plot*

This is the presentation of the intersection of dip planes with given vertical azimuth planes (Fig. 19-24). The azimuth values can be chosen by the user, if not the six values of 360°, 30°, 60°, 90°, 120°, 150° are arbitrarily kept. These sections allow a geologist to trace the dips on a non-deviated cross-section of the well. Schlumberger is now studying a method of vertically representing the deviation of a well on which will be drawn the stick plot intersections according to the different azimuth planes. The use of this diagram which could be drawn on different

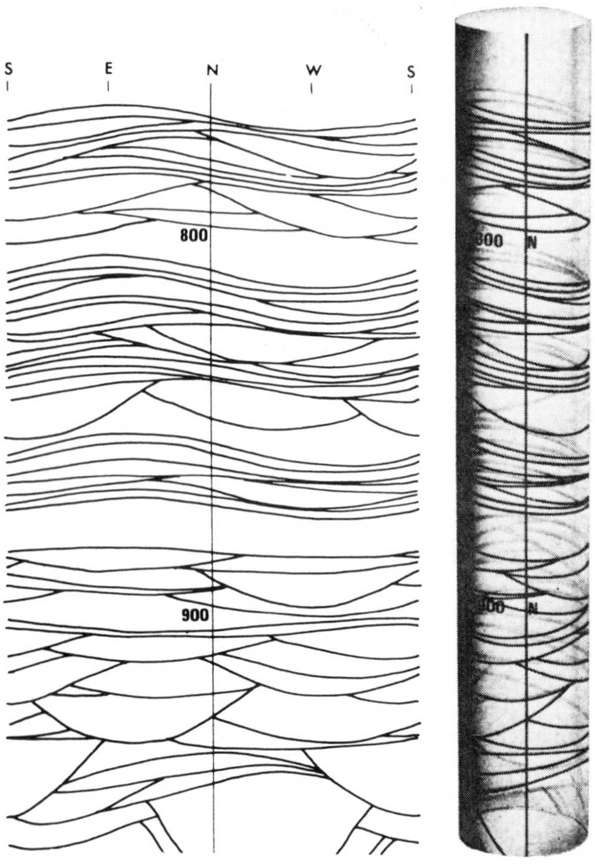

Fig. 19-23. Example of FAST plot.

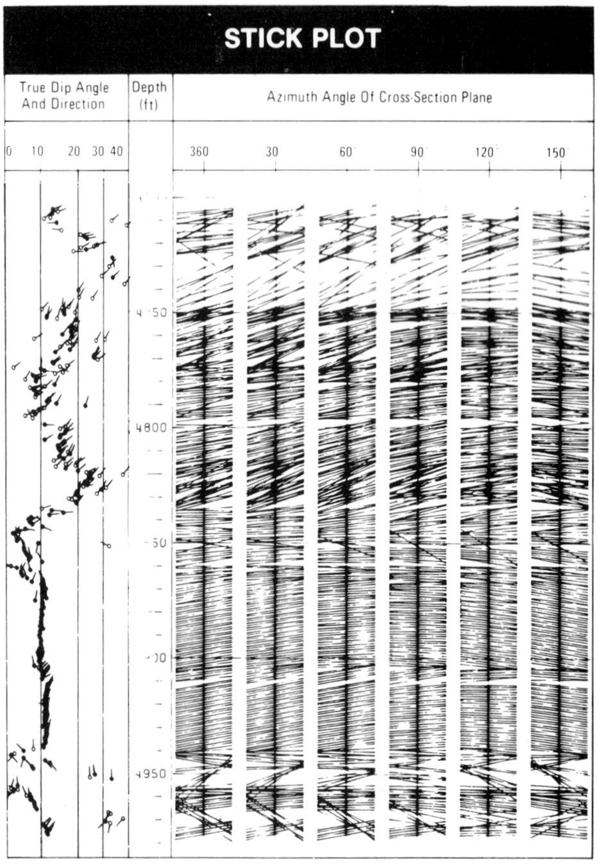

Fig. 19-24. Example of Stick plot.

* FAST means: Formation Anomaly Simulation Trace.

TABLE 19-5

Listing of a stick plot (courtesy of Schlumberger)

ZONE 1–   PAGE 1–7

DEPTH	DIP	DIP	Q	360	30	60	90	120	150
4748.9	9.9	43	C	7.3	9.7	6.5	6.8	2.2	−2.5
4748.4	15.4	5	C	15.3	14.0	9.0	1.4	−6.6	−12.7
4747.9	17.1	34	D	14.3	17.1	15.5	9.8	1.2	−7.7
4744.3	25.1	35	D	21.0	25.0	23.0	15.0	2.3	−11.2
4743.8	19.5	31	D	17.2	19.0	17.6	10.6	0.4	−10.0
4740.8	19.7	36	D	16.2	15.6	18.1	11.9	2.1	−8.3
4738.2	37.6	41	D	30.2	37.1	36.1	26.8	8.4	−14.1
4735.4	33.0	45	D	24.7	32.1	32.1	24.7	9.5	−9.5
4739.4	28.7	48	D	20.1	27.5	28.2	22.1	9.6	−6.5
4730.6	31.7	44	D	24.0	30.9	30.7	23.2	8.5	−9.7
4726.8	32.9	167	D	−32.2	−25.1	−10.7	8.3	23.8	31.7
4724.0	18.4	92	H	−1.7	8.9	15.8	18.4	16.4	10.0
4723.3	22.7	74	A	6.6	16.7	22.1	21.9	16.2	5.8
4722.5	20.6	62	F	10.0	17.7	20.6	18.4	11.3	0.8
4722.0	25.7	66	B	11.1	21.1	25.6	23.7	15.8	2.9
4721.3	27.0	73	A	8.5	20.4	26.4	26.0	19.2	6.5
4718.6	23.0	79	A	4.6	15.6	21.9	22.6	17.8	7.9
4717.6	21.9	78	A	4.8	15.1	20.9	21.5	16.6	7.1
4716.9	20.5	77	H	4.8	14.8	19.7	20.0	15.3	6.2
4714.9	15.2	279	O	2.4	−5.6	−11.9	−15.0	−14.2	−9.7
4713.3	42.4	60	D	24.5	38.1	42.4	38.3	24.5	0.0
4712.8	38.4	53	D	25.5	36.1	38.2	32.3	17.2	−5.5
4711.5	14.0	268	O	−0.5	−7.5	−12.4	−14.0	11.6	−6.7
4710.8	32.6	48	D	28.2	31.3	32.0	25.4	11.2	−7.6
4709.8	10.6	47	D	7.2	10.1	12.3	7.5	3.1	−2.4
4709.3	12.3	18	D	11.7	12.0	9.2	3.9	−2.6	−8.1
4707.5	12.8	52	O	8.0	11.9	12.7	10.2	14.9	−1.4
4706.7	14.3	70	D	5.0	11.0	14.1	12.5	9.3	2.5
4705.7	12.9	68	D	4.9	10.2	12.8	12.0	8.0	1.8

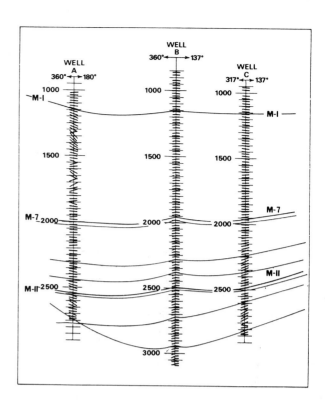

Fig. 19-25. Example of cross-section plot.

horizontal and vertical scales and adapted to each case, would allow easier preparation of cross-sections between deviated wells, notably in development fields. A listing of the apparent dips in each plane is given (Table 19-5).

19.6.2.7. *Cross-section plot*

This is an adaption of the Stick plot allowing the representation of the dips on two cross-sectional planes of different azimuth (Fig. 19-25).

19.6.2.8. *Borehole geometry plot*

This plot presents the two caliper curves (1-3 and 2-4), the position of the sonde, the position of pad 1 and the hole orientation (Fig. 19-26). Hole volume integration (expressed in cubic meters) is added.

19.6.3. **Results given as punched cards or magnetic tapes**

19.6.3.1. *Punched cards*

Each punch card corresponds to one level and has the same information as a listing and in the same order. This service is only rarely given and is tending to disappear.

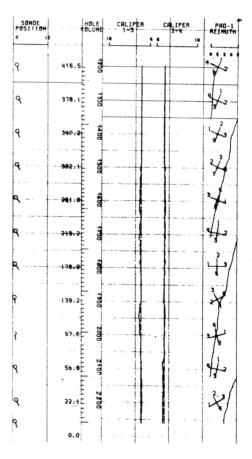

Fig. 19-26. Example of borehole geometry plot.

19.6.3.2. *Magnetic tape*

This is a tape of either 7-track 556 BPI or 9-track 800 or 1600 BPI. It has the same data as a listing and adds the displacements found in correlation between curves.

Cards are given free but tapes are invoiced at cost price. These allow the making of different or further playbacks of results in the period after the three months that Schlumberger itself keeps a result tape. The result tape can be used to make up composite logs (at 1/2000 scale for example).

If the result tape does not have the edited field tape data on it then it cannot be used to recompute with new parameters. In general the result tape should be completed with the basic field data (perhaps on request) for this reason. The field data is recorded on the result tape in a format similar to that produced by Schlumberger CSU field units (i.e. modified LIS).

19.6.4. Horizontal or vertical projections of the hole deviation

This presentation is supplied by Schlumberger with each deviation survey (Fig. 19-27).

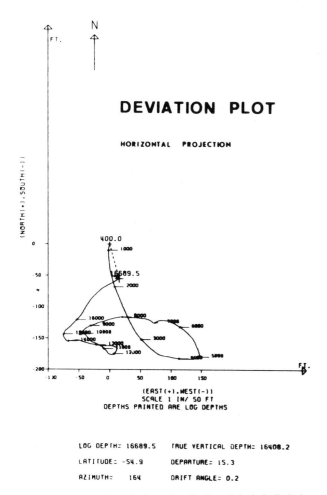

Fig. 19-27. Example of horizontal projection of the hole deviation.

19.6.5. Special presentation of GEODIP results

Given the considerable increase in the density of data these can rarely be made at scales of 1/1000 or

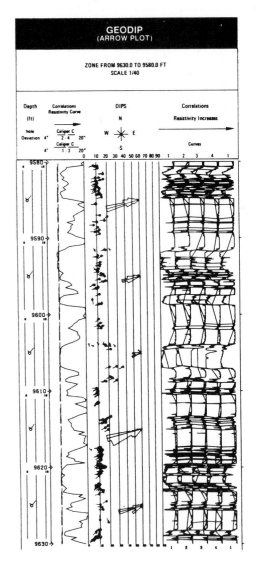

Fig. 19-28. Example of Geodip arrow plot.

1/200 for detailed studies, and so arrow plots are usually made at scales of 1/40 or 1/20.

The four resistivity curves are reproduced and the pattern recognition correlation drawn on the curves (more or less horizontal lines and vertical lines joining inflection points). This presents a considerable improvement since the validity of individual correlation can be checked (Fig. 19-28) and so the quality of the dip computed.

In addition, an azimuth frequency plot appears at regular intervals so giving an idea of the average azimuth. This is taken off for 1/200 scale plots.

19.7. TREATMENT OF RESULTS

19.7.1. The Cluster method

This method was introduced by Schlumberger in 1975 in an interpretation program using cross correlation (MARK IV program). It seeks to eliminate false results that occur because of mathematical or statistical problems.

19.7.1.1. *Basic principle*

The basic principle is; "Only those correlations which are still found after a slight movement of the correlation interval (half, a third or a quarter of the interval length) are valid".

The Cluster method therefore tries to obtain all possible correlations, by using a sufficient overlap between successive correlation intervals, or levels, and by keeping only those results that repeat from one level to the next, defining adjacent dips (Fig. 19-29).

19.7.1.2. *Procedure*

This is recapped in the flow chart in Table 19-6. As well as information on deviation, the azimuth and calipers the Cluster method uses all the dips found by taking two by two all the correlations obtained by the

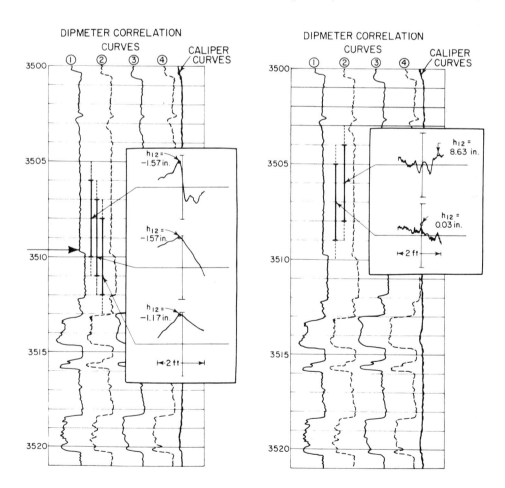

Fig. 19-29a. Illustration of dominant anomaly, included here within three overlapping correlation intervals. Corresponding correlograms at right show small differences between the locations of their maxima (-1.57, -1.57, -1.17).

Fig. 19-29b. Illustration of case where there is no dominant anomaly in the two correlation intervals. The correlograms show a wide difference in the locations of their maxima (from Hepp et al., 1975).

TABLE 19-6

HDT cluster logic--flow diagram (courtesy of Schlumberger)

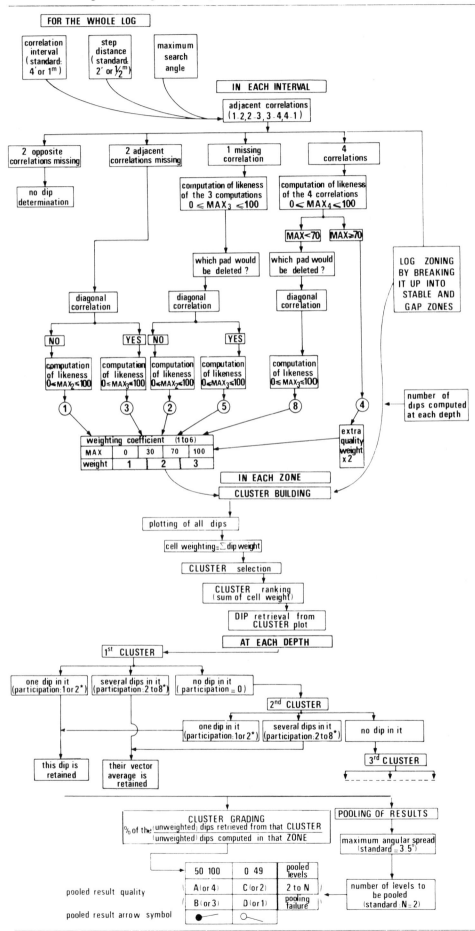

TABLE 19-7

Example of results from an existing dipmeter program used as input to cluster (from Hepp et al., 1975)

DEPTH (FEET)	CALIPERS		|——— CURVE DISPLACEMENTS ———|							MAX DEV	WELL DEVIATION	TOOL ORIENTATION		
	|D13	D24 (INCHES)	H12	H23	H34 (INCHES)	H41	H13	H24	MAX	DEV (DEGREES)	DVAZ	PAZ (DEGREES)	RB
3836	8.9	8.4	-0.34	-0.99	0.39	0.79	✲	✲	57	2.3	7	202	195
3834	8.8	8.4	-0.29	4.20	-0.29	1.26	-1.91		46	2.4	7	203	196
3832	8.8	8.4	3.58	1.46			-0.45		10	2.4	5	203	198
3830	8.8	8.4	-2.97	1.42	-0.06	1.07	-2.22		24	2.4	3	202	199
3828	8.8	8.4	-3.24		1.77	0.74	1.52		31	2.5	2	200	198
3826	8.7	8.4	-3.21	-0.40	1.21	0.49	-3.35		36	2.5	3	201	198
3824	8.6	8.4	-0.90	-0.28	1.10	0.37		0.73	38	2.5	4	204	200
3822	8.5	8.4	-1.04	-0.04	1.12	0.51		0.94	35	2.6	4	202	198
3820	8.6	8.2	-4.45	5.10	1.21	3.22	-3.20		40	2.7	3	196	193
3818	8.5	8.1	0.63	0.11	6.88	-0.07		-0.20	22	2.6	3	191	188
3816	8.5	8.4	0.70	0.04			0.60		27	2.5	5	191	185
3814	8.6	8.5	-0.90	1.04	-0.80		0.11		30	2.4	4	188	183
3812	8.6	8.4	-0.57	0.97			0.62		25	2.3	360	185	184
3810	8.5	8.3	0.16	0.99	0.87	-1.92	✲	✲	23	2.2	359	182	183
3808	8.5	8.2			0.79	-1.16	-0.17		24	2.1	360	177	177
3806	8.6	8.2	-2.42	1.56	1.27	-0.11			37	2.1	359	172	173
3804	8.6	8.3		0.19	-4.47			-4.16	14	2.1	358	171	172
3802	8.6	8.5	NO CORRELATION							2.0	1	174	172
3802	8.6	8.5	NO CORRELATION							2.0	1	174	172
3800	8.4	8.6	-5.80	-6.00	0.38	-9.13			35	2.0	360	183	182
3798	8.0	8.3		5.49	0.93	-2.31			37	2.0	359	192	192
3796	8.2	8.2	0.81	0.11	-0.78		1.16		14	2.0	2	194	192

TABLE 19-8

Example of zoning defined from the displacements (from Hepp et al., 1975)

	ADJACENT-CURVE DISPLACEMENTS				
DEPTH (FEET)	H12	H23 (INCHES)	H34	H41	PAZ (DEGR.)
OPEN ZONE					
3858	3.71	-0.69	3.43	-3.27	181
3856	0.81	-0.65	-0.90	0.78	192
3854	0.97	-0.43	-1.24	-0.70	189
3852	0.92	-0.46	-1.39		182
3850	0.77	-0.20	-1.45	0.59	178
3848	0.70	-0.12	-4.46		176
3846		-0.31	-4.82		180
STABLE ZONE					
3844	0.32	-0.58	-0.37	0.45	195
3842	0.11	-0.79	-0.34	0.36	205
3840	-0.04	-0.93		0.65	208
3838	-0.25	-1.03	0.44	0.73	205
3836	-0.34	-0.99	0.39	0.79	202
OPEN ZONE					
3834	-0.29	4.20	-0.29	1.26	203
3832	3.58	1.46			203
3830	-2.97	1.42	-0.06	1.07	202
3828	-3.24		1.77	0.74	200
3826	-3.21	-0.40	1.21	0.49	201
3824	-0.90	-0.28	1.10	0.37	204
3822	-1.04	-0.04	1.12	0.51	202
3820	-4.45	5.10	1.21	3.22	196
3818	0.63	0.11	6.88	-0.07	191
3816	0.70	0.04			191
OPEN ZONE					
3814	-0.90	1.04	-0.80		188
3812	-0.57	0.97			185
3810	0.16	0.99	0.87	-1.92	182
3808			0.79	-1.16	177
3806	-2.42	1.56	1.27	-0.11	172
3804		0.19	-4.47		171
3802					174
3800	-5.80	-6.00	0.38	-9.13	183
3798		5.49	0.93	-2.31	192
3796	0.81	0.11	-0.78		194

cross-correlation programs (2, 3 or 4 adjacent correlations and one diagonal). Then the displacement listing (Table 19-7) is divided into zones corresponding to given depth intervals.

The consecutive displacements between the same two electrodes are studied with the objective of identifying stable intervals and so grouping levels into stable or unstable zones.

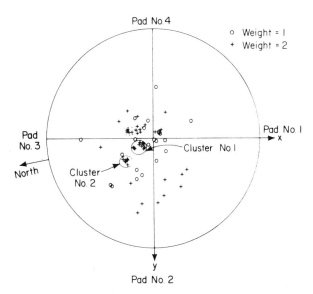

Fig. 19-30. Plot of dip results on two-dimensional map (scalings removed). Two clusters are circled, and a near cluster is indicated in a dashed circle. (from Hepp et al., 1975).

TABLE 19-9

Cluster output listing (from Hepp et al., 1975)

DEPTH	DIP	DIP AZM	DEV	DEV AZM	D13	D24	LOGIC	QRC	CE	PAR	MAX	SPD COR
3896	6.4	222	2.1	1	8.5	8.9		C	10	1	30	
3894	8.8	158	2.1	2	8.5	8.9		C	10	1	18	
3892	8.5	190	2.1	4	8.5	9.0		C	10	2	27	
3890	11.4	224	2.1	3	8.5	9.0		A	10	4	33	
3888	5.1	181	2.2	3	8.5	9.0		A	10	7	29	
3886	3.9	200	2.1	4	8.5	9.0		A	10	8	56	
3884	7.7	206	2.1	2	8.5	9.1	**	A	10	8	31	
3882	7.1	196	2.2	0	8.5	9.3		A	10	8	53	
3880	7.6	164	2.2	2	8.5	9.4		A	10	8	51	
3876	18.6	143	2.4	5	8.5	9.2		D	10	1	22	
3874	14.9	146	2.4	4	8.5	9.1	**	B	10	8	8	
3872	13.5	151	2.3	3	8.5	9.1		B	10	5	26	
3870	13.8	152	2.3	1	8.5	9.1		B	10	4	31	
3868	29.7	297	2.4	1	8.5	8.9		D	20	2	30	
3866	12.8	153	2.4	1	8.5	8.8		B	10	5	25	
3864	15.0	165	2.4	1	8.3	8.8		D	10	1	38	*
3862	11.3	177	2.4	1	8.2	8.7		D	10	1	31	*
3856	10.2	119	2.1	9	8.5	8.4	**	A	10	8	40	
3854	10.3	139	2.0	12	8.5	8.4		A	10	5	17	
3852	10.6	129	2.0	14	8.5	8.4		A	10	5	18	
3850	10.1	122	2.0	16	8.5	8.5		A	10	8	35	
3848	6.5	135	2.0	17	8.5	8.5		C	10	3	23	
3844	7.7	112	2.0	16	8.5	8.4		A	10	8	56	
3842	6.1	115	2.1	14	8.5	8.4		A	10	8	49	
3840	6.0	85	2.2	11	8.6	8.4		A	10	5	35	*
3838	6.4	62	2.3	7	8.8	8.4	**	A	10	8	49	
3836	6.4	59	2.3	7	8.9	8.4	**	A	10	8	57	
3830	10.7	24	2.4	3	8.8	8.4		D	10	1	24	
3828	23.8	346	2.5	2	8.8	8.4		D	20	1	31	
3826	8.4	354	2.5	3	8.7	8.4		D	10	2	36	
3824	6.6	353	2.5	4	8.6	8.4		B	10	8	38	
3822	7.3	345	2.6	4	8.5	8.4		B	10	8	35	
3814	12.0	262	2.4	4	8.6	8.5		D	10	3	30	
3812	10.8	243	2.3	360	8.6	8.4		D	10	3	25	
3810	11.8	258	2.2	359	8.5	8.3	**	D	10	2	23	
3808	12.5	257	2.1	360	8.5	8.2		D	10	3	24	
3806	18.1	250	2.1	359	8.6	8.2		D	10	1	37	
3790	4.6	244	2.0	360	8.5	8.5	**	A	10	8	36	
3788	5.2	257	2.0	0	8.5	8.5		A	10	5	42	
3782	0.5	346	2.0	0	8.3	8.4		A	10	5	26	

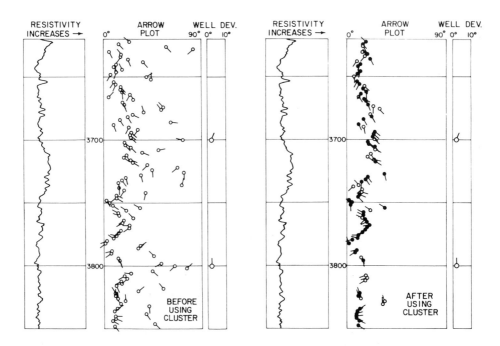

Fig. 19-31. Dipmeter results before and after using Cluster (from Hepp et al., 1975).

By a stable zone we mean an interval in depth in which successive displacements (one level to the next) do not differ by more than a certain percentage (Fig. 19-29 and Table 19-8) (if not the zone is open or unstable).

For each level of each zone so defined as stable or open, each of the intermediate dips is calculated by grouping two by two the displacements (the number of combinations is limited to eight).

Each intermediate dip so determined is given a weight (from 1 to 3) according to the quality of the correlation. The weight is doubled if the level has a good planarity (and thus a good closure as well).

For a given zone, each weighted intermediate dip is plotted on a Schmidt plot. When dips fall at the same place their weights are added.

The plot is analyzed to identify dip concentrations or groupings * of sufficient weight or density (Fig. 19-30).

The final dip retained for a level is selected only from those that fall in the grouping or near to it. This selection is not very restrictive in order not to prevent the production of red or blue patterns.

Each final dip selected is given with a classification that depends on the quality of the cluster or grouping in which it falls and on the number of dips at the level that fall in the group (which depends on the closure, the planarity and stability of the adjacent levels).

These dips are set out in a listing with their quality (Table 19-9).

Simultaneously the program puts together the dips from adjacent stable levels in one average dip vector that is output in a separate file. The arrow plot is made from this file. This reduces the large density of arrows on the log and eliminates duplication of the same dip result arising from the presence of the same dominant anomaly in successive correlation intervals that overlap. The file also serves as the basis for the results listing given to a client (see example in Table 19-9).

19.7.1.3. *Critique of the Cluster method*

It has the advantage of eliminating noise as the comparison of processing with and without the Cluster method show (Fig. 19-31).

However, there is the risk of eliminating dips considered as noise when they really do correspond to an actual phenomenon and so are likely to help in interpretation of the arrow plot.

19.7.2. **Determination of structural dip (Diptrend)**

A statistical study of the dips determined by Cluster, made in intervals more or less large and homogeneous, showing dips constant in magnitude and azimuth corresponding to parallel boundaries without current influence, permits the determination of structural dip for these intervals.

19.7.3. **Dip-removal**

Knowing the structural dip it can be removed from the dip values (vector subtraction). This leads to

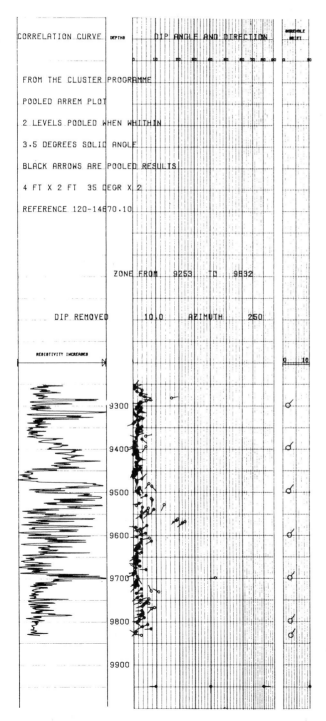

Fig. 19-32. Example of result after dip-removal.

* i.e. the participation of each level to the grouping (column PAR on the results listed in Table 19-9).

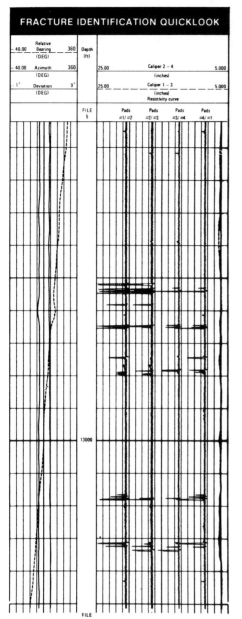

Fig. 19-33. Example of Fracture Identification Log (F.I.L.) (courtesy of Schlumberger).

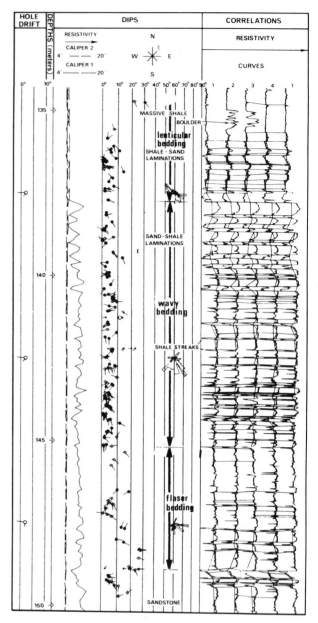

Fig. 19-34. Example of Sedimentary Features seen on a Geodip (courtesy of Schlumberger).

a new arrow plot that can be used to make a study of sedimentary features and to define the direction of transport (Fig. 19-32).

19.8. APPLICATIONS

These are of three types.

19.8.1. Tectonic or structural applications

Dip measurements are fundamentally used to define regional or structural dip, but also to detect structural dip anomalies associated with structural deformation such as faults, folding, etc.

The resistivity curves can be used to detect fractures. (F.I.L. Presentation, Fig. 19-33.)

19.8.2. Sedimentary applications

A dipmeter, especially when it has been processed with GEODIP, gives an indication of sedimentary features present in the interval and consequently an idea of the depositional processes: bedding type (flaser, wavy, lenticular bedding, parallel boundaries), current features (cross-bedding), and graded bedding.

In addition, a study of the original resistivity curves and of the dips related to each correlation

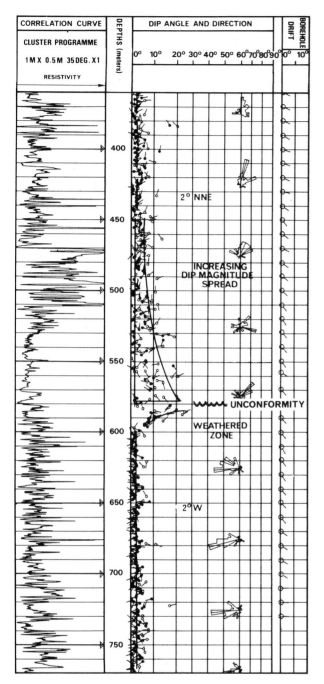

CORRELATION CURVE	DEPTHS (meters)	DIP ANGLE AND DIRECTION	BOREHOLE DRIFT

CLUSTER PROGRAMME
1 M X 0.5 M 35 DEG. X1
RESISTIVITY

2° NNE

INCREASING
DIP MAGNITUDE
SPREAD

~~~~ UNCONFORMITY

WEATHERED
ZONE

2° W

Fig. 19-35. Example of unconformity detected by the dipmeter (courtesy of Schlumberger).

supplies information on bed thickness, the repetition of phenomenon, vertical evolution in thickness, in grain size, in composition, in bedding type, and on the nature and origin of bed boundaries, the homogeneity or heterogeneity of each unit. (Fig. 19-34).

19.8.3. Stratigraphic applications

Analysis of the dipmeter may bring out information on stratigraphic phenomenon such as discontinuities or angular unconformities (Fig. 19-35).

These three types of applications will be studied further in a second volume on log interpretation.

19.9. REFERENCES

Allaud, L.A. and Ringot, J., 1969. The high resolution dipmeter tool. Log Analyst, 10 (3).

Autric, A., 1967. Erreurs dans les mesures de pendagemétrie-Effet de la rotation de la sonde (communication SAID, Paris).

Bemrose, J., 1957. Component dip nomogram. World Oil.

Bricaud, J.M. and Poupon, A., 1959. Le pendagemètre continu à potéclinomètre et à microdispositifs focalisés. 5e Cong. mond. Pétrole, New York.

Campbell, R.L., Jr., 1968. Stratigraphic Application of dipmeter data in Mid-Continent. Bull. Am. Assoc. Pet. Geol., 52 (9).

Cox, J.W., 1968. Interpretation of dipmeter data in the Devonian Carbonates and evaporites of the Rainbow and Zama Areas. 19th Annu Techn. Meet. Pet. Soc. CIM., paper no. 6820.

Cox, J.W., 1972. The high resolution dipmeter reveals dip-related borehole and formation characteristics. 11th Annu. SPWLA Logging Symp. Trans., Pap. No. D.

de Chambrier, P., 1953. The microlog continuous dipmeter. Geophysics, 18 (4).

de Witte, A.J., 1956. A graphic method of dipmeter interpretation using the stereo-net. Pet. Trans. AIME, 207, and J. Pet. Technol., 8 (8).

Doll, H.G., 1943. The S.P. dipmeter. J. Pet. Technol., 6.

Dresser-Atlas. Relating Diplogs to practical geology.

Franks, C. and White, B., 1957. New method of interpretation gives you more from your dip-log surveys. Oil Gas J.

Gartner, J.E., 1966. Application de la pendagemétrie. Communication AFTP.

Gilreath, J.A., 1960. Interpretation of dipmeter surveys in Mississippi. Ann. Symp. SPWLA.

Gilreath, J.A. and Maricelli, J.J., 1964. Detailed stratigraphic control through dip computations. Bull. Am. Assoc. Pet. Geol., 48 (12).

Gilreath, J.A. and Stephens, R.W., 1971. Distributary front deposits interpreted from dipmeter patterns. Gulf Coast Assoc. Geol. Soc., Trans. 21.

Gilreath, J.A., Healy, J.S. and Yelverton, J.N., 1969. Depositional environments defined by dipmeter interpretation. Gulf Coast Assoc. Geol. Soc., Trans. 19.

Goetz, J.F., Prins, W.J. and Logar, J.F., 1977. Reservoir delineation by wireline techniques. 6th Annu. Conv. of Indonesian Petrol. Assoc.

Hammack, G.W., 1964. Diplog. Techn. Bull. Lane Wells.

Hepp, V. and Dumestre, A.C., 1975. Cluster- A method for selecting the most probable dip results from dipmeter surveys. Soc. Pet. Eng. AIME, Paper 5543.

McRobertson, J., 1972. Deficiencies of computer correlated dip logs. 13th Annual Symp. SPWLA.

Moran, J.H., Coufleau, M.A., Miller, G.K. and Timmons, J.P., 1963. Automatic computation of dipmeter logs digitally recorded on magnetic tapes. J. Pet. Technol., 14 (7).

Payre, X. and Serra, O., 1979. A case history-Turbidites recognized through dipmeter. SPWLA, 6th Europ. Logging Symp. Trans., paper K.

Phillips, F.C., 1960. Utilisation de la projection stéréographique en géologie structurale. Edward Arnold, London, 2nd ed.

Pirson, S.J., 1970. Geologic well log analysis. Gulf Publ. Co., Houston.

Schlumberger. Schlumberger continuous dipmeter. Schlumberger Engineering Report #8.

Schlumberger. Schlumberger Computer-Processed dipmeter presentations.

Schlumberger. (Surenco et L.A. Allaud) Use of stereographic projection in dipmeter analysis.

Schlumberger, 1970. Fundamentals of dipmeter interpretation. New York.

Schlumberger, 1981. Dipmeter interpretation. Fundamentals.

Selley, R.C., 1978. Ancient Sedimentary Environments. Chapman and Hall, London.

Selley, R.C., 1979. Dipmeter and log motifs in North Sea submarine-fan sands. Bull. Am. Assoc. Pet. Geol., 63 (6).

Vincent, Ph., Gartner, J.E. and Attali, G., 1977. Geodip: an approach to detailed dip determination using correlation by pattern recognition. 52nd Annu. Fall Techn. Conf. SPE of AIME, Paper 6823.

Welex, 1970. Dip log Interpretation.

20. WIRELINE SAMPLING

In order to confirm or have a more precise idea of the lithology of a rock or its fluid content suggested by the interpretation of the logs already run, a geologist may require samples taken after drilling. This is the reason why logging companies offer various tools which can be run on wireline to the required depth to take samples of rock or fluid, with depth positioning (correlation) made by an SP or a gamma-ray log run simultaneously.

20.1. ROCK SAMPLING

In 1937 Schlumberger introduced for the first time a rock sampling device, known since by the name of sidewall sampling, and using bullets. This tool consists of a series of hollow "bullets" which are projected into the formation by a charge exploded behind the bullet using a small detonator fired on command from the surface (Fig. 20-1a, b). The bullets and charges are loaded into a gun attached to the wireline.

The bullet is recovered by means of two steel cables fixed at one end to the sides of the bullet and at the other to the gun. When the gun is pulled upwards the steel cables pull the bullet free of the formation (Fig. 20-2).

The strength of the explosive charge is chosen according to the hardness of the formation to be

Fig. 20-1a. Schlumberger's Sidewall sample-taker for large and small hole diameters (courtesy of Schlumberger).

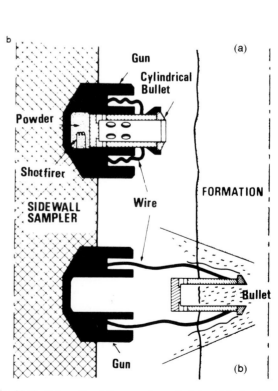

Fig. 20-1b. Schematic of the sampling principle. (a) bullet in the gun; and (b) bullet fired into the formation (courtesy of Schlumberger).

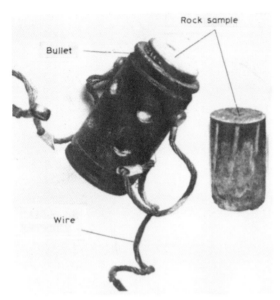

Fig. 20-2. Bullet and sidewall core (courtesy of Schlumberger).

cored. Too large a charge in a soft formation will risk losing the bullet by breaking the retaining cables, whereas too small a charge in a hard formation means that little or no core is recovered. Some previous tests are sometimes necessary or a use of local knowledge. Correction depth positioning is ensured by the use of an SP or GR log.

20.2. FLUID SAMPLING AND PRESSURE MEASUREMENTS

It was around 1952 that Schlumberger introduced the first formation fluid sampler that could be lowered on a logging cable.

This sampling technique constitutes a fast, economical and sure method of testing the production potential of a zone and without great risk. Although other companies have introduced similar tools we shall look at the Schlumberger types namely the FT, FIT and RFT.

20.2.1. Formation tester (FT)

This was the first tool proposed by Schlumberger. The tool is shown in the photograph of Fig. 20-3 and its principle by the diagram in Fig. 20-4. The tool includes a rubber pad about 15 cm wide and 70 cm long in the centre of which is a block with a shaped charge perforator. This pad is pushed strongly against the borehole wall using another pad on the reverse side of the tool, with the expansion or distance between the pads achieved using hydraulic pressure. The tool is positioned in depth by recording an SP or GR log. After its positioning and the opening of the pads a flowline valve is opened in order to let forma-

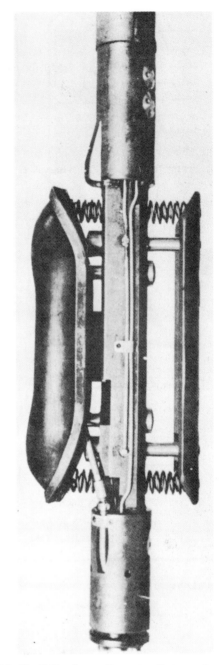

Fig. 20-3. The Schlumberger Formation Tester packer assembly (courtesy of Schlumberger).

tion fluids flow into the sample chamber. If a flow is observed the shaped charge is not used. On the other hand, if the test pressure or the flow is low and it is not possible to recover a sample the charge is shot. After a sufficient time, the close valve is activated, shutting the sample chamber. If a formation pressure build up is required in a low permeability formation the tool has to be left in place long enough to obtain a final shut-in pressure. Finally the tool is released from the formation by equalizing internal and external pressures, causing the pistons and arms to retract. The tool with its sealed sample chamber is then pulled out of the well.

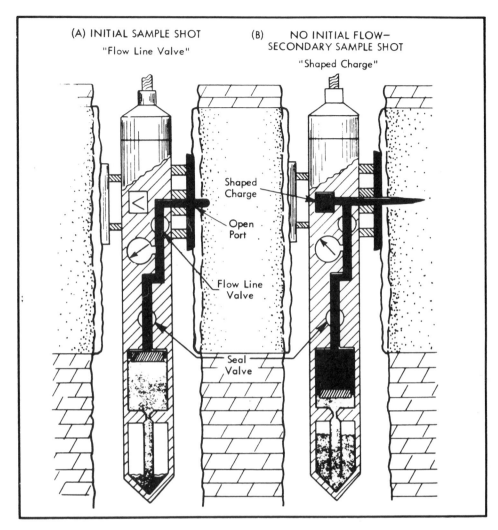

Fig. 20-4. Operation of the Formation Tester (courtesy of Schlumberger).

For the duration of the test the recorded curves (Fig. 20-5) show the firing of the charge and allow pressure measurements: (a) the pad hydraulic circuit pressure (this is an internal tool measurement); (b) the isolation pressure; (c) the build-up and static pressure; and (d) the mud column pressure. This information allows the test to be followed and controlled throughout its duration.

We should note that this tool does not allow any measurement of the virgin formation pressure.

Different sample chamber sizes are available: 4, 10, or 20 litres. The FT tool has been almost entirely replaced by the FIT and RFT.

20.2.2. Formation Interval Tester (FIT)

As the FT pad is relatively large it can become differentially stuck to the borehole wall. Equalizing pressures to release the pad have no effect and the whole tool remains stuck. To limit this possibility a tool with a smaller pad has been designed. This is illustrated in the photograph in Fig. 20-6. Its func-

tioning is explained in the diagrams of Fig. 20-7 and the measurement process explained in Fig. 20-8.

The recording of the surface controls and of the pressures is illustrated in Fig. 20-9. The recording sequence is explained below (the letters refer to the figure).

(A) A calibration signal is sent to the circuits measuring the flowline and hydraulic pressures.

(B) Hydrostatic pressure measurement (i.e. mud column pressure).

(C) The opening of the mud valve. The pressure rises as the tool is opened at the chosen test depth (positioning by GR or SP).

(D) The opening of the flow valve allowing formation fluid to enter the sample chamber.

(E) Firing of the shaped charge.

(F) The sample chamber is full and the pressure begins to rise.

(G) The static reservoir pressure is reached.

(H) The sample chamber is closed.

(I) The rise of pressure above the static pressure is due to an over compression of the fluid in the flow

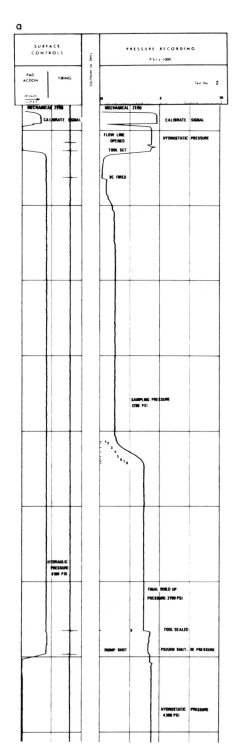

Fig. 20-5a. Example of pressure-time record in a low permeability sandstone.

lines (pseudo-closing pressure).

(J) The hydraulic pressure is released and the pads retract.

(K) Hydrostatic pressure measurement.

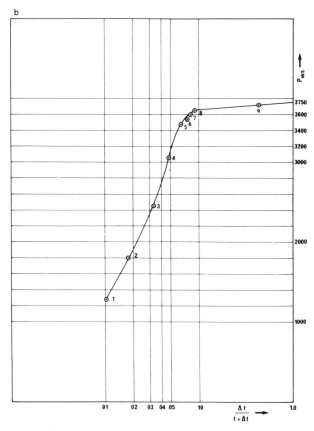

Fig. 20-5b. Interpretation of the measurement for pressure determination.

Fig. 20-6. The Schlumberger Formation Interval Tester (courtesy of Schlumberger).

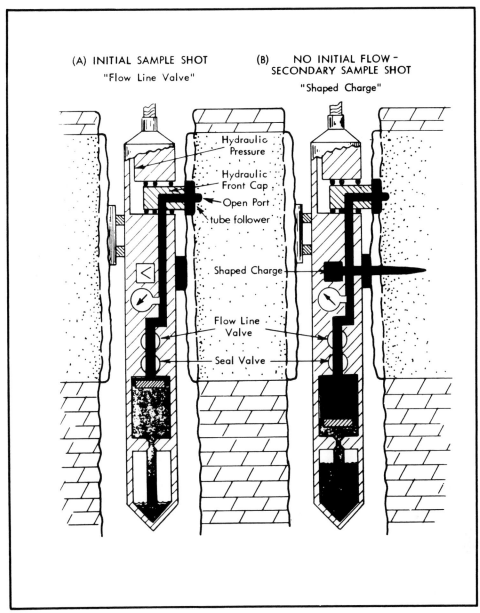

(A) INITIAL SAMPLE SHOT
"Flow Line Valve"

(B) NO INITIAL FLOW –
SECONDARY SAMPLE SHOT
"Shaped Charge"

Hydraulic Pressure

Hydraulic Front Cap

Open Port

tube follower

Shaped Charge

Flow Line Valve

Seal Valve

Fig. 20-7. Operation of the Formation Interval Tester (courtesy of Schlumberger).

20.2.3. Repeat Formation Tester (RFT)

The preceeding tools allow only one sample to be collected and one pressure test to be made on each descent in the hole. Normally if additional tests are required the tool has to be brought out of the hole, cleaned, the sample chamber emptied and some parts replaced (shaped charge for example) leading to a loss of rig time. This is true even if the test is dry.

Several companies have introduced tools that are much more flexible in their use. The Schlumberger version is shown in the photograph of Fig. 20-10. Its characteristics are as follows:

(a) In one trip in the hole it is possible to set the tool any number of times and take a formation pressure measurement. The tool does not have to be brought to surface after each measurement so saving a large amount of time.

(b) It is possible to pretest a formation to choose the most permeable zones on which to take fluid samples. If the pretest is satisfactory the sample valve is opened. If the test is not satisfactory, the tool is closed, the pretest chambers emptied and the equalization valve opened.

(c) Since there are two sample chambers it is possible either to sample twice or to collect a larger sample on one test.

(d) The precision on measurement is higher (error less than 0.49% and even less than 0.29% with special calibration procedures).

(e) An internal filter prevents sand flow and fluid loss (Fig. 20-11).

308

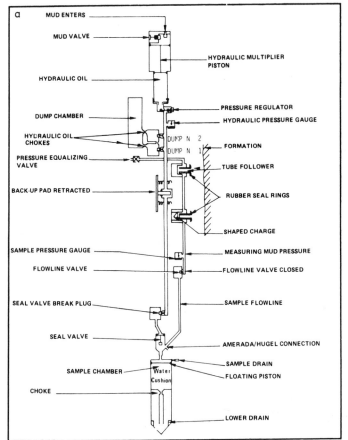

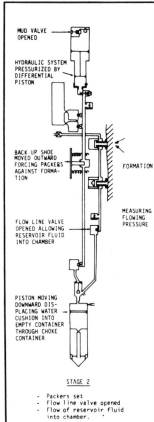

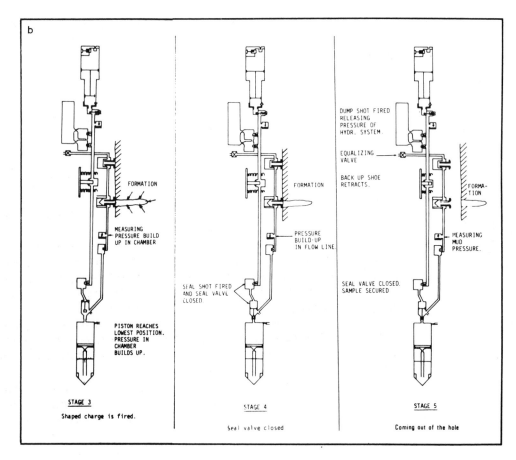

Fig. 20-8. Schematic of the different stages of a test with the FIT.

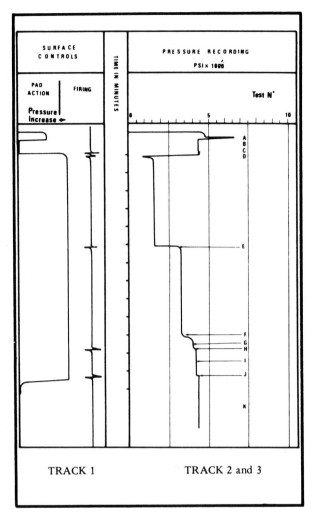

Fig. 20-9. Example of pressure recording with FIT (courtesy of Schlumberger).

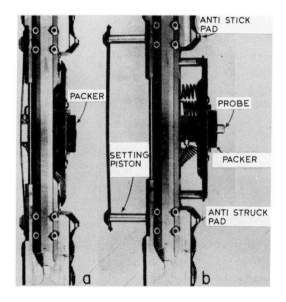

Fig. 20-10. The Schlumberger Repeat Formation Tester (RFT) (a) pad retracted; and (b) pad opened. (courtesy of Schlumberger).

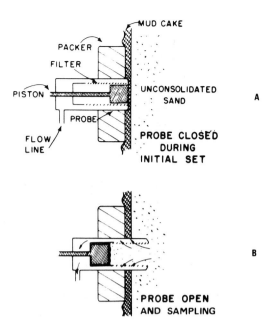

Fig. 20-11. Schematics showing details of the probe with piston and filter (courtesy of Schlumberger).

(f) Anti-stick pads practically eliminate the chance of the tool sticking.

The tool's sampling system is shown diagramatically in Fig. 20-12. A direct digital measurement is made by the panel and the log recording is made both digitally and in analog form. An example of a log is shown in Fig. 20-13.

20.2.4. Fluid sampler applications

The development of tools that allow multiple pressure tests to be made as well as one or two samples on the same trip in the well has meant that the use of wireline fluid testing has increased considerably.

Both rock and fluid sampling offer considerable benefits. Among these are:

(a) The collection of a large number of samples in a short time, so saving rig time that might be spent in conventional casing or drill stem testing.

(b) The precise positioning of samples relative to the standard formation evaluation logs. This means that very detailed sampling may be made in the zones of interest or where there is some doubt as to the log interpretation.

(c) The sampling of zones that might otherwise be left untested. This is a result of the speed of sampling and its relatively low cost compared to other sampling methods.

Some uses of wireline testing can be listed.

(1) To predict or confirm the productivity of a formation by identification of formation fluids and by an analysis of the pressure measurements.

(2) To establish the main fluid characteristics such as:

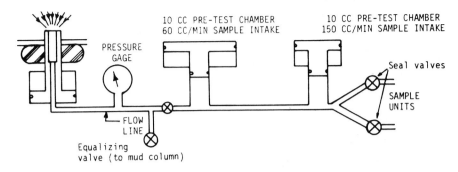

Fig. 20-12. Schematic of the sampling system of the RFT (courtesy of Schlumberger).

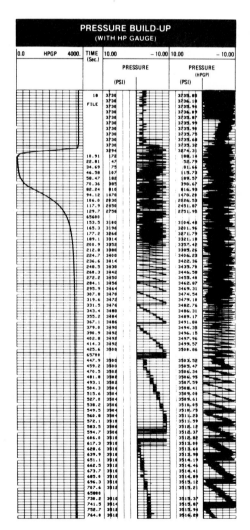

Fig. 20-13. Example of pressure recording (courtesy of Schlumberger).

(a) oil density;

(b) the gas/oil ratio (GOR) by use of an empirical chart (Fig. 20-14) as a function of the volumes collected during the test;

(c) the water cut equal to the ratio of the volume of formation water to the sum of the volumes of oil and water:

Water cut % =

$$\frac{\text{Volume of formation water}}{\text{Volume of formation water} + \text{volume of oil}} \times 100$$

(3) To evaluate bottom hole pressures (often using special charts as plots):

(a) virgin formation pressure, initial shut-in pressure;

(b) flowing pressures;

(c) the pressure build-up and final shut-in pressure if necessary by extrapolation (Fig. 20-15);

(d) hydrostatic pressure.

(4) To determine reservoir parameters: (a) a productivity index; and (b) permeabilities.

Considerable progresses in recent years in the analysis of pressure build up curves and associated pressures have allowed RFT formation pressure readings to be used more accurately in predicting reservoir behaviour.

(5) To determine fluid contacts by use of depth versus pressure plots. Pressure measurement in, for example, the oil and water sections allows the gradients in each section to be detected, and by identifying the point in depth where the gradient changes, the actual fluid contact is found (see Fig. 20-16).

(6) To correlate between wells. This is essentially a correlation of pressures where the same reservoir is penetrated by both wells and there is communication through the reservoir.

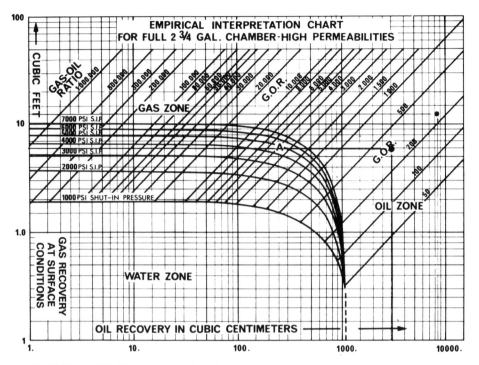

Fig. 20-14. Empirical interpretation chart for determining the GOR (courtesy of Schlumberger).

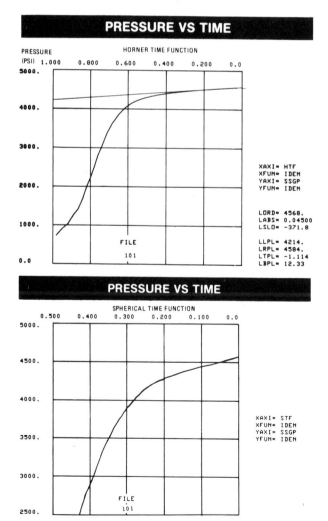

Fig. 20-15. Two of the pressure vs. time plots obtained at the well-site with CSU (courtesy of Schlumberger).

312

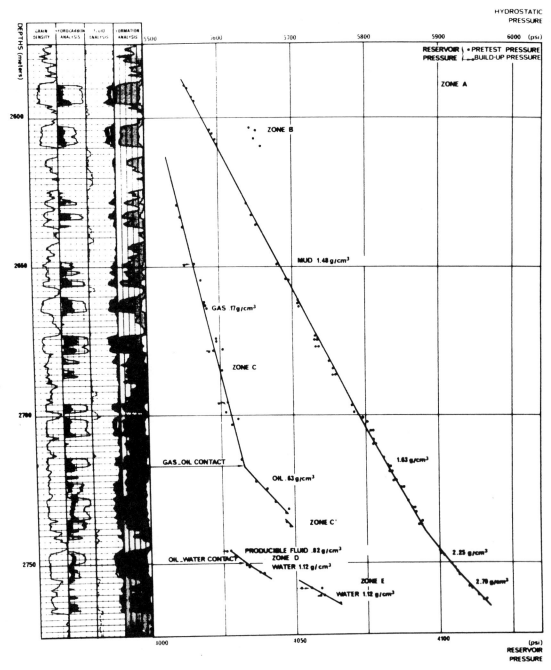

Fig. 20-16. Determination of G-O and O-W contacts with pressure measurements (courtesy of Schlumberger).

20.3. REFERENCES

Banks, K.M., 1963. Recent achievements with the formation tester in Canada. J. Can. Pet. Technol., 2 (2).

Moran, J.H. and Finklea, E.E., 1962. Theoretical analysis of pressure phenomena associated with the wireline formation tester. J. Pet. Technol., 14 (8).

Schlumberger. The Essentials of Wireline formation tester.

Schlumberger, 1981. RFT. Essentials of Pressure Test Interpretation.

Schultz, A.L., Bell, W.T. and Urbanovsky, H.J., 1974. Advancements in uncased-hole wireline-formation-tester techniques. Soc. Pet. Eng. of AIME, Paper 5035.

Smolen, J.J. and Litsey, L.R., 1977. Formation evaluation using wireline-formation-tester pressure data. Soc. Pet. Eng., Paper 6822.

Stewart, G. and Wittmann, M., 1979. Interpretation of the pressure response of the repeat formation tester. SPE of AIME, paper SPE 8362.

Tison, J., 1977. Obtention et traitement des mesures de pression en géologie des fluides-Utilisation de l'ordinateur. DES Fac. Sci. Bordeaux.

21. OTHER MEASUREMENTS

21.1. BOREHOLE TELEVIEWER (BHTV)

The concept of this tool was introduced in 1969 by the Mobil Oil Company and commercialized a year or two later by several logging companies which obtained licences from Mobil. But, due to some limitations and problems in obtaining a good picture of the wall, the industry did welcome accept this new tool, so it was removed from the market a few years later by most of the service companies.

Over the last two or three years, we have observed a renewal of interest for this technique. Improvements made on the data acquisition, presentation and processing by companies like Amoco which obtained a licence from Mobil, give a new impulse to this tool.

21.1.1. Principle

The basic principal of the BHTV is quite similar to that of sonar. This is fundamentally an acoustic logging tool that produces an acoustic "picture" of the borehole wall. This is achieved by recording part of the initial acoustic energy, sent by a transmitter (the transducer), reflected back by the formation to the same transducer working now as a receiver.

21.1.2. Tool description

The tool (Fig. 21-1) contains a piezo-electric transducer, which is the sensitive part of the tool, a motor, a magnetometer and associated electronics. The sonde diameter is 3 3/8", its length 12 ft. The transducer emits bursts of acoustic energy at a rate of 1500 pulses/s. It comprises a disk (1/2" diameter and 0.045" thickness) of piezo-electric crystal lead methaniobate for Amoco or lithium niobate for Sandia Nat. Lab. It operates in a thickness mode with a dominant frequency of 1.3 MHz.

An excitation voltage pulse is applied to the transducer at a rate of 1500 times per second to produce the bursts, and a synchronization pulse is generated such that every reflected and received signal contains amplitude information. A trigger pulse is provided to initiate a scan line containing 485 data points.

The transducer acts as a receiver. This is possible because the low quality of the transducer reduces the ringdown time such that the transducer has stopped reverberating, due to the excitation voltage, before the reflected signal arrives. The motor rotates the transducer at 3 revolutions per second.

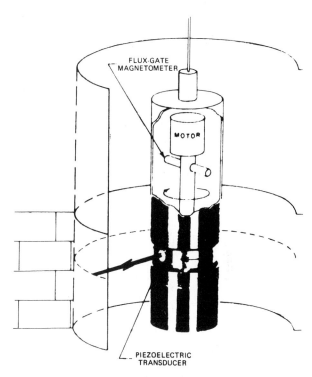

Fig. 21-1. Borehole televiewer (from Zemanek et al., 1968).

The magnetometer provides the orientation of the tool with respect to the Earth's magnetic field. The recommended recording speed is 300 ft/hour. The tool is centralized.

21.1.3. Data recording and display

The intensity (Z-axis) of an oscilloscope is modulated by the reflected signal and is recorded on film and tape. With the recording speed of 300 ft/hour and 3 revolutions per second, a 5-ft interval corresponds to 180 scan lines, where each scan line contains 485 data points.

The quality of the prints can be improved by controlling the range of the initial amplitude, initial setting of the oscilloscope beam intensity and camera focus.

21.1.4. Parameters affecting the measurement

The amount of reflected acoustic energy received at the transducer depends on rock reflectance, wall roughness, borehole attenuation and hole geometry.

(a) Rock reflectance is higher for hard rock, such

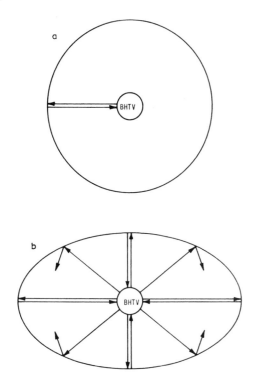

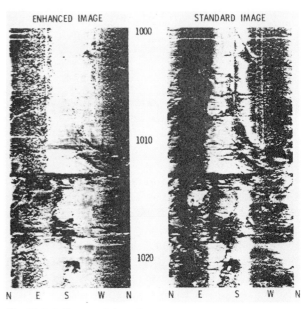

Fig. 21-2. Direction of reflected sound for: (a) circular borehole; and (b) elliptical borehole (from Wiley, 1980).

as compacted carbonates or quartzite, than for soft rocks (sand and shale) or mud-cake.

(b) Wall roughness depends on the texture of the rock: fractures and vugs will create a rough wall so more energy dispersion. On the contrary a well sorted sand or homogeneous shale will give a smooth wall which will reflect more energy, but not so much as a well consolidated compact carbonate.

(c) A circular borehole (Fig. 21-2a) has a reflecting surface normal to the pressure wave and consequently will send more energy back towards the transducer. In an elliptical hole the incident wave will be normal to the surface only when the transducer is opposite each major and minor axis. Between these positions the majority of the energy will be reflected away from the transducer (Fig. 21-2b). However, the intensity will be reduced at the major axis.

(d) Like all acoustic logs, it can be run only in liquid-filled holes. The borehole fluid affects the measurement. Too many suspended particles in the mud will create a speckled log due to dispersion and reflection on particles.

21.1.5. Processing of the data

To enhance the picture the digitized data are processed using a square-root operator, instead of linear, and Fourier filtering.

21.1.6. Interpretation

It consists of analyzing the "picture" which is like that presented in Fig. 21-3. On such a picture it is easy to recognize lamination, vugs and fractures, these features appearing as dark lines or spots due to dispersion of the acoustic energy on the edges of these features. The orientation is easy to define and the dip easy to compute (Fig. 21-4).

21.1.7. Application

The main application is the detection of fractures and the definition of their orientation. But this can be used also to define some textural features in carbonates as vugs or bird's eyes.

21.2. VERTICAL SEISMIC PROFILING (VSP)

Vertical Seismic Profiling is a high-resolution seismic method similar to the check-shot survey technique but in which the entire seismic trace is digitally recorded. This allows a more detailed analysis of the different wave trains and provides information in the vicinity of the well-bore both in spatial and time domains.

21.2.1. Principle

A velocity geophone anchored to the borehole wall receives information coming from two opposite directions: the downgoing waves and the upgoing or reflected waves (Fig. 21-5).

Fig. 21-3. Influence of the enhancement technique on the picture of the borehole wall (from Wiley, 1980).

315

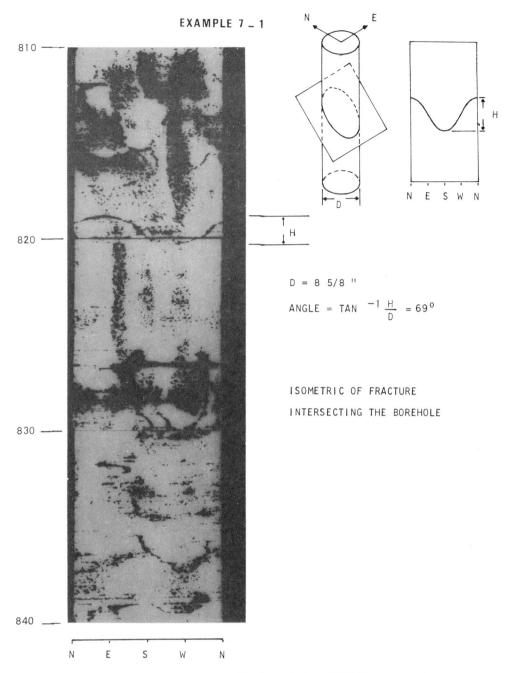

EXAMPLE 7 – 1

810

820

830

840

N E S W N

D = 8 5/8 ''

ANGLE = TAN $^{-1}\dfrac{H}{D}$ = 69°

ISOMETRIC OF FRACTURE
INTERSECTING THE BOREHOLE

N E S W N

Fig. 21-4. Example of a fracture and computation of its dip angle (from WEC Libya, courtesy of Schlumberger).

To allow a detailed analysis of the downgoing wave propagation and to permit a precise separation of the up- and downgoing signals, it is necessary to record a large number of levels in the well (50 to 400).

Moreover, the precise knowledge of the downgoing wavetrain at all depths allows the computation of a very powerful deconvolution operator that will be applied to the upgoing wavetrains. This allows high-resolution processing of VSP data with a minimum of assumptions concerning the earth response.

21.2.2. **Tool description**

The equipment is composed of:

(a) a source which can be an air or water gun or even Vibroseis;

(b) a downhole sensor which is a hydraulically operated anchored geophone and features low-noise and high-gain amplifiers;

(c) a set of specialized modules for the CSU.

316

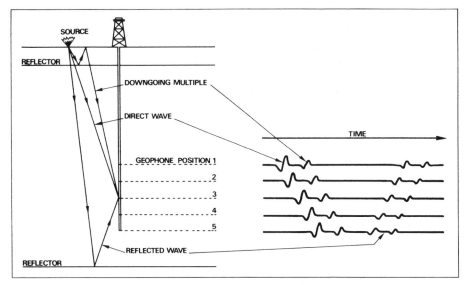

Fig. 21-5. A VSP is made of two wavetrains: the upgoing and the downgoing waves. The multiples can clearly be seen on the displays (from Mons and Babour, 1981).

21.2.3. Data recording

Data are sampled every millisecond and digitized with a dynamic range of 90 db in an instantaneous floating-point converter. The storage is done on 9-track compatible tapes. VSP recording consists of the acquisition of 200 to 2000 shots at up to 400 levels. The well must be sampled every 10 to 50 m in order to allow a full separation of the up and downgoing wavetrains.

21.2.4. Data processing

21.2.4.1. *At the well-site (Fig. 21-6)*

With the CSU capability the following "quick-look" products can be displayed:

(a) correction of the measured transit times taking account of the time vertical depth of the measurements (TVD corrections and Surface Reference Datum (SRD) corrections);

(b) time to depth relationship;

(c) interval velocities between shooting levels;

(d) sonic drift.

In order to remove high-frequency noise, a band-pass filter is applied. The true amplitude recovery is performed to correct for spherical divergence.

21.2.4.2. *At the computing center*

The following steps are taken in the processing:

(a) The shots are retrieved from the tape, edited and stacked.

(b) Bandpass filtering is applied to restrict the bandwidth from 1 Hz/250 Hz (acquisition band-

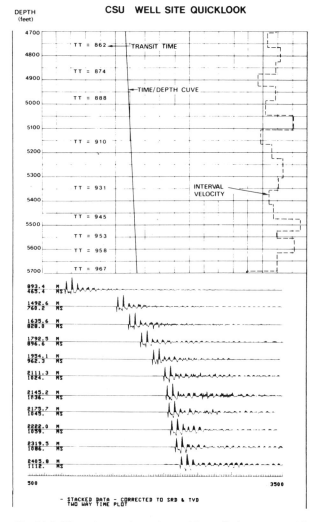

Fig. 21-6. The cyber service unit provides well site answers. All times and velocities are corrected to vertical and to the SRD (from Mons and Babour, 1981).

width) to the usable bandwidth. This varies from case to case, but is generally of the order of 10 to 80 Hz.

(c) Velocity filtering is then applied to separate the up- and downgoing wavetrains. The trace is considered as the sum of an up- and a downgoing wavetrain and is assumed that the signal shift is equal but of opposite sign from one level to the following. Next, the energy along one of the two wavepaths is minimized in the least-square sense. It is important to notice that the transit time difference between levels is taken into account exactly. This allows the placement of the downhole sensor at levels where the borehole conditions are favourable. It is also possible to reject noisy levels during the processing without affecting the output.

Predictive deconvolution is applied to remove the multiples. The operators are computed level by level from the autocorrelation function of the downgoing wavetrain. The operators are then applied to both the down- and the upgoing wavetrains.

The next step is to apply waveshaping deconvolution to tailor the signal signature to the seismic section, and to ultimately improve the signal-to-noise ratio.

The last step of the processing consists of the stacking of all the upgoing waves in a window after the first break. This produces a trace which has the appearance of a synthetic seismogram but is in fact made of real seismic data (Fig. 21-7).

21.2.5. Applications

The main applications are listed below.

(a) *High-resolution time-depth curve.* As the levels are separated by only 3 to 7 milliseconds, an accurate velocity analysis can be made. This is very useful when the sonic log is of poor quality due to borehole effects and also when no sonic log can be recorded. For example in the vicinity of the surface the sonic log cannot be acquired if the velocity of the formation is lower than 1600 m per second. In this case the velocity derived from the VSP is the only one available.

(b) The analysis of the downgoing wavetrain (seismic wave versus depth) provides information on the filtering of the earth. The attenuation of the energy and the phase rotation can be evaluated.

(c) The acoustic impedance is computed by the evaluation in the frequency domain of the transfer function from one level to the next. The effect of noise or of tool coupling is minimized by the computation of interspectra and not only of ratios of amplitude. Further, from the downgoing wavetrain the multiples can be studied and their genesis deduced from the evolution of their pattern versus depth.

(d) However, one of the most important applications of VSP remains the *analysis of reflected signals from below the sensor.* This provides an accurate seismogram in the vicinity of the well bore. The lateral depth of investigation of a VSP is intermediate between surface seismic and logs. If one considers

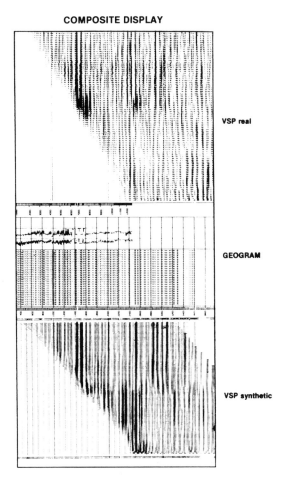

Fig. 21-7. Composite display of the different geophysical answer products: the logs, the Geogram, the VSP and the synthetic VSP (from Mons and Babour, 1981).

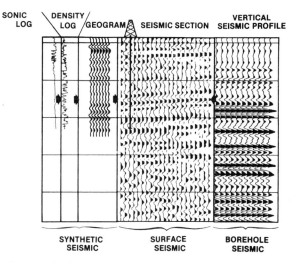

Fig. 21-8. Simultaneous interpretation of log data and VSP allows precise calibration of surface seismics (from Mons and Babour, 1981).

that half a Fresnel zone of the reflector is providing information, this is a radius of the order of 40 m that is investigated in the case of an average VSP.

When recorded in deep wells, the VSP provides information on the *reflectors below the bottom of the well*, and its depth of investigation is often significantly larger than that which can be achieved with surface seismic (Fig. 21-8).

One of the most striking properties of VSP is to be vertical i.e. the moveout effects are minimized. This simplifies greatly the analysis of highly dipping reflectors, and also the interpretation of data recorded in faulted areas.

Vertical Seismic Profiling presents a strong potential as a high-resolution technique because the shallow layers (WZ) are crossed only once by the seismic wavetrain, causing the frequency content of VSP to be generally higher than that which can be achieved by surface to surface seismic.

21.3. NUCLEAR MAGNETIC LOG (NML)

21.3.1. Introduction

The Nuclear Magnetic Logging tool measures the free precession of proton nuclear magnetic moments in the earth's magnetic field.

21.3.2. Principle

Some nuclei, e.g. hydrogen, have a spin, I, which corresponds to one macroscopic angular momentum or magnetic moment, μ. In this case the nuclei behave like magnets which rotate on themselves. Following the quantum theory nuclei having a spin I, can take $2I + 1$ possible states. For instance if the nucleus has a value of $I = 1/2$ it will take two opposite values. In this case, with a population of N nuclei, n_{up} will have a negative momentum, $-\mu$, and n_{down} a positive one, $+\mu$. In the absence of a magnetic field the equilibrium state is obtained when n_{up} is equal to n_{down}, and the total momentum is nil. When a magnetic field, H_0, is applied, a new equilibrium state is reached after a certain time. In this new state more spins are parallel to the magnetic field than antiparallel (Fig. 21-9). The presence of the magnetic field implies an energy shift of the spins, equal to the product of the angular momentum μ by the amplitude of the field H_0.

$$E = \mu H_0$$

According to quantum statistics the probability that a spin is in one state is proportional to:

$$e^{-E/kT}$$

where: E is the energy of this state; k is the Boltz-

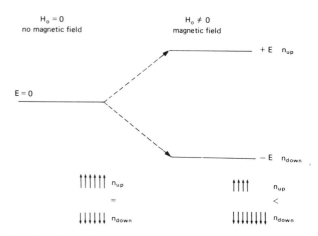

Fig. 21-9. Energy levels of nuclei in the absence and presence of magnetic field (courtesy of Schlumberger).

mann constant; and T is the absolute temperature. Hence we have:

n_{up} is proportional to $e^{+\mu H_0/kT}$

n_{down} is proportional to $e^{-\mu H_0/kT}$

More spins are now in the lower energy level and the total momentum, M_0, of the nuclei is no longer nil but equal to:

$$M_0 = -n(_{down} - n_{up})\mu$$

The earth's magnetic field itself polarizes the nuclei. But the net polarization is too small and the magneti-

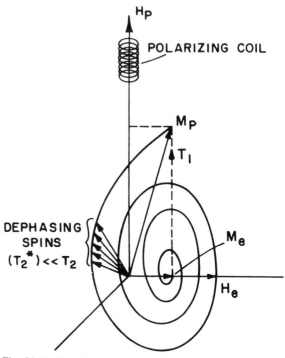

Fig. 21-10. Free induction decay in the earth's magnetic field following application of a DC polarizing magnetic field (from Herrick et al., 1979).

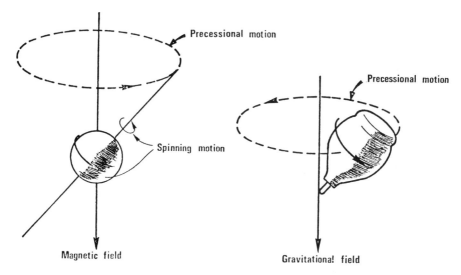

Fig. 21-11. Examples of precessional motions.

zation of the sample difficult to observe.

To increase the magnetization, or the difference of populations of opposite spins, it is possible to use an external large field H_p, induced by a polarizing coil (Fig. 21-10), which more efficiently polarizes the observed sample and induces a magnetization M_0. The effect of H_e (earth's magnetic field) is negligible during this polarization period.

After a sufficient time, T_1, called longitudinal or spin-lattice relaxation time the equilibrium is reached and the polarizing magnetic field H_0 is removed.

The spins are now acted upon only by the earth's magnetic field. Their behaviour can be described classically as analogous to the motion of a gyroscope in the earth's gravitational field at a frequency proportional to the field strength, the Larmor frequency, $f = \gamma H_e$, where γ is the gyromagnetic ratio of the proton ($\gamma = 4.2576 \times 10^3$ Hz/G).

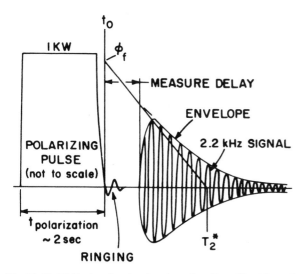

Fig. 21-12. NML signal and estimation of ϕ_f (from Herrick et al., 1979).

The precession of the spins results in a varying magnetic field which induces an AC voltage (at the precession frequency) in the measuring coil positioned perpendicular to the earth's magnetic field (Fig. 21-11). It is obvious that the measuring coil can be the same as the polarizing coil as polarization and signal measurement are made separately.

The signal in the coil is an exponentially decaying sine wave (Fig. 21-12). It is characterized by its decay time constant or transversal or spin-spin relaxation time T_2, its amplitude at the beginning of the precession and its frequency. Transitions from one equilibrium state to another (when an external magnetic field is applied or removed) are not instantaneous. They take a certain amount of time which depends on the structure of the material containing the hydrogen nuclei and the concentration of impurities in the material.

T_1, the spin-lattice relaxation time characterizes the time taken by the magnetic field to align the spins in the direction of the field.

T_2, the spin-spin relaxation time characterizes the time taken by the ordered spins in a magnetic field to become randomly oriented. T_2 is shorter than T_1.

Figure 21-9 shows that the alignment of spins corresponds to a loss of energy $2E$ from the upper to the lower level. This loss of energy must be communicated to the surrounding medium. The way this energy is dissipated commands the rate of relaxation T_1.

In the same manner the way the spin system gains energy from the surrounding medium commands T_2.

Three cases are considered:

(a) Solids. In solids the hydrogen nuclei are tightly bound to the surrounding medium. Energy can be easily dissipated or gained through thermal vibrations. The spins can flip rapidly to a different energy level. Consequently, relaxation times are short, gener-

320

ally a few hundred microseconds.

(b) Liquids. Molecules in a liquid are independent. They do not interfere with other molecules. Each molecule with hydrogen nuclei keeps unchanged spins as there is little or no support for an energy transfer. Aligning or dephasing spins is a lengthy process. Relaxation times are long, hundreds of milliseconds.

(c) Liquids with paramagnetic ions. The behaviour of paramagnetic media in the presence of an external magnetic field is more complex: electron spins are also split into two energy levels. But because the electron magnetic moment is much larger than the nucleus magnetic moment, the energy of separation of the two electronic energy levels is much larger. The difference of populations in the lower and upper energy levels consequently is also larger.

In a liquid with paramagnetic ions there is a coupling of the nuclear spin system and the electron spin system. Generally when an electron flips from one spin position to the opposite one, a nucleus flips the complementary way. Because of a large number of flipping electrons a large number of nuclear transitions is achieved and the relaxation times are considerably reduced. An increase in the concentration of paramagnetic ions is followed by a reduction in the relaxation time (Fig. 21-13).

It is for that reason that the addition of ferromagnetic materials is recommended.

Because of their large magnetic moment, magnetite grains dispersed in fluids in random motion, create large varying magnetic fields which flip over the spins of the nuclei in the fluids.

The relaxation time is shortened to the time T_2^*. The difference between T_2^* and T_2 involves microscopic magnetic interactions. The reciprocal of T_2^* is proportional to the concentration of ferromagnetic material:

$$\frac{1}{T_2^*} \sim 1.4 \, M$$

(M = concentration of magnetite in $\mu g/cm^3$)

21.3.3. **Tool description**

The downhole tool consists of a sonde, an electronic cartridge, an auxiliary caliper and two centralizers (Fig. 21-14). On the surface, a tool module interfaces to the CSU. The sonde contains two coils which are wound in a series opposing ("bucking") configuration. This arrangement provides cancellation of most environmental noise, allows the sonde to be calibrated in a standard test pit on the surface.

The coils perform two functions: (a) production of a polarizing magnetic field from the DC current supplied by the cartridge, approximately 1 kw is dissipated in the coils during polarization; and (b) detection of the proton precession signal using the same set of coils. The induced signal across the coil is of the order of 0.1 microvolt.

The CSU supplies AC power to the cartridge where it is transformed, rectified and applied to the sonde coil during the polarization period. Tuning pulses from the CSU are received by the cartridge and used to drive switches. The signal induced in the

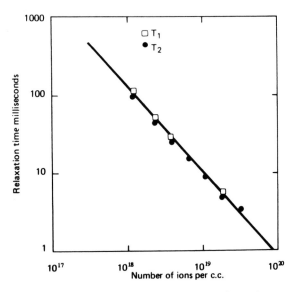

Fig. 21-13. Longitudinal and transverse relaxation times of protons at room temperature in paramagnetic solution of ferric ions of different concentration (courtesy of Schlumberger).

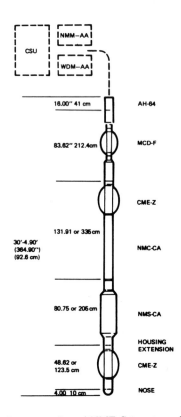

Fig. 21-14. Nuclear magnetic tool NMT-C (courtesy of Schlumberger).

coil is amplified to a level that can be transmitted up the logging cable.

The coil is parallel tuned to the local Larmor frequency by a capacitor network. The tuning range is from 1300 Hz to 2600 Hz. The proper capacitors for a given Larmor frequency are determined by the CSU unit and switched in automatically.

A calibrated circuit is included in order to select the proper tuning capacitors as well as to calibrate the gain of the entire system.

The ratings are: temperature—175°C or 340°F; pressure—20,000 psi or 1350 atm. The recommended logging speed is: 700 ft/hr.

21.3.4. Method of measurement

Three operating modes have been implemented for the NML tool:

(a) Normal mode. The log is run continuously with polarizing pulses synchronized to each depth sample (every 6″). Maximum power is applied with sufficient duration to reach saturation (get maximum magnetization of the proton spins). The free-fluid index is computed from the envelope and displayed in real time along with T_2^* and the caliper. T_2^* does not characterize the formation but is useful for log quality control.

(b) Continuous mode. This mode is also run continuously versus depth. Three non-saturating pulses of polarization are applied, each twice the duration of the preceeding pulse. Three free-fluid indices are derived with the same extrapolation technique as in normal mode. These three values lie on an increasing exponential if plotted versus time. The time constant of this curve is a good approximation of T_1.

(c) Stationary mode. Multiple polarizing pulse durations are applied to the formation while the tool remains stationary in front of the formation. With an approach similar to the previous mode (but now with many more free-fluid index values versus time) T_1 is determined more accurately.

21.3.5. Signal processing

(a) CSU signal processing.

The measurement does not begin until after a preselected delay, to discriminate against tool transients and residual mud signal. The signal coming up-hole is fully rectified, then digitized in fifty 2-ms samples and stored as envelope values. During the downhole calibration, the signal without polarization is measured several times and averaged to give noise. The envelope values are corrected for noise by root-mean-square subtraction for any value larger than noise. The corrected envelope is fitted to a single exponential using a weighted least square or gate estimation algorithm to determine amplitude of T_2.

The outputs of the curve fittings are:
- AE: the amplitude of the fitted exponential at the beginning of the estimation period.
- T_2: the decay constant of the envelope in milliseconds.

(b) Reprocessing

This processing uses the full waveform sampled in 1024 intervals. Examples of raw and reprocessed waveforms are shown in Figs. 21-15 and 21-16.

21.3.6. Geological factor influencing the measurement

(a) Mineral composition of the rocks, shale.

Most of the time they have very short relaxation times (solids or liquids bound to solids). The amplitude of their contribution becomes negligible before the instrumentation delay expires.

(b) Fluids

(1) Bound water (irreducible water saturation). It has a very short relation time the water being bound to grains.

(2) Oil: most common oils contain hydrogen protons with T_1 comparable to water. But their relaxation time depends on their physical characteristics and decreases with viscosity. Oil with viscosities above 600 cp do not contribute to the NML signal.

(3) Gas: because of its low hydrogen concentration, low pressure gas makes little or no contribution to the NML signal.

(4) Free water: In general, the only signal observed after a certain delay is due to the free fluids (Fig. 21-17). Therefore, the amplitude of this signal extrapolated at the beginning of the precession reflects the total free-fluid porosity ϕ_f or FFI (Free Fluid Index).

(c) Texture

Due to the influence of textural parameters on the porosity and the permeability, its influence on the measurement is very high. In fact the measurement reflects the textural properties of the rock.

(d) Temperature

The magnetization is a function of the temperature. This influence must be taken into account and corrected before the interpretation.

21.3.7. Environmental factors

(a) Angular factor

The signal depends on the direction of the earth's magnetic field. For a straight hole the correction to apply is proportional to $1 + \cos^2$ (MINC), where MINC is the inclination of the earth's magnetic field measured from the vertical.

For a deviated hole the correction is proportional to $1 + \alpha^2$ with $\alpha =$ Cos (MINC) Cos (DEVI) + Cos (AZIM) Sin (MINC) Sin (DEVI) where DEVI is the deviation angle and AZIM is the deviation azimuth.

322

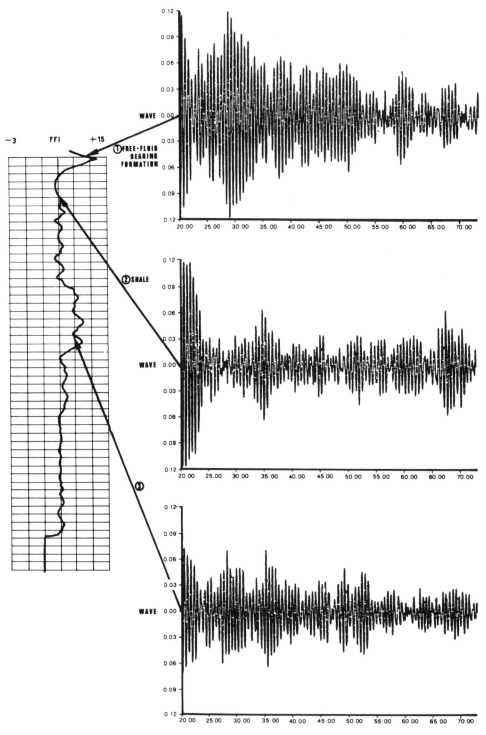

Fig. 21-15. Raw waveforms.

(b) Larmor frequency

The signal is proportional to the Larmor pulsation:

$$\omega_L = \gamma H_e \text{ (in Hz)}$$

where: γ is the gyromagnetic ratio ($= 2.675 \times 10^4$ S^{-1} gauss^{-1}) and H_e is the amplitude of the earth's magnetic field (in kilogammas)

(c) Borehole influences

The mud signal should be eliminated by doping with magnetite. Nevertheless the diameter of the hole has a strong influence which must be corrected before interpretation.

The nuclear magnetism signal is of little value in holes larger than 13″.

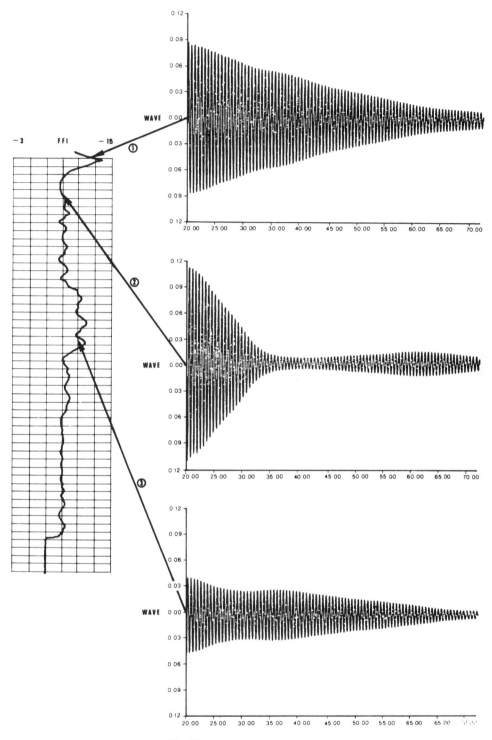

Fig. 21-16. Reprocessed waveforms.

21.3.8. **Interpretation–Application**

21.3.8.1. *Irreducible water saturation determination*

The NML porosity, ϕ_f, can be used in conjunction with other porosity data to determine irreducible water saturation, S_{wirr}. The quantity $\phi_e - \phi_f$, rock effective porosity less free fluid saturated porosity, is the bulk amount of fluid bound to the formation. Thus irreducible water saturation is simply:

$$S_{wirr} = (\phi_e - \phi_f)/\phi_e$$

For clean formations at irreducible water saturation this equation can be combined with the Archie water saturation equation to yield:

$$\phi_e - \phi_f = \sqrt{R_w/R_t}$$

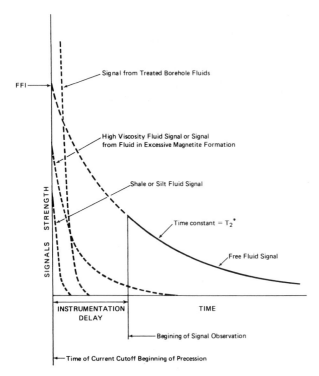

Fig. 21-17. Pictorial representation of fluid signals in a nuclear magnetism log.

The quantity $\phi_e - \phi_f$ may be crossplotted against $\sqrt{1/R_t}$ (Fig. 21-18). All levels at or near irreducible water saturation will plot on a straight line through the origin with a slope of $\sqrt{R_w}$. If the formation is shaly and requires a more complicated water saturation equation, S_{wirr} from the NML Log and S_w can simply be compared to determine whether the formation is at irreducible water saturation.

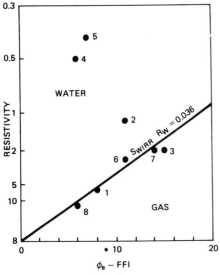

Fig. 21-18. Cross-plot of $\sqrt{1/R_t}$ vs ($\phi_e - $FFI) for eight zones (from Herrick et al., 1979).

21.3.8.2. Permeability estimation (Fig. 21-19)

Several equations have been proposed to compute permeability from porosity and S_{wirr}; most are of the form:

$$k = A\phi^B / (S_{wirr})^c$$

where k is permeability and A, B and C are empirical constants which are determined for a specific area or formation and may not be widely applicable.

An expression for use in sandstones has been proposed by Timur (1969) and has the form:

$$k = A\left[\frac{\phi^{4.4}}{10^4(S_{wirr})^2}\right]^B$$

where A and B are empirical constants to be determined for a given lithology. Either expression will probably require some initial calibration with core data after which they should be applicable to subsequent wells drilled in the same formation.

21.3.8.3. Porosity determination

In carbonate formations, which have little surface activity, the NML tool measures essentially total porosity and can therefore supplement existing commercial porosity tools. Since the NML log does not respond at all to hydrogen protons in the matrix, it can make a valuable contribution in unusual lithologies containing considerable water of hydration, such as gypsum, where nuclear tools often give erroneously high readings.

21.3.8.4. Determination of hydrocarbon characteristics

As previously discussed formation fluid viscosity (and consequently hydrocarbon density) is related to the polarization time constant T_1. The NML log can thus identify intervals bearing heavy crude or tar (little or no NML signal) and permeable water bearing zones (large signal).

21.3.8.5. Residual oil saturation

The NML survey is the only logging technique which directly measures the saturation of residual oil rather than inferring it from other measurements. In addition, the NML technique does not depend on the validity or choice of saturation equation. The procedure involves the addition of paramagnetic ions to the drilling fluid in order to inhibit the filtrate signal in the NML measurement by greatly reducing its relaxation time. In this case the free-fluid porosity is equal to the volume of residual oil in the flushed zone, $\phi_e S_{or}$ (Fig. 21-20).

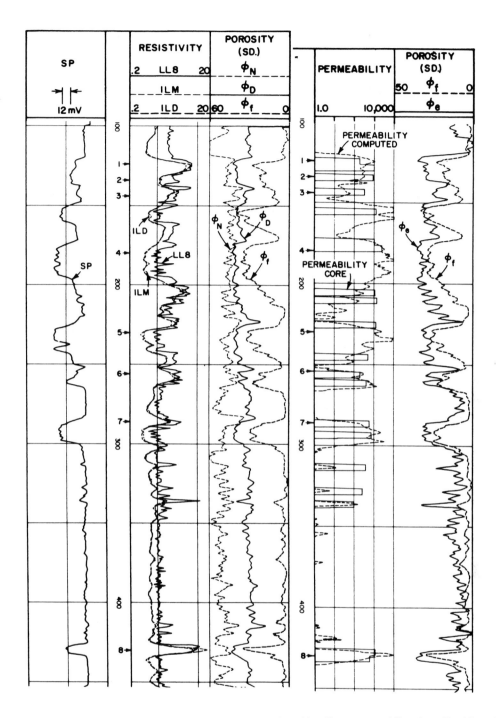

Fig. 21-19. Example of permeability computed from the NML compared to sidewall core permeability (from Herrick et al., 1979).

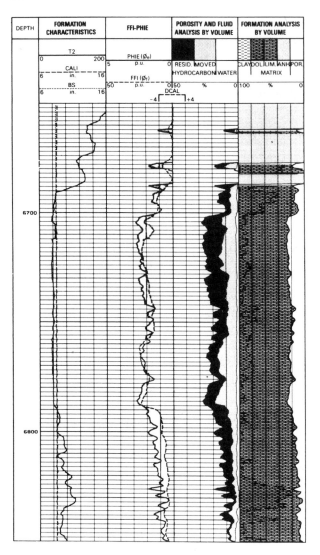

Fig. 21-20. Comparison of NML field data and CPI results (from Well Evaluation Conference, U.A.E./Qatar, 1981).

21.4. REFERENCES

Artus, D.S., 1967. Nuclear Magnetism Logging in the Rocky Mountain Area. The Log Analyst.

Desbrandes, R., 1968. Theorie et Interpretation des Diagraphies. Technip, Paris.

Herrick, R.C., Couturie, S.H., and Best, D.L., 1979. An improved nuclear magnetism logging system and its application to formation evaluation. SPE of AIME, 54th Annu. Techn. Conf. of SPE of AIME. Paper. No. SPE 8361.

Mons, F. and Babour, K., 1981. Vertical seismic profiling. Recording, Processing, Applications.

Neuman, C.H., 1978. Log and core measurements of oil in place, San Joaquin Valley, California. SPE 7146 presented at the California Regional Meeting of the SPE.

Robinson, J.D., Loren, J.D., Vajnar, E.A., and Hartman, D.E., 1974. Determining residual oil with the nuclear magnetism log. J. Pet. Technol., pp. 226–236.

Schlumberger Well Services, 1978. Log Interpretation Charts.

Timur, A., 1969. Pulsed nuclear magnetic resonance studies of porosity, Movable fluid and permeability of sandstones. J. Pet. Technol., pp. 775–786.

Timur, A., 1972. Nuclear Magnetic Resonance Study of Carbonate Rocks. SPWLA, 13th Annu. Log. Symp. Trans., paper N.

Wiley, R., 1980. Borehole televiewer—revisited. SPWLA, 21st Annu. Log. Symp., Trans., Paper HH.

Zemanek, V. et al., 1968. The Borehole Televiewer—A new logging concept for fracture location and other types of Borehole Inspection. SPE Trans., 246.

22. THE PLACE AND ROLE OF LOGS IN THE SEARCH FOR PETROLEUM

Let us suppose that it was possible to core a well continuously to total depth. We could conclude that in this case there would be no real use in running logs, since the geologists, reservoir engineers and drilling engineers apparently have on hand all the samples necessary for their particular studies. However, they still lack certain information:

(a) Data that are difficult or expensive to obtain (measurements of natural gamma-ray radiation, neutron hydrogen index, sonic transit time, bulk density, etc.).

(b) An analysis of a larger volume of formation than the core, especially when we consider that core measurements are themselves made on thin sections or plugs of material.

(c) The possibility of making a computer analysis of quantitative data as with logs.

(d) Information which is permanent.

Everyone, therefore realizes that continuous coring without running logs is not a reasonable economic way of evaluating a formation. In any case since there is almost always a risk of a less than 100 percent core recovery then there is also a risk of having gaps in the well information.

These are some of the reasons why the last 50 years has seen almost continuous research into techniques that attempt to cover deficiencies in core data. Logs make up the majority of these techniques. The number of types of logs and their reliability has increased and new interpretation methods have been developed simultaneously.

Logs, having become in some ways the geologist's eye—an eye that is imperfect and sometimes distorted but nevertheless not blind—and an instrument for the reservoir engineer, occupy a special place and play an important role in petroleum research by the economies that they bring and the amount of information they contain.

Desbrandes (1968) estimated that the cost of logging was 2% of the total drilling cost whereas the information that they supplied on the content of reservoirs was obtained at some one fifth of the cost of any other comparable method.

From a statistical study made by the drilling department of the Elf-Aquitaine group on 43 wells operated by the group in 1973, it turns out that the cost of logging operations, measured by the total bills paid to service companies, was 5% of the drilling costs, leaving out the immobilisation time of the necessary equipment (the rig) or production operations. If this time is added, the total cost comes to about 10%.

The amount of money that is saved is more difficult to determine but we can estimate that continuous coring would triple the drilling costs. In the same way, formation tests that are made without reliable information on their positioning in depth, or their necessity, raise costs very strongly.

All the same even if logging can be shown to be an economical and practical method of obtaining data we can always conceive of methods of reducing overall logging costs. Several issues need to be considered before doing this:

(1) All geological or other information either comes from samples collected during drilling (cores, cuttings, wireline cores) and testing (fluid samples) or from logs. While the cost of coring and testing is often less than 5% of the well cost and geological monitoring less than 2% it is evident that good quality samples are rare and relatively expensive. If the logging program is limited we risk losing part of our only other source of information. This is a false economy since we risk losing information that is more valuable than the costs saved.

(2) The economy may show up in the long term to be disastrous. When a study of a basin or field is made the lack of information may lead to the non-identification of a potential well or zone.

(3) In recording logs we include information that may not be of immediate use but which may show up as useful later on.

(4) Reducing the frequency of logging may lead to the loss of information entirely as in the case where a well is lost before the next logging point is reached or where further drilling deteriorates the well conditions to the point where good quality logs are unobtainable (well caving in or deep invasion).

(5) Redundant information should not be eliminated since it acts as a control of the quality of the recording.

(6) Logs are obtained rapidly after the completion of drilling, are easily transported, easily compared and are a relatively indestructable form of information.

We have to then measure the economies made in terms of the actual information given by each log and the risk involved when this information is not available.

We can obtain an idea of the information given by each log by looking at Table 1-4 of this volume and the paragraphs on geological factors influencing measurements and applications.

Table 22-1 gives a more complete idea by looking at the information obtained from each log singly or in combination with others and by classifying them as more or less quantitative or qualitative.

We need to clearly establish the areas of application of logs. These and the corresponding interpretation methods will be the subject of volume 2 which will cover the following subjects:

  (a) Description of sedimentary series:
     (1) their mineral composition and lithology;
     (2) their texture;
     (3) their sedimentary structure
     (4) the petrophysical characteristics of the reservoirs;
     (5) the nature of the reservoir fluids.
  (b)   The formation of sedimentary series (sedimentology) attempting to reconstruct the depositional environment from:

     (1) a facies type analysis;
     (2) analysis of sequences;
     (3) facies correlation.
  (c) Transformation of sediments. The effects of diagenesis, their identification, compaction studies.
  (d) The organisation of sedimentary series (stratigraphy)
     (1) study of their order and relative age;
     (2) the significance of ruptures (discontinuities);
     (3) stratigraphic correlation.
  (e) The deformation of sedimentary series (structural geology)
     (1) folds;
     (2) faults;
     (3) fractures.
  (f) Geophysical, geothermal and geochemical applications.
  (g) Role and importance of logs in geological synthesis.

TABLE 22-1

Information provided by basic measurements or their combination

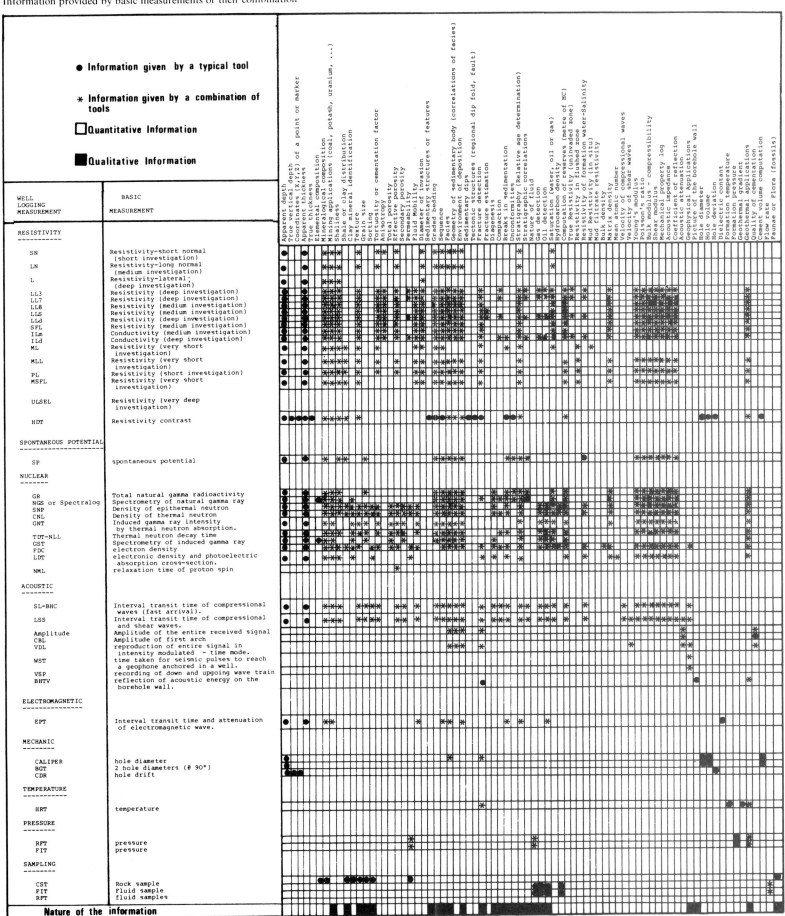

APPENDIX 1

TABLE A1-1

List of open hole logging tools and their symbol

OPEN HOLE LOGGING—Formation Evaluation Logs

*Electrical Survey *

Birdwell	Electric Log	ES
	Single Point Resistivity Log	
Dresser Atlas	Electrolog	EL
GO International	Electrical Log	EL
Schlumberger	Electrical Survey	ES
	Salt Dome Profiling (ULSEL *)	ES-ULS
Welex	Electric Log	
	Drill Pipe Electric Log *	

*Induction Logs * (focused conductivity)*

Birdwell	Induction Log	IS
	Induction Electric Log	IES
Dresser Atlas	Induction Electrolog	IEL
	Dual Induction Focused Log	DIFL
GO International	Induction Log	IL
	Induction-Electrical Log	IEL
Schlumberger	Induction Log	
	Induction-Electrical Survey	IES
	ISF Induction Log	ISF
	Dual Induction-Laterolog	DIL
	Dual Induction-SFL	
Welex	Induction Electric Log/Long Lateral	IEL
	Dual Induction Guard	DIGL

Focused Resistivity Log

Birdwell	Guard Log *	GDS
Dresser Atlas	Laterolog *	IL
	Dual Laterolog	
GO International	Guard Log *	GL
Schlumberger	Laterolog *	L
	Dual Laterolog *	DLL
	Spherically Focused Log *	SFL
Welex	Guard Log *	
	Guard-FoRxo	

*Micro-resistivity Log (non-focused, Microlog * type)*

Birdwell	Micro-Contact Caliper	EMC
Dresser Atlas	Minilog	ML
GO International	Micro-Electrical Log	ML
Schlumberger	Microlog	ML
Welex	Contact Caliper Log	

*Micro-resistivity Log (focused, Microlaterolog * type)*

Dresser Atlas	Micro-Laterolog	MLL
	Proximity *-Minilog	PML
GO International	Micro-Guard	MGL
Schlumberger	Microlaterolog	MLL
	Proximity *-Microlog	PLL
	Micro-SFL	
Welex	FoRxo-Caliper Log	

*Gamma-Ray Log * (natural gamma)*

Birdwell	Gamma-Ray Log	
Dresser Atlas	Gamma-Ray Log	
GO-International	Gamma-Ray Log	GR
McCullough	Gamma-ray	

Schlumberger	Gamma-Ray Log	GR
Welex	Gamma-Ray Log	

*Spectral Gamma - Ray Log ***
Dresser Atlas	Spectralog	
Schlumberger	Natural Gamma-Ray Spectrometry	NGS

*Neutron Logs ***
Birdwell	Neutron Log (n-γ, n-n)	NL,NNL
	Epithermal Neutron Log	ENP
Dresser Atlas	Neutron Log (n-t)	
	Epithermal Neutron Log (n-e)	
	Sidewall Epithermal Neutron * (n-e)	SWN
	Compensated Neutron Log * (hybrid)	CNL
GO International	Neutron Log (n-γ, n-e)	NL
McCullough	Neutron Log	
Schlumberger	Neutron Log (n-γ)	GNT
	Sidewall Neutron Log * (n-e)	SNP
	Compensated Neutron Log * (n-t)	CNL
Welex	Neutron Log (n-γ, n-t)	
	Sidewall Neutron Log * (n-e)	SWN

*Pulsed Neutron Logs ***
Dresser Atlas	Neutron Lifetime Log *	NLL
	Dual Space Neutron Lifetime Log	DNLL
	Carbon/Oxygen Log *	
Schlumberger	Thermal Decay Time Log	TDT
	Dual-Spacing Thermal Decay Time Log *	TDT-K

*Density Log * (gamma - gamma)*
Birdwell	Formation Density	FDL
Dresser Atlas	Compensated Densilog	CDL
GO International	Density Log	DL
	Borehole Compensated Density Log	CDL
McCullough	Density	
Schlumberger	Formation Density Log (Compensated)	FDC
Welex	Compensated Density Log	CDL
	Articulated Density Log	

Litho - density Log
Schlumberger	Lithodensity Log	LDT

*Acoustic Log * — Transit Time * Logs*
Birdwell	Continuous Velocity Log *	V2S
Dresser Atlas	BHC Acoustilog	BHC
	Sidewall Acoustic Neutron Log *	SWAN
GO International	Borehole Compensated Sonic Log *	CSVL
	Sonic Log *	SVL
Schlumberger	BHC Sonic Log	BHC
	Long Spacing Sonic	LSS
Welex	Acoustic Velocity Log	AVL
	Compensated Acoustic Velocity Log	

*Acoustic Log * — Amplitude Logs ***
Dresser Atlas	Signature (or Half Signature) Log *	
GO International	Sonic Formation Amplitude Log	SFAL
Schlumberger	Amplitude Log	A-BHC
Welex	Fracture Finder	

*Acoustic Log * — Wave Train * Logs*
Birdwell	3-D Velocity Log	V3D
Dresser Atlas	Variable Density Log *	VDL
Go International	Seismic Spectrum	SS
	Signature * Curve	SC
Schlumberger	Variable Density Log *	BHC-VD
Welex	Micro-Seismogram *	MSG

Electromagnetic propagation
Schlumberger	Electromagnetic propagation	EPT

Spectrometry of Induced Gamma - Rays
Dresser Atlas	Carbon-Oxygen Log	C/O
Schlumberger	Gamma Spectrometry	GST

OPEN HOLE LOGGING—Special Purpose Logs and Services

Beta Log
    Birdwell   Beta Log   BL

Borehole Camera
    Birdwell   Borehole Camera *   BC

Category	Company	Service	Code
Beta Log	Birdwell	Beta Log	BL
Borehole Camera	Birdwell	Borehole Camera *	BC
Borehole Gravimeter *	Gravilog	Esso Vibrating String	
Borehole Televiewer *	Birdwell	Seisviewer	SVS
	Schlumberger	Borehole Televiewer (not avail. 1975)	TVT
	Amoco	Borehole Televiewer	
	Sandia	Acoustic borehole Televiewer	
Caliper Log *	Birdwell	Caliper Logging	CA3, CA6, CAL
		Seiscaliper (sonar caliper *)	CSC
	Dresser Atlas	Caliper	
	GO International	Caliper Log	CL
	Schlumberger	Caliper	CAL
	Welex	Caliper	
Computed Analysis Logs *	Birdwell	COM-PRO LOGS	
		Elastic-Properties Log	
	Dresser Atlas	EPILOG SYSTEMS	
	Schlumberger	SYNERGETIC LOG SYSTEMS (SARABAND, CORIBAND, GLOBAL, VOLAN)	
		Mechanical Properties Log	
	Welex	CAL SYSTEMS	
Core Slicer *	Dresser Atlas	Tricore	
	Schlumberger	Diamond Core Slicer (not avail.)	DCS
Dipmeter *	Dresser Atlas	Focused 3-Arm Diplog	FDIP
		Hi Resolution 4-Arm Diplog	HRDIP
	GO International	Dipmeter Log	
	Schlumberger	4-Arm High Resolution Digital Dipmeter	HDT-D
		Sedimentary Dipmeter	SHDT
	Welex	Dip Log	
Directional Surveys *	Birdwell	Continuous Directional Inclinometer Log	CDI
	Dresser Atlas	Directional Survey	
	GO International	Continuous Directional Survey	CDS
	Schlumberger	Continuous Directional Survey	CDR
	Welex	Drift Log	
Formation Tester *	Dresser Atlas	Formation Tester	FT
	GO International	Formation Tester	FT
	Schlumberger	Formation Tester	FT
		Formation Interval Tester	FIT
		Multiple Fluid Sampler	MFS
		Repeat Formation Tester	RFT
	Welex	Formation Tester	
		Jet-Vac Tester	
Nuclear Magnetism Log *	Dresser Atlas	Nuclear Magnetism Log (not avail. 1975)	NML
	Schlumberger	Nuclear Magnetism Log	NML
Sidewall Sample Taker * (percussion)	Birdwell	Core Gun	SWD
	Dresser Atlas	Corgun	
	GO International	Sidewall Coring	SWC
	Schlumberger	Sidewall Coring	CST
	Welex	Sidewall Coring	

Temperature Logs *

Birdwell	Temperature Log	TL
	Differential Temperature Log *	TLD
Dresser Atlas	Temperature Log	
	Differential Temperature Log *	
GO International	Temperature Log	TL
	Differential Temperature Log *	DTL
Schlumberger	High Resolution Thermometer	HRT
Welex	Precision Temperature Log	

Well Seismic

Schlumberger	Well Seismic Survey	WST
	Vertical Seismic Profile	VSP

TABLE A1-2

Specifications of open hole logging tools [1]

Tool type	Company	Equipment type	Tool diameter (inch)	Tool length (ft)	Hole Diameter Min (inch)	Max	Pressure rating (PSI)	Temperature rating (°F)
E	Dresser Atlas	1702	$3\frac{1}{4}$	$11\frac{1}{16}$	5	12	20000	450
	GO International	EL	$3\frac{1}{2}$	7, 6	$4\frac{1}{2}$		20000	400
	Schlumberger	S54 & 55	$3\frac{5}{8}$	$9\frac{1}{2}$	$5\frac{1}{8}$	12	20000	350
	Welex		$3\frac{5}{8}$	$19\frac{2}{3}$	$4\frac{3}{4}$	–	20000	350
IL	Dresser Atlas	806	$3\frac{7}{8}$	$17\frac{5}{12}$	$5\frac{3}{4}$	16	18000	350
	Dresser Atlas	1502 (DIF)	$3\frac{5}{8}$	$25\frac{2}{3}$	$5\frac{1}{2}$	16	18000	350
	GO International	IEL	$3\frac{7}{8}$	14.11	6		15000	325
	GO International	IL	$3\frac{7}{8}$	14.11	6		15000	325
	GO International	DIL	4	6.25			20000	350
	Schlumberger	IRT-J	$2\frac{3}{4}$	$27\frac{3}{4}$	$4\frac{1}{8}$	12	20000	350
	Schlumberger	DIL	$3\frac{7}{8}$	$24\frac{1}{2}$	6	12	20000	350
	Schlumberger	IRT-J	$2\frac{3}{4}$		$4\frac{1}{8}$	12	20000	350 [a]
	Schlumberger	IRT-M	$3\frac{3}{8}$		$4\frac{7}{8}$		25000	500
	Schlumberger	IRT-F	$3\frac{7}{8}$		$5\frac{3}{8}$		20000	350
	Schlumberger	IRT-L	$3\frac{7}{8}$		$5\frac{3}{8}$		20000	400
	Schlumberger	IRT-K	$3\frac{1}{2}$		5		20000	350
	Schlumberger	IRT-Q	$3\frac{1}{2}$		5		20000	400
	Schlumberger	IRT-R	$3\frac{1}{2}$	28	5		20000	350
	Schlumberger	DIT-A	$3\frac{7}{8}$		$5\frac{3}{8}$		20000	350
	Schlumberger	DIT-B	$3\frac{3}{8}$		$4\frac{7}{8}$		18000	340
	Schlumberger	DIT-D	$3\frac{5}{8}$		$5\frac{1}{8}$		20000	350
	Welex	42	$3\frac{3}{8}$	23	$4\frac{3}{4}$	–	20000	325
	Welex	DIG	$3\frac{3}{8}$	24	$4\frac{3}{4}$	–	20000	400
	Welex	42A	$2\frac{3}{4}$	23	$4\frac{5}{8}$	–	20000	325
LL	Dresser Atlas	1209	$3\frac{5}{8}$	$17\frac{3}{4}$	6	16	20000	400
	GO International	GL	$3\frac{1}{2}$	12.2	$5\frac{1}{2}$		15000	300
	GO International	DLL	3.5	28			20000	400
	Schlumberger	SLT	$3\frac{7}{8}$	18	6	12	20000	350
	Schlumberger	LGT-D	$3\frac{5}{8}$		$5\frac{1}{8}$		20000	350
	Schlumberger	DLT-A/B	$3\frac{5}{8}$		$5\frac{1}{8}$		20000	350
	Schlumberger	DST-A/B	$5\frac{1}{4}$		$6\frac{3}{4}$	22	20000	350
	Schlumberger	DLT-C	$4\frac{1}{2}$		6		25000	500
	Schlumberger	DLT-D	$3\frac{5}{8}$		$5\frac{1}{8}$		20000	350
	Welex		$3\frac{5}{8}$	10	$4\frac{3}{4}$	–	20000	325

TABLE A1-2 (continued)

Tool type	Company	Equipment type	Tool diameter (inch)	Tool length (ft)	Hole Diameter (inch) Min	Max	Pressure rating (PSI)	Temperature rating (°F)
SFL	Schlumberger	SRS	$5\frac{1}{4}$		$6\frac{5}{8}$	22	20000	350
ML	Dresser Atlas	1106	5	7	$5\frac{3}{4}$	16	20000	350
	Dresser Atlas	1110	$3\frac{3}{4}$	$7\frac{1}{4}$	5	16	20000	350
	GO International	MEL	4	6.25	6		15000	350
	Schlumberger	1 pad	$5\frac{1}{16}$	10	$6\frac{5}{8}$	20	20000	350
	Schlumberger	1X pad	$4\frac{5}{8}$	10	6	16	20000	350
	Schlumberger	MPT-C/D	$5\frac{1}{2}$	7		20	20000	350
	Welex	6000-3	$5\frac{1}{2}$	20	6	16	20000	350
MLL	Dresser Atlas	1211	$5\frac{1}{2}$	$11\frac{7}{12}$	6	16	20000	375
	Dresser Atlas	3101	$5\frac{3}{4}$	16	$6\frac{1}{2}$	16	20000	350
	GO International	MGL	$4\frac{1}{2}$	?	6		15000	300
	GO International	MLL	5	6.33			15000	350
	Schlumberger	PML-B	$5\frac{1}{16}$	$18\frac{1}{2}$	$6\frac{1}{2}$	20	20000	250
	Schlumberger	MPS	$5\frac{1}{16}$	$18\frac{1}{2}$	$6\frac{1}{2}$	20	20000	350
	Schlumberger	MPT-C/D[b]	$4\frac{5}{8}$		6	20	20000	350
	Schlumberger	MLT-A	$5\frac{1}{16}$		$6\frac{5}{8}$	20	20000	350
	Welex		$5\frac{1}{2}$	20	6	16	20000	325
MSFL	Schlumberger	SRT-B	4.25	20	$5\frac{1}{2}$	22	20000	350
Cal	Dresser Atlas		4	4	5	36	20000	350
	Dresser Atlas		$1\frac{11}{16}$	6	2	36	15000	300
	GO International	CL	$1\frac{1}{4}$	$6.3\frac{1}{2}$	2		10000	300
	GO International		$3\frac{1}{2}$	6.11	$4\frac{1}{2}$		15000	300
	McCullough		$3\frac{1}{2}$	$3\frac{1}{2}$	4	36	18200	350
	Schlumberger	SHC	$2\frac{6}{8}$	8	6	16	20000	350
	Schlumberger	TIC	$1\frac{11}{16}$	5	2	12	10000	285
	Schlumberger	BGT-A/B	4		$5\frac{1}{2}$	30	20000	350
	Welex		3	6	4	36	20000	350
GR	Dresser Atlas	310	$1\frac{3}{4}$	$11\frac{2}{3}$	2	16	18000	300
	Dresser Atlas	414	$3\frac{5}{8}$	$14\frac{7}{12}$	$4\frac{1}{2}$	16	20000	300
	Dresser Atlas	420	$3\frac{3}{8}$	$10\text{-}9\frac{5}{12}$	$4\frac{1}{2}$	16	20000	400
	Dresser Atlas	421	$1\frac{11}{16}$	11	2	16	20000	400
	Dresser Atlas	2402	$3\frac{5}{8}$	12	$4\frac{1}{2}$	16	20000	400
	Dresser Atlas	2404	$1\frac{11}{16}$	11	2	16	17000	400
	Dresser Atlas	2406	$1\frac{3}{8}$	$8\frac{1}{2}$	$1\frac{3}{4}$	16	20000	350

	Dresser Atlas	1305	3^{5}_{x}	4^{5}_{6}	4^{1}_{2}	16	20 000	400
	Dresser Atlas	1306	3^{5}_{x}	4^{5}_{6}	4^{1}_{2}	16	20 000	400
	GO International		1	7	1^{1}_{4}	—	15 000	325
	GO International		1^{11}_{16}	6.11	2^{1}_{2}	—	15 000	300
	GO International		3^{1}_{2}	7	4^{1}_{2}	—	15 000	300
	GO International		3^{1}_{2}	3.08	4^{1}_{2}		15 000	350
	McCullough		3^{1}_{3}	11	3.75		27 000	500
	McCullough		3^{1}_{3}	11	3.920		18 200	350
	McCullough		1^{3}_{3}	14	1.995		18 200	350
	McCullough		1^{3}_{4}	5	1.5		22 000	300
	McCullough		1^{3}_{4}	9'7"	1.995		18 000	425
	McCullough		1^{7}	9'7"	2.25		23 000	425
	Schlumberger	SGT-D	3^{6}_{x}	7^{7}_{x}	6	14	20 000	400
	Schlumberger	SGT-EA with SGD-SB/SA	2^{5}_{x}	4^{1}_{x}			25 000	300
	Schlumberger	SGT-EB w/SGD-TAA/TB/TAB	3^{3}_{x}	4^{7}_{x}			25 000	350
	Schlumberger	SGT-EC with SGD-UAA	3^{3}_{x}	4^{7}_{x}			25 000	400
		with SGD-UB/UAB	3^{3}_{x}	4^{7}_{x}			25 000	350
	Schlumberger	SGT-G with SGD-P	1^{11}_{16}	3^{1}_{x}			17 000	300
	Schlumberger	SGT-GB with SGD-R	1^{11}_{16}	3^{1}_{4}			17 000	300
	Schlumberger	SGT with SGH-A	3^{5}_{x}	5^{5}_{x}			20 000	350
		with SGH-E	3^{5}_{x}	5^{5}_{x}			25 000	350
	Schlumberger	SGT-H	2^{3}_{x}	4^{4}_{4}			25 000	500
	Schlumberger	SGT-JA	1^{11}_{16}	3^{1}_{4}			20 000	275
	Schlumberger	SGT-JB	1^{11}_{16}	3^{1}_{4}			20 000	325
	Schlumberger	GPT-A	3^{3}_{x}	4^{7}_{x}			20 000	350
	Schlumberger	GFT-A	3^{5}_{x}	5^{1}_{x}			20 000	350
	Welex		3^{5}_{x}	15	4^{3}_{4}	16	15 000	300
	Welex		2^{1}_{x}	9.76	3		15 000	300
	Welex		1.66	12.75	1.782		15 000	300
	Welex		1^{5}_{x}	20.25	1.75		20 000	400
Spectral GR	Dresser Atlas	Spectralog						
	Schlumberger	NGT-A/B	3^{5}_{x}				20 000	300
FD	Dresser Atlas	2207	4^{3}_{4}	10^{3}_{4}	5^{1}_{2}	16	20 000	400
	GO International	CDL	4	9.5	5		20 000	350
	GO International	DL	3^{1}_{2}	6.2	4^{1}_{2}		17 000	300
	GO International	DL/CL	3^{1}_{2}	11.9	4^{1}_{2}		17 000	300
	McCullough		3^{1}_{2}	11	3.92		18 200	350
	Schlumberger	PGT	4^{3}_{x}	18^{1}_{6}	6	21^{c}	20 000	350
	Schlumberger	PGT-E	4^{5}_{x}		6	16^{c}	20 000	350

TABLE A1-2 (continued)

Tool type	Company	Equipment type	Tool diameter (inch)	Tool length (ft)	Hole Diameter (inch) Min	Hole Diameter (inch) Max	Pressure rating (PSI)	Temperature rating (°F)
	Schlumberger	PGT-FA	$4\frac{5}{8}$		6	16^c	20000	400
	Schlumberger	PGT-FB	$4\frac{5}{8}$		6	16^c	20000	400
	Schlumberger	PGT-G	$3\frac{1}{2}$		5	8	20000	350
	Schlumberger	PGT-HA	$3\frac{1}{2}$		5	8	20000	400
	Schlumberger	PGT-HB	$3\frac{1}{2}$		5	8	20000	400
	Schlumberger	PGT-J	$4\frac{3}{4}$	16.6	6	22	20000	350
	Schlumberger	FGT-A	$1\frac{11}{16}$		$3\frac{1}{4}$	8	16500	325
	Schlumberger	FGT-B	$2\frac{3}{4}$		$4\frac{1}{4}$	12	25000	500
	Welex	Model 121	4	25	$5\frac{1}{4}$	19	20000	350
	Welex	Model 125	4	27	$5\frac{1}{4}$	19	20000	400
LDT	Schlumberger	LDT-A	$3\frac{7}{8}$	24	6	22	20000	350
N-Th N	Dresser Atlas	402	$3\frac{5}{8}$	16	$4\frac{1}{2}$		20000	300
	GO International		1	7	$1\frac{1}{4}$		15000	325
	GO International		$1\frac{11}{16}$	6.11	$2\frac{1}{2}$		15000	300
	GO International		$3\frac{1}{2}$	7	$4\frac{1}{2}$		15000	300
	McCullough		$3\frac{3}{8}$	11	3.75		27000	500
	McCullough		$3\frac{1}{2}$	11	3.920		18000	350
	McCullough		$1\frac{3}{4}$	14	1.995		18000	350
	McCullough		$1\frac{3}{4}$	$9\frac{7}{12}$	1.995		18200	425
	McCullough		$1\frac{7}{8}$	$9\frac{7}{12}$	2.25		23000	425
	Welex		$3\frac{5}{8}$	18	$4\frac{3}{4}$		15000	300
	Welex		$3\frac{5}{8}$	11.3	$4\frac{3}{4}$		20000	400
	Welex		$3\frac{5}{8}$	11.3	$4\frac{3}{4}$		15000	300
	Welex		$1\frac{3}{4}$	12.25	$3\frac{1}{2}$		15000	300
	Welex		$1\frac{5}{8}$	20.25	2		20000	400
N-ep N	Dresser Atlas	SWN	$5\frac{1}{2}$	9	6	16	20000	350
	Dresser Atlas	REH-10	$3\frac{3}{8}$	11	4	16	20000	400
	Dresser Atlas	2402	$3\frac{1}{2}$	$10\frac{1}{2}$	4	16	15000	400
	Dresser Atlas	2402	$3\frac{5}{8}$	$10\frac{1}{2}$	$4\frac{1}{2}$	16	20000	400
	GO International	SNL	4	9.5			20000	350
	Schlumberger	SNP	$4\frac{3}{8}$	18'2''	6		20000	350
	Welex	SWN	4.1	18.2	$6\frac{1}{4}$	19	20000	350
	Welex	102	$3\frac{5}{8}$	12	$4\frac{3}{4}$	16	20000	400

Type	Company	Tool	OD (in)	Length	Weight (in)		Max. pressure (psi)	Max. temp (°F)
N–γ	Dresser Atlas	301	2 5/8, 3 5/8	14	4 1/2, 4		18000, 20000	300, 400
	Schlumberger	GNT-K (STD)	1 11/16	7 1/2	3 1/4		12000	300
		GNT-K (H. PRESS)	2		3 1/2		20000	350[d]
		GNT-N	2 5/8		4 7/8		20000	500
		GNT-Q (F) with GNH-F/SGD-J	3 5/8		5 5/8		20000	400
		with GNH-F/SGD-F	3 5/8		5 5/8		20000	350
		with GNH-F/SGD-L	3 5/8		5 5/8		20000	300
		with GNH-P	3 5/8		5 1/8		25000	400
		GNT-R (G.H.) with GNH-G/SGD-J	3 7/8		5 3/8		20000	350
		with GNH-G/SGD-F	3 7/8		5 3/8		20000	300
		with GNH-G/SGD-L	3 7/8		5 3/8		20000	300
		with GNH-Q	3 7/8				25000	300
	Welex	CNL	3 5/8	12 1/2	4 1/2		20000	300
Compensated neutron	GO International	CNS	4.0	9.5	4 7/8		20000	350
	Schlumberger	CNT-A (Bare)	3 3/8		6 1/2	16	20000	400
	Schlumberger	CNT-A (with Spring ecc.)	5		3 1/4		20000	400
	Schlumberger	CNT-B	1 11/16				16500	350
	Schlumberger	CNT-D	2 3/4				25000	500
NLL-TDT	Dresser Atlas	2701	3 5/8	24 5/12	5		20000	275
	Dresser Atlas	2702	2 3/16	20 7/12	2 7/8		10000	275
	Schlumberger	TDT	1 11/16	19'2"	3 1/4		14000	300
	Schlumberger	TDT	3 5/8	30'6"	4 1/2		20000	275
	Schlumberger	TDT-K	1 11/16		3 3/4		17000	325
	Schlumberger	TDT-M	1 11/16		3 1/4		20000	350
SL	Dresser Atlas	1610	4 1/4	23 1/3	6		20000	300
	Dresser Atlas	623	3 5/8	14 3/4	4 1/2		20000	400
	Dresser Atlas	3001	1 11/16	21 1/2	2	12	15000	300
	Dresser Atlas	1408	3 5/8	9 2/3	4 3/4	12	20000	350
	GO International	SVL	3 1/2	11.6	6		15000	300
	GO International	CSVL	3 1/2	11.6	6		15000	300
	GO International	BCS	3 1/2	15.96			15000	350
	McCullough	Cement Bond	1 3/4, 3 1/2	15, 17	1.995, 3.920		18200, 18200	350, 350
	Schlumberger	SLT-J	1 11/16	16	3 1/4	12	16500	300
	Schlumberger	BHC	3 5/8	27	6	12	20000	350 (21)
	Schlumberger	SLH-A w/SLS D/H/NA/QA	3 5/8		5 1/8		20000	350
	Schlumberger	SLH-C with SLS-K/KB	3 3/8		4 7/8		20000	350

TABLE A1-2 (continued)

Tool type	Company	Equipment type	Tool diameter (inch)	Tool length (ft)	Hole Diameter Min (inch)	Max	Pressure rating (PSI)	Temperature rating (°F)
	Schlumberger	SLT-M	$3\frac{3}{8}$		$4\frac{7}{8}$		25 000	500
	Schlumberger	SLT-N	$3\frac{3}{8}$				20 000	350
	Schlumberger	SLT-L	$3\frac{5}{8}$				20 000	350
	Schlumberger	VCD Caliper	$3\frac{5}{8}$		$5\frac{1}{8}$		20 000	350
	Schlumberger	MCD Caliper	$3\frac{3}{8}$		$4\frac{7}{8}$		20 000	350
LSS	Schlumberger	LSS	$3\frac{5}{8}$				20 000	350
	Welex	24A	$3\frac{5}{8}$	21	$4\frac{3}{4}$	12	20 000	325
	Welex	304	$2\frac{1}{8}$	15	3	12	20 000	350
Temp.	Dresser Atlas	Diff.Temp.	1.44	2.4			15 000	390
	GO International	HRT	$1\frac{11}{16}$				15 000	350
WST	Schlumberger	WST-A or B	$5\frac{5}{8}$		$5\frac{1}{2}$	$13\frac{1}{2}$	20 000	350
					$8\frac{1}{2}$	$19\frac{1}{2}$	20 000	350
Form Tester	Dresser Atlas	SFT	5.5	7.5–24.0			15 000	350
	GO International	FIT-A	5.1		6	$14\frac{3}{4}$	20 000	250
	Schlumberger		5.1				20 000	350
		RFT-A	5.1		6	$14\frac{3}{4}$	20 000	350
Sidewall Sampler	Dresser Atlas	SWC	4.0	6.67			20 000	350
	GO International	CST	5.1	$14\frac{3}{4}$			20 000	430
	Schlumberger	CST-C	$5\frac{1}{4}$	14			20 000	280
		CST-U, V	$4\frac{3}{8}$	$12\frac{1}{4}$			20 000	280
		CST-W	$4\frac{3}{8}$	$12\frac{1}{4}$			20 000	280
DIP	Dresser Atlas	FED	4.5	16			20 000	300
	GO International	HDT-D			$4\frac{1}{2}$	18	20 000	350
	Schlumberger	HDT-E			4	22	20 000	350
		HDT-F			4	21	20 000	400

a Two special tools rated at 400°F.
b With type IX Microlog Pad and type XVII Microlaterolog Pad.
c Special back-up arm permits 21″.
d With hi-temp. P.M. tube.
¹ This table lists the open hole logging tools of the principal logging companies as available at the end of 1980. A more up-to-date list can be obtained from the logging companies.

APPENDIX 2

MATHEMATICS OF MANUAL DIP COMPUTATION (reproduced from "Fundamentals of dipmeter interpretation" by courtesy of Schlumberger)

Figure A2-1 shows a section of the borehole traversed by bedding plane B. Hole axis OA is supposed vertical in this section, leading to the determination of *apparent dip* θ and *apparent azimuth* ϕ. A reference plane DOF is drawn perpendicular to OA. As electrodes 1, 2 and 3 travel upward, they encounter plane B at elevations l_1, l_2 and l_3 above DOF. Axis OD is in the reference plane containing Axis OA and electrode 1. The sonde is assumed not to rotate as it crosses plane B.

A plane drawn perpendicular to B through axis OA cuts B along its apparent line of greatest slope, and DOF through line OM, where M is in the downdip direction. Angle (OMC) is the apparent dip angle θ and angle (DOM) is the apparent azimuth ϕ, counted positive clockwise from D to M.

Let a be the borehole radius ($a = d_h/2$):

$$l_1 = a \tan\theta [1 - \cos\phi]$$
$$l_2 = a \tan\theta \left[1 - \cos\left(\frac{2\pi}{3} - \phi\right)\right] \tag{1}$$

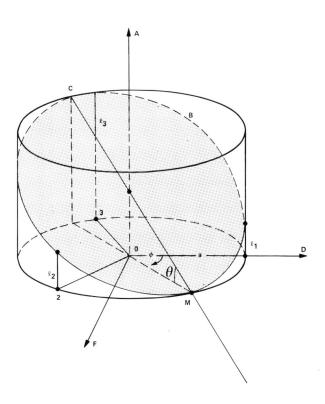

$$l_3 = a \tan\theta \left[1 - \cos\left(\frac{4\pi}{3} - \phi\right)\right]$$

The displacements measured between dip curves are, therefore:

$$h_{1\text{-}2} = l_2 - l_1 = a \tan\theta \left[\cos\phi - \cos\left(\frac{2\pi}{3} - \phi\right)\right]$$
$$h_{1\text{-}3} = l_3 - l_1 = a \tan\theta \left[\cos\phi - \cos\left(\frac{4\pi}{3} - \phi\right)\right]$$

Noting that $\sin(2\pi/3) = -\sin(4\pi/3) = \sqrt{3}/2$ and $\cos(2\pi/3) = \cos(4\pi/3) = -1/2$ this transforms to:

$$h_{1\text{-}2} = \frac{3}{2} a \tan\theta \left[\cos\phi - \frac{1}{\sqrt{3}} \sin\phi\right] \tag{2}$$

$$h_{1\text{-}3} = \frac{3}{2} a \tan\theta \left[\cos\phi + \frac{1}{\sqrt{3}} \sin\phi\right] \tag{3}$$

Knowing a, this system can be solved for θ and ϕ.

The method of solution used is graphical and uses the concept of *combined displacement*.
We can write:

$$K = \tfrac{3}{2} a \tan\theta = \text{combined displacement.}$$

$$I_2 = \cos\phi - \frac{1}{\sqrt{3}} \sin\phi$$

$$I_3 = \cos\phi + \frac{1}{\sqrt{3}} \sin\phi$$

and

$$h_{1\text{-}2} = K \times I_2 \tag{2a}$$
$$h_{1\text{-}3} = K \times I_3 \tag{3a}$$

Equations 2a and 3a are the equations of a set of ellipses as shown in Fig. A3-1a. Each ellipse corresponds to a value of the combined displacement.

Carry displacements $h_{1\text{-}2}$ and $h_{1\text{-}3}$ on their respective axes. If electrode 2 is "up" with respect to electrode 1, $h_{1\text{-}2}$ is positive, and similarly for electrode 3 and $h_{1\text{-}3}$ *. The representative point ($h_{1\text{-}2}$, $h_{1\text{-}3}$) falls between two ellipses. Combined displacement K and apparent azimuth ϕ are read.

Note that whenever one displacement is equal to zero, (or both displacements are equal), the other displacement is (or both displacements are) equal to the combined displacement.

The combined displacement is now carried into

---

* In another approach, the sign convention is the opposite, "up" is negative, "down" is positive. The reasoning remains the same.

Fig. A3-1b which solves the equation:

$$K = \tfrac{3}{2} a \tan \theta \qquad \text{for } \theta, \text{ the apparent dip}$$

Example: data of Appendix 3, Method 1.
Displacement between curves:
Curves I and 2 2 up = 3.0 mm
Curves I and 3 3 up = 1.5 mm
Figure A3-1a gives $K = 2.65$ mm on 1/20 scale So on
scale 1: $K = 2.65 \times 20 = 53$ mm and $a = 4.125 \times 25.4$

= 104.775 mm. So we have:

$$\tan \theta = \frac{2K}{3a} = \frac{106}{314.325} = 0.33723$$

which gives:

$$\theta = 18°38'$$

as apparent dip.
Compare with graphical methods explained in Appendix 3.

APPENDIX 3.

AN EXAMPLE OF A MANUAL DIPMETER CALCULATION

Method 1 (Table A3-1)

Data

Azimuth of electrode 1: AZM1 = 120°
Relative bearing   : β = 60°
Well deviation   : δ = 6°30′
Hole diameter   : d_h = $8\frac{1}{4}$ inches
Magnetic declination : 10°W.
Displacement between curves:

1 and 2, #2 up 3.0 mm
1 and 3, #3 up 1.5 mm

Calculation

(1) Calculation of the magnetic azimuth of the deviation

AZM d = AZM1 − β = 120° − 60° = 60°

(2) Determination of the apparent dip angle relative to electrode (pad) 1:

(a) The chart at the top of Fig. A3-1 gives for the azimuth AZMa = 330° and for the combined displacement 2.65 mm (point A).

TABLE A3-1

First method (Schlumberger, 1960)

Data from the log (CDM)	magnetic azimuth of electrode 1 relative bearing deviation of the well bore hole diameter	: AZM 1 : β : δ : d_h	Magnetic azimuth of deviation AzMd = AzM1 − β
Computation of displacements between curves	Displacement between curves I and II = x Displacement between curves I and III = y		
Use of a chart, Fig. A3-1a	Introduce x and y in the chart → (up if 2 above 1)	Magnetic azimuth of apparent dip: AzMa Combined displacement	
Use of a chart, Fig. A3-1b	Introduce: combined displacement → hole diameter in the chart	Apparent dip: θ	
Computation of the azimuth of the apparent dip relative to the hole deviation	AzMa + β	F	
Use of a chart, Fig. A3-2	Introduce: F θ → δ	true dip: α Azimuth of true dip relative to the deviation: G	
Computation of the magnetic azimuth of the dip	G + AzMd →	AzM	
Correction for declination	AzM + magnetic declination →	Geographic azimuth of the true dip: AzG	

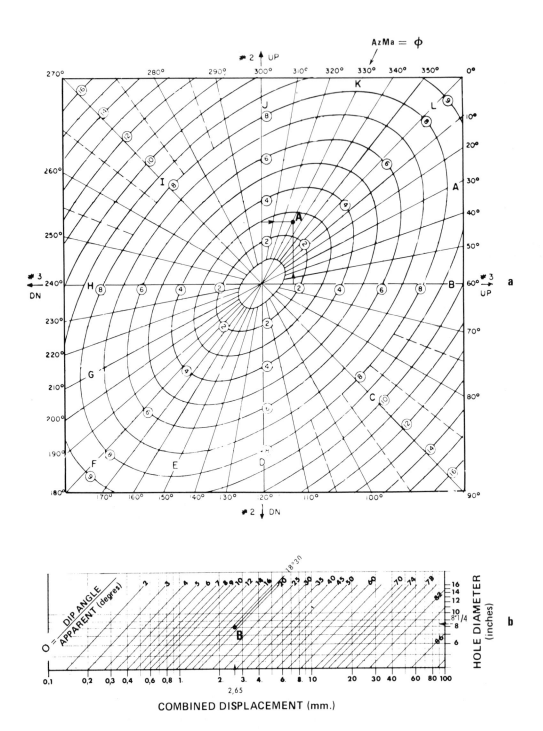

Fig. A3-1

(b) The chart at the bottom of Fig. A3-1 gives for a combined displacement of 2.15 mm and a hole diameter of $8\frac{1}{4}$ inches, an apparent dip angle θ equal to 18°30′ (point B).

(3) Calculation of the azimuth of the apparent dip, F relative to the hole deviation:

$$F = \text{AzMa} + \text{relative bearing} = 330° + 60° - 360° = 30°$$

(4) Calculation of the actual dip α and the azimuth of the actual dip G relative to the hole deviation:

Using the chart in Fig. A3-2 gives $\alpha = 13°$. F is entered (the perimeter's graduations are from 0° to 360°) and θ (on the radii graduated from 0° to 90°) and from the point obtained (C) we go down the length of a small circle by a value equal to the deviation. The real dip angle $\alpha = 13°$ (Point D) and its azimuth relative to the hole deviation, $G = 45°$

(5) Calculation of the magnetic azimuth of the dip:

G + magnetic azimuth of the deviation

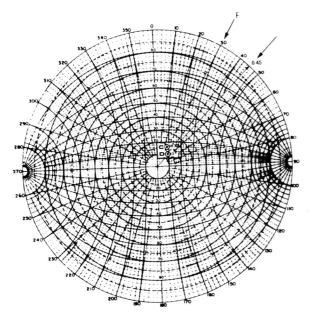

Fig. A3-2

$AZM = 45° + 60° = 105°$

(6) Calculation of the geographic azimuth

$AZM = 105° - 10° = 95°$

Method 2 (Table A3-2)

Data

Azimuth of electrode 1: AZM 1 = 120°

Relative bearing	$:\beta$	$= 60°$
Well deviation	$:\delta$	$= 6°30'$
Hole diameter	$:d_h$	$= 8\frac{1}{4}$ inches

Magnetic declination : 10°W.

Displacement between curves:

1 and 2 = 3.0 mm (2 above 1)

1 and 3 = 1.5 mm (3 above 1)

2 and 3 = 1.5 mm (2 above 3)

Calculation

(1) Place the displacements to the scale of the well (multiply by the inverse of the log scale, 20 if 1/20 and 40 if 1/40).

(2) Use the nomogram in Fig. A3-3 to calculate the three apparent dip angles $\theta_1 = 18°30'$, $\theta_2 = 9°15'$, and $\theta_3 = 9°15'$.

(3) Place the guide line of the disc on N of the net (see Fig. A3-4a).

(4) Mark on the disc in a clockwise direction:

the azimuth of electrode 1: Point I (120°)

the azimuth of electrode 2: Point II (120° + 120° = 240°)

the azimuth of electrode 3: Point III (240° + 120° = 360°)

the deviation of the azimuth: Point BH (60°)

(5) Turning the disc, place I and II on the 60° line of the net (hemisphere N): Fig. A3-4b.

(6) Find the side of the net where the lowest

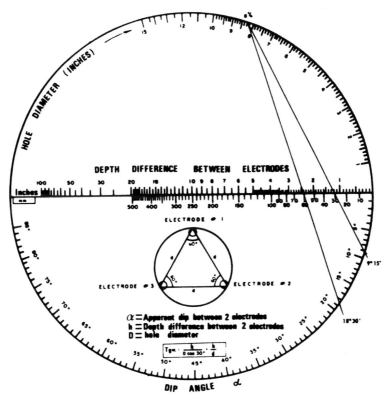

Fig. A3-3

348

TABLE A3-2

Second method using a stereogram

Computation of displacements between curves	(1) displacement between:	curves I and II	x
		curves I and III	y
		curves II and III	z
	(2) Conversion of the displacements x, y and z to the scale of the well (multiply by the inverse of the log (scale)		a
			b
			c
Use of a nomogram, Fig. A3-3	Introduce: a, b and c hole diameter in the nomogram		
Use of a stereogram	see Fig. A3-4		

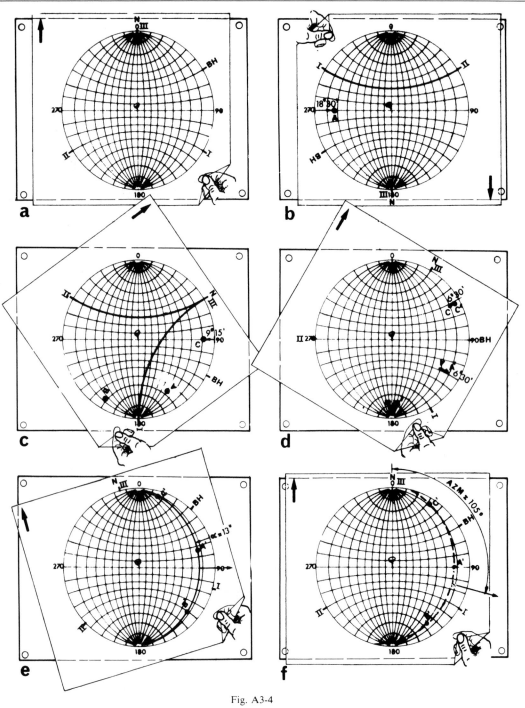

Fig. A3-4

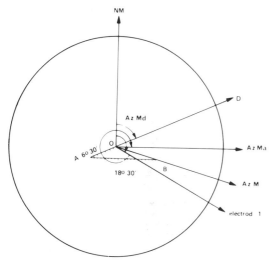

Fig. A3-5

electrode is located and measuring across the E-W axis of the net mark the Point A corresponding to a distance θ: Fig. A3-4b.

(7) Repeat steps 5 and 6 for the set I-III and θ_2 giving Point B and the set II-III and θ_3 giving Point C (Fig. A3-4c).

(8) Turning the disc bring BH to the E-W axis (Fig. A3-4d).

(9) On the disc move the points A, B and C along small circles of the net in the direction of BH by quantities equal to the well deviation so giving points A', B', and C' (Fig. A3-4d).

(10) Turn the disc to bring A', B' and C' onto a great circle of the net (Fig. A3-4e).

(11) Draw an arrow on the disc along the E-W axis from the common great circle and towards the edge of the net.

(12) Read off the dip angle α by counting the number of degrees from the edge of the net to the common great circle:

$\alpha = 13°$

(13) Turn the disc to bring the guide line to N and read off the position of the azimuth arrow giving the magnetic azimuth of the actual dip (Fig. A3-4f):

AZM = 105°

(14) Bring the guideline onto the declination NM/NG by turning the disc in the direction of the declination (towards the west in this case) by a value equal to the angle of declination and read off the position of the arrow the geographical azimuth of the actual dip:

AZG = 95°

Method 3 (Table A3-3)

The same data as in Method 1 are used.

Calculation

(1) and (2) the same as in Method 1.

Graphical construction (Fig. A3-5)

(1) We plot the magnetic azimuth of the deviation:

AZMd = 60° $(\overline{OD})$

(2) Plot the magnetic azimuth of the apparent dip

AZMa = 330° + 120° = 90°

(3) Plot the value of the deviation δ ($\delta = 6°30'$) in the opposite direction to the deviation azimuth and using an arbitary scale (Point A).

(4) From this point A trace a vector parallel to the magnetic azimuth of the apparent dip AZMa on which we plot using the same scale the value of the apparent dip: point B (18°30').

(5) Trace the vector $\overline{OB}$ which gives the magnetic azimuth of the actual dip (105° on the circle).

(6) Converting the length OB, using the same scale as above, into degrees gives the actual dip angle (13°).

(7) We calculate the geographical azimuth AZG of the dip by subtracting the magnetic declination

AZG = 105° − 10° = 95°

TABLE A3-3

Third method (Ref. SPE 7067, Nov. 1955; Ref. SPE 7071, Nov. 1955)

Define	AzMa and θ as in first method
Use a simplified graphical construction. Fig. A3-5 to determine AzM and α	
Correction for declination	AzM + magnetic declination → AZG

APPENDIX 4

QUICK-LOOK METHOD TO DETERMINE THE
DIP AZIMUTH

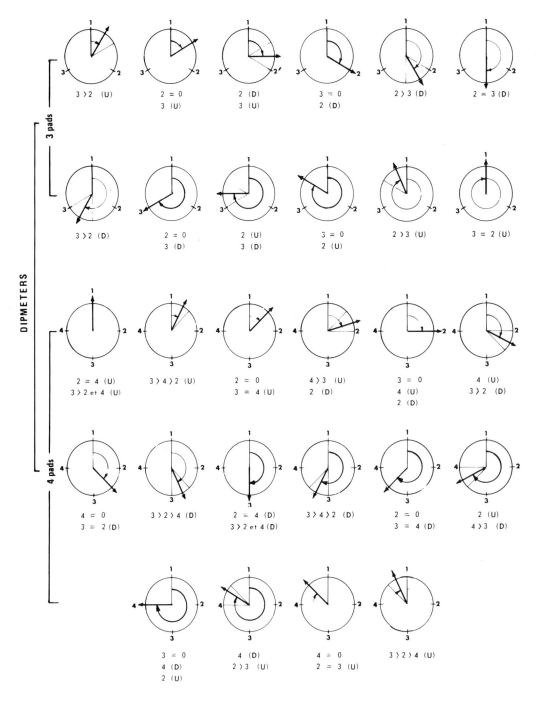

0 means no shift relative to curve 1.

U means "up" relative to curve 1.

D means "down" relative to curve 1.

To determine the dip azimuth add azimuth of electrode 1, the angle between electrode 1 and arrow, and the magnetic declination.

APPENDIX 5

QUALITY CONTROL OF LOG MEASUREMENTS

As logs are used as the basis for a certain number of applications and are often used as a reference document for a well by the addition of other data it is important that their quality should be as high as possible, both in their technical values and in their presentation. There is therefore a need for a systematic method of *log quality control*. It is obvious that any log interpretation and action decided upon using that interpretation will be false if the logs are bad.

An aid to the maintenance of good quality logs is to perform a series of calibrations before and after the actual logging operation. These calibrations (and their recording) help to check that the tools are functioning properly and that the various circuits and sensors are calibrated to read properly under test conditions. We should note that it is quite possible to have good calibrations but the log itself may be bad. This results from the position in the chain of sensors-circuit-recording where the calibration check occurs.

In general there are two kinds of calibration:

(a) "Shop" or "Master" calibrations usually made at the Service company workshops and involving the use of primary calibration standard or representing a primary A.P.I. standard.

(b) Wellsite calibrations. These generally include: (1) surface calibrations made before entering the well; (2) downhole calibrations; and (3) after-survey calibrations.

In each case the following points should be checked: (1) the mechanical position of the galvanometers corresponding to no signal being fed to the camera; (2) the electrical zero usually corresponding to no signal being fed to the measurement circuits; and (3) the calibration value resulting from a known signal source or a known calibration standard.

These calibration measurements are recorded on film at the beginning and at the end of the logging operation. In the case of computers controlled logging units (e.g. the Schlumberger CSU) the presentation of calibration data is entirely different and separate quality controls need to be established.

As a further check to log quality a repeat section is recorded over a section of the main log. This is designed to check the repeatability of and to detect erroneous shifts in the measurement. Repeat sections should:

(a) include a section covering an interval of 100 feet or 30 meters minimum;

(b) include a depth mark;

(c) be recorded to include both a section with memorisation and a section without (this checks that the memorizer panel does not introduce an electrical shift);

(d) be recorded with the same parameters as the main log including scale, logging speed, time constant etc;

(e) be identical with the main log except for statistical variation occurring in radioactivity logging.

Repeat sections are generally recorded at the bottom of the hole but another zone can be chosen either because it is a zone of interest or because it includes good contrasts in log values.

Service and oil companies usually give their engineers general guidelines for quality control for each tool. It is a good idea to have a look at these. However, in order to help the geologist, it has seemed useful to gather an example of each repeat section and calibration for each of the Schlumberger tools. (These are non-CSU examples).

Finally we note some general recommendations.

GENERAL RECOMMENDATIONS COVERING ALL OPERATIONS

The "Company" man's wellsite responsibility

For anyone given the job of watching logging operations the following responsibilities apply: company representation, control, coordination and information.

In *representing* the company it is not only necessary to know the reasons why logs are being run, the problems that can arise and how to solve them, but also to be aware of and to use with some discretion his powers of decision throughout the entire operation (day and night).

General or specific guidelines for each log are not intended to reduce the material responsibilities for control and coordination of operations. Instead they are supposed to provide a framework to which are added those human qualities that should be displayed by an observer in these circumstances: adaptability, common sense, etc.

A wellsite observer is considered to have fulfilled

352

his obligations when his company both locally and centrally has information on the operation that is both complete and objective. Just filling in the boxes on a form does not immediately imply that the information is objective; some other added remarks may make the situation clearer.

Problems of security and safety of personnel depend on the regulations in force at the wellsite. For equipment this is the responsibility of the oil company or operator in charge.

One "established" rule puts this in perspective—all contractor equipment lost or damaged at a wellsite unless it is clearly the fault of the contractor's personnel may be charged to the oil company or operator.

When a tool or cable is stuck in the well the contractor furnishes a "fishing" kit with suitable adaptors and recommends a method to follow. A fishing kit should be kept in each logging unit. The drilling department actually takes charge of the fishing operations.

Starting a logging operation

(a) *Call-out of the logging team*

The logging contractor should arrive sufficiently well ahead of time to avoid any downtime of the rig.

(1) Defining the operation: this implies stipulating the logs to be run and the additional equipment needed (blowout-preventors, riser, etc.) with reference to the established well program. At this stage it should be decided which are strictly essential logs and which are optional.

(2) Keep in touch with the availability of logging equipment (via the contractor).

(3) Take into account the limitations of each tool by consulting the technical specifications given by the contractor. In· particular, the contractor should be notified if the temperature exceeds 250°F (120°C), the pressure is above 10,000 PSI (700 kg/cm) or if the well is highly deviated.

(4) Take into account the economic implications of each request: decide on the particular combination of logging tools to save time; and avoid duplication of curves unless they are necessary to provide quality control or indicate that the tool is functioning properly (for example the caliper where a pad tool is used).

(b) *Coordination with drilling or production departments*

(1) Ensure that the drillers and the logging company are co-ordinated.
(a) Make sure that the well is drilled deep enough so that the top-most sensor in the logging string can reach the bottom-most zone of interest. For this the logging company should provide the measured distance from the sensors to the bottom of the tool string.

(b) Keep in touch with any changes in mud properties that will affect the logging operation. In the case of electrical logs of all kinds (ES, IES, LL, ML, MLL, SFL, HDT) the mud type and salinity will be critical. In addition changes in mud properties that occur after drilling a section but before logging might mean that the fluid in the invaded zone bears no relationship to that measured by the logging engineer just before logging. Changes in the SP and other logs will result.

(c) Make decisions with the drillers and logging engineers on the condition of the hole and the need for wiper trips and so on. It is particularly important not to run some types of tools in adverse hole conditions.

(2) In all operations where a well is to be completed someone from the Production Service should be present to ensure the safety of the operation, the opening and closing of flowline valves, pressure measurements and so on. The logging "observer" can play the role of an "intermediary" if necessary.

Operations preceeding logging

(1) Documents to prepare:
–a control record of logs and calibrations;
–a price list (for invoicing);
–logs at 1/200 scale already recorded in the same well;
–logs from nearby wells;
–information required to complete the log headings:
. name of the field or country;
. well name;
. well-co-ordinates;
. depths and their source (driller, logging etc.);
. mud properties (density, pH, viscosity, water loss);
. bit size;
. casing sizes and depths, and casing weight;
. tubing (if any) sizes, weights, depths;
. deviation expected.
–relevant data on the well, drilling history, changes in mud, lost circulation zones, zones that flowed, hydrocarbon shows, lithological description from cuttings or coring, test results, etc.

(2) Request a sample, of mud, mud filtrate and mudcake, taken using mud from the flowline just before circulation ceases before logging. These samples should be given to the logging engineer to measure R_m, R_{mf} and R_{mc}.

(3) In the case of a floating rig ensure that a wave-motion-compensating device has been correctly installed.

(4) Give the orders of the logs to be run not forgetting that:

    (a) Some logs can only be run once results from other logs have been obtained. In this case it is best to run these logs first and then run other logs which give time to interpret the first set.

    (b) A tool containing a radioactive source should never be run first in open hole because of the risk of getting stuck.

    (c) When hole conditions are poor it is best to run the more difficult tools first (i.e. long tools, tools with centralizers) in order to avoid making a wiper trip in the middle of the logging operation.

    (d) When the bottom hole temperature is high run the most heat sensitive tools first.

    (e) In order to correlate one log set with a previous set in the same well it is often a good idea to run a gamma ray tool in combination with the first tools run in the hole so that a correlation gamma-ray log can be run through casing.

    (f) To save time run in combination as much as possible those tools that need to be run slowly.

(5) Choose depth scales for logging, generally 1/200 or 1/500 with 1/40 for microtools or HDT.

General checks to be made before and during running in the hole

(a) *Before running in*

(1) Some tools are calibrated on surface before running in. Each of these tools has specific calibration guidelines given by the logging companies and these guidelines should be verified. Check the quality of the calibration.

(2) Make sure that a maximum-reading thermometer is attached to each tool.

(b) *While running in*

(1) Choice of scale

Where scales are not fixed, check while running in the maximum and minimum log values seen and choose a scale accordingly. In some cases a back-up galvanometer is required on a particular log.

Where a first log is run on a logarithmic scale the probable position of minimum and maximum values on other logs can be produced. Note that with many tools it is not possible to see or read the progressing values while running in the hole, for example, a pad tool that is closed up.

A correctly chosen scale should show detail in readings of low amplitude. Generally a log should be kept within the film track or tracks chosen. Try to avoid frequent use of back-ups as this makes readings difficult.

Record all resistivity logs so as to have the same scale in the low resistivity track. Generally resistivity logs are all recorded on the same scale and more and more a logarithmic scale is chosen for them all.

(2) Depth correlation

The first log run serves as the depth reference for all other logs. When possible correlate the first log with other logs run in preceeding log operations. If not, depth correlates with the casing shoe. In any case check the casing shoe and the total drilling depths.

If there is a difference in the log depth and the drilling depths greater than one in a thousand, check if the error is connected with the logging operation or arises from some other local source (for example missing out one stand or joint of drill pipe in the drill-pipe tally). Check that logging depths are measured with the cable under tension (logging up not down).

(3) Other checks while running in. Keep a close eye on the cable tension meter to detect when the tools hold up while running in. Check that the depth measuring wheels running on the cable turn normally and evenly with cable movement. Watch the galvanometers to see that there is movement indicating that the tool is moving.

(4) When the well is very hot record logs while running in where possible. As some tools are particularly temperature sensitive this is often the only way to record a log.

General checks at bottom and while logging

(a) *At bottom and before recording*

(1) Downhole calibrations: this should correspond exactly with the surface or shop (master) calibration. Check that the calibration is recorded at least on a scale of 1/200.

(2) Record zeros before recording the logs.

(3) Record a repeat section (see above comments).

(b) *During logging*

Record zeros before logging.

Note the pick-up point of each tool from the bottom (and in the case of nuclear tools record the tension on film).

Check the logging speed.

Watch the tension meter and note "sticky" points during logging.

Check the cable magnetic marks.

Check that the magnetic mark is consistent and that no changes are greater than 1m in 100m.

Check correlation markers, ie. beds that are easily identified as depth marks, and that these are appearing at the correct depth.

Check memorization of one log to the reference.

Check that all curves are present (back-up curves in particular).

Watch that the film turns in the film cassettes and in the film take-up cassettes. It can happen that the film does not turn or becomes jammed inside the camera.

Note points at which scale changes or mechanical shifts are made.

Watch out for strange effects such as galvanometer vibration, imposed periodic parasitic effects, etc.

Where possible record at least one hundred feet of overlap with previous logs.

Record all logs at least twenty meters into the casing.

Record all calibrations after logging.

Develop all films immediately after the calibrations are recorded.

In this way if something is wrong the operation can be restarted without further loss of time so long as the problem arises from an operator error (film not turning, recording started at the wrong depth or stopped too early, etc.).

General quality checks on recorded logs

Logging Speed Checks: a blank marker appears every one minute in the black edge of one track.

The main log and repeat section should be identical except for statistical variation arising on nuclear logs.

The calibrations and zeros recorded before and after logging should be within the tolerances prescribed for each tool.

Check that the memorisation is correct.

Compare overlap sections with previous logging runs. These should match in log readings and should be perfectly matched in depth.

Overlay the different logs to check depth correlation. In overlap sections the correlation should be good for porosity tools (FDC, SL, etc.). For resistivity tools changes in invasion diameters may influence readings. However, in front of completely impermeable zones the logs should read the same. The same problems arise for neutron logs in front of gas-bearing formations.

Correlate the new logs with those recorded in nearby wells checking local markers (salt, anhydrite, coal, etc.) and also readings in shale beds and water-bearing formations.

Ensure that there are no parasitic effects by way of superimposed periodic signals, kicks on the logs or galvanometer vibrations.

Check that all curves are present.

Check that all depth and grid lines are present.

Check that the film is correctly developed and that the traces are clearly seen, particularly in the case of micro-resistivity logs.

If there is any doubt about the quality of a log the logging engineer should try to find its cause and following consultation with the operation supervisor, rerun the log if necessary.

If the rerun is still unsatisfactory or the original fault is traced to a tool fault immediate action should be taken to bring in a replacement tool or to repair the tool in such a way that the loss of rig time is minimized. Generally one tool can be repaired or replaced while other services are being run.

When the tool is brought out of the hole ensure that the pads, back-up arms, centralizers, etc. are all in place and that nothing has been left in the hole.

Checks on log headings and prints

Check that log headings are correctly filled in with hole information, equipment types, stand-off or centralizer information, mud characteristics, mention of other logs run in the same series of generations, the maximum temperature, R_m at that temperature and general comments.

Check that all scale changes and mechanical shifts are noted on the film.

Make sure that a note is made on the heading or film of all data relating to tool sticking, problems with the recording camera, loss of one or more curves on the film for whatever reason, etc.

In the case of some logging operations (e.g. production logging) make sure that the exact sequence of operations is noted on the heading.

Check that the headings of composite logs made up of several logging runs contains all the information relating to each individual run.

Check print quality.

Reports at the end of logging

This report is made up of two parts:

(a) A sheet detailing operations. This document should be completed by the engineer and the operation supervisor. It should: (b) allow the proper invoicing of the operation; and (c) give a recap of the logs noting any anomalies in recording or quality.

(b) A complete interpretation report of the operation in four parts: (1) a detailed qualitative study; (2) the lithological report of the logged section; (3) correlations with neighbouring wells; and (4) a quantitative evaluation of the reservoirs.

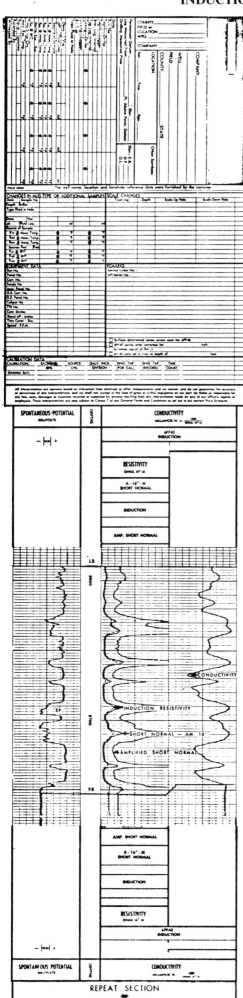

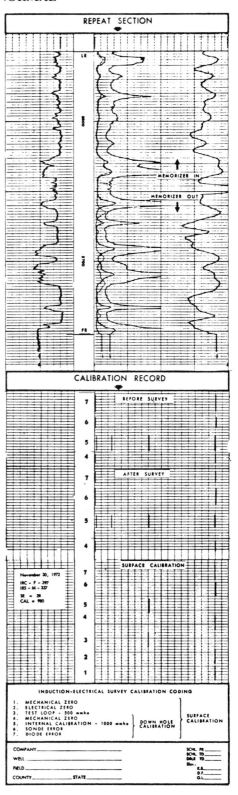

APPENDIX 5A

Examples of logs with repeat section and calibrations for the most common Schlumberger logs

DUAL INDUCTION-LL8

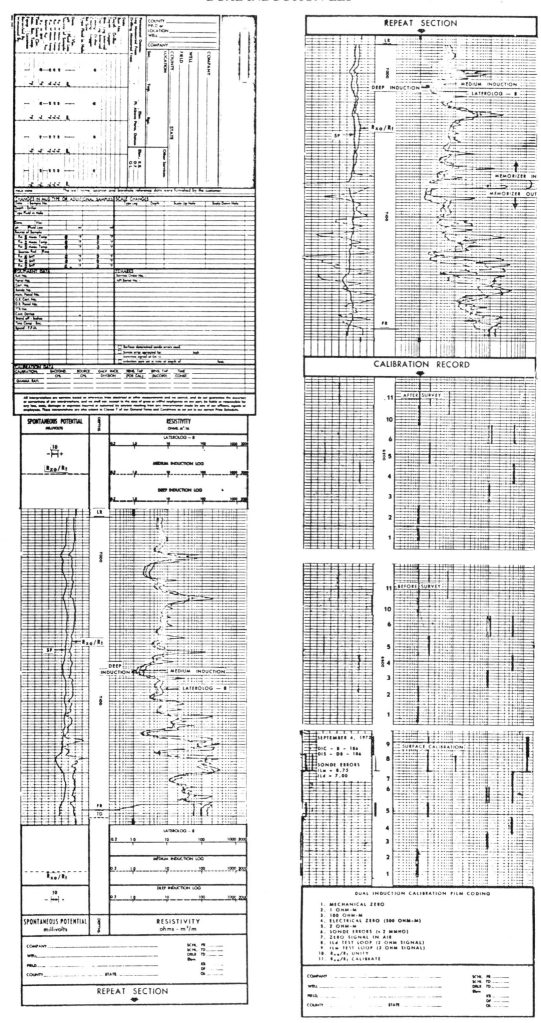

INDUCTION — SPHERICALLY FOCUSED LOG SONIC

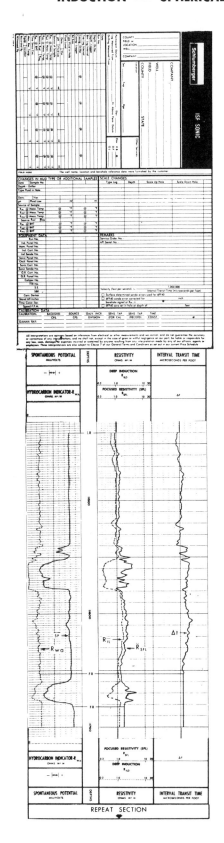

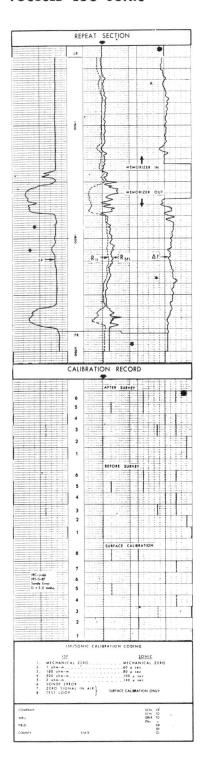

DUAL LATEROLOG

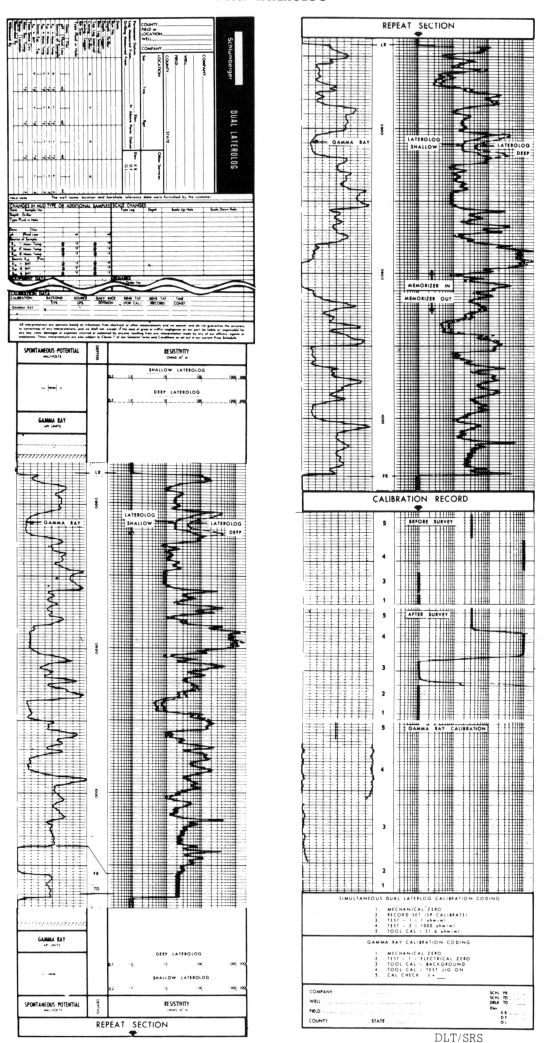

DLT/SRS

DUAL LATEROLOG — MICROSPHERICAL LOG

Schlumberger

SIMULTANEOUS
DUAL LATEROLOG
MICRO SPHERICAL LOG

	COUNTRY			
FIELD or LOCATION				
WELL				
COMPANY				

COMPANY
FIELD
WELL
COUNTRY
LOCATION

Set

Top

Elev.
Ft. Above Perm. Datum

Btm

Elev. K.B
D.F
G.L

Other Services

The well name, location and borehole reference data were furnished by the customer.

EQUIPMENT DATA

| Run N° | Panel | Laterolog | | | S. Resistivity Panel | Tape Recorder | Memory Panel | G.R Cartridge |
| | | Cartridge | Sonde | Electrode | | | | |

LOGGING DATA

| Run N° | Central | Stand off | SP Electr | Pad Type | Averaging | T.C. | Ø Mud Cake |

REMARKS   Beam width 24

| CALIPER | Depths | RESISTIVITY |
| hole diameter in inches | | ohms - m²/m |

LATEROLOG DEEP Rld
0.2   1.0   10   100   1000 2000

SPONTANEOUS POTENTIAL
millivolts

LATEROLOG SHALLOW Rls
0.2   1.0   10   100   1000 2000

MICROSPHERICAL LOG
0.2   1.0   10   100   1000 2000

GAMMA RAY
api units

GAMMA RAY
api units

MICROSPHERICAL LOG

SPONTANEOUS POTENTIAL
millivolts

LATEROLOG SHALLOW Rls

LATEROLOG DEEP Rld

CALIPER
hole diameter in inches

RESISTIVITY
ohms m²/m

REPEAT SECTION

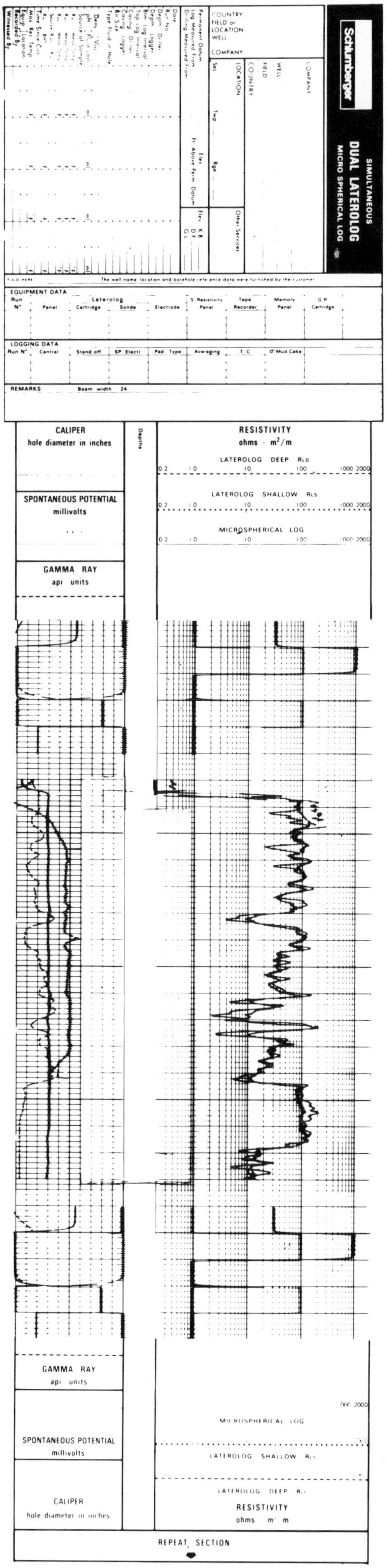

DLT/SRS

MICROLOG-MICROLATEROLOG

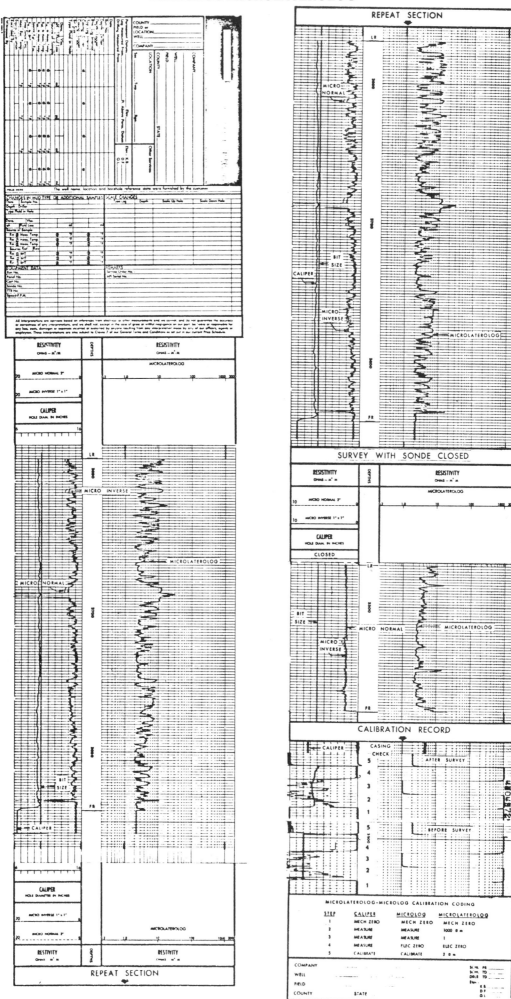

MICROLOG-PROXIMITY LOG

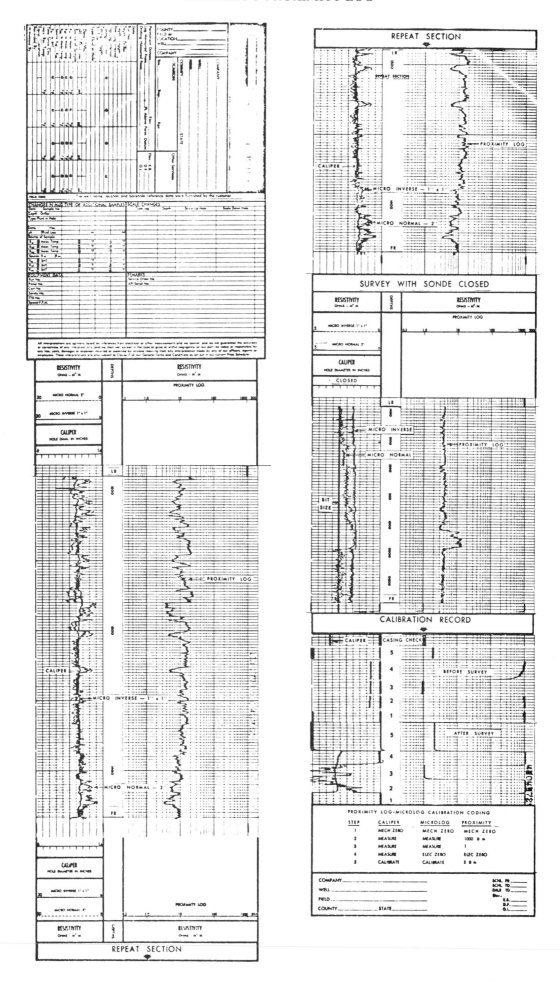

GAMMA RAY-NEUTRON GNT

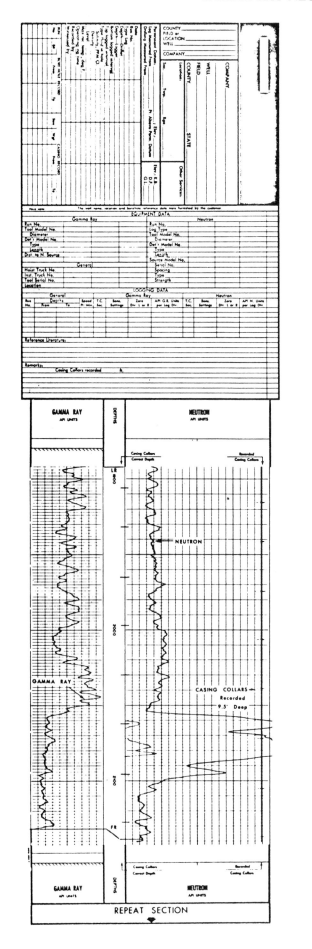

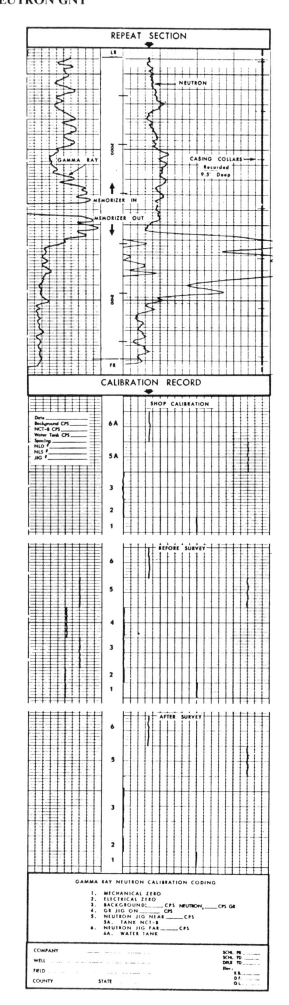

GAMMA RAY-SIDEWALL NEUTRON (SNP)

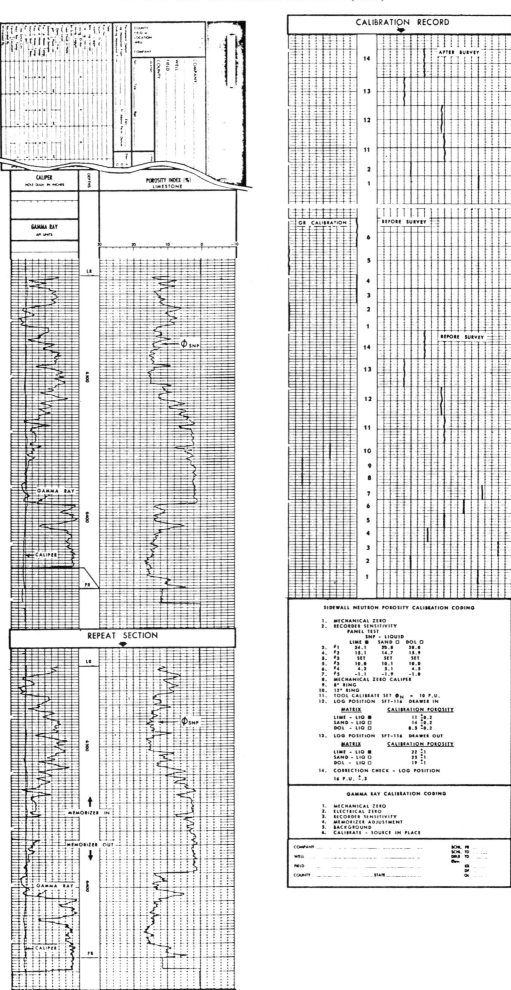

COMPENSATED NEUTRON LOG

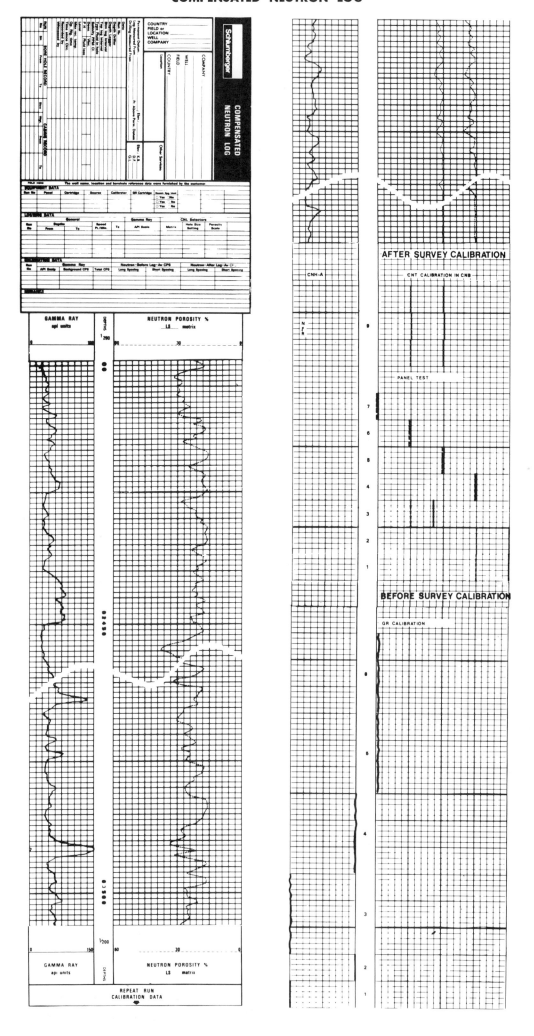

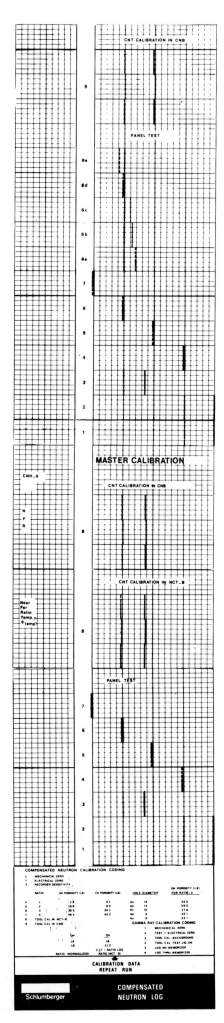

CNT CALIBRATION IN CNB

PANEL TEST

9
8e
8d
6c
6b
8a
7
6
5
4
3
2
1

MASTER CALIBRATION

CNT CALIBRATION IN CNB

CNH_A

N
F
R

9

CNT CALIBRATION IN NCT_B

Near
Far
Ratio
Temp
Φ temp

8

PANEL TEST

7
6
5
4
3
2
1

COMPENSATED NEUTRON CALIBRATION CODING

			DH POROSITY (LS) FOR RATIO 3
1	MECHANICAL ZERO		
2	ELECTRICAL ZERO		
3	RECORDER SENSITIVITY		

	RATIO	DH POROSITY (LS)	CH POROSITY (LS)	HOLE DIAMETER	DH POROSITY (LS) FOR RATIO 3
4	1	1.5	0.1	8a 16	26.0
5	2	15.6	9.0	8b 12	26.6
6	3	30.5	24.1	8c 10	27.8
7	4	46.4	43.2	8d 8	30.1
8	TOOL CAL IN NCT_B			8e 6	33.7
9	TOOL CAL IN CNB				

GAMMA RAY CALIBRATION CODING

1	MECHANICAL ZERO
2	TEST + ELECTRICAL ZERO
3	TOOL CAL BACKGROUND
4	TOOL CAL TEST JIG ON
5	LOG NO MEMORIZER
6	LOG THRU MEMORIZER

	LS	CH	LS
	1.9		11.2

RATIO (NORMALIZED)   2.17 - RATIO LOG   RATIO (NCT_B)

CALIBRATION DATA
REPEAT RUN

Schlumberger

COMPENSATED
NEUTRON LOG

CNT-A

COMPENSATED FORMATION DENSITY LOG

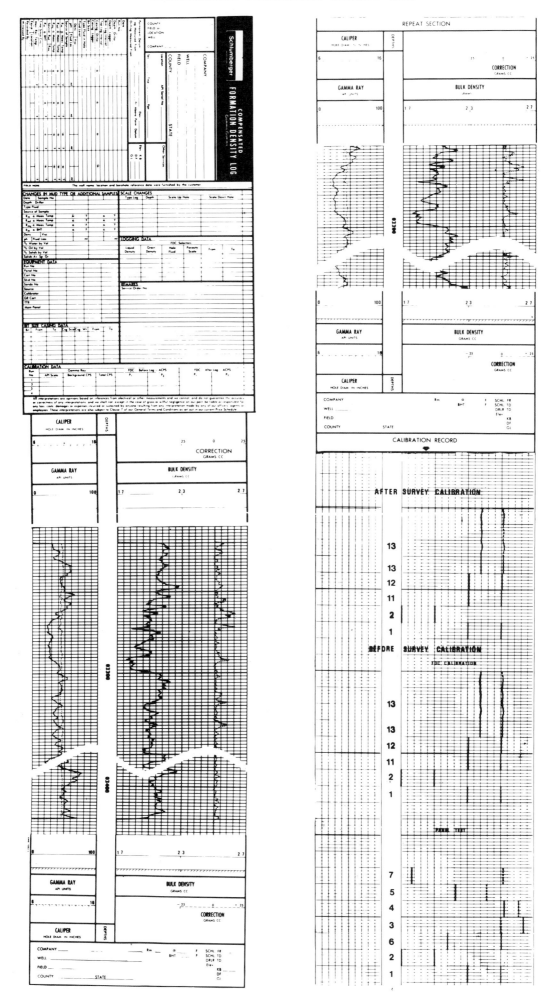

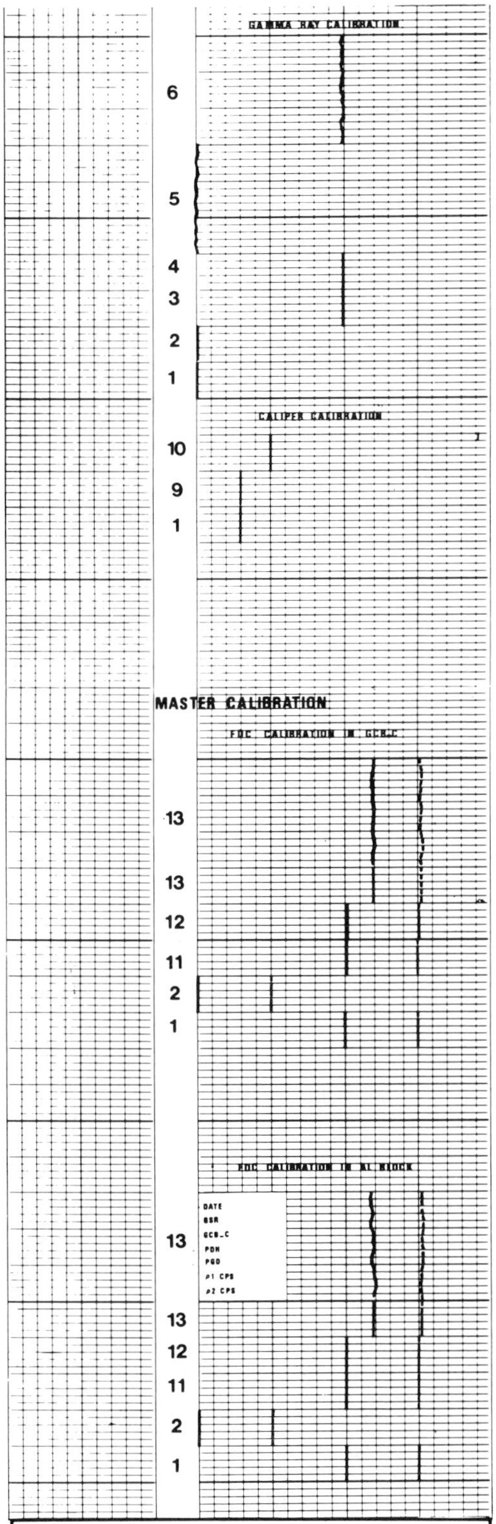

GAMMA RAY CALIBRATION

6

5

4
3
2
1

CALIPER CALIBRATION

10
9
1

MASTER CALIBRATION

FDC CALIBRATION IN GCRR

13

13

12

11

2

1

FDC CALIBRATION IN AL BLOCK

DATE
GSR
GCB_C
PDN
PGD
#1 CPS
#2 CPS

13

13
12
11
2
1

FORMATION DENSITY COMPENSATED CALIBRATION CODING

		FDC	LIQUID	
1	MECHANICAL ZERO			
2	RECORDER SENSITIVITY			
	PANEL TEST			
	POS			
3	# 1	2.92	.00	
4	# 2	2.78	-.14	
5	# 3	2.42	.10	
6	# 4	2.35	.00	
7	# 5	2.08	.01	
8	MECHANICAL ZERO CALIPER			
9	8 RING			
10	12 RING			
11	TOOL CALIBRATE # 1 SET	2.50		
12	TOOL CALIBRATE # 2 SET	.00		
13	LOG POSITION	2.59	.015	

GAMMA RAY CALIBRATION CODING

1 MECHANICAL ZERO
2 ELECTRICAL ZERO
3 RECORDER SENSITIVITY
4 MEMORIZER ADJUSTMENT
5 BACKGROUND
6 CALIBRATE SOURCE IN PLACE

CALIBRATION RECORD

COMPANY _____   Rm ___ ℩ ___ F   SCHL FR ____
                           BHT ____   F   SCHL TD ____
WELL _____   DRLR TD ____
                                           Elev.
FIELD   KB ____
COUNTY _____   STATE   DF ____
                                              GL ____

PGT

GAMMA RAY-BOREHOLE COMPENSATED SONIC

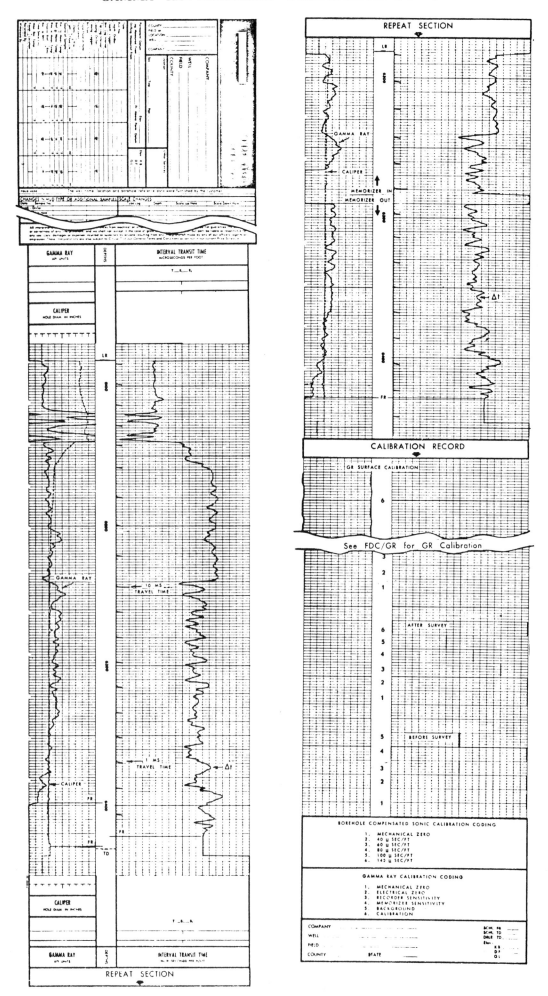

HDT-D/E

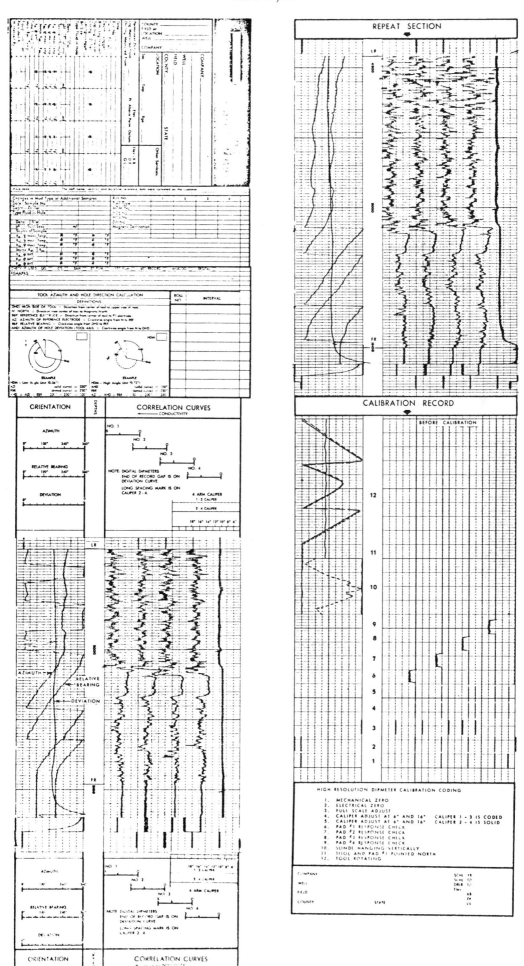

TDT — DUAL SPACING

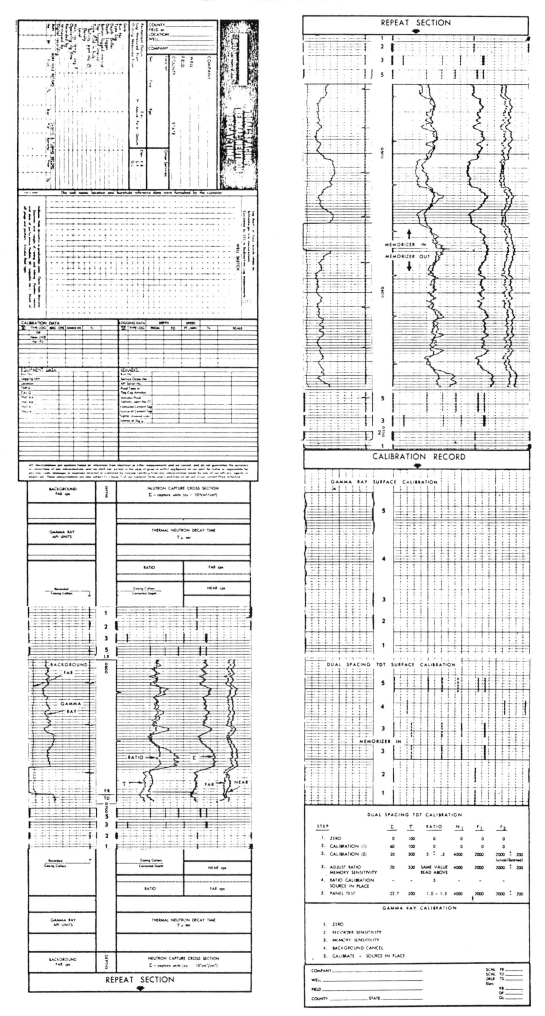

DUAL SPACING TDT CALIBRATION

STEP		Σ	T	RATIO	N_1	F_1	F_3
1.	ZERO	0	100	0	0	0	0
2.	CALIBRATION (1)	60	100	0	0	0	0
3.	CALIBRATION (2)	20	300	5 ± .5	4000	2000	2000 ± 200
							(uncalibrated)
3.	ADJUST RATIO MEMORY SENSITIVITY	20	330	SAME VALUE READ ABOVE	4000	2000	2000 ± 200
4.	RATIO CALIBRATION SOURCE IN PLACE	-	-	5	-	-	-
5.	PANEL TEST	22.7	200	1.0 - 1.5	4000	2000	2000 ± 200

GAMMA RAY CALIBRATION

1. ZERO
2. RECORDER SENSITIVITY
3. MEMORY SENSITIVITY
4. BACKGROUND CANCEL
5. CALIBRATE - SOURCE IN PLACE

COMPANY_____   SCHL PR_____

WELL_____   SCHL TD_____

FIELD_____   DRLR TD_____

COUNTY_____ STATE_____   Elev._____

                KB_____

                DF_____

                GL_____

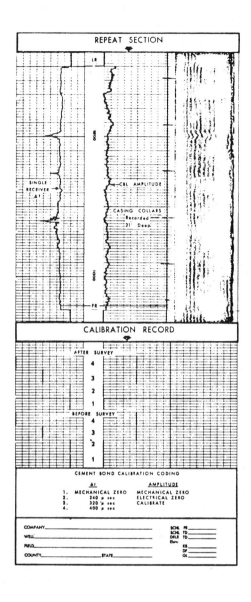

REPEAT FORMATION TESTER

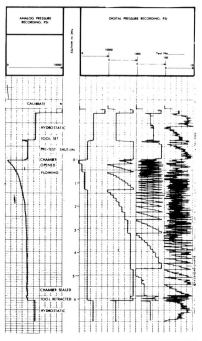

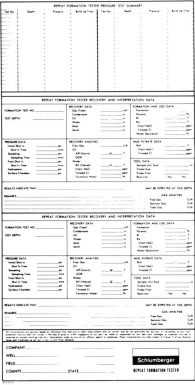

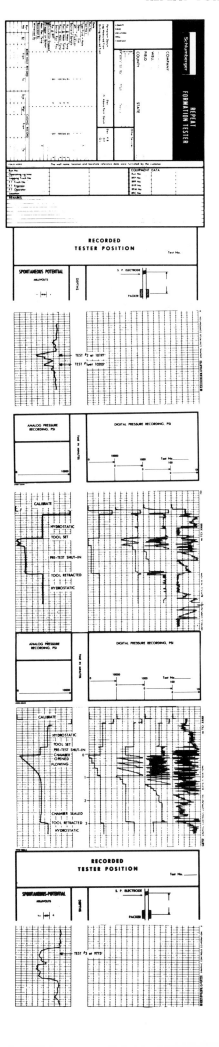

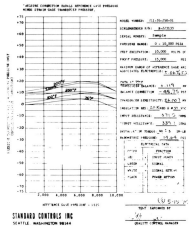

CALIBRATION CERTIFICATE

APPENDIX 5B

CSU calibrations (reproduced by courtesy of Schlumberger)

DETAILED CALIBRATION SUMMARIES

This section is designed to clarify the meaning of the various calibration summaries that are generated by the CSU system. Only one example will be given of each general log type; for instance, the Dual Induction-SFL * printout illustrates the significance of all induction log calibration summaries.

The CSU programs which govern the log formats have been improved somewhat, and will undoubtedly continue to change as the need arises. The calibration summaries may therefore follow a different format than those shown here. However, the principle remains the same, and the various entries will remain recognizable.

INDUCTION LOGS DISF

Shop summary

Induction calibration begins with a shop check, because the primary calibration standard is a precision test loop which is impractical for wellsite use.

The shop summary (Fig. A5B-1) begins as do all sections of the calibration summary, with identification of the date, file, and program (numbers ①, ②, ③).

All dates are listed by year/month/day. ④ gives the tool code.

Electronics calibration summary ⑤

Induction calibrations are done as follows: the "Measured zero" signal ⑥ is recorded with the sonde electronically disconnected. Then a precision calibration resistor in the tool is switched into the circuit, giving a "Measured plus" value ⑦. Offset and gain factors are computed to correct these to the calibrated values, zero ⑧ and 500 millimhos/m ⑨.

Then, with these correction factors in play, the "sonde error corr." ⑩ and "Test loop" ⑪ readings are taken in a zero-conductivity area. Sonde error represents the small residual signal, contributed by the sonde, which is removed from the log readings by CSU system. The test loop reading must fall within a small tolerance range on 500 mmho/m, the absolute equivalent value of the test loop. If either figure is outside the specified tolerances, maintenance is required. The SFL * and SP measurements do not require shop calibration. ⑫ gives the tool serial numbers.

Before survey calibration summary (Fig. A5B-2)

This calibration sequence is normally performed at the wellsite, as confirmed by the data ①.

```
                              SHOP SUMMARY

   PERFORMED:       ①78/08/07
   PROGRAM FILE: ②SHOP      (VERSION    ③10.2          78/ 6/27)

DITD④                  ELECTRONICS CALIBRATION SUMMARY

              ⑤ELECTRONICS CALIBRATION              SONDE
               MEASURED             CALIBRATED       ERROR      TEST
          ⑥ZERO    ⑦PLUS      ⑧ZERO       ⑨PLUS    ⑩CORR.    ⑪LOOP       UNITS
ILD         1.7     514.6       0.0        499.9     5.33      511.2       MM/M
ILM         3.4     497.4       0.0        499.9     5.06      498.7       MM/M

                      ⑫(IS:119   , IC:141   )
```

Fig. A5B-1. Shop summary, DISF log.

* Mark of Schlumberger.

The "Measured" section corresponds with numbers ⑥ and ⑦ on Fig. A5B-1, except for the addition of the SFL. The ILd and ILm values ⑤ and ⑥ may differ slightly from the shop summary values due to drift, but the differences are normally insignificant. Again, these values are used to compute offsets and gains which produce "Calibrated" readings of 0 and 500 mmho/m ⑦.

The SFL tool is an electrode device which is calibrated by precisely measuring the zero ⑧, and the plus values of the electronics ⑨. Again, computed offset and gain convert the tool actual response in these configurations to the known values, 0 and 500 mmho/m.

The ILm and ILd "Sonde error" listings ⑩ are normally the same as those shown on the shop summary, ⑩ on Fig. A5B-1. However, they may be set differently by the engineer to compensate for borehole effect. When this is the case, it should be explained on the log heading.

The gamma-ray calibration begins with tool identification ⑪. The "measured" readings are taken with the tool reading background radiation only ⑫, then with a jig producing a known radiation field ⑬. The difference between those readings is defined for this tool as 165 API gamma-ray units, printed under the "calibrated" heading ⑭. This conversion is, of course, accomplished by the computer.

The calibration value varies according to tool size and type: all gamma-ray calibrations are based on the response of that tool type in the API Gamma Ray Test Pit in Houston.

SP measurements, consisting as they do of direct recordings of the measured quantity (millivolts), require no conversion factors. No calibration summaries are printed for the SP. The log measurement is in fact calibrated against an accurately regulated voltage, but the measuring device is in effect a digital voltmeter which has no drift.

After survey tool check summary (Fig. A5B-3)

This printout indicates at a glance whether any drift has occurred in the "Zero" ① or "Plus" ② readings, and if so, how much. The tolerances specified for reasonable logging conditions are shown in Table A5B-1.

RESISTIVITY TOOLS DLL *-MSFL *

Before survey calibration summary (Fig. A5B-4)

The identifier and column headings are the same as those for the induction tools. However, a shop summary is not used, because the primary calibration standard is an internal resistor in each measuring circuit, permitting primary calibrations at the wellsite.

Dual Laterolog * tool calibrations are straightforward. The "Zero measure" ① values are made with the downhole tool electrically disconnected, and the "Plus" values ② are produced by downhole resistors which represent formation resistivities of 31.6 Ωm. The actual readings usually differ from this figure because of downhole circuit variables, but are ad-

---

* Mark of Schlumberger.

```
            BEFORE SURVEY CALIBRATION SUMMARY

    PERFORMED:   ①78/09/18
    PROGRAM FILE:②NUC     (VERSION  ③10.1      78/ 5/12)

  DITD④            ELECTRONICS CALIBRATION SUMMARY

                MEASURED              ⑦CALIBRATED
              ZERO   PLUS          ZERO      PLUS        UNITS
     ⑤ILD     2.7    514.3          0.0       499.9       MM/M
     ⑥ILM     4.4    496.3          0.0       499.9       MM/M
      SFL    ⑧0.1  ⑨539.0          0.0       499.9       MM/M

  ILM⑩SONDE ERROR CORRECTION :        4.5     MM/M
  ILD⑩SONDE ERROR CORRECTION :        4.5     MM/M

  SGTE⑪            DETECTOR CALIBRATION SUMMARY

                MEASURED
             ⑫BKGD  ⑬JIG     ⑭CALIBRATED       UNITS
       GR      19    200         164            GAPI
```

Fig. A5B-2. Before survey calibration summary, DISF.

```
              AFTER SURVEY TOOL CHECK SUMMARY

    PERFORMED:     78/09/18
    PROGRAM FILE:  MUC     (VERSION    10.1     78/ 5/12)

    DITD                   TOOL   CHECK

                    ① ZERO                ② PLUS
                  BEFORE     AFTER       BEFORE     AFTER     UNITS
          ILD     0.0        0.0         499.9      494.4     MM/M
          ILM     0.0        -0.0        499.9      494.8     MM/M
          SFL     0.0        0.1         499.9      496.7     MM/M

    ILM  SONDE ERROR CORRECTION :    5.0    MM/M
    ILD  SONDE ERROR CORRECTION :    8.5    MM/M
```

Fig. A5B-3. After survey tool check summary, DISF.

TABLE A5B-1

Log	Curve	Tolerance Zero	Plus	Units
IES	ILd	±2	±20	mho/m
	SFL	±2	±20	mho/m
Slim Hole IES	ILd	±2	±50	mho/m
	16″ N	±0.08	±1.05	Ω.m
DIL	ILm	±2	±20	mho/m
	ILd	±2	±20	mho/m
	LL8	±2	±20	mho/m
DISF	ILm	±2	±20	mho/m
	ILd	±2	±20	mho/m
	SFL	±2	±20	mho/m

```
              BEFORE SURVEY CALIBRATION SUMMARY
    PERFORMED:     78/10/10
    PROGRAM FILE:  MILL    (VERSION    10.2     78/ 6/27)

    DLT           ELECTRONICS CALIBRATION SUMMARY

                MEASURED                 CALIBRATED
             ①ZERO  ②PLUS          ZERO      PLUS      UNITS
       LLD   0.0    34.5           0.0       31.6      OHMM
       LLS   0.0    33.9           0.0       31.6      OHMM

    MSFL          ELECTRONICS CALIBRATION SUMMARY

                MEASURED            ⑤CALIBRATED
             ③ZERO  ④PLUS          ZERO      PLUS      UNITS
      MSFL   0.0    1003.          0.0       999.9     MMHO
        I1   5.8    206.0          0.0       199.9     MMHO

    SGTE          DETECTOR CALIBRATION SUMMARY

                MEASURED
             BKGD   JIG        CALIBRATED        UNITS
        GR   43     195           164            GAPI

    MSFL          CALIPER CALIBRATION SUMMARY

                MEASURED                 CALIBRATED
             SMALL  LARGE          SMALL     LARGE     UNITS
      CALI   9.4    13.6           8.0       12.0      IN
```

Fig. A5B-4. Before survey calibration summary, DLL-MSFL.

justed to read the exact 31.6 by CSU-computed factors.

The MSFL is calibrated somewhat similarly, but uses the reciprocal units, mmho/m, for calibration. The "Zero" end ③ is again read with downhole circuits disconnected, but the calibrating resistor gives a "Plus" response equal to that of a 1000-mmho/m formation ④. Deviations caused by circuit variables are adjusted by CSU to give precise "Calibrated" readings ⑤.

Gamma-ray calibrations have already been described.

After survey tool check summary (Fig. A5B-5)

These comparisons of "Before" and "After" readings are self-explanatory. Tolerances are shown in Table A5B-2.

NUCLEAR TOOLS FDC-CNL*-GR

Shop summary (Fig. A5B-6)

The nuclear tools are shop calibrated with primary field standards which are impractical for wellsite use. Once again, the shop summary is identified as to date, file, and program. Then the neutron tool is identified, CNT-A, in this case ①.

CNL and FDC tool calibrations may be made in either of two ways. In the prevalent method, the engineer enters known calibration values into the CSU system during the "Before Survey" calibration. These values are found during prior shop checks, and recorded on the shop summary under the "Tank measured" (CNL) and "Block measured" (FDC) columns. The operation and stability of the measuring systems are verified by radioactive jigs which produce predetermined responses in the tools.

The other method requires that these tool-check jigs be set to reproduce as accurately as possible the "Tank" and "Block" responses. Then the jigs become secondary calibration standards which can be used at the wellsite. The "Before survey calibration summary" reveals which method was used, as will be explained.

CNT calibration

Detector calibration summary for the neutron tool recaps the "Tank" calibration. The neutron calibrating tank is the primary field standard for neutron tools. It consists of a cylindrical water jacket around the tool, which should produce a fixed neutron ratio (NRAT, ②) between the near and far detector count rates. If the "Measured NRAT" differs from the correct value (2.15 here), the CSU system computes a gain factor which restores it to the standard ③.

A special radioactive jig is used to verify the detector sensitivity during prejob checkout. During the shop check the jig is set to reproduce approximately the tank ratio ④. This check can then be repeated in the field to verify that detector sensitivities have not changed. On this example, the CNT component numbers are also listed ⑤ and ⑥.

FDC calibration

The FDC calibration is similar to that of the CNT, except that raw count rates are recorded instead of ratios. Again, the primary standard calibration is done using a special aluminium block ⑦. Then the FDC jig is set to match that response approximately ⑧. The "Block calibrated" figures are in this case arbitrary numbers that produce the correct bulk density value; scaling factors are computed to convert the actual block count rates to these standards. Finally, the FDC component numbers are listed ⑨.

Before survey calibration summary (Fig. A5B-7)

After dating and identifying, we proceed to calibrate the gamma ray as described under the induction tool section. Then comes the CNL calibration.

The Schlumberger engineer inserts into the CSU system the count-rate ratio recorded during the "Tank" calibration. The "Calibrated" value for NRAT will appear as shown at ⑩. The jig will be used to verify the stability of the detectors, but the "Jig calibrated" figure will usually differ somewhat from the precise value of 2.15 ⑪. As mentioned earlier, it is possible to calibrate the CNL tool by use of the jig alone, in which case the figure "0.0" will appear at ⑩, and the "Jig calibrated" figure will be exactly 2.15.

The individual near and far neutron count rates do not appear on the calibration summaries since they are not calibrated and are not used to compute porosity.

FDC tool calibration is similar, except that the actual count rates are calibrated instead of a ratio.

The calibration is done by inserting the "Block" count rates, recorded during shop check, into the CSU system, as shown at ⑫. Then the tool operation and stability are checked by means of the jig ⑬.

If the jig is used to calibrate the tool, the "Block calibrated" column will show only zeroes.

---

* Mark of Schlumberger.

```
            AFTER SURVEY TOOL CHECK SUMMARY

    PERFORMED:     78/10/10
    PROGRAM FILE:  MILL    (VERSION    10.2      78/ 6/27)

    DLT                    TOOL   CHECK

                ZERO                      PLUS
             BEFORE     AFTER        BEFORE     AFTER    UNITS
       LLD     0.0       0.0          31.6      31.6     OHMM
       LLS     0.0       0.0          31.6      31.6     OHMM

    MSFL                   TOOL   CHECK

                ZERO                      PLUS
             BEFORE     AFTER        BEFORE     AFTER    UNITS
      MSFL     0.0      -0.1         999.9     1001.     MMHO
        I1     0.0       0.0         199.9      205.3    MMHO
```

Fig. A5B-5. After survey tool check summary, DLL-MSFL.

TABLE A5B-2

Curve	Tolerance Zero	Tolerance Plus	Units
LLs	± 0.1	± 2%	Ωm
LLd	± 0.1	± 2%	Ωm
MSFL	± 2	±20	mmho/m

```
                SHOP  SUMMARY

    PERFORMED:     78/09/14
    PROGRAM FILE:  SHOP    (VERSION    10.2      78/ 6/27)

    PGTK            DETECTOR CALIBRATION SUMMARY

              ⑦ BLOCK              ⑧ JIG
          MEASURED   CALIBRATED   MEASURED   CALIBRATED   UNITS

    FFDC    396         336         398         338        CPS
    NFDC    697         527         698         528        CPS

           ⑨ (PGS:20   ,  PGC:35   , SFT:1308 )


                SHOP  SUMMARY

    PERFORMED:     78/09/21
    PROGRAM FILE:  SHOP    (VERSION    10.2      78/ 6/27)

    CNTA①            DETECTOR CALIBRATION SUMMARY

                TANK                  JIG
          MEASURED  CALIBRATED    MEASURED   CALIBRATED

    NRAT  ②2.37   ③2.15           2.38      ④2.16

           ⑤ (CNC:1262 ,⑥CNB:1290  )
```

Fig. A5B-6. Shop summary, FDC-CNL-GR log.

```
                BEFORE  SURVEY  CALIBRATION  SUMMARY


    PERFORMED:      78/10/10
    PROGRAM FILE:   NUC      (VERSION    10.2      78/ 6/27)

  SGTE              DETECTOR  CALIBRATION  SUMMARY

              MEASURED
             BKGD    JIG        CALIBRATED          UNITS
      GR      19    171            164               GAPI

  CNTA              DETECTOR  CALIBRATION  SUMMARY

                  TANK                  JIG
               CALIBRATED     MEASURED      CALIBRATED
      NRAT     ⑩2.15           2.39         ⑪2.18

  PGTK              DETECTOR  CALIBRATION  SUMMARY

                 BLOCK            ⑬JIG
              CALIBRATED      MEASURED      CALIBRATED    UNITS
      FFDC     ⑫336             402            342         CPS
      NFDC      527             695            526         CPS

  PGTK              CALIPER  CALIBRATION  SUMMARY

              MEASURED                  CALIBRATED
             SMALL   LARGE           SMALL    LARGE       UNITS
      CALI    8.6    12.1             8.0     12.0         IN
```

Fig. A5B-7. Before survey calibration summary, FDC-CNL-GR.

TABLE A5B-3

Curve	Tolerance	Units
CNL (NRAT)	±0.04	(Ratio)
SNP	±13	cps
FDC (F)	±14	cps
FDC (N)	±22	cps

```
                AFTER  SURVEY  TOOL  CHECK  SUMMARY


    PERFORMED:      78/10/10
    PROGRAM FILE:   NUC      (VERSION    10.2      78/ 6/27)

  CNTA                      TOOL   CHECK

                      JIG
                  BEFORE      AFTER
      NRAT         2.18        2.19

    POROSITY CHANGE  (LIME):  0.002

  PGTK                      TOOL   CHECK

                      JIG
                  BEFORE      AFTER    UNITS
      FFDC         342         341      CPS
      NFDC         526         525      CPS
```

Fig. A5B-8. After survey tool check summary, FDC-CNL-GR.

TABLE A5B-4

Curve	Zero	Tolerance Plus	Units
Proximity	±1	±15	mmho/m
MLL	±1	±15	mmho/m

After survey tool check summary (Fig. A5B-8)

As before, this shows, the changes in various parameters from before to after the survey. The tolerances quoted in Table A5B-3 are realistic values which can be expected under reasonable logging conditions.

MICRORESISTIVITY ML-MLL *-PROXIMITY *

(Micro SFL calibration described in the DLL section)

Before survey calibration summary (Fig. A5B-9)

The microresistivity tool runs either a Proximity-Microlog or a Microlaterolog *-Microlog, each with a caliper. Calibration is similar to that for the Dual Laterolog tool, by means of internal calibration resistors.

The operation is identical to that described for the Dual Laterolog tools.

After survey tool check summary (Fig. A5B-10)

The presentation is very similar to that for the Dual Laterolog tool. Reasonable-condition tolerances are shown in Table A5B-4.

* Mark of Schlumberger.

* Mark of Schlumberger.

```
              BEFORE SURVEY CALIBRATION SUMMARY

   PERFORMED:    78/09/20
   PROGRAM FILE: MPT     (VERSION    10.1     78/ 5/12)

   MPT        ELECTRONICS CALIBRATION SUMMARY

              MEASURED              CALIBRATED
              ZERO   PLUS           ZERO   PLUS         UNITS
   MINV      -0.0    4.4            0.0    4.9          OHMM
   MNOR       0.0    3.4            0.0    4.9          OHMM
   PROX      -0.2  517.5            0.0  499.9          MM/M

   MPT         CALIPER CALIBRATION SUMMARY

              MEASURED              CALIBRATED
              SMALL  LARGE          SMALL  LARGE        UNITS
   CALI       9.3    12.7           8.0    11.9         IN
```

Fig. A5B-9. Before survey calibration summary, Proximity-Microlog.

```
              AFTER SURVEY TOOL CHECK SUMMARY

   PERFORMED:    78/09/20
   PROGRAM FILE: MPT     (VERSION    10.1     78/ 5/12)

   MPT               TOOL  CHECK

                ZERO                 PLUS
              BEFORE  AFTER        BEFORE  AFTER     UNITS
   MINV       0.0     0.0          4.9     4.9       OHMM
   MNOR       0.0     0.0          4.9     4.9       OHMM
   PROX       0.0     0.1          499.9   497.5     MM/M
```

Fig. A5B-10. After survey tool check summary, Proximity-Microlog.

SONIC LOGGING

There are no CSU system calibration summaries for the sonic logs. Like the SP measurement, sonic transit-time logs are direct measurements of a basic parameter, namely time. The only calibration of conventional analog sonic recordings were of the galvanometers and associated circuitry, not of the measurement itself. In the CSU system, time intervals are keyed to a crystal-controlled oscillator, and processed in digital form; the surface equipment does not vary in sensitivity. The accuracy of the crystal clock is verified before each job.

APPENDIX 6

SI BASE AND SUPPLEMENTARY UNITS

Quantity	Unit Name	Unit Symbol	Remarks
BASE UNITS			
length	meter, or metre	m	U.S. spelling is "meter". Canadian and ISO (International Organization for Standardization) spelling is "metre".
mass	kilogram	kg	This is the only base unit having a prefix. In SI the "kilogram" is always a unit of mass, never a unit of weight or force.
time	second	s	The "second" is the base unit, but in practice other time units are allowable.
electric current	ampere	A	
thermodynamic temperature	kelvin	K	Note lower-case k in "kelvin", but cap K for unit symbol. No degree sign is used with "kelvin".
amount of substance	mole	mol	
luminous intensity	candela	cd	Pronounced can dell' uh.
SUPPLEMENTARY UNITS			
plane angle	radian	rad	These angular units are designated by ISO to be dimensionless with respect to the base quantities.
solid angle	steradian	sr	

TABLE A6-2

SI Fundamental Units

Unit	Quantity	Dimension	Unit name SI	Unit symbol	Other units & their correspondence
Space	Plane angle	rad	Radian	rad	degree $1°$ minute $1'$ second $1''$ grade 1 gr
	Solid angle	sr	steradian	sr	
	Length	m	metre	m	foot ft = 0.305m inch in = 2.54 cm
	Surface	m^2	square metre	m^2	
	Volume	m^3	cubic metre	m^3	litre l = 1dm^3 = 10^{-3} m^3
Time	Time	s	second	s	minute mn = 60s hour h = 60mn = 3600s day
	Frequency	s^{-1}	Hertz	Hz	cycle/second c/s = 1 Hz
	Velocity	$m\,s^{-1}$	m/second	m/s	
	Acceleration, linear	$m\,s^{-2}$	m/second squared	m/s^2	Gal Gal = 1 cm/s^2
Mass	Mass	kg	kilogram	kg	tonne t = 10^3 kg = 10^6 g
	Density (mass)	$kg\,m^{-3}$	kilogram per cubic metre	kg/m^3	gram per cm^3 g/cm^3 = 10^3 kg/m^3
Force	Force	$m\,kg\,s^{-2}$	newton	N	dyne dyn = 10^{-5} N kg-force kgf = 9.8 N
	Pressure	$m^{-1}\,kg\,s^{-2}$	pascal	Pa	bar bar = 10^6 dyne/cm^3 = 10^5 Pa atmosphere atm = 101.3 kPa
	Dynamic viscosity	$m^{-1}\,kg\,s^{-1}$	pascal.second or poiseuille	Pa.s	poise Po = 1 dyn s/cm^2 = 0.1 Pa s
	Kinematic viscosity	$m^2\,s^{-1}$		m^2/s	stokes St = 1 cm^2/s
Energy	Work	$m^2\,kg\,s^{-2}$	Joule	J	erg erg = 0.1 J watt.second = newton metre = J
	Power	$m^2\,kg\,s^{-3}$	Watt	W	W = J/s Horsepower Hp = 735.498 W
Temperature	Thermodynamic temperature	K	kelvin	K	Degree -Celsius °C = K + 273.15 -Fahrenheit °F = 1.8°C + 32
	Quantity of heat	$m^2\,kg\,s^{-2}$	joule	J	calorie cal = 4.1868 J
	Heat capacity	$m^2\,kg\,s^{-2}\,K^{-1}$	joule per kelvin	J/K	
Electricity	Electrical potential	$m^2\,kg\,s^{-3}\,A^{-1}$	Volt	V	V = W/A
	Quantity of electricity	s A	Coulomb	C	ampere.second As = 1C
	Capacitance	$m^{-2}\,kg^{-1}\,s^4\,A^2$	Farad	F	F = C/V
	Intensity	A	Ampere	A	
	Conductance	$m^{-2}\,kg^{-1}\,s^3\,A^2$	Siemens	S	mho mho = 1S = $1\Omega^{-1}$
	Resistance	$m^2\,kg\,s^{-3}\,A^{-2}$	Ohm	Ω	Ω = V/A
	Inductance	$m^2\,kg\,s^{-2}\,A^{-2}$	Henry	H	H = Wb/A
	Resistivity	$m^3\,kg\,s^{-3}\,A^{-2}$		Ω, m	
	Conductivity	$m^{-3}\,kg^{-1}\,s^3\,A^2$	Siemens per m	S/m	
Magnetism	Magnetic potential or magnetomotive force		Gilbert	Gb	Gb = $(10/4\pi)$ A turns
	Magnetic flux strength or magnetic field intensity	$A\,m^{-1}$	Ampere per m	A/m	Oersted Oe = $(10^3/4\pi)$ A turns/m
	Magnetic flux	$m^2\,kg\,s^{-2}\,A^{-1}$	Weber	Wb	Maxwell Mx = 10^{-8} Wb
	Magnetic flux density	$kg\,s^{-2}\,A^{-1}$	Tesla	T	Gauss Gs = 10^{-4} T T = Wb/m^2
Light	Luminous intensity	cd	Candela	cd	
	Luminous flux	cd sr	Lumen	lm	
	Illuminance	$m^{-2}\,cd\,sr$	lux	lx	phot ph = 10 klx
	Luminance	$cd\,m^{-2}$	Candela per metre squared	cd/m^2	stilb sb = 1 cd/m^2

TABLE A6-3

UNITS FOR COMMON LOGGING MEASUREMENTS

Quantity	Customary Unit	Metricated Unit	Unit Symbol
angle of dip	degree	degree	°
Caliper, bit and casing size, mud-cake thickness, microspacing, tool diameter	inch	millimeter	mm
conductivity	millimho per meter	millisiemens per meter	mS/m
density	gram per cubic centimeter	kilogram per cubic meter*	kg/m³
depth, bed thickness, macrospacing, invasion depth	foot	meter	m
gamma-ray intensity	A.P.I. Unit	A.P.I. Unit	A.P.I. Unit
interval transit time (Sonic)	microsecond per foot	microsecond per meter	μs/m
macroscopic thermal neutron capture cross section	capture unit (= 10^{-3} cm^{-1})	per meter	m^{-1}
Neutron-log reading	A.P.I. Unit	A.P.I. Unit	A.P.I. Unit
porosity	fraction or percent	cubic meter per cubic meter or porosity unit	m³/m³
resistivity	ohm meter, or ohm meter squared per meter	ohm meter	$\Omega \cdot$m
spontaneous potential	millivolt	millivolt	mV
temperature	degree Fahrenheit	degree Celsius	°C

*"Megagram per cubic meter" (Mg/m³) was used in metric Chart Book.

TABLE A 6-4
SI COHERENT DERIVED UNITS

Quantity	Unit Name	Unit Symbol	Expressed in Terms of Other Derived SI Units	Expressed in Terms of Base- and Supplementary-Unit Symbols
absorbed dose (of ionizing radiation)	gray (replaces the rad)	Gy	J/kg	$m^2 \cdot s^{-2}$
acceleration, linear	meter per second squared	m/s^2		$m \cdot s^{-2}$
activity (of radionuclides)	becquerel (replaces the curie)	Bq		s^{-1}
angular acceleration	radian per second squared	rad/s^2		$rad \cdot s^{-2}$
angular velocity	radian per second	rad/s		$rad \cdot s^{-1}$
area	square meter	m^2		m^2
capacitance (electrical)	farad	F	C/V	$m^{-2} \cdot kg^{-1} \cdot s^4 \cdot A^2$
charge (electrical)	coulomb	C	$A \cdot s$	$s \cdot A$
conductance (electrical)	siemens* (replaces the mho)	S	A/V	$m^{-2} \cdot kg^{-1} \cdot s^3 \cdot A^2$
conductivity (electrical)	siemens per meter	S/m		$m^{-3} \cdot kg^{-1} \cdot s^3 \cdot A^2$
current density	ampere per square meter	A/m^2		$A \cdot m^{-2}$
density (mass)	kilogram per cubic meter	kg/m^3		$kg \cdot m^{-3}$
electromotive force	volt	V	W/A	$m^2 \cdot kg \cdot s^{-3} \cdot A^{-1}$
energy	joule*	J	$N \cdot m$ or $W \cdot s$	$m^2 \cdot kg \cdot s^{-2}$
entropy	joule per kelvin	J/K		$m^2 \cdot kg \cdot s^{-2} \cdot K^{-1}$
field strength (electrical)	volt per meter	V/m		$m \cdot kg \cdot s^{-3} \cdot A^{-1}$
force	newton	N		$m \cdot kg \cdot s^{-2}$
frequency	hertz	Hz		s^{-1}
heat capacity	joule per kelvin	J/K		$m^2 \cdot kg \cdot s^{-2} \cdot K^{-1}$
heat, quantity of	joule*	J		$m^2 \cdot kg \cdot s^{-2}$
illuminance	lux	lx	lm/m^2	$m^{-2} \cdot cd \cdot sr$
inductance	henry	H	$V \cdot s/A$ ($= Wb/A$)	$m^2 \cdot kg \cdot s^{-2} \cdot A^{-2}$
luminance	candela per square meter	cd/m^2		$cd \cdot m^{-2}$
luminous flux	lumen	lm		$cd \cdot sr$
magnetic field strength	ampere per meter	A/m		$A \cdot m^{-1}$
magnetic flux	weber	Wb	$V \cdot s$	$m^2 \cdot kg \cdot s^{-2} \cdot A^{-1}$
magnetic flux density	tesla	T	Wb/m^2	$kg \cdot s^{-2} \cdot A^{-1}$
magnetic permeability	henry per meter	H/m		$m \cdot kg \cdot s^{-2} \cdot A^{-2}$
neutron capture cross section	per meter (i.e., square meter per cubic meter)	1/m	m^2/m^3	m^{-1}
permittivity	farad per meter	F/m		$m^{-3} \cdot kg^{-1} \cdot s^4 \cdot A^2$
potential, potential difference (electrical)	volt	V	W/A	$m^2 \cdot kg \cdot s^{-3} \cdot A^{-1}$
power	watt	W	J/s	$m^2 \cdot kg \cdot s^{-3}$
pressure	pascal*	Pa	N/m^2	$m^{-1} \cdot kg \cdot s^{-2}$
quantity of electricity (charge)	coulomb	C		$s \cdot A$
radiant flux	watt	W	J/s	$m^2 \cdot kg \cdot s^{-3}$
radiant intensity	watt per steradian	W/sr		$m^2 \cdot kg \cdot s^{-3} \cdot sr^{-1}$
resistance (electrical)	ohm	Ω (cap omega)	V/A	$m^2 \cdot kg \cdot s^{-3} \cdot A^{-2}$
resistivity (electrical)	ohm meter or ohm-meter**	$\Omega \cdot m$		$m^3 \cdot kg \cdot s^{-3} \cdot A^{-2}$
specific heat capacity	joule per kilogram kelvin	J/(kg. K)		$m^2 \cdot s^{-2} \cdot K^{-1}$
stress	pascal*	Pa	N/m^2	$m^{-1} \cdot kg \cdot s^{-2}$
thermal conductivity	watt per meter kelvin	W/(m. K)		$m \cdot kg \cdot s^{-3} \cdot K^{-1}$
velocity	meter per second	m/s		$m \cdot s^{-1}$
viscosity, dynamic	pascal second or pascal-second	$Pa \cdot s$	$N \cdot s/m^2$	$m^{-1} \cdot kg \cdot s^{-1}$
viscosity, kinematic	square meter per second	m^2/s		$m^2 \cdot s^{-1}$
voltage	volt	V	W/A	$m^2 \cdot kg \cdot s^{-3} \cdot A^{-1}$
volume	cubic meter	m^3		m^3
wave number	(cycles) per meter	1/m		m^{-1}
work	joule*	J	$N \cdot m$	$m^2 \cdot kg \cdot s^{-2}$

*Pronounce "siemens" like "seamen's", pronounce "pascal" to rhyme with "rascal", pronounce "joule" to rhyme with "pool".
**The "ohm meter squared per meter", sometimes used to designate the resistivity unit in the past, is definitely discarded.

TABLE A6-5

ALLOWABLE UNITS AND CONVERSIONS

Quantity	Coherent SI Units	Allowable Units for Logging and Related Use	Comments and Conversions
Acceleration, linear	m/s² (meter per second squared)	m/s² mm/s² Gal (gal)	1 ft/s² = 0.3048* m/s² 1 Gal = 1 cm/s² The "gal" and "milligal" are special units used in geodetic and gravity work to express the acceleration due to gravity. The internationally accepted value of acceleration due to gravity is 9.806 65 m/s² = 32.1740 ft/s². Actual value will vary with latitude, densities of surrounding rocks, and depth.
Angle, plane	rad (radian)	rad mrad (milliradian) μrad (microradian) ° (degree) ′ (minute) ″ (second)	1° = 0.017 453 29 rad (ANSI prefers the "unit degree" with decimal divisions.)
Area	m² (square meter)	km² ha (hectare) dm² cm² mm²	1 ha = 10,000 m² = 1 hm² The "hectare" is used for land measure.
Conductance	S (siemens) (1 S = 1 A/V)	S	1 mho = 1 Ω⁻¹ = 1* S The "mho" is replaced by the "siemens".
Conductivity	S/m (siemens per meter)	S/m mS/m	"mS/m" replaces "mmho/m" on induction-log conductivity curves.
Density	kg/m³ (kilogram per cubic meter)	kg/m³ Mg/m³	1 lbm/ft³ = 16.085 kg/m³ ("lbm" is "pound mass") 1 g/cm³ = 1000 kg/m³ = 1 Mg/m³
Depth, bed thickness, tool length, macro-spacing (and invasion depth)	m (meter)	m	1 ft = 0.3048* m 1 yd = 0.9144* m
Diameter of hole, bit or casing size, mud-cake thickness, microspacing, tool diameter	m (meter)	mm	1 in. = 25.4* mm

Quantity	Coherent SI Units	Allowable Units for Logging and Related Use	Comments and Conversions
Distance	m (meter)	km	1 mi = 1.609 344* km 1 naut. mi = 1.852* km
Energy	J (joule) Pronounce "joule" to rhyme with "pool"	J	1 Btu = 1.055 056 kJ 1 eV (electronvolt = 1.602 19 × 10⁻¹⁹ J = 16.0219 aJ
Flow Rate, mass	kg/s (kilogram per second)	kg/s	1 lbm/s = 0.453 59 kg/s ("lbm" is "pound mass")
Flow Rate, volumetric	m³/s (cubic meter per second)	m³/s m³/min m³/h m³/d L/s	1 BPD = 0.158 987 m³/d (For standard conditions, see "Gas Volume" and "Oil Volume".)
Force	N (newton) 1 N = 1 kg/s²	N	1 lbf (pound force) = 4.448 22 N 1 kgf (kilogram force) = 9.806 65* N Note: The kilogram is NEVER used as a unit of force in SI.
Gamma Ray Intensity		API Unit	
Gas-Oil Ratio	m³/m³ (dimensionless)	Std. m³/m³ at specified standard conditions	1 scf/bbl (standard cubic foot per barrel) = 0.180 117 5 std. m³/m³. (See Gas Volume.)
Gas volume	m³ (cubic meter)	m³ at specified standard conditions	1 scf (standard cubic foot at 60° F and 14.65 psi) = 2.817 399 × 10⁻² m³ (at 15°C and 1 atm = 101.325 kPa)
Gravity: See Relative Density			
Interval transit time	s/m (second per meter)	μs/m (microsecond per meter.	1 μs/ft = 3.280 840 μs/m
Length (see Depth, Diameter, Distance)			
Mass	kg (kilogram)	t (metric ton or tonne)** Mg kg g mg ** In Canadian French "tonne" may refer to 2000-lb ton.	1 t = 1 Mg (megagram) 1 lbm avoir. (pound mass avoirdupois) = 0.453 592 4 kg.
Mud Weight (see also Density)	kg/m³ (kilogram per cubic meter)	kg/m³ Mg/m³	1 lbm/U.S. gal = 119.826 4 kg/m³ 1 lbm/U.K. gal = 99.776 33 kg/m³

*Exact value

TABLE A6-5 (CONTINUED)

Quantity	Coherent SI Units	Allowable Units for Logging and Related Use	Comments and Conversions
Neutron log count rate		API Unit	
Neutron capture cross section (macroscopic)	m^{-1} (per meter)	m^{-1} c.u. $=10^{-3}$ cm^{-1} (U.S. only)	1 c.u. (capture unit) $=$ 10^{-3} $cm^{-1} = 10^{-1}$ m^{-1}
Oil Volume	m^3 (cubic meter)	m^3 at 15°C	1 bbl $=$ 1 barrel $=$ 42 U.S. gal. 1 bbl at 60°F $= 0.158\ 91$ m^3 at 15°C.
Permeability (hydrodynamic)	m^2 (square meter)	μm^2 (square micrometer)	1 darcy $= 0.986\ 923\ 2\ \mu m^2$. Since the μm^2 differs by only 1.3% from the darcy, it has been suggested to define a metric darcy equal to the μm^2. The unit symbol for the metric darcy would be "D", and for the metric millidarcy would be "mD".
Porosity	m^3/m^3 (cubic meter per cubic meter, dimensionless)	p.u.	1 p.u. $= 0.01^*$ m^3/m^3
Power	W (Watt) 1 W $=$ 1 J/s	MW kW W etc.	
Pressure	Pa (pascal) 1 Pa $=$ 1 N/m²	Pa kPa MPa	1 psi $= 6.894\ 757$ kPa 1 bar $= 100^*$ kPa atmospheric pressure: 1 atm $= 101.325$ kPa $=$ 760 mm Hg$= 14.696$ psi
Radioactivity of radio-nuclides	Bq (becquerel)	Bq	1 Ci (curie) $= 37^*$ GBq $=$ $37 \times 10^9{}^*$ Bq
Radioactivity, gamma ray. See Gamma Ray API Units			
Relative Density (see also Gravity, Specific, and Gravity, Gas.)	m^3/m^3 (dimensionless)		The term "relative density" may replace "specific gravity". The reference conditions (temperature and pressure) should be stated. Water is the implied reference substance for solids and liquids and air for gases unless otherwise stated.
Resistance	Ω (ohm) 1 Ω $=$ 1 V/A	$M\Omega$ $k\Omega$ Ω etc.	No change.
Resistivity	$\Omega \cdot m$ (ohm-meter)	$\Omega \cdot m$ various multiples	No change.

Quantity	Coherent SI Units	Allowable Units for Logging and Related Use	Comments and Conversions
Rotational Speed		r/s (revolution per second)	
Specific gravity or specific weight			These terms are indicated as obsolescent. See "Relative Density".
Spontaneous Potential	V (volt)	mV (millivolt)	No change.
Temperature	K (kelvin)	°C (degree Celsius) (No longer called "degree centigrade")	$T_{°C} = T_K - 273.15$ $T_{°F} = 32 + 1.8 \times T_{°C}$
Temperature gradient	°C/m (degree Celsius per meter)	°C/m	1°F/100 ft $= 0.018\ 227$ °C/m
Time	s (second)	s min (minute) h (hour) d (day) a (year)	
Velocity	m/s (meter per second)	m/s	1 ft/s $= 0.3048^*$ m/s
Viscosity, dynamic	Pa·s (pascal second) 1 Pa·s $=$ 1 N·s/m²	Pa·s mPa·s N·s/m² (U.S.)	1 P (poise) $=$ 1 dyn·s/ cm² $= 0.1$ Pa·s
Viscosity, kinematic	m^2/s (square meter per second)	m^2/s mm^2/s	1 St (stokes) $=$ 1 cm²/s $=$ 100 mm²/s
Volume, very large, such as volume of a sedimentary basin	m^3 (cubic meter)	km^3 (cubic kilometer)	1 cubem (cubic mile) $=$ $4.168\ 182$ km³ 1 km³ $= 10^9{}^*$ m³
Volume, large, such as volume of a local reservoir	m^3	ha·m (hectare meter, U.S.)	1 acre ft $= 0.123\ 348\ 2$ ha·m 1 ha·m $= 10^4$ m³
Volume, tank size, hole volume	m^3	m^3	1 yd³ $= 0.764\ 554\ 9$ m³ 1 ft³ $= 0.028\ 316\ 85$ m³ 1 U.S. gal $= 3.785\ 412 \times$ 10^{-3} m³ 1 U.K. gal $= 4.546\ 092 \times$ 10^{-3} m³ 1 bbl (42 U.S. gal) $=$ $0.158\ 987\ 3^*$ m³
Smaller volumes	m^3	dm³ or L (liter)	1 L $= 1^*$ dm³ $=$ $10^{-3}{}^*$ m³
See also "Gas Volume"			

*Exact value

TABLE A6-6

ROUNDUP OF UNIT SYMBOLS FOR
UNITS WITH NAMES

Symbol	Unit	Status
a, ha	are, hectare	allowable
a	year (annum)	allowable
atm	atmosphere	limited allowability
A	ampere	base unit
Å	angstrom	limited allowability
b	barn	limited allowability
bar	bar	limited allowability
B, dB	bel, decibel	allowable
Bq	becquerel	derived unit
cd	candela	base unit
C	coulomb	derived unit
Ci	curie	limited allowability
°C	degree Celsius	allowable
d	day	allowable
D	darcy (metric)	allowable
eV	electronvolt	allowable
F	farad	derived unit
g	gram	derived unit
Gal	gal	limited allowability
Gy	gray	derived unit
h	hour	allowable
H	henry	derived unit
Hz	hertz	derived unit
J	joule	derived unit
kg	kilogram	base unit
kn	knot	limited allowability
K	kelvin	base unit
lm	lumen	derived unit

Symbol	Unit	Status
lx	lux	derived unit
L, l, ℓ	liter	allowable
m	meter	base unit
min	minute (time)	allowable
mol	mol	base unit
N	newton	derived unit
naut. mi. (no symbol)	nautical mile	limited allowability
ps	parsec	allowable
Pa	pascal	derived unit
r	revolution (as r/s)	allowable
rad	radian	supplementary unit
rd	rad	limited allowability
R	roentgen	limited allowability
s	second (time)	base unit
sr	steradian	supplementary unit
S	siemens	derived unit
t	metric ton (tonne)	?
T	tesla	derived unit
u	unified mass unit (formerly amu)	allowable
V	volt	derived unit
W	watt	derived unit
Wb	weber	derived unit
Ω	ohm	derived unit
°	degree (angle)	allowable
'	minute (angle)	allowable
"	second (angle)	allowable

TABLE A6-7

Metricated values of some constants

Atmosphere:	1 atm = 101.325 kPa
Density of dry air at 0°C and 1 atm:	1.2929 kg/m^3
Molar gas constant, R:	$R = 8.314 J/(mol \cdot K)$
Avogadro constant, N_A:	$N_A = 6.0220 \times 10^{23}$ mol^{-1}
Velocity of light, c:	$c = 299.792\ 46$ Mm/s
Faraday constant, F:	$F = 96.484$ kC/mol
Permeability of free space, μ_0:	$\mu_0 = 1.256\ 637$ H/m
Standard acceleration due to gravity, g:	$g = 9.806\ 65$ m/s^2
Volume of 1 mol of ideal gas under standard conditions (at 15°C and 1 atm):	23.64 L
Planck constant, h:	$h = 6.626176 \times 10^{-34}$ j s
Boltzmann constant, k:	$k = 1.38054 \times 10^{-23}$ j K^{-1}
Electron rest mass, m_e:	$m_e = 9.10956 \times 10^{-31}$ kg
Proton rest mass, m_p:	$m_p = 1.67261 \times 10^{-27}$ kg
Neutron rest mass, m_n:	$m_n = 1.67492 \times 10^{-27}$ kg

TABLE A6-8

SI prefixes

Multiplication factor	SI prefix for unit name	SI prefix for unit symbol
10^{18}	hexa	H
10^{15}	penta	P
10^{12}	tera	T
10^{9}	giga	G
10^{6}	mega	M
10^{3}	kilo	k
10^{2}	hecto	h
10	deca	da
10^{-1}	deci	d
10^{-2}	centi	c
10^{-3}	milli	m
10^{-6}	micro	μ
10^{-9}	nano	n
10^{-12}	pico	p
10^{-15}	femto	f
10^{-18}	atto	a

TABLE A6-9

CONVERSION FACTORS BETWEEN METRIC, A.P.I., AND U.S. MEASURES

multiply	by	to find	multiply	by	to find
acres	0.4047	hectares	cm per second	0.03281	ft/sec
"	43,560	sq ft	" " "	0.6	m/min
"	4,047	sq m	cubic centimeters	3.531×10^{-5}	cu ft
acre-feet	7,758	bbl	" "	6.102×10^{-2}	cu in.
" "	43,560	cu ft	" "	10^{-6}	cu m
" "	3.259×10^5	gals	" "	2.642×10^{-4}	gals
atmospheres	76	cms of Hg	" "	10^{-3}	liters
"	29.92	in. of Hg	" "	6.2897×10^{-6}	bbl
"	33.93	ft of water	cubic feet	0.1781	bbl
"	1.033	kg/cm²	" "	2.832×10^4	cc
"	14.70	psi	" "	7.481	gals
barrels (API)	1.289×10^{-4}	acre-ft	" "	1,728	cu in.
" "	158,987	cc	" "	0.02832	cu m
" "	5.615	cu ft	" "	28.32	liters
" "	42	gals	cubic feet/day	1.18	liters/hr
" "	9,702	cu in.	" " "	1.18×10^{-3}	cu m/hr
" "	1,590	liters	" " "	0.02832	cu m/day
" "	0.1590	cu m	" " "	0.1781	bbl/day
barrels/day	5.615	cu ft/day	cubic feet/minute	10.686	bbl/hr
" "	0.02917	gal/min	" " "	256.5	bbl/day
" "	6.625	liters/hr	" " "	472	cc/sec
" "	0.1590	cu m/day	" " "	7.481	gal/min
" "	0.006625	cu m/hr	" " "	0.472	liters/sec
barrels/hour	0.0936	cu ft/min	cubic inches	16.39	cc
" "	0.700	gal/min	" "	5.787×10^{-4}	cu ft
" "	2.695	cu in./sec	" "	1.639×10^{-5}	cu m
bars	0.9869	atm	" "	4.329×10^{-3}	gals
"	1.020	kg/cm²	" "	1.639×10^{-2}	liters
"	14.50	psi	cubic meters	6.2897	bbl
British Thermal Units	778.57	ft-lbs	" "	10^6	cc
" " "	0.2520	kg-cal	" "	264.2	gals
" " "	0.2930	watt-hrs	" "	6.102×10^4	cu in.
Btu/minute	0.02357	hp	" "	35.31	cu ft
" "	0.01758	kw	" "	10^3	liters
" "	12.97	ft-lbs/sec	cubic meters/hour	151.0	bbl/day
centimeters	3.281×10^{-2}	ft	" " "	847.8	cu ft/day
"	0.3937	in.	" " "	10^3	liters/hr
"	0.01	meters	" " "	24	cu m/day
"	10	mm	cubic meters/day	0.2621	bbl/hr
cm of mercury	0.01316	atm	" " "	6.2897	bbl/day
" " "	0.4461	ft of water	" " "	1.471	cu ft/hr
" " "	0.01360	kg/cm²	" " "	35.31	cu ft/day
cm per second	0.1934	psi	" " "	41.67	liters/hr
" " "	1.969	ft/min			

CONTINUED →

TABLE A6-9 (CONTINUED)

multiply	by	to find
kgs/square cm	0.9807	bars
,, ,, ,,	32.84	ft of water
,, ,, ,,	28.96	in. of Hg
,, ,, ,,	14.22	psi
kilowatts	56.88	Btu/min
,,	4.427×10^4	ft-lbs/min
,,	737.8	ft-lbs/sec
,,	1.341	hp
,,	10^3	watts
kilowatt-hours	3,413	Btu
,, ,,	2.656×10^6	ft-lbs
,, ,,	1.341	hp-hrs
,, ,,	860	kg-cal
,, ,,	3.672×10^5	kg-m
liters	10^3	cc
,,	6.2897×10^{-3}	bbl
,,	0.03531	cu ft
,,	0.2642	gals
,,	61.02	cu in.
,,	10^{-3}	cu m
liters/hour	0.1509	bbl/day
,, ,,	6.289×10^{-3}	bbl/hr
,, ,,	5.885×10^{-4}	cu ft/min
,, ,,	0.8475	cu ft/day
,, ,,	10^{-3}	cu m/hr
,, ,,	0.02400	cu m/day
meters	3.281	ft
,,	39.37	in.
,,	10^3	mm
,,	6.214×10^{-4}	mile
meters/minute	1.667	cm/sec
,, ,,	3.281	ft/min
,, ,,	196.9	ft/hr
,, ,,	0.05468	ft/sec
mile	5,280	ft
,,	1.609	km
mile/hour	44.70	cm/sec
,, ,,	88	ft/min
,, ,,	26.82	m/min
millimeters	0.1	cm
,,	3.281×10^{-3}	ft
,,	0.03937	in.
minutes	6.944×10^{-4}	days
,,	1.667×10^{-2}	hrs

multiply	by	to find
parts/million	0.05835	grains/gal
,, ,,	8.337	lbs/10^6 gals
pound	7,000	grains
,,	453.6	gm
,,	0.4536	kg
pounds/cubic ft	0.1337	lbs/gal
,, ,, ,,	0.01602	gm/cc
,, ,, ,,	16.02	kg/cu m
,, ,, ,,	5.787×10^{-4}	lbs/cu m
pounds/square in.	0.06805	atm
,, ,, ,,	2.309	ft of water
,, ,, ,,	2.036	in. of Hg
,, ,, ,,	51.70	mm of Hg
,, ,, ,,	0.07031	kg/cm^2
,, ,, ,,	144	lbs/cu ft
seconds	1.157×10^{-5}	days
,,	2.778×10^{-4}	hrs
,,	1.667×10^{-2}	min
square cm	1.076×10^{-3}	sq ft
,, ,,	0.1550	sq in.
,, ,,	10^{-4}	sq m
,, ,,	100	sq mm
square feet	2.296×10^{-5}	acres
,, ,,	929.0	sq cm
,, ,,	144	sq in.
,, ,,	0.09290	sq m
square inches	6.452	sq cm
,, ,,	6.944×10^{-3}	sq ft
,, ,,	645.2	sq mm
square meters	10.76	sq ft
,, ,,	2.471×10^{-4}	acres
,, ,,	1,550	sq in.
°Cent. + 273	1	°K (abs)
°Fahr. + 460	1	°R (abs)
°Cent + 17.8	1.8	°F
°Fahr. −32	5/9	°C
°Cent./100 meters	0.5486	°F/100 ft
°F/100 ft	1.823	°C/100 ft
tons (long)	1,016	kg
,, ,,	2,240	lbs
tons (metric)	10^3	kg
,, ,,	2,205	lbs
tons (short)	2,000	lbs

CONTINUED –

TABLE A6-9 (CONTINUED)

multiply	by	to find
cubic meters/day	0.04167	cu m/hr
days	1,440	min
''	86,400	sec
feet	30.48	cm
''	12	in.
''	0.3048	meters
feet of water	0.02950	atm
'' '' ''	0.8826	in. of Hg
'' '' ''	0.03048	kg/sq cm
'' '' ''	62.43	lbs/sq ft
'' '' ''	0.4335	psi
feet/hour	0.008467	cm/sec
'' ''	5.086×10^{-3}	m/min
'' ''	0.01667	ft/min
feet/minute	0.5080	cm/sec
'' ''	0.01667	ft/sec
'' ''	0.01829	km/hr
'' ''	0.3048	m/min
feet/second	30.48	cm/sec
'' ''	18.29	m/min
foot-pounds	1.285×10^{-3}	Btu
'' ''	3.238×10^{-4}	kg-cal
foot-pounds/minute	3.030×10^{-5}	hp
'' '' ''	2.260×10^{-5}	kw
foot-pounds/second	1.818×10^{-3}	hp
'' '' ''	1.356×10^{-3}	kw
gallons (US)	0.02381	bbl
'' ''	3,785	cc
'' ''	0.1337	cu ft
'' ''	231	cu in.
'' ''	3.785×10^{-3}	cu m
'' ''	3.785	liters
gallons (Imperial)	1.2009	gal (U.S.)
gallons/minute	1.429	bbl/hr
'' ''	34.286	bbl/day
'' ''	0.1337	cu ft/min
'' ''	192.5	cu ft/day
'' ''	3.785	liters/min
'' ''	90.84	liters/hr
grain (avoir)	0.06480	gm
grains/gal	17.12	ppm
'' ''	142.9	lbs/10^6 gals
'' ''	0.01714	gm/liter
grams	15.432	grains

multiply	by	to find
grams	10^{-3}	kg
''	0.3215	oz
''	2.205×10^{-3}	lbs
grams/cc	62.43	lb/cu ft
'' ''	8.344	lbs/gal
'' ''	0.03613	lbs/cu in.
grams/liter	58.42	grains/gal
hectares	2.471	acres
''	1.076×10^5	sq ft
''	0.010	sq km
horsepower	42.40	Btu/min
''	33,000	ft-lbs/min
''	550	ft-lbs/sec
''	1.014	metric hp
''	10.68	kg-cal/min
''	0.7457	kw
''	745.7	watts
horsepower-hour	2,544	Btu
'' ''	641.1	kg-cal
'' ''	2.737×10^5	kg-m
'' ''	0.7455	kw-hr
inches	2.540	cm
''	8.333×10^{-2}	ft
in. of mercury	0.03342	atm
'' '' ''	1.133	ft of water
'' '' ''	0.03453	kg/sq cm
'' '' ''	0.4912	psi
in. of water	0.002458	atm
'' '' ''	0.07349	in. of Hg
'' '' ''	0.002538	kg/cm^2
'' '' ''	0.03609	psi
kilograms	10^3	gm
''	2.205	lbs
''	1.102×10^{-3}	tons (short)
kg-calories	3.986	Btu
'' ''	3,088	ft-lbs
'' ''	1.560×10^{-3}	hp-hrs
'' ''	427	kg-meters
'' ''	1.163×10^{-3}	kw-hrs
kg-calories/min	0.09358	hp
'' '' ''	0.06977	kw
kgs/cubic meter	10^{-3}	gm/cc
kgs/square cm	0.9678	atm

CONTINUED →

TABLE A6-9 (CONTINUED)

multiply	by	to find
viscosity, lb-sec/sq in.	6.895×10^6	viscosity, cp
viscosity, lb-sec/sq ft	4.78×10^4	viscosity, cp
viscosity, centistokes	density	viscosity, cp
watts	0.05688	Btu/min
"	44.27	ft-lbs/min
"	0.7378	ft-lbs/sec
"	1.341×10^{-3}	hp
"	0.01433	kg-cal/min

multiply	by	to find
watts	10^{-3}	kw
watt-hours	3.413	Btu
" "	2,656	ft-lbs
" "	1.341×10^{-3}	hp-hrs
" "	0.860	kg-cal
" "	367.2	kg-m
" "	10^{-3}	kw-hrs

TABLE A6-10

Schlumberger

CONVERSIONS

DEPTH | TEMPERATURE | PRESSURE | MUD WEIGHT

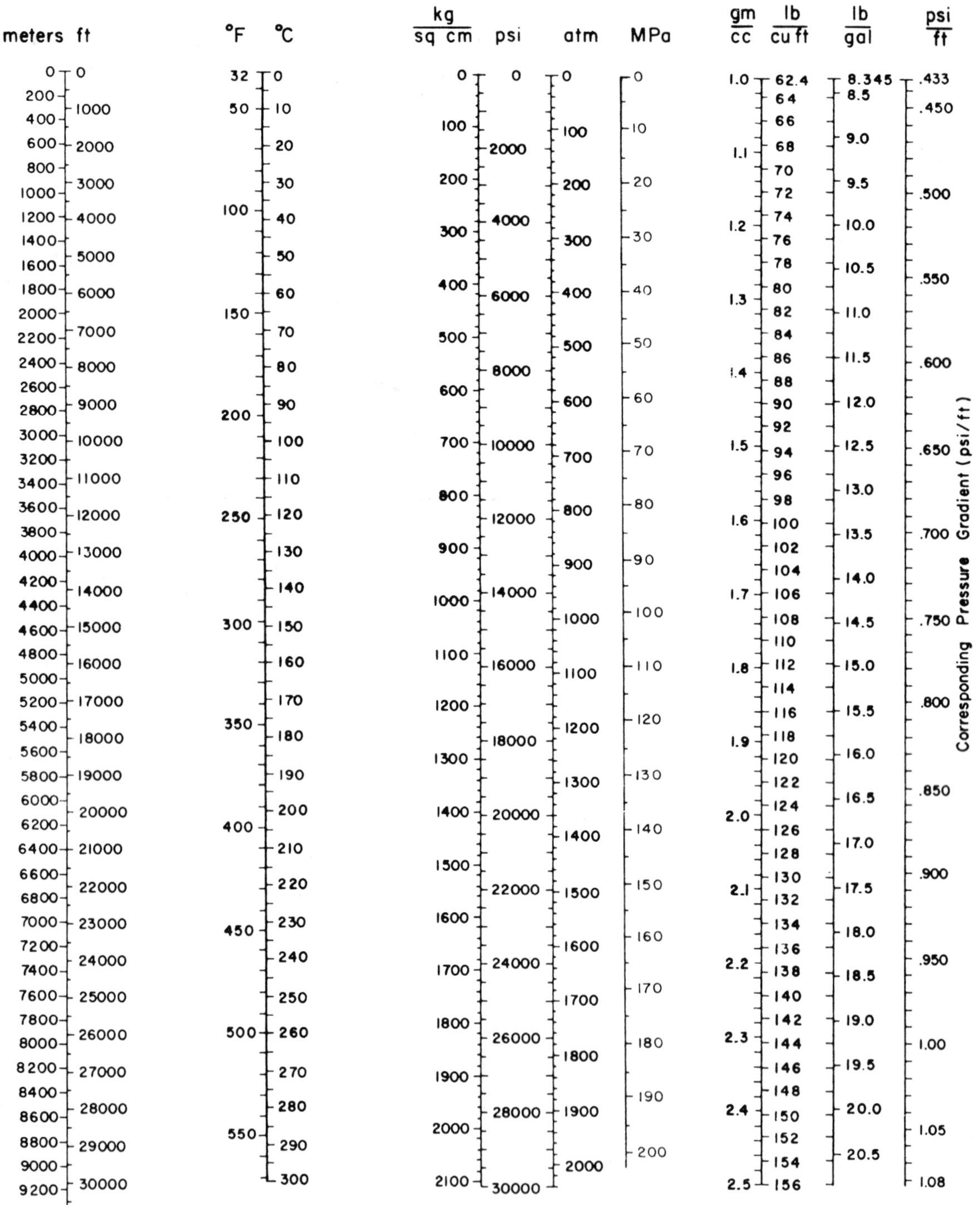
meters ft
0 200 400 600 800 1000 1200 1400 1600 1800 2000 2200 2400 2600 2800 3000 3200 3400 3600 3800 4000 4200 4400 4600 4800 5000 5200 5400 5600 5800 6000 6200 6400 6600 6800 7000 7200 7400 7600 7800 8000 8200 8400 8600 8800 9000 9200 9400 9600 9800 10000
0 1000 2000 3000 4000 5000 6000 7000 8000 9000 10000 11000 12000 13000 14000 15000 16000 17000 18000 19000 20000 21000 22000 23000 24000 25000 26000 27000 28000 29000 30000 31000 32000 33000
°F °C
32 50 100 150 200 250 300 350 400 450 500 550
0 10 20 30 40 50 60 70 80 90 100 110 120 130 140 150 160 170 180 190 200 210 220 230 240 250 260 270 280 290 300
kg/sq cm psi atm MPa
0 100 200 300 400 500 600 700 800 900 1000 1100 1200 1300 1400 1500 1600 1700 1800 1900 2000 2100
0 2000 4000 6000 8000 10000 12000 14000 16000 18000 20000 22000 24000 26000 28000 30000
0 100 200 300 400 500 600 700 800 900 1000 1100 1200 1300 1400 1500 1600 1700 1800 1900 2000
0 10 20 30 40 50 60 70 80 90 100 110 120 130 140 150 160 170 180 190 200
gm/cc lb/cu ft lb/gal psi/ft
1.0 1.1 1.2 1.3 1.4 1.5 1.6 1.7 1.8 1.9 2.0 2.1 2.2 2.3 2.4 2.5
62.4 64 66 68 70 72 74 76 78 80 82 84 86 88 90 92 94 96 98 100 102 104 106 108 110 112 114 116 118 120 122 124 126 128 130 132 134 136 138 140 142 144 146 148 150 152 154 156
8.345 8.5 9.0 9.5 10.0 10.5 11.0 11.5 12.0 12.5 13.0 13.5 14.0 14.5 15.0 15.5 16.0 16.5 17.0 17.5 18.0 18.5 19.0 19.5 20.0 20.5
.433 .450 .500 .550 .600 .650 .700 .750 .800 .850 .900 .950 1.00 1.05 1.08
Corresponding Pressure Gradient (psi/ft)

1 m = 3.28 ft

°C = (°F − 32) × $\frac{5}{9}$

°F = (°C × $\frac{9}{5}$) + 32

psi = kg/sq cm × 14.22

atm = kg/sq cm × .968

atm = psi × .068

psi = MPa × 145.038

1 gm/cc = 62.43 lb/ft^3

= 8.345 lb/gal (U.S.)

psi/ft = .433 × gm/cc

= lb/ft^3/144

= lb/gal/19.27

TABLE A6-11

Schlumberger

CONVERSIONS AND EQUIVALENTS

CONCENTRATION of NaCl SOLUTIONS

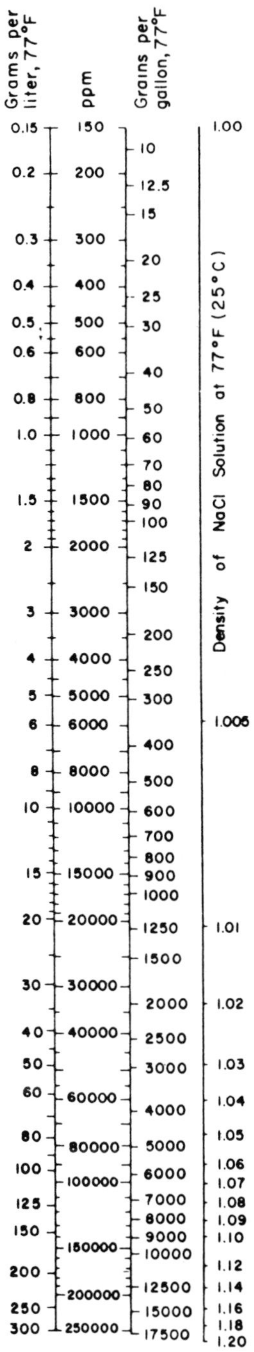

Grams per liter, 77°F | ppm | Grains per gallon, 77°F

TEMP. GRADIENT CONVERSION

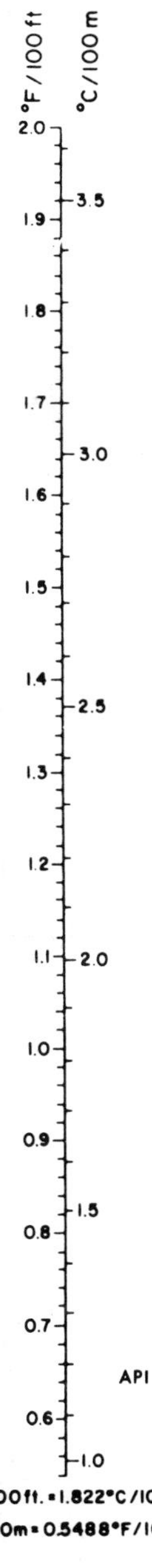

°F/100 ft | °C/100 m

Density of NaCl Solution at 77°F (25°C)

OIL GRAVITY

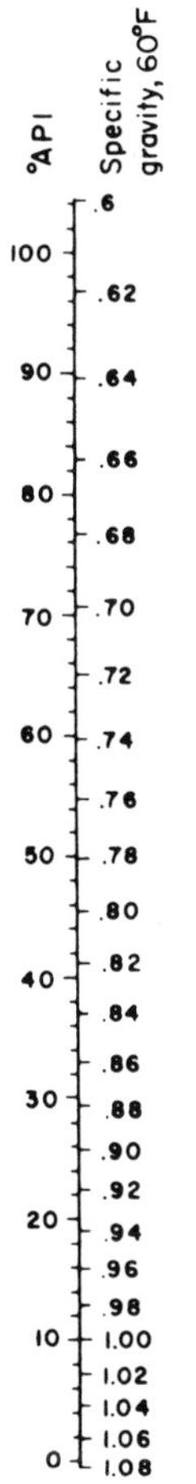

°API | Specific gravity, 60°F

$$API = \frac{141.5}{sp.\ gr.\ @\ 60°F} - 131.5$$

1°F/100ft.=1.822°C/100m
1°C/100m=0.5488°F/100ft.

USEFUL EQUIVALENTS

LENGTH
1 foot (ft)	30.48 cm
1 inch (in.)	2.540 cm
1 meter (m)	3.281 ft
"	39.37 in.

VOLUME
1 acre-foot	7,758 bbls
"	43,560 cu ft
1 barrel (bbl) of oil	42 U.S. gal
"	5.6154 cu ft
"	158.98 liters
1 cubic foot (cu ft)	7.481 U.S. gal
"	28.32 liters
1 U.S. gallon (gal)	231.00 cu in.
"	0.1337 cu ft
"	3.785 liters
1 imperial gallon (England, Canada, Australia, etc.)	1.2009 U.S. gal
"	4.5460 liters
1 liter (1000 cc)	0.03532 cu ft
"	0.2642 U.S. gal

MASS
1 grain	0.0001429 (or 1/7000 lb)
"	0.6480 g
1 pound (lb) avoirdupois	0.4536 kg
1 metric ton (1000 kg)	2205 lb

DENSITY, SPECIFIC GRAVITY, etc.
1 gram per cubic centimeter (g/cc)	62.43 lb/cu ft
"	8.345 lb per U.S. gal
1 U.S. gallon of liquid weighs	(in pounds avoir.) 8.345 multiplied by density in g/cc
1 imperial gallon of water at 62°F weighs 10 lb	
1 barrel of oil weighs	(in pounds avoir.) 350 multiplied by density of oil in g/cc

Oil gravity in degrees API is computed as:

$$°API = \frac{141.5}{Spec.\ Grav.\ 60/60F} - 131.5$$

where "Spec. Grav. 60/60F" means specific gravity of oil at 60°F referred to water at 60°F

PRESSURE
1 atmosphere (atm)	14.70 psi
	1.0332 kg/sq cm
1 kilogram per square centimeter pressure (kg/sq cm)	14.22 psi
1 pound per square inch (psi)	0.07031 kg/sq cm
	0.06805 atm

Pressure Gradient
psi/ft. = 0.433 x g/cc
= lb/ft³/144
= lb/gal/19.27
kg/sq cm/meter = 0.1 x g/cc
= psi/ft x 0.231

TEMPERATURE CONVERSIONS
°F = 1.8°C + 32   °R(Rankine) = °F + 459.69
°C = 5/9(°F − 32)   K(kelvins) = °C + 273.16

CONCENTRATION
1 grain/U.S. gallon	0.017118 g/liter
1 grain/U.S. gallon (in ppm)	17.118 divided by the density in g/cc
1 gram/liter	58.417 grains/gal
1 gram/liter (in ppm)	1000 divided by the density in g/cc

APPENDIX 7

TABLE A7-1

Letter and computer symbols for well logging and formation evaluation (from: Society of Petroleum Engineers and Society of Professional Well Log Analysts (1975))

Quantity	Letter, symbol	Computer symbol			Reserve SPE letter symbols	Dimensions *
		Operator field	Quantity symbol field	Subscript field		
acoustic velocity	v	VAC			V, u	m s^{-1}
acoustic velocity, apparent (measured)	v_a	VAC	A		V_a, u_a	m s^{-1}
acoustic velocity, fluid	v_f	VAC	F		V_f, u_f	m s^{-1}
acoustic velocity, matrix	v_{ma}	VAC	MA		V_{ma}, u_{ma}	m s^{-1}
acoustic velocity, shale	v_{sh}	VAC	SH		V_{sh}, u_{sh}	m s^{-1}
activity	a	ACT				
amplitude	A	AMP				various
amplitude, compressional wave	A_c	AMP	C			various
amplitude, relative	A_r	AMP	R			various
amplitude, shear wave	A_s	AMP	S			various
angle	α (alpha)	ANG			β (beta)	
	θ (theta)	AGL			γ (gamma)	
angle of dip	Θ (theta$_{cap}$)	ANG	D		α_d (alpha)	
anisotropy coefficient	K_{ani}	COE	ANI		M_{ani}	
area	A	ARA			S	m^2
atomic number	Z	ANM				
atomic weight	A	AWT				kg
attenuation coefficient	α (alpha)	COE	A		M_α	m^{-1}
azimuth of dip	Φ (phi$_{cap}$)	DAZ			β_d (beta)	
azimuth of reference on sonde	μ (mu)	RAZ			M (mu$_{cap}$)	
bearing, relative	β (beta)	BRG	R		γ (gamma)	
bottom-hole pressure	p_{bh}	PRS	BH		P_{BH}	$\text{kg m}^{-1}\text{ s}^{-2}$
bottom-hole temperature	T_{bh}	TEM	BH		θ_{BH} (theta)	K
bubble-point (saturation) pressure	p_b	PRS	B		p_s, P_b, P_s	$\text{kg m}^{-1}\text{ s}^{-2}$
bulk modulus	K	BKM			K_b	$\text{kg m}^{-1}\text{ s}^{-2}$
bulk volume	V_b	VOL	B		v_b	m^3
capacitance	C	ECQ				$\text{A}^2\text{ s}^4\text{ kg}^{-1}\text{ m}^{-2}$
capillary pressure	P_c	PRS	CP		P_C, p_C	$\text{kg m}^{-1}\text{ s}^{-2}$
cementation (porosity) exponent	m	MXP				
charge	Q	CHG			q	A s
coefficient, anisotropy	K_{ani}	COE	ANI		M_{ani}	
coefficient, attenuation	α (alpha)	COE	A		M_α	m^{-1}
coefficient, electrochemical	K_c	COE	C		M_c, K_{ec}	$\text{kg m}^2\text{ s}^{-3}\text{ A}^{-1}$
coefficient, formation resistivity factor	K_R	COE	R		M_R, a, C	
coefficient or multiplier	K	COE			M	various
compressibility	c	CMP			k, κ (kappa)	$\text{m kg}^{-1}\text{ s}^2$
concentration (salinity)	C	CNC			c, n	various
conductivity, apparent	C_a	ECN	A		γ_a (gamma)	$\text{m}^{-3}\text{ kg}^{-1}\text{ s}^3\text{ A}^2$
conductivity, electric	C	ECN			γ (gamma)	$\text{m}^{-3}\text{ kg}^{-1}\text{ s}^3\text{ A}^2$
conductivity, thermal	k_h	HCN			λ (lambda)	$\text{m kg s}^{-3}\text{ K}^{-1}$
constant, decay $(1/\tau_d)$	λ (lambda)	LAM			C	s^{-1}
constant, dielectric	ϵ (epsilon)	DIC				$\text{A}^2\text{ s}^4\text{ kg}^{-1}\text{ m}^{-3}$
correction term or correction factor (either additive or multiplicative)	B	COR			C	
cross-section (area)	A	ARA			S	m^2
cross-section, macroscopic	Σ (sigma$_{cap}$)	XST			S	m^{-1}
cross-section of a nucleus, microscopic	σ (sigma)	XNL			s	m^2
current, electric	I	CUR			i, ι (script i)	A
decay constant $(1/\tau_d)$	λ (lambda)	LAM			C	s^{-1}
decay time $(1/\lambda)$	τ_d (tau)	TIM	D		t_d	s
decrement	δ (delta)	DCR			Δ (delta$_{cap}$)	various
density	ρ (rho)	DEN			D	kg m^{-3}

TABLE A7-1 (continued)

Quantity	Letter, symbol	Computer symbol			Reserve SPE letter symbols	Dimensions *
		Operator field	Quantity symbol field	Subscript field		
density, apparent	ρ_a (rho)		DEN	A	D_a	kg m^{-3}
density, bulk	ρ_b (rho)		DEN	B	D_b	kg m^{-3}
density, fluid	ρ_f (rho)		DEN	F	D_f	kg m^{-3}
density, flushed zone	ρ_{xo} (rho)		DEN	XO	D_{xo}	kg m^{-3}
density (indicating "number per unit volume")	n		NMB		N	m^{-3}
density, matrix [1]	ρ_{ma} (rho)		DEN	MA	D_{ma}	kg m^{-3}
density (number) of neutrons	n_N		NMB	N		m^{-3}
density, true	ρ_t (rho)		DEN	T	D_t	kg m^{-3}
depth	D		DPH		γ, H	m
depth, skin	δ (delta)		SKD		r_s	m
deviation, hole	δ (delta)		ANG	H		m
dew-point pressure	p_d		PRS	D	P_d	kg m^{-1} s^{-2}
diameter	d		DIA		D	m
diameter, hole	d_h		DIA	H	d_H, D_h	m
diameter, invaded zone (electrically equivalent)	d_i		DIA	I	d_I, D_i	m
dielectric constant	ϵ (epsilon)		DIC			A^2 s^4 kg^{-1} m^{-3}
difference	Δ (delta$_{cap}$)	DEL				[X]
dip, angle of	Θ (theta$_{cap}$)		ANG	D	α_d (alpha)	
dip, apparent angle of	θ (theta)		ANG	DA	α_{da} (alpha)	
dip, apparent azimuth of	ϕ (phi)		DAZ	A	β_{da} (beta)	
dip, azimuth of	Φ (phi$_{cap}$)		DAZ		β_d (beta)	
distance, length, or length of path	L		LTH		s, l	m
distance, radial (increment along radius)	Δr	DEL	RAD		ΔR	m
drift angle, hole (deviation)	δ (delta)		ANG	H		
electric current	I		CUR		i, ι (script i)	A
electrochemical coefficient	K_c		COE	C	M_c	kg m^2 s^{-3} A^{-1}
electrochemical component of the SP	E_c		EMF	C	Φ_c (phi$_{cap}$)	kg m^2 s^{-3} A^{-1}
electrokinetic component of the SP	E_k		EMF	K	Φ_k (phi$_{cap}$)	kg m^2 s^{-3} A^{-1}
electromotive force	E		EMF		V	kg m^2 s^{-3} A^{-1}
energy	E		ENG		U	kg m^2 s^{-2}
exponent, porosity (cementation)	m		MXP			
exponent, saturation	n		SXP			
factor	F		FAC			various
flow rate, heat	$\dot{Q}$		HRT		q, Φ (phi$_{cap}$)	kg m^2 s^{-3}
force, electromotive	E		EMF		V	kg m^2 s^{-3} A^{-1}
force, mechanical	F		FCE		Q	kg m s^{-2}
formation resistivity factor	F_R		FAC	HR		
formation resistivity factor coefficient ($F_R \phi^m$)	K_R		COE	R	M_R, a, C	
formation volume factor	B		FVF		F	
fraction	f		FRC		F	
fraction of bulk (total) volume	f_V		FRC	VB	f_{Vb}, V_{bf}	
fraction of intergranular space ("porosity") occupied by all shales	$f_{\phi sh}$		FIG	SH	ϕ_{igfsh} (phi)	
fraction of intergranular space ("porosity") occupied by water	$f_{\phi w}$		FIG	W	ϕ_{igfw} (phi)	
fraction of intermatrix space ("porosity") occupied by nonstructural dispersed shale	$f_{\phi shd}$		FIM	SHD	ϕ_{imfshd} (phi), q	
fracture index	I_f		FRX		i_f, I_F, i_F	
free fluid index	I_{Ff}		FFX		i_{Ff}	
frequency	f		FQN		ν (nu)	s^{-1}
gamma ray count rate	N_{GR}		NGR		N_γ, C_G	s^{-1}
gamma ray [usually with identifying subscript(s)]	γ (gamma)		GRY			various
gas compressibility factor	z		ZED		Z	
gas constant, universal (per mole)	R		RRR			kg m^2 s^{-2} K
gas-oil ratio, producing	R		GOR		F_g, F_{go}	
gas specific gravity	γ_g (gamma)		SPG	G	s_g, F_{sg}	

TABLE A7-1 (continued)

Quantity	Letter, symbol	Computer symbol			Reserve —SPE letter symbols	Dimensions *
		Operator field	Quantity symbol field	Subscript field		
geometrical fraction (multiplier or factor)	G		GMF		f_G	
geometrical fraction (multiplier or factor), annulus	G_{an}		GMF	AN	f_{Gan}	
geometrical fraction (multiplier or factor), flushed zone	G_{xo}		GMF	XO	f_{Gxo}	
geometrical fraction (multiplier or factor), invaded zone	G_i		GMF	I	f_{Gi}	
geometrical fraction (multiplier or factor), mud	G_m		GMF	M	f_{Gm}	
geometrical fraction (multiplier or factor), pseudo-	G_p		GMF	P	f_J	
geometrical fraction (multiplier or factor), true	G_t		GMF	T	f_{Gtr}	
gradient	g		GRD		γ (gamma)	various
gradient, geothermal	g_G		GRD	GT	g_g	K m^{-1}
gradient, temperature	g_T		GRD	T	g_h	K m^{-1}
gravity, specific	γ (gamma)		SPG		s, F_s	
half life	$t_{1/2}$		TIM	H		s
heat flow rate	$\dot{Q}$		HRT		q, Φ (phi$_{cap}$)	kg m^2 s^{-3}
height, or fluid head	Z		ZEL		D, h	m
hold-up (fraction of the pipe volume filled by a given fluid: y_0 is oil hold-up, y_w is water hold-up; sum of all the hold-ups at a given level is one)	y		HOL		f	
hydrocarbon resistivity index	I_R		RSX	H	i_R	
hydrogen index	I_H		HYX		i_H	
impedance, acoustic	Z_a		MPD	A		kg m^{-2} s^{-1}
impedance, electric	Z_e		MPD	E	Z_E, η (eta)	kg m^2 s^{-3} A^{-2}
index, fracture	I_f		FRX		i_f, I_F, i_F	
index, free fluid	I_{Ff}		FFX		i_{Ff}	
index, (hydrocarbon) resistivity	I_R		RSX	H	i_R	
index, hydrogen	I_H		HYX		i_H	
index, porosity	I_ϕ		PRX		i_ϕ	
index, primary porosity	$I_{\phi 1}$		PRX	PR	$i_{\phi 1}$	
index, secondary porosity	$I_{\phi 2}$		PRX	SE	$i_{\phi 2}$	
index, shaliness gamma-ray $(\gamma_{LOG} - \gamma_{cn})/(\gamma_{sh} - \gamma_{cn})$	I_{shGR}		SHX	GR	i_{shGR}	
index (use subscripts as needed)	I		–X		i	
injection rate	i		INJ			m^3 s^{-1}
interfacial tension	σ (sigma)		SFT		y, γ (gamma)	kg s^{-2}
intergranular "porosity" (space) $(V_b - V_{gr})/V_b$	ϕ_{ig} (phi)		POR	IG	f_{ig}, ϵ_{ig} (epsilon)	
intermatrix "porosity" (space) $(V_b - V_{ma})/V_b$	ϕ_{im} (phi)		POR	IM	f_{im}, ϵ_{im} (epsilon)	
interval transit time	ℓ (script t)		TAC		Δt	s m^{-1}
interval transit time, apparent	ℓ_a (script t)		TAC	A	Δt_a	s m^{-1}
interval transit time-density slope (absolute value)	M		SAD		$m_{\theta D}$	s m^2 kg^{-1}
interval transit time, fluid	ℓ_f (script t)		TAC	F	Δt_f	s m^{-1}
interval transit time, matrix	ℓ_{ma} (script t)		TAC	MA	Δt_{ma}	s m^{-1}
interval transit time, shale	ℓ_{sh} (script t)		TAC	SH	Δt_{sh}	s m^{-1}
length, path length, or distance	L		LTH		s, l	m
lifetime, average (mean life)	$\bar{\tau}$ (tau)		TIM	AV	$\bar{t}$	s
macroscopic cross-section	Σ (sigma$_{cap}$)		XST	MAC	S	m^{-1}
magnetic permeability	μ (mu)		PRM	M	m	kg m s^{-2} A^{-2}
magnetic susceptibility	k		SUS	M	κ (kappa)	kg m s^{-2} A^{-2}
magnetization	M		MAG		I	kg s^{-2} A^{-1}
magnetization, fraction	M_f		MAG	F		
mass	m		MAS			kg

TABLE A7-1 (continued)

Quantity	Letter, symbol	Computer symbol			Reserve SPE letter symbols	Dimensions *
		Operator field	Quantity symbol field	Subscript field		
mass flow rate	w		MRT		m	kg s^{-1}
matrix (framework) volume (volume of all formation solids except nonstructural clay or shale)	V_{ma}		VOL	MA		m^3
mean life	$\bar{\tau}$ (tau)		TIM	AV	$\bar{t}$	s
microscopic cross-section	σ (sigma)		XST	MIC		m^2
modulus, bulk	K		BKM		K_b	kg m^{-1} s^{-2}
modulus of elasticity (Young's modulus)	E		ELM	Y	Y	kg m^{-1} s^{-2}
modulus, shear	G		ELM	S	E_s	kg m^{-1} s^{-2}
multiplier (fraction), geometrical	G		GMF		f_G	
multiplier (fraction), geometrical, invaded zone	G_i		GMF	I	f_{Gi}	
multiplier (fraction), geometrical, true	G_t		GMF	T	f_{Gtr}	
multiplier or coefficient	K		COE		M	various
neutron count rate	N_N		NEU	N	N_n, C_N	s^{-1}
neutron lifetime	t_N		NLF		τ_N (tau), t_n	s
neutron porosity-density slope (absolute value)	N		SND		$m_{\phi ND}$	m^3 kg^{-1}
neutron [usually with identifying subscript(s)]	N		NEU			various
nucleus cross-section	σ (sigma)		XNL		s	m^2
number, atomic	Z		ANM			
number, dimensionless, in general (always with identifying subscripts) [2]	N		NUM	Q		
number of moles, total (see also moles, number of, Supplement II)	n		MOL		n_t, N_t	
number (quantity)	n		NMB		N	
oil specific gravity	γ_o (gamma)		SPG	O	s_o, F_{so}	
period	T		PER		Θ (theta$_{cap}$)	s
permeability, absolute (fluid flow)	k		PRM		K	m^2
permeability, magnetic	μ (mu)		PRM	M	m	kg m A^{-2} s^{-2}
Poisson's ratio	μ (mu)		PSN		ν (nu), σ (sigma)	
porosity $(V_b - V_s)/V_b$	ϕ (phi)		POR		f, ϵ (epsilon)	
porosity, apparent	ϕ_a (phi)		POR	A	f_a, ϵ_a (epsilon)	
porosity (cementation) exponent	m		MXP			
porosity, effective (interconnected) (V_{pe}/V_b)	ϕ_e (phi)		POR	E	f_e, ϵ_e (epsilon)	
porosity index	I_ϕ		PRX		i_ϕ	
porosity index, primary	$I_{\phi 1}$		PRX	PR	$i_{\phi 1}$	
porosity index, secondary	$I_{\phi 2}$		PRX	SE	$i_{\phi 2}$	
porosity, noneffective (noninterconnected) (V_{pne}/V_b)	ϕ_{ne} (phi)		POR	NE	f_{ne}, ϵ_{ne} (epsilon)	
"porosity" (space), intergranular $(V_b - V_{gr})/V_b$	ϕ_{ig} (phi)		POR	IG	f_{ig}, ϵ_{ig} (epsilon)	
"porosity" (space), intermatrix $(V_b - V_{ma})/V_b$	ϕ_{im} (phi)		POR	IM	f_{im}, ϵ_{im} (epsilon)	
porosity, total	ϕ_t (phi)		POR	T	f_t, ϵ_t (epsilon)	
potential difference (electric)	V		VLT		U	kg m^2 s^{-3} A^{-1}
pressure	p		PRS		P	kg m^{-1} s^{-2}
primary porosity index	$I_{\phi 1}$		PRX	PR	$i_{\phi 1}$	
production rate or flow rate	q		RTE		Q	m^3 s^{-1}
productivity index	J		PDX		j	m^4 s kg^{-1}
pseudo-geometrical fraction (multiplier or factor)	G_p		GMF	P	f_J	
pseudo-SP	E_{pSP}		EMF	P	Φ_{sp} (phi$_{cap}$)	kg m^2 s^{-3} A^{-1}
radial distance (increment along radius)	Δr	DEL	RAD		ΔR	m
radius	r		RAD		R	m
reactance	X		XEL			kg m^2 s^{-3} A^{-2}
reduction or reduction term	α (alpha)		RED			
reduction, SP, due to shaliness	α_{SP} (alpha)		RED	SP		
relative amplitude	A_r		AMP	R		
relaxation time, free-precession decay	t_2		TIM	AV	τ_2 (tau)	s

TABLE A7-1 (continued)

Quantity	Letter, symbol	Computer symbol			Reserve SPE letter symbols	Dimensions *
		Operator field	Quantity symbol field	Subscript field		
relaxation time, proton thermal	t_1		TIM	RP	τ_1 (tau)	s
resistance	r		RST		R	$kg\ m^2\ s^{-3}\ A^{-2}$
resistivity	R		RES		ρ (rho), r	$kg\ m^3\ s^{-3}\ A^{-2}$
resistivity, annulus	R_{an}		RES	AN	ρ_{an} (rho), r_{an}	$kg\ m^3\ s^{-3}\ A^{-2}$
resistivity, apparent	R_a		RES	A	ρ_a (rho), r_a	$kg\ m^3\ s^{-3}\ A^{-2}$
resistivity, apparent, of the conductive liquids mixed in invaded zone	R_z		RES	Z	ρ_z (rho), r_z	$kg\ m^3\ s^{-3}\ A^{-2}$
resistivity factor coefficient, formation ($F_R\phi^m$)	K_R		COE	R	M_R, a, C	
resistivity factor, formation	F_R		FAC	HR		
resistivity, flushed zone (that part of the invaded zone closest to the wall of the hole, where flushing has been maximum)	R_{xo}		RES	XO	ρ_{xo} (rho), r_{x0}	$kg\ m^3\ s^{-3}\ A^{-2}$
resistivity, formation 100 percent saturated with water of resistivity R_w (zero hydrocarbon saturation)	R_0		RES	ZR	ρ_0 (rho), r_0	$kg\ m^3\ s^{-3}\ A^{-2}$
resistivity, formation, true	R_t		RES	T	ρ_t (rho), r_t	$kg\ m^3\ s^{-3}\ A^{-2}$
resistivity index, hydrocarbon	I_R		RSX	H	i_R	
resistivity, invaded zone	R_i		RES	I	ρ_i (rho), r_i	$kg\ m^3\ s^{-3}\ A^{-2}$
resistivity, mud	R_m		RES	M	ρ_m (rho), r_m	$kg\ m^3\ s^{-3}\ A^{-2}$
resistivity, mud-cake	R_{mc}		RES	MC	ρ_{mc} (rho), r_{mc}	$kg\ m^3\ s^{-3}\ A^{-2}$
resistivity, mud-filtrate	R_{mf}		RES	MF	ρ_{mf} (rho), r_{mf}	$kg\ m^3\ s^{-3}\ A^{-2}$
resistivity, shale	R_{sh}		RES	SH	ρ_{sh} (rho), r_{sh}	$kg\ m^3\ s^{-3}\ A^{-2}$
resistivity, surrounding formation	R_s		RES	S	ρ_s (rho), r_s	$kg\ m^3\ s^{-3}\ A^{-2}$
resistivity, water	R_w		RES	W	ρ_w (rho), r_w	$kg\ m^3\ s^{-3}\ A^{-2}$
Reynolds number (dimensionless number)	N_{Re}		REY	Q		
salinity	C		CNC		c, n	various
saturation	S		SAT		ρ (rho), s	
saturation exponent	n		SXP			
saturation, gas	S_g		SAT	G	ρ_g (rho), s_g	
saturation, hydrocarbon	S_h		SAT	H	ρ_h (rho), s_h	
saturation, oil	S_o		SAT	O	ρ_o (rho), s_o	
saturation, oil, residual	S_{or}		SAT	OR	ρ_{or} (rho), s_{or}	
saturation, water	S_w		SAT	W	ρ_w (rho), s_w	
secondary porosity index	$I_{\phi2}$		PRX	SE	$i_{\phi2}$	
self potential (see SP quantities)						
shaliness gamma-ray index ($(\gamma_{LOG} - \gamma_{cn})/(\gamma_{sh} - \gamma_{cn})$)	I_{shGR}		SHX	G	i_{shGR}	
shear modulus	G		ELM	S	E_s	$kg\ m^{-1}\ s^{-2}$
skin depth (logging)	δ (delta)		SKD		r_s	m
slope, interval transit time vs density (absolute value)	M		SAD		$m_{\theta D}$	$s\ m^2\ kg^{-1}$
slope, neutron porosity vs density (absolute value)	N		SND		$m_{\phi ND}$	$m^3\ kg^{-1}$
solid(s) volume (volume of all formation solids)	V_s		VOL	S	v_s	m^3
SP (measured SP)	E_{SP}		EMF	SP	Φ_{SP} (phi$_{cap}$)	$kg\ m^3\ s^{-3}\ A^{-1}$
SP, pseudo-	E_{pSP}		EMF	PSP	Φ_{pSP} (phi$_{cap}$)	$kg\ m^2\ s^{-3}\ A^{-1}$
SP reduction due to shaliness	α_{SP} (alpha)		RED	SP		
SP, static (SSP)	E_{SSP}		EMF	SSP	Φ_{SSP} (phi$_{cap}$)	$kg\ m^2\ s^{-3}\ A^{-1}$
spacing	L_s		LEN	S	s_s, l_s	m
specific gravity	γ (gamma)		SPG		s, F_s	
spontaneous potentials (see SP quantities)						
SSP (static SP)	E_{SSP}		EMF	SSP	Φ_{SSP} (phi$_{cap}$)	$kg\ m^2\ s^{-3}\ A^{-1}$
summation (operator)	Σ (sigma$_{cap}$)	SUM				
superficial phase velocity (flux rate of a particular fluid phase flowing in pipe; use appropriate phase subscripts)	u		VEL	V	ψ (psi)	$m\ s^{-1}$
surface production rate	q_{sc}		RTE	SC	q_σ, Q_{sc}	$m^3\ s^{-1}$
susceptibility, magnetic	k		SUS	M	κ (kappa)	$kg\ m\ A^{-2}\ s^{-2}$
temperature	T		TEM		θ (theta)	K

TABLE A7-1 (continued)

Quantity	Letter, symbol	Computer symbol			Reserve SPE letter symbols	Dimensions *
		Operator field	Quantity symbol field	Subscript field		
temperature gradient	g_T		TEM	GR	g_h	K m^{-1}
thermal conductivity	k_h		HCN		λ (lambda)	$\text{kg m s}^{-3}\,\text{K}^{-1}$
thickness	h		THK		$d.\,e$	m
thickness, mud-cake	h_{mc}		THK	MC	d_{mc}, e_{mc}	m
thickness, pay, gross (total)	h_t		THK	T	d_t, e_t	m
thickness, pay, net	h_n		THK	N	d_n, e_n	m
time	t		TIM		τ (tau)	s
time constant	τ (tau)		TIM	C	τ_c (tau)	s
time, decay ($1/\lambda$)	τ_d (tau)		TIM	D	t_d	s
time difference	Δt	DEL	TIM		$\Delta\tau$ (tau)	s
time, interval transit	t (script t)		TAC		Δt	s m^{-1}
tortuosity	τ (tau)		TOR			
tortuosity, electric	τ_e (tau)		TOR	E		
tortuosity, hydraulic	τ_H (tau)		TOR	HL		
transit time, interval	t (script t)		TAC		Δt	s m^{-1}
valence	z		VAL			
velocity	v		VEL		V, u	m s^{-1}
velocity, acoustic	v		VAC		V, u	m s^{-1}
viscosity (dynamic)	μ (mu)		VIS		η (eta)	$\text{m}^{-1}\text{ kg s}^{-1}$
volume	V		VOL		v	m^3
volume, bulk	V_b		VOL	B	v_b	m^3
volume fraction or ratio (as needed, use same subscripted symbols as for "volumes"; note that bulk volume fraction is unity and pore volume fractions are ϕ)	V		VLF		f_V, F_V	various
volume, grain (volume of all formation solids except shales)	V_{gr}		VOL	GR	v_{gr}	m^3
volume, intergranular (volume between grains; consists of fluids and all shales) ($V_b - V_{gr}$)	V_{ig}		VOL	IG	v_{ig}	m^3
volume, intermatrix (consists of fluids and dispersed shale) ($V_b - V_{ma}$)	V_{im}		VOL	IM	v_{im}	m^3
volume, matrix (framework) (volume of all formation solids except dispersed shale)	V_{ma}		VOL	MA	v_{ma}	m^3
volume, pore ($V_b - V_s$)	V_p		VOL	P	v_p	m^3
volume, pore, effective (interconnected) (interconnected pore space)	V_e		VOL	E	V_{pe}, v_e	m^3
volume, pore noneffective (noninterconnected) (noninterconnected pore space) ($V_p - V_e$)	V_{ne}		VOL	NE	V_{pne}, v_{ne}	m^3
volume, shale, dispersed	V_{shd}		VOL	SHD	v_{shd}	m^3
volume, shale, laminated	V_{shl}		VSH	LAM	v_{shl}	m^3
volume, shale, structural	V_{shst}		VOL	SHS	v_{shs}	m^3
volume, shale(s) (volume of all shales: structural and dispersed)	V_{sh}		VOL	SH	v_{sh}	m^3
volume, solid(s) (volume of *all* formation solids)	V_s		VOL	S	v_s	m^3
volumetric flow rate	q		RTE		Q	$\text{m}^3\,\text{s}^{-1}$
volumetric flow rate down hole	q_{dh}		RTE	DH	q_{wf}, q_{DH}, Q_{dh}	$\text{m}^3\,\text{s}^{-1}$
volumetric flow rate, surface conditions	q_{sc}		RTE	SC	q_σ, Q_{sc}	$\text{m}^3\,\text{s}^{-1}$
wavelength	λ (lambda)		WVL			m
weight, atomic	A		AWT			kg
weight (gravitational)	W		WGT		w, G	kg m s^{-2}
work	W		WRK		w	$\text{kg m}^2\,\text{s}^{-2}$
Young's modulus	E		ELM	Y	Y	$\text{kg m}^{-1}\,\text{s}^{-2}$

* Dimensions: m = length, kg = mass, A = electric current, s = time, K = temperature.

[1] See subscript definitions of "matrix" and "solid(s)".

[2] Dimensionless numbers are criteria for geometric, kinematic and dynamic similarity between two systems. They are derived by one of three procedures used in methods of similarity: integral, differential, or dimensional. Examples of dimensionless numbers are Reynolds number (N_{Re}) and Prandtl number (N_{Pr}). For a discussion of methods of similarity and dimensionless numbers, see "Methods of Similarity," by R.E. Schilson, J. Pet. Tech. (Aug. 1964) 877.

TABLE A7-2

Subscripts in alphabetical order

Subscript definition	Letter subscript	Computer subscript Subscript field	Reserve SPE letter subscripts
acoustic	a	A	A, a alpha
amplitude log (derived from or given by)	A	A	a
anhydrite	anh	AH	
anisotropic	ani	ANI	
annulus	an	AN	AN
apparent (from log readings: use tool-description subscripts)			
apparent (general)	a	A	ap
borehole televiewer log (derived from or given by)	TV	TV	tv
bottom hole	bh	BH	w, BH
bulk	b	B	B, t
calculated	C	CA	calc
caliper log (derived from or given by)	C	C	c
capillary	c	CP	C
capture	cap	C	
casing	c	CS	cg
cement bond log (derived from or given by)	CB	CB	cb
chlorine log (derived from or given by)	CL	CL	cl
clay	cl	CL	cla
clean	cn	CN	cln
coil	C	C	c
compaction	cp	CP	
compensated density log (derived from or given by)	CD	CD	cd
compensated neutron log (derived from or given by)	CN	CN	cn
compressional wave	c	C	C
conductive liquids in invaded zone	z	Z	
contact log, microlog, minilog (derived from or given by)	ML	ML	ml
corrected	cor	COR	
critical	c	CR	cr
decay	d	D	
deep induction log (derived from or given by)	ID	ID	id
deep laterolog (derived from or given by)	LLD	LLD	llD
density log, compensated (derived from or given by)	CD	CD	cd
density log (derived from or given by)	D	D	d
differential temperature log (derived from or given by)	DT	DT	dt
dip	d	D	
diplog, dipmeter (derived from or given by)	DM	DM	dm
directional survey (derived from or given by)	DR	DR	dr
dirty (clayey, shaly)	dy	DY	dty
dolomite	dol	DL	
down-hole	dh	DH	DH
dual induction log (derived from or given by)	DI	DI	di
dual laterolog (derived from or given by)	DLL	DLL	d
effective	e	E	
electric, electrical	e	E	E
electrochemical	c	C	ec
electrode	E	E	e
electrokinetic	k	K	ek
electrolog, electrical log, electrical survey (derived from or given by)	EL	EL	el, ES
electron	el	E	$\ell\,\ell$ (script)
empirical	E	EM	EM
epithermal neutron log (derived from or given by)	NE	NE	ne
equivalent	eq	EV	EV
experimental	E	EX	EX
external, outer	e	E	o
fast neutron log (derived from or given by)	NF	NF	nf
fluid	f	F	fl
flushed zone	xo	XO	
formation (rock)	f	F	fm
formation, surrounding	s	S	
fraction	f	F	r

402

TABLE A7-2 (continued)

Subscript definition	Letter subscript	Computer subscript Subscript field	Reserve SPE letter subscripts
fracture	f	FR	F
free	F	F	f
free fluid	Ff	FF	
gamma-gamma ray log (derived from or given by)	GG	GG	gg
gamma ray log (derived from or given by)	GR	GR	gr
gas	g	G	G
geometrical	G	G	
geothermal	G	GT	T
grain	gr	GR	
gross (total)	t	T	T
gypsum	gyp	GY	
half	$1/2$	H	
heat or thermal	h	HT	T, θ (theta)
heavy phase	HP	HP	hp
hole	h	H	H
hydraulic	H	HL	
hydrocarbon	h	H	H
hydrogen nuclei or atoms	H	HY	
induction log, deep investigation (derived from or given by)	ID	ID	id
induction log (derived from or given by)	I	I	i
induction log, medium investigation (derived from or given by)	IM	IM	im
inner	i	I	ι (iota), ι (script i)
intergranular	ig	IG	
intermatrix	im	IM	
internal	i	I	ι (iota), ϑ (script i)
intrinsic	int	I	
invaded zone	i	I	I
irreducible	i	IR	ir, ϑ (script i), ι (iota)
junction	j	J	
laminar	l	LAM	L
laminated, lamination	l	LAM	L
lateral (resistivity) log (derived from or given by)	L	L	l
laterolog (derived from or given by)	LL	LL	ll
	LL_3	LL_3	ll_3
	LL_7	LL_7	ll_7
	LL_8	LL_8	ll_8
	LLD	LLD	llD
	LLS	LLS	llS
	etc.	etc.	etc.
light phase	LP	LP	lp
limestone	ls	LS	lst
limiting value	lim	LM	
liquid	L	L	l
liquids, conductive, in invaded zone	z	Z	
log (derived from or given by)	LOG	L	log
lower	l	L	L
matrix [solids except dispersed (nonstructural) clay or shale]	ma	MA	
maximum	max	MX	
medium investigation induction log (derived from or given by)	IM	IM	im
microlaterolog (derived from or given by)	MLL	MLL	mll
microlog, minilog, contact log (derived from or given by)	ML	ML	ml
micro-seismogram log, signature log, variable density log (derived from or given by)	VD	VD	vd
microspherically focused log (derived from or given by)	$MSFL$	MSFL	
minimum	min	MN	
mixture	M	M	z, m
mud	m	M	
mud cake	mc	MC	
mud filtrate	mf	MF	
net	n	N	

TABLE A7-2 (continued)

Subscript definition	Letter subscript	Computer subscript Subscript field	Reserve SPE letter subscripts
neutron	N	N	n
neutron activation log (derived from or given by)	NA	NA	na
neutron lifetime log, TDT (derived from or given by)	NL	NL	nl
neutron log (derived from or given by)	N	N	n
noneffective	ne	NE	
normal (resistivity) log (derived from or given by)	N	N	n
nuclear magnetism log (derived from or given by)	NM	NM	nm
oil	o	O	N
outer (external)	e	E	o
pipe inspection log, electromagnetic (derived from or given by)	EP	EP	ep
pore	p	P	P
porosity	ϕ (phi)	PHI	f, ϵ (epsilon)
porosity data (derived from or given by)	ϕ (phi)	P	f, ϵ (epsilon)
primary	1 (one)	PR	p, pri
proximity log (derived from or given by)	P	P	p
pseudo	p	P	
pseudo-critical	pc	PC	
pseudo-reduced	pr	PRD	
pseudo-SP	pSP	PSP	
radial	r	R	R
reduced	r	RD	
relative	r	R	R
residual	r	R	R
resistivity	R	R	
resistivity log (derived from or given by)	R	R	r, ρ (rho)
Reynolds	Re	–	
sand	sd	SD	sa
sandstone	ss	SS	sst
scattered, scattering	sc	SC	
secondary	2	SE	s, sec
shale	sh	SH	sha
shallow laterolog (derived from or given by)	LLS	LLS	llS
shear wave	s	S	τ (tau)
sidewall	S	SW	SW
sidewall neutron log (derived from or given by)	SN	SN	sn
signature log, micro-seismogram log, variable density log (derived from or given by)	VD	VD	vd
silt	sl	SL	slt
skin	s	S	S
slip or slippage	s	S	σ (sigma)
slurry ("mixture")	M	M	z, m
solid(s) (*all* formation solids)	s	S	σ (sigma)
sonde, tool	T	T	t
sonic velocity log (derived from or given by)	SV	SV	sv
SP (derived from or given by)	SP	SP	sp
spacing	s	L	
spherically focused log (derived from or given by)	SFL	SFL	
SSP	SSP	SSP	
standard conditions	sc	SC	σ (sigma)
static or shut-in conditions	ws	WS	s
stock-tank conditions	st	ST	
structural	st	ST	s
surrounding formation	s	S	
TDT log, neutron lifetime log (derived from or given by)	NL	NL	nl
temperature	T	T	h, θ (theta)
temperature log (derived from or given by)	T	T	t, h
temperature log, differential (derived from or given by)	DT	DT	dt
thermal	h	HT	T, θ (theta)
thermal decay time (TDT) log (derived from or given by)	NL	NL	nl
thermal neutron log (derived from or given by)	NT	NT	nt
tool-description subscripts: see individual entries, such as "amplitude log", "neutron log", etc. *			

TABLE A7-2 (continued)

Subscript definition	Letter subscript	Computer subscript Subscript field	Reserve SPE letter subscripts
tool, sonde	T	T	t
total (gross)	t	T	T
true (opposed to apparent)	t	T	tr
upper	u	U	U
variable density log, micro-seismogram log, signature log (derived from or given by)	VD	VD	vd
volume or volumetric	V	V	v
water	w	W	W
water-saturated formation, 100 percent (*zero* hydrocarbon saturation)	0 (zero)	ZR	zr
weight	W	W	w
well flowing conditions	wf	WF	f
well static conditions	ws	WS	s
zero hydrocarbon saturation	0 (zero)	ZR	zr

* If service-company identification is needed for a tool, it is recommended that the appropriate following capital letter be added to the tool-description subscript: Birdwell, B; Dresser Atlas, D; GO International, G; Lane-Wells, L; PGAC, P; Schlumberger, S; Welex, W.

TABLE A7-3

Symbols

Traditional symbols	Standard SPE of AIME and SPWLA [a]	Standard computer symbols [a]	Description	Customary units or relation	Standard reserve symbols [b]
A	A	AMP	amplitude		γ (gamma)
A	A	AWT	atomic weight	atomic mass units (amu)	
C	C	ECN	conductivity, electric	millimhos per meter (mmhos/m)	C_{cp}
C_p	B_{cp}	CORCP	sonic compaction correction factor	$\phi_{SVcor} = B_{cp}\phi SV$	
C		CNC	concentration (salinity)		
C			electronic density coefficient	$C = 2Z/A$	z, H
D	D	DPH	depth	feet (ft) or meters (m)	
E	E	ELMY	Young's modulus		
E	E	EMF	electromotive force	millivolts (mV)	V
F	F_R	FACHR	formation resistivity factor	$F_R = K_R/\phi^m$	
G	G	ELMS	shear modulus		
G		GMF	geometrical factor (function of d_i)		f_G
H	I_H	HYX	hydrogen index		i_H
I	I	X	index		i
FFI	I_{Ff}	FFX	free fluid index		i_{Ff}
SI	I_{sl}	SLX	silt index		I_{slt}, i_{sl}, i_{slt}
	I_ϕ	PRX	porosity index		i_ϕ
SPI	$I_{\phi2}$	PRXSE	secondary porosity index		$i_{\phi2}$
J	G_p	GMFP	pseudo geometrical factor		f_J
K	K_c	COEC	electrochemical SP coefficient	$E_c = K_c \log(a_w/a_{mf})$	M_c, K_{cc}
K	K	BKM	bulk modulus	$M = K + \tfrac{4}{3}\mu$	
M	M		space modulus		
M	M	SAD	slope, sonic interval transit time vs Density$\times$0.01, in M-N Plot	$M = [(t_f - t_{LOG})/(\rho_b - \rho_1)]\times0.01$	$m_{\theta D}$
N	N	NUMQ	number, dimensionless		
N	N	SND	slope, Neutron porosity vs Density, in M-N Plot	$N = (\phi_{Nf} - \phi_N)/(\rho_b - \rho_f)$	$m_{\phi ND}$
C	C	NEU	neutron		
P	P	CNC	salinity	grams per gram, parts per million	c, n
p	p	PRS	pressure	pounds/sq. inch (psi), kilograms per sq cm [c], atmospheres	P
P_c	P_c	PRSCP	capillary pressure	(same units as for "pressure")	P_c, p_c
Q_v			shaliness (CEC per ml water)	milliequivalents per milliliter	
R	R	RES	resistivity	ohm-meters (ohm-m)	ρ (rho), r
S	S	SAT	saturation	fraction or percent of pore volume	ρ (rho), s
T	T	TEM	temperature	degrees (°F or °C), or kelvin (K)	θ (theta)
BHT, T_{bh}	T_{bh}	TEMBH	bottom-hole temperature	(same units as Temperature)	$T_{BH}, \theta_{bh}, \theta_{BH}$
FT, T_{fm}	T_f	TEMF	formation temperature	(same units as Temperature)	$T_{fm}, \theta_f, \theta_{fm}$
V		VLT	potential difference (electric)	volt	υ
V	V	VOL	volume	cubic centimeters (cc), cubic feet, etc.	
V	V	VOL	volume fraction		f_v, F_v

406

TABLE A7-3 (continued)

Traditional symbols	Standard SPE of AIME and SPWLA [a]	Standard computer symbols [a]	Description	Customary units or relation	Standard reserve symbols [b]
Z	Z	ANM	atomic number		
a	a	ACT	electrochemical activity	equivalents per liter, moles per liter	
a	K_R	COER	coefficient in $F_R - \phi$ relation	$F_R = K_R/\phi^m$	M_R, a, C
c		CMP	compressibility		
d	d	DIA	diameter	inches (in.)	D
f		FRC	fraction		
f		FQN	frequency	Hz	
g		GRD	gradient		
h	h	THK	thickness (bed, mud-cake, etc.)	feet, meters, inches	d, e
k	k	PRM	permeability, absolute (fluid flow)	millidarcies (md)	K
k		SUSM	susceptibility, magnetic		
m		MAS	mass		
m	m	MXP	porosity (cementation) exponent	$F_R = K_R/\phi^m$	
n		MOL	number of moles		
n	n	SXP	saturation exponent	$S_w^n = F_R R_w / R_t$	
p		PRS	pressure		
q	$f_{\phi shd}$	FIMSHD	dispersed-shale volume fraction of intermatrix porosity		$\phi_{imf,shd}, q$
r	r	RAD	radial distance from hole axis	inches	R
r		RST	resistance	ohms	
t	t	TIM	time	microseconds (microsec), seconds (sec) minutes	τ (tau)
v	v	VAC	velocity (acoustic)	feet per second, meters per second	V, u
Δ	Δ	DEL	difference		
Δr	Δ_r	DELRAD	distance, radial (increment along radius)		
Δt	t	TAC	sonic interval transit time	microseconds per foot	
$\Delta \phi_{Nex}$	(e)		excavation effect	porosity units (p.u.)	
Θ		ANGD	angle of dip		

Symbol	Code	Quantity	Units
Φ	DAZ	azimuth of dip	
Σ	XST	neutron capture cross-section	
Σ	XSTMAC	macroscopic cross-section	capture units (c.u.), cm^{-1}
Ω		ohm	ohm
α	COEA	coefficient, attenuation	
α_{SP}	REDSP	SP reduction factor due to shaliness	
α	ANG	angle	
β	BRGR	relative bearing	
γ	GRY	gamma ray	
γ	SPG	specific gravity (ρ/ρ_{w} or ρ_{g}/ρ_{air})	
δ	DCR	decrement	
δ	SKD	skin depth	
δ	ANGH	hole deviation (drift angle)	
δ		sound attenuation	
ϵ	DIC	dielectric constant	
θ	AGL	angle	
η		viscosity	
λ	LAM	decay constant	$\lambda = 1/\tau_{d}$
λ (K_{ani})	COEANI	coefficient of anisotropy	
λ	WVL	wavelength	
μ (G)	ELMS	shear modulus	
μ	PRMM	magnetic permeability	
μ	RAZ	azimuth of reference on sonde	
μ	PSN	Poisson's ratio	
ν	VIS	viscosity (dynamic)	
ρ		frequency	
ρ	DEN	density	g cm^{-3}
σ	SFT	interfacial tension	
σ	XNL	microscopic cross-section	
σ	PSN	Poisson's ratio	
τ (τ_{dN})	TIMD	decay time (thermal neutron)	microseconds
ϕ	POR	porosity	

TABLE A7-4

Subscripts

Traditional subscripts	Standard SPE of AIME and SPWLA [a]	Standard Computer Subscripts [a]	Explanation	Example	Standard Reserve Subscripts [b]
a	LOG	L	apparent from log reading (or use tool-description subscript)	R_{LOG}, R_{LL}	log
a	a	A	apparent (general)	R_a	ap
abs	cap	C	absorption, capture	Σ_{cap}	
anh	anh	AH	anhydrite		B, t
b	b	B	bulk	ρ_b	B, t
bh	bh	BH	bottom hole	T_{bh}	w, BH
clay	cl	CL	clay	V_{cl}	cla
cor, c	cor	COR	corrected	t_{cor}	
c	c	C	electrochemical	E_c	ec
cp	cp	CP	compaction	B_{cp}	
dis	shd	SHD	dispersed shale	V_{shd}	
dol	dol	DL	dolomite	t_{dol}	
e, eq	eq	EV	equivalent	R_{weq}, R_{mfeq}	EV
f, fluid	f	F	fluid	ρ_f	fl
fm	f	F	formation (rock)	T_f	fm
g, gas	g	G	gas	S_g	G
	gr	GR	grain	ρ_{gr}	
gxo	gxo	GXO	gas in flushed zone	S_{gxo}	GXO
gyp	gyp	GY	Gypsum	ρ_{gyp}	
h	h	H	hole	d_h	H
h	h	H	hydrocarbon	ρ_h	H
hr	hr	HR	residual hydrocarbon	S_{hr}	HR
i	i	I	invaded zone (inner boundary)	d_i	I
ig	ig	IG	intergranular (incl. disp. and str. shale)	ϕ_{ig}	
im, z	im	IM	intermatrix (incl. disp. shale)	ϕ_{im}	
int	int	I	intrinsic (as opposed to log value)	Σ_{int}	
irr	i	IR	irreducible	S_{wi}	ir, i (script)
J	j	J	liquid junction	E_j	ι (iota)
j	j		invaded zone (outer boundary)	d_j	
k	k	K	electrokinetic	E_k	ek
lam	ℓ (script)	LAM	lamination, laminated	$V_{sh}l$	L
lim	lim	LM	limiting value	ϕ_{lim}	
liq	L	L	liquid	ρ_L	ℓ (script)
log	LOG	LOG	log values	t_{LOG}	log
ls	ls	LS	limestone	t_{ls}	1st
m	m	M	mud	R_m	
max	max	MX	maximum	ϕ_{max}	
ma	ma	MA	matrix	t_{ma}	
mc	mc	MC	mud cake	R_{mc}	
mf	mf	MF	mud filtrate	R_{mf}	
mfa	mfa	MFA	mud filtrate, apparent	R_{mfa}	
min	min	MN	minimum value		

ni		non-invaded zone	R_{ni}	N
o	o	oil (except with resistivity)	S_o	
or	or	residual oil	S_{or}	
o, O	OR	100-percent water saturated	F_0	zr
PSP	ZR	pseudo-static SP	E_{pSP}	
pri	PSP	primary	ϕ_1	p, pri
r	PR	relative	$k_{ro},\, k_{rw}$	R
r	R	residual	$S_{or},\, S_{hr}$	R
s	R	adjacent (surrounding) formation	R_s	
sd	S	sand		sa
ss	SD	sandstone		sst
sec	SS	secondary	ϕ_2	s, sec
sh	SE	shale	V_{sh}	sha
silt	SH	silt	I_{sl}	slt
SP	SL	spontaneous potential	E_{SP}	sp
SSP	SP	static SP	E_{SSP}	
str	SSP	structural shale	V_{shst}	s
t	SH ST	true (as opposed to apparent)	R_t	tr
T	T	total	C_t	T
w	T	water, formation water	S_w	W
wa	W	formation water, apparent	R_{wa}	WA
wf	WA	well-flowing conditions	p_{wf}	f
ws	WF	well-static conditions	p_{ws}	s
xo	WS	flushed zone	R_{xo}	
z, im	XO	intermatrix	ϕ_{im}	
0 (zero)	IM	100-percent water saturated	R_0	zr
IL	ZR	from Induction Log	R_I	i
ILd	I	from Deep Induction Log	R_{ID}	id
ILm	ID	from Medium Induction Log	R_{IM}	im
LL (Also LL3, LL8, etc.)	IM	from Laterolog (Also LL3, LL7, LL8, LLd, LLs.)	R_{LL}	$\ell\ell$ (script)
	LL (Also LL3, LL7, LL8, LLd, LLs.)			
6FF40		from 6FF40 IL	R_{6FF40}	
MLL	MLL	from Microlaterolog	R_{MLL}	mll
PL	P	from Proximity Log	R_P	p
N	N	from normal resistivity log	R_N	n
16″, 16″N		from 16″-normal log	$R_{16''}$	
1″×1″		from 1″×1″ microinverse (ML)	$R_{1''\times1''}$	
2″		from 2″ micronormal (ML)	$R_{2''}$	
D	D	from Density Log	ϕ_D	d
GG	GG	from gamma-gamma Log	ϕ_{GG}	gg
N	N	from Neutron Log	ϕ_N	n
SNP	SN	from Sidewall Neutron Log	$\phi_{SN},\, \phi_{SNP}$	sn
CNL	CN	from Compensated Neutron Log	ϕ_{CN}	cn
TDT	PNC	from Thermal Decay Time Log	$\phi_{PNC},\, \phi_{TDT}$	
S	SV	from Sonic Log	ϕ_{SV}	sv
ND		from Neutron and Density Logs	ϕ_{ND}	
GR	GR	from Gamma Ray Log	ϕ_{GR}	gr

[a] References: Supplement V to 1965 Standard—"Letter and Computer Symbols for Well Logging and Formation Evaluation", in *Journal of Petroleum Technology* (October, 1975), pages 1244–1261, and in *The Log Analyst* (November-December 1975), pages 46–59.

[b] Reserve symbols are to be used only if conflict arises between standard symbols used in the same paper.

TABLE A7-5

Abbreviations

acre	spell out	megahertz (mega = 10^6)	MHz
acre-foot	acre-ft	meter	m **
alternating-current (as adjective)	AC	mhos per meter	mho/m
ampere	A, amp	microsecond (micro = 10^{-6})	microsec
ampere-hour	amp-hr	mile	spell out
angstrom unit (10^{-8} cm)	Å	miles per hour	mph
atmosphere	atm	milliamperes (milli = 10^{-3})	milliamp
average	avg	millicurie	mC
barrel	bbl	millidarcy, millidarcies	md
barrels of liquid per day	BLPD	milliequivalent	meq
barrels of oil per day	BOPD	milligram	mg
barrels of water per day	BWPD	milliliter	ml
barrels per day	B/D	millimeter	mm
barrels per minute	bbl/min	millimho	mmho
billion cubic feet (billion = 10^9)	Bcf	million cubic feet (million = 10^6)	MMcf
billion cubic feet per day	Bcf/D	million cubic feet per day	MMcf/D
billion standard cubic feet per day	Bscf/D	million electron volts	meV
bottom-hole pressure	BHP	million standard cubic feet per day	MMscf/D
bottom-hole temperature	BHT	milliseconds	millisec
British thermal unit	Btu	millivolt	mV
centimeter (centi = 10^{-2})	cm	minimum	min
centipoise	cp	minutes	spell out *
centistoke	cstk	mole	mol
cosecant	cosec	nanosecond (nano = 10^{-9})	nsec
cosine	cos	ohm	spell out
cotangent	cot	ohm − centimeter	ohm-cm
coulomb	C	ohm-meter	ohm-m
cubic	cu	ounce	oz
cubic centimeter	cc, cm^3	outside diameter	OD
cubic foot	cu ft, ft^3	parts per million	ppm
cubic feet per barrel	cu ft/bbl	picofarad (pico = 10^{-12})	pF
cubic feet per day	cu ft/D	pint	pt
cubic feet per minute	cu ft/min	pore volume	PV
cubic feet per pound	cu ft/lb	pound	lb
cubic feet per second	cu ft/sec	pounds per cubic foot	lb/cu ft
cubic inch	cu in.	pounds per gallon	lb/gal
cubic meter	cu m, m^3	pounds per square inch	psi
cubic millimeter	cu mm, mm^3	pounds per square inch absolute	psia
cubic yard	cu yd	pounds per square inch gauge	psig
darcy, darcies	spell out	pressure-volume-temperature	PVT
day	spell out *	productivity index	PI
dead-weight ton	DWT	quart	qt
decibel (deci = 10^{-1})	dB	reservoir barrel	res bbl
degree (American Petroleum Institute)	°API	reservoir barrel per day	RB/D
degree Celsius	°C	revolutions per minute	rpm
degree Fahrenheit	°F	secant	sec
degree Kelvin	(see "kelvin")	seconds	spell out *
degree Rankine	°R	self-potential	SP
direct-current (as adjective)	DC	sine	sin
electromotive force	emf	specific productivity index	SPI
electron volt	eV	square	sq
farad	F	square centimeter	sq cm, cm^2
feet per minute	ft/min	square foot	sq ft, ft^2
feet per second	ft/sec	square inch	sq in.
feet square	ft^2	square meter	sq m, m^2
foot	ft	square millimeter	sq mm, mm^2
foot-pound	ft-lb	standard	std
gallon	gal	standard cubic feet per day	scf/D
gallons per minute	gal/min	standard cubic foot	scf
gas-oil ratio	GOR	stock-tank barrel	STB
gigawatt (giga = 10^9)	GW	stock-tank barrels per day	STB/D
gram	gm	stoke	spell out
hertz	Hz	tangent	tan
horsepower	hp	teragram (tera = 10^{12})	Tg
horsepower-hour	hp-hr	thousand cubic feet	Mcf

hour	spell out *	thousand cubic feet per day	Mcf/D
hyperbolic sine, cosine, etc.	sinh, cosh, etc.	thousand standard cubic feet per day	Mscf/D
inch	in.	trillion cubic feet (trillion = 10^{12})	Tcf
inches per second	in./sec	trillion cubic feet per day	Tcf/D
inside diameter	ID	versus	vs
kelvin	K	volt	V
kilogram (kilo = 10^3)	kg	volume per volume	vol/vol
kilohertz	kHz	water-oil ratio	WOR
kilovolt	kV	watt	W
kilowatt	kW	yard	yd
kilowatt-hour	kW-hr		
liquefied petroleum gas	LPG		
liter	spell out		
logarithm	log		
logarithm (natural)	ln		
maximum	max		

* Except in combinations such as ft/D, cc/sec, ft/hr.
** Except with the number 1: 1 meter, not 1 m.

INDEX AND GLOSSARY

Here below are listed the most common terms or expressions used in well logging, followed by a short definition or explanation.

When they are treated more fully in this book, only the relevant chapter is referred to.

Absorption:
1. the process of taking up by capillary, osmotic, chemical or solvent action
2. the process by which energy such as that of electromagnetic or acoustic waves, is converted into other forms of energy.
– neutron absorption: 8.2.1.4.
– gamma-ray absorption: 12.1
Accelerator:
a device which permits the acceleration of electrons or nuclear particles. For example, an accelerator is used to produce neutrons by accellerating deuterium before striking a tritium target, in pulsed neutron logging. 10.3
Acoustic:
of or pertaining to sound.
– acoustic impedance: the product of acoustic velocity and density, 13.6.
– acoustic log: 14.
– acoustic signal: 13.1.
– acoustic travel time: time required for an acoustic wave to travel from one point to another.
– acoustic wave: a sound wave in which the disturbance propagated through a medium is an elastic deformation of the medium. 13.2
Activation:
the process of making a substance radioactive by bombarding it with nuclear particles. For example irradiation by neutrons which transform some nuclei into radio-isotopes, which are characterized by the energy of the induced gamma ray and by their decay time schemes. 9.2.2.
Activity:
1. for a solution the thermodynamic equivalent concentration ie. the ion concentration corrected for the deviation from ideal behaviour due to the interionic attraction of ions. 4.3.
2. the intensity of radioactive emission of a substance, measured as the number of atoms decaying per unit of time. 6.2.6.
Adsorption:
Adherence of gas molecules or of ions of molecules in solutions to the surfaces of solids with which they are in contact.
A-electrode:
the current-emitting electrode in a resistivity-measuring device. 3.2.1.
Alpha:
the ratio of the pseudo-static SP (PSP) to the static SP (SSP). 4.6.1.2.1.
Alpha particle:
6.2.1.
AM:
notation used to refer to the distance between the current electrode (A) and the potential measuring electrode (M). 3.2.1.1.
Amplified curve:
a curve recorded on a more sensitive scale.

Amplitude:
the maximum value of the displacement in an oscillatory motion. 14. and 15.
Analog system:
system in which the information is represented as a continuous flow of the quantity constituting the signal.
Anisotropy:
1.3.3.2
Annulus:
1. that space between a drill pipe and the formation through which the drilling fluid returns to the surface.
2. the space between casing and formation or between tubing and casing.
3. a ring of interstitial water sometimes produced by invasion processes in hydrocarbon-bearing beds. When $R_{xo} > R_o$, the annulus will be more conductive than the flushed zone (R_{xo}) or the virgin zone (R_t).
AO:
notation used to refer to the distance between the current electrode (A) and the point (O) midway between the potential measuring electrodes. 3.2.1.2.
API:
abbreviation for American Petroleum Institute.
API log grid:
is the standard format used by all logging companies for recording well logging measurement. This grid has:
– one left-hand track, 2.5 inches wide,
– the depth track or column, 0.75 inch wide,
– two right-hand tracks, 2.5 inches wide each.
The tracks may be divided into a linear or logarithmic scale. 2.4.
API test pits:
calibration pits at the University of Houston
– for gamma ray: 6.11.
– for neutron: 8.2.5.
API unit:
(1) for gamma-ray curves: the difference in curve deflection between zones of low and high radiation in the API gamma-ray calibration pit is 200 API gamma-ray units. 6.2.6. and 6.11.
(2) for neutron curves: the difference between electrical zero and the curve deflection opposite a zone of Indiana limestone (19% porosity) in the API neutron calibration pit is 1,000 API units. 8.2.5.
Apparent:
as recorded, before correction for environmental influence.
Archie:
engineer of the Shell Company who found the relationship between the formation resistivity factor, the porosity and the water saturation. 1.4.2.
Arm:
a bow spring or lever connected to a logging sonde which presses against the borehole wall to centralize the tool, to push the tool to the opposite side of the borehole, or to hold a sensor pad to the borehole wall.
Arrow plot:
a display of dipmeter data. 19.6.2.1.
Atomic mass unit:
symbol AMU: a measure of atomic mass, defined as equal to 1/12 mass of a carbon of mass 12.

Atomic number:
> symbol Z: number of protons within an atomic nucleus, or the number of orbital electrons in a neutral atom.

Atomic weight:
> symbol A: the relative weight of an atom on the basis that carbon is 12. Equal to the total number of neutrons and protons in the atomic nucleus.

Attenuation:
> 15.

Azimuth:
> 1. In the horizontal plane, it is the clockwise angle of departure from magnetic north.
> 2. Curve recorded in dipmeter survey. It is the clockwise angle from magnetic north to the reference electrode (no.1) on the sonde. 19.5.

Azimuth frequency plot:
> a diagram on polar chart paper which presents only a count of how many dip azimuth measurements fall within each ten degree sector, within a given group of dips. Dip magnitude is ignored. 19.6.2.3.

Back-up:
> a curve recorded by a back-up galvanometer, which begins to record when the primary galvanometer has reached the limit of available track width or goes off scale.

Barn:
> 10^{-24} cm^2/nucleus: a unit for measuring capture cross-sections of elements.

Base-line shift:
> 1. generally refers to a naturally occurring shift of the base line of any specific curve (SP or GR...). Usually the base line referred to is the shale base line.
> 2. sometimes refers to a manual shift made by the logging engineer.

Beta particle:
> a high-speed disintegration electron spontaneously emitted from an atomic nucleus as a form of radiation. 6.2.2.

BHC:
> Borehole Compensated Sonic log. 14.2

BHT:
> Bottom hole temperature.

Bimetallism:
> 4.5.8.1.

Blind zone:
> shadow zone. Commonly observed on curves recorded by a lateral device. 3.2.6.2.2.

Blue pattern:
> a convention used in dipmeter interpretation. It corresponds to an increasing dip magnitude with decreasing depth with nearly uniform azimuth.

Bond index:
> the ratio of attenuation in zone of interest (db/ft) to attenuation in well-cemented section (db/ft). It is an indicator of the quality of cement bond. 15.3.

Borehole effect:
> the spurious influence on a well-logging measurement due to the influence of the borehole environment which includes: hole diameter, shape of the borehole wall (rugosity), type of borehole fluid, mud-cake.

Borehole televiewer:
> 21.1.

Bound water:
> 1. water which has become adsorbed to the surfaces of solid particles or grains. Under natural conditions this water tends to be viscous and immobile but might not have lost its electrolytic properties.
> 2. water which is chemically bound by becoming part of a crystal lattice. This water has lost its electrolytic properties.

Bridle:
> 2.3.2.4.

Bulk density:
> symbol ρ_b: it is the value of the density of rock as its occurs in nature. 11.3

Bulk modulus:
> symbol K: 13.3

Button:
> a small disc-shaped, button-like electrode used on micro-resistivity pads (ML, MLL, HDT...).

Cable:
> 2.3.2.

Calibration:
> the process wherein the scale and sensitivity of the measuring circuit is adjusted to meaningful units. 2.5.

Caliper:
> 17.

Camera:
> 1. recorder. An instrument which records traces of light which have been beamed on film by galvanometers responsive to logging tool measurements. 2.3.4.1.
> 2. borehole camera: downhole instrument which photographs the interior of the borehole or casing.

Capture cross-section:
> 1. the nuclear capture cross section for neutrons is the effective area within which a neutron has to pass in order to be captured by an atomic nucleus. It is often measured in barns.
> 2. macroscopic capture cross section, symbol Σ, is the effective cross-sectional area per unit volume of material for capture of neutrons. 10.2.1.2.

Capture unit:
> symbol c.u. $= 10^{-3}$ cm^{-1}. A unit of measure of macroscopic capture cross section; equivalent to sigma unit. 10.6.

Carbon/oxygen:
> 9.2.6. C/O is a Dresser Atlas trademark.

Cartridge:
> a package which contains electronic modules or hardware for the down hole instrument. The package is carried in a protective housing. 2.3.3.

Casing collar locator:
> symbol CCL: a coil and magnet system used to locate casing collars.

Cation exchange capacity:
> symbol CEC: a measure of the extent to which a substance will supply exchange cations. Compensating cations serve to compensate the excess of charge (usually negative) in clay lattice. It is related to the concentration of compensating cations near clay-layer surfaces, which in the presence of water can be exchanged for other cations available in solution. The CEC is expressed in terms of milliequivalents of exchangeable ions per 100 grams of dry clay. 1.3.3.3.

Cationic membrane:
> a membrane which permits the passage of cations but not of anions. Clay acts as a membrane. 4.2.1.

CBL:
> Cement Bond Log. 15.3.1.

Cementation factor:
> symbol m: the porosity exponent in Archie's formula. 1.4.1.

Cement bond log:
> symbol CBL: used to determine the presence of cement behind casing and the quality of cement bond to casing or formation wall. 15.3.1.

Centralizer:
> a device which positions the logging tool in the center or near the center, of the well bore, aligned with the well bore axis.

Channel:
1. a defect in cement quality which prevents zone isolation.
2. in a pulse height analyzer, an energy gate in which only pulses occurring within a specific energy range are registered. The width of the channel corresponds to the difference between the upper and lower limits.
3. a path along which digital or other information may flow in a computer.

Chlorine log:
9.1.1. and 9.1.3.

Clean:
containing no appreciable amount of clay or shale.

Combination logging tool:
a single assembly of logging tools capable of performing more than one general type of logging service with a single trip into the well bore, so saving rig time. 2.3.5.

Compaction correction:
symbol C_p or $c\Delta t_{sh}$: an empirical correction applied to porosity derived from the sonic log in uncompacted formations. 14.8.

Compatible scale:
the Quick-Look Interpretation of well logs often requires a direct comparison of log responses (i.e. ρ_b vs. ϕ_N). In order to facilitate this comparison, the same grid type and equal scale sensitivities must be used. Overlay techniques require the use of compatible scales.

Compensated:
corrected out unwanted effects associated with the borehole.
- Compensated formation density log: symbol FDC: 11.7. FDC is a Schlumberger registered trademark.
- Compensated neutron log: symbol CNL: 8.2.6. CNL is a Schlumberger registered trademark.
- Compensated sonic log: symbol BHC: 14.2

Composite log:
several well logs on the same interval of a well, spliced together to form a single continuous record for correlation purposes.

Compressibility:
symbol C: the volumetric change in a unit volume of fluid when the pressure on that volume is increased. 13.3

Compressional wave:
symbol P-Wave: 13.2.1.

Compton scattering:
11.1.2.

Conductivity:
the property of a solid or fluid medium which allows the medium to conduct a form of energy (electricity, temperature ...). 1.3.3.2.

Consolidated:
pertains to a rock framework provided with a degree of cohesiveness or rigidity by cementation or other binding means.

Contact log:
a genetic term referring to the log produced by any logging tool which uses pad or skid devices to make direct contact with the formation wall.

Continuous velocity log:
symbol CVL: see Sonic Log 14.

Correlation interval:
19.5.4.2.

Crossplot:
a plot of one parameter versus another.

Crosstalk:
4.5.8.2.

Curie:
a standard measure of the rate of radioactive decay. 6.2.6.

Cycle skip:
14.9.2.

Darcy:
a unit of permeability: 1.4.5.

Dead-time:
it corresponds to the recovery period, required by the system to prepare itself for counting each successive event. Events occurring during dead-time are not counted. 5.4.

Decay:
the spontaneous disappearance of an effect
radioactive decay: 6.2.4.
thermal decay time: 10.

Deep investigation:
the measurement of formation properties far enough from the well bore that the effects of the invaded zone become minimal.

Deflection:
1. the internal movement in a galvanometer, in response to an impressed voltage, which produces the excursion on a logging trace or curve.
2. the lateral movement or excursion of a curve.

Delaware:
an anomalous effect on early Laterolog first observed in the Delaware Basin. 3.3.2.5.4.

Delay panel:
a memorizer panel.

Density:
symbol ρ: mass per unit volume

Density log:
11.

Departure curves:
graphs which show the influence of various conditions on the basic measurement.

Depth column:
the depth track (see API log grid).

Depth datum:
the zero depth reference for well logging. Usually kelly bushing, but could be ground level, derrick floor....

Depth matching:
to put in depth different well-logging measurements.

Depth of invasion:
the radial depth from the wellbore to which mud filtrate has invaded porous and permeable rock. 2.2.1.2.

Depth of investigation:
radius of investigation.

Detector:
a sensor used for the detection of some form of energy.

Deviation:
1. departure of a borehole from vertical
2. angle measured between tool axis and vertical

DF:
derrick floor

Diameter of invasion:
symbol d_i: the diameter to which mud filtrate has invaded porous and permeable rock. 2.2.1.2.

Dielectric:
a material having low electrical conductivity compared to that of a metal.

Dielectric constant:
relative permittivity. 16.1.

Diffusion:
1. of ions. The spontaneous flow of ions from a more concentrated solution into a more dilute solution. 4.2.2.
2. of thermal neutrons. 8.2.1.3. and 10.2.2.

Diffusion potential:
liquid-junction potential. 4.2.2.

Digital:
representation of quantities in discrete (quantized) units. The information is represented as a series of discrete numbers.

Digitize:
convert data from analog trace records to binary computer useable numbers.

Diode error:

an unwanted portion of the total electrical conductivity signal sent to the surface from downhole induction logging instruments. The diode error is produced by the measure circuit electronics of the induction cartridge. It is isolated and measured during the calibration operation at a step where the sonde output is zero. Once evaluated, it is cancelled during the survey operation.

Dip:

the angle that a surface makes with the horizontal. 19.

Dipmeter:

19.

Directional survey:

measurement of drift, azimuth and inclination of a borehole with vertical.

Dirty:

said of a formation which contains appreciable amount of clay or shale.

Dispersed:

a term used to refer to particles (clays) distributed within the interstices of the rock framework. 1.4.4.2.

Drift:

the attitude of a borehole, the drift angle or deviation is the angle between the borehole axis and the vertical.

Dual induction:

combination of a deep reading with a medium investigating. 3.3.1.8.5.

Dual Laterolog:

combination of a deep and shallow laterologs. 3.3.2.1.4. Dual Laterolog is a Schlumberger registered trademark.

Dual spacing:

tool with two detectors with different spacings from the source. This configuration allows to decrease the borehole influence.

Eccentering arm:

eccentralizer. A device which presses the sonde body against the borehole wall.

Effective permeability:

symbol k_e: The ability of a rock to leave a fluid flow through it in the presence of fluid immiscible with the first. 1.4.5.1.3.

Effective porosity: symbol ϕ_e:

1. interconnected pore volume occupied by free fluids. Hydrodynamically effective pore volume. 1.3.3.1.
2. also electrically effective porosity.

Eh:

oxidation-reduction potential.

Elastic constants:

13.3

Electrical coring:

the name given in 1932 by C. and M. Schlumberger and E.G. Leonardon to a series of well surveying operations in open hole.

Electrical survey:

a genetic term used to refer to the combination SP, short and long normals, lateral.

Electrical zero:

the recorded output of the electronic measure circuit when no signal is being measured,

Electrobed:

corresponds to an interval of depth in which log response is near constant. 2.2.2.3.

Electrochemical potential:

4.2.1.

Electrofacies:

the set of log responses that characterizes a sediment and permits the sediment to be distinguished from other.

Electrofiltration potential:

4.1.

Electrokinetic potential:

4.1.

Electrolyte:

1. material in which the flow of electric current is accompanied by a movement of ions.
2. any chemical compound which when dissolved in a solvent will conduct electric current.

Electronic density:

symbol ρ_e: electron population per unit volume. 11.2.

Electron volt:

symbol eV: a unit of energy equal to the kinetic energy acquired by a charged particle carrying unit electronic charge when it is accelerated a potential difference of one volt. Equivalent to 1.602×10^{-19}, joule.

Epithermal neutron:

a neutron which has been slowed down to a low energy level, in the range 100 eV–0.1 eV. 8.2.1.2.

Excavation effect:

a decrease in the neutron log apparent porosity reading below that expected on the basis of the hydrogen indices of the formation component. 8.2.10.

Fast neutron:

a neutron of 100 KeV or greater energy.

Fast-plot:

19.6.2.5.

FFI:

free fluid index: other symbol ϕ_f: the percent of the bulk volume occupied by fluids which are free to flow as measured by the nuclear magnetism log (NML). 21.3.

FIL:

fracture identification log. 19.8.

First reading:

symbol FR: refers to the depth of the first useable reading or value recorded on a curve.

Fish:

a foreign object lost in the borehole.

FIT:

formation interval tester: 20.2.2.

Floating pad:

a term used to refer to a pad that does not make good contact with the borehole wall to record quality information.

Flushed zone:

the zone at a relatively short radial distance from the borehole, immediately behind mud cake, which is considered to be flushed by mud filtrate.

Focused log:

refers to a well log produced by any well logging device in which survey-signal sent by the tool is focused. 3.3. and 3.5.

Formation:

a general term applied in well logging industry to the external environment of the drilled well bore without stratigraphic connotation.

Formation factor:

symbol F: factor derived from the porosity following law $F = a/\phi^m$

Formation interval tester:

symbol FIT: 20.2.2.

Formation resistivity factor:

symbol F_R: 1.4.1.

Formation signal:

the signal related to the formation.

Formation tester:

symbol FT: 20.2.1.

Formation water:

water present on the virgin formation.

FR:

first reading.

Free fluid index:
symbol FFI or ϕ_f, see FFI.

Free pipe:
pipe in a well bore which is free to vibrate

Free point:
the deepest depth in the well bore that stuck casing or drill pipe is free and can be salvaged.

Frequency:
1. the number of cycles or waves in a unit of time: symbol f: 13.
2. the number of occurrences or events over a specified period of time or length of borehole.

Fresh:
very low in dissolved salt.

FT:
formation tester: 20.2.1.

Galvanometer:
a small voltmeter which has a miniature mirror fastened to the moving coil. Light from a source is directed onto the mirror from which it is reflected onto a moving photographic film where it traces a curve. 2.3.4.1.

Gamma-gamma:
equivalent to density log

Gamma ray:
electromagnetic radiation emitted from an atomic nucleus during radioactive decay. 6.2.3.

Gamma-ray detector:
Geiger-Mueller counter, ionization chamber or scintillation counter are gamma-ray detectors.

Gamma-ray source:
an encapsulated radioactive material which emits gamma rays. Usually ^{137}Cs. 11.4.

Gate:
a window or opening, usually in time, during which certain measurements are made. 6.5.

Geiger-Mueller counter:
a type of gamma-ray detector. 6.5.

Geometrical factor:
symbol G: 3.3.1.2.

GL:
ground level.

GOR:
gas-oil ratio

Gradient:
the change in any parameter per unit change of another parameter.

Grain density:
the density of a unit volume of a mineral or other rock matter at zero porosity.

Grand slam:
a combination of logs or a computation procedure for calculating the depth of invasion and the resistivity of both invaded and virgin zones.

Green pattern:
a convention used in dipmeter interpretation. It represents a succession of dips of relatively constant azimuth and magnitude.

Grid:
see API log grid

Groningen effect:
3.3.2.5.5.

Guard electrode:
electrode which carries electrical current which tends to confine the survey portion of the current flow to a thin horizontal layer.

Half-life:
the time required for any amount of radioactive nuclide to lose one-half of its original activity by decay. 6.2.4.

Half-thickness:
the thickness of an absorbing material necessary to reduce the intensity of incident radiation by one-half 6.2.

Head:
the threaded connector end of downhole logging tool (sonde-cartridge set). 2.3.3.

Heading:
the form attached to the top of a well log which displays all relevant information about the well, the survey, and the well-bore condition. 2.4.

Hertz:
symbol Hz: the measure of frequency, defined as equal to 1 cycle per second.

Horn:
spurious high-resistivity anomalies found on induction-derived curves inside the upper or lower boundary of a resistive bed. A result of improper boundary compensation for the level of formation resistivity being logged.

Hostile environment:
any of the following criteria, found in a well bore, which will severely affect or restrict the logging operations:
- depth greater than 20,000 ft
- temperature greater 325 °F
- pressure greater than 20,000 psi
- hole deviation greater than 50
- H_2S or gas-cut mud.

Housing:
a cylindrical metal case which protects the electronic cartridge of the downhole logging instrument from damage by pressure and moisture. 2.3.3.

Humble formula:
a modified form of the Archie's relationship proposed by Humble Company. $F = 0.62/\phi^{2.15}$

Hybrid scale:
a resistivity scale used with laterolog, below midscale, the scale is linear in resistivity; above midscale, the scale is still in resistivity units but is linear with conductivity.

Hydraulic pad:
an articulating, liquid-filled pad used to place current and measuring electrodes in direct contact with the borehole wall. It allows better electrode contact by improving pad conformity with the wall.

Hydrocarbon saturation:
symbol S_h: fraction of the pore volume filled with hydrocarbon.

Hydrogen index:
symbol I_H: the ratio of the number of hydrogen atoms per unit volume of a material to that number in pure water at 75°F. 8.2.10.

Hydrostatic load:
the weight of formation fluid filling the pores of the rock and in communication with the water table at the well site, or sea surface.

IL:
induction log. 3.3.1.

Impedance:
13.6.

Inclinometer:
a device for measuring hole inclination and azimuth.

Induced:
term related to any provocated phenomenon produced by excitation.
- induced polarisation
- induced gamma ray
- induced gamma-ray spectrometry. 9.

Induction:
3.3.1.

Interval transit time:
symbol Δt or t: the travel time of a wave over a unit distance.
Invaded zone:
the portion of formation surrounding a well bore into which drilling fluid has penetrated, displacing some of the formation fluids. 2.2.1.2.
Invasion:
2.2.1.2.
Ion:
atom or group of atoms which has either taken on or given up one or more orbital electrons.
Ionization chamber:
a type of gamma-ray detector. 6.5.
Irreducible saturation:
it corresponds to the minimum saturation of a fluid when the fluid is displaced from a porous medium by another fluid immiscible with the first.
Isotope:
atoms of a single element which have differing masses. It is either stable or unstable (radioisotope).
Isotropy:
the property of homogeneity of a rock which allows it to show the same responses when measured along different axes.

Junction:
refers to liquid junction potential. 4.2.2.

KB:
Kelly bushing: rotary drilling bushing. Often taken as a depth datum.

Lag:
corresponds to the distance the detector moves, during one time constant.
Last reading:
symbol LR: refers to the depth of the last useable reading or value recorded on a curve.
Lateral device:
3.2.1.2.
Laterolog:
3.3.2. Laterolog is a Schlumberger registered trademark.
Line:
1. ground line
2. survey cable
Liquid-junction potential:
4.2.2.
Lithology:
refers to the mineralogical composition and to the texture of a rock.
Lithostatique load:
the weight of the overlying rock column without the fluid contained in the pore volume of the rock.
LL:
laterolog. 3.3.2.
Log:
1. a record containing one or more curves related to some property in the formation surrounding the well bore. 2.4.
2. to run a survey.
Log zero:
depth datum for the survey.
Longitudinal wave:
symbol P-Wave: see compressional wave. 13.2.1.
Long normal curve:
a resistivity curve recorded with a normal electrode configuration in which AM spacing is 64 inches.
LR:
last reading

Macroscopic anisotropy:
1.3.3.2.
Magnetism:
4.5.8.3.
Magnetostriction:
the change in dimension of a body when subjected to a magnetic field.
Magnetostrictive transducer:
converts electro-magnetic energy to mechanical energy and vice versa.
Manual shift:
an intentional electrical or mechanical shift given to a curve by the engineer during a survey in order to keep the curve within a given track.
Mark:
1. a magnetic mark or metal shin on a survey cable for depth-control.
2. one of the marks on the left part of Track-1 of a log used to compute logging speed. See Minute Mark.
Matrix:
1. for log analyst the solid framework, except shale, of rock which surrounds pore volume. 1.3.1.
2. for geologist the smaller or finer-grained, continuous material of a sediment or sedimentary rock.
Maximum pressure rating:
corresponds to the maximum pressure for satisfactory operation of the downhole instrument.
Maximum temperature rating:
corresponds to the maximum temperature for satisfactory operation of a downhole instrument.
md:
millidarcy. 1.4.5.
Measured depth:
corresponds to the depth measured along the drilled hole not corrected for hole deviation.
Measured point:
a depth reference point, on a logging sonde, at which measurements are taken.
Mechanical zero:
the reading of a galvanometer at rest. With no potential applied, this is the zero reference for galvanometer deflections.
M electrode:
the potential-measuring electrode nearest the A electrode in a resistivity-measuring device. 3.2.1.
Memorizer:
an electronic storage device which delays the measured signal to record at depth the different measuring points of a combined device. 2.3.6.
mho:
a unit of electrical conductance.
mho per meter:
a unit of electrical conductivity. 3.4.1.
Microinverse:
a very short lateral electrode arrangement.
Microlaterolog:
symbol MLL: 3.5.1. MLL is a Schlumberger registered trademark.
Microlog:
symbol ML: 3.4.1. ML is a Schlumberger registered trademark.
Micronormal:
a very short normal electrode arrangement. 3.4.1.
Microresistivity log:
a log of the resistivity of the flushed zone around a borehole (see ML, MLL, MSFL, PML).
Microscopic anisotropy:
1.3.3.2.
Microspherically focused log:
symbol MSFL: 3.5.3. MSFL is Schlumberger registered trademark.

Mineral:

a naturally occurring material having a definite chemical composition and, usually, a characteristic crystal form, entering in the composition of a rock (Table 1-1).

Minitron:

a neutron generator.

Minute-mark:

a mark or grid line interruption which is placed on the film every 60 seconds during the survey, to control the logging speed. It is usually found near the outside edge of track 1. 2.2.3.

mmho:

millimho

MN spacing:

the distance between the two potential measuring electrodes in a lateral device. 3.2.1.2.

Mobility:

of a fluid. 2.2.1.4.

Modified Schmidt diagram:

a plot of dipmeter information on polar chart paper where 0° dip is represented on the circumference and 90° at the center. 19.6.2.4.

Monitor curve:

a curve recorded on a well log which is a measure of some aspect of a tool performance or stability. An indicator to the quality of measurements being made by the instrument.

Monocable:

an armoured single conductor cable. 2.3.2.2.

Mono electrode:

a single electrode for measuring formation resistance.

MOP:

movable oil plot

Movable oil plot:

symbol MOP: a computed log, based on several logging operations, prepared for the purpose of determining the presence and quality of movable hydrocarbon.

Mud:

drilling mud.

Mud cake:

the residue deposited on the borehole wall as the mud looses filtrate into porous, permeable formations. 2.2.1.2.

Mud density:

symbol ρ_m: the density of the drilling mud. Measured in pounds per US gallon (lb/gl) or in g/cm^3.

Mud filtrate:

the effluent of the continuous phase liquid of drilling mud which penetrates porous and permeable rock, leaving a mud cake on the drilled face of the rock. 2.2.1.2.

Mud log:

a log recorded with Microlog or microlaterolog sonde with the arms collapsed, so that the measuring pad loses contact with the formation wall and "reads" the mud resistivity at downhole conditions.

Negative separation:

a term used in reference to microlog curves to describe the condition where the longer spaced resistivity curve ($R_{2''}$) reads a lower value than the shorter space curve ($R_{1'' \times 1''}$). 3.4.3.

Neutron:

an electrically neutral, elementary nuclear particle having a mass of 1.0089 atomic mass unit (very close to that of a proton). 8.

Neutron generator:

an electromechanical device operating at high voltage (125–130 kV) which focuses a beam of high-energy deuterons on a target surface containing tritium. Nuclear fusion of the deuteron ions and target atoms produces high-energy (14 MeV) neutrons. 10.3.

Neutron lifetime log:

symbol NLL: 10.2.4. NLL is Dresser Atlas registered trademark.

Neutron log:

a log of a response related to the interaction of neutrons with matter. 8.

Neutron source:

1. an encapsulated radioactive material which produces neutrons. 8.2.4.

2. a neutron generator. 10.3.

Noise:

sudden spurious readings on a curve.

Normal device:

3.2.1.1.

Normalize:

to adjust two log curves in order that one value may be compared with the other.

Nuclear magnetic resonance:

a phenomenon exhibited by atomic nuclei which is based on the existence of nuclear magnetic moments associated with quantized nuclear spins, 21.3.

Nuclear magnetism log:

symbol NML: a well log which is dependent on the alignment of the magnetic moment of protons (hydrogen nuclei) with an impressed magnetic field. 21.3.

Ohm:

symbol Ω: a unit of electrical resistance.

Ohm-meter:

symbol Ω-m: a unit of electrical resistivity.

Open hole:

uncased hole or portion of the hole.

Osmosis:

the spontaneous flow of molecules of the solvent of a more dilute solution into a more concentrated solution when separated from one another by a suitable semi-permeable membrane.

Overburden:

geostatic load.

Oxidation-reduction potential:

symbol Eh: redox potential.

Overlay:

to place one recorded curve over another to provide specific information with regard to lithology and fluid saturation.

Pad:

see sidewall pad.

Pair production:

11.1.1.

Permeability:

symbol k: a measure of the ability of a rock to conduct a fluid through its connected pores. 1.4.5.

pH:

an expression representing the negative logarithm of the effective hydrogen-ion concentration or hydrogen-ion activity.

Photoelectric absorption:

11.1.3.

Photomultiplier:

used with a scintillation crystal to make up a scintillation counter. The flash of light produced in a scintillating crystal strikes the sensitive surface of a photocathode in the photomultiplier causing the emission of a number of primary electrons. These electrons are drawn to an anode maintained at a higher potential whereupon a number of secondary electrons are emitted for each impinging electron. The secondary electrons are drawn to a second anode, maintained at a higher potential than the first, whereupon additional multiplication occurs. This process is repeated until the initial current has been multiplied about a million fold. 6.5.

Photon:

a quantity of energy emitted in the form of electromagnetic radiation.

Poisson's ratio:

symbol σ or ν: 13.3

Polar plot:

plots on polar coordinate paper usually used to aid dipmeter interpretation. 19.6.2.4.

Porosity:

symbol ϕ: 1.3.3.1.

Porosity exponent:

symbol m: the exponent of the porosity term in Archie's formula. 1.4.1.

Porosity overlay:

a log of porosity values computed from different logs plotted on top of each other.

Positive separation:

a term usually used in reference to Microlog curves to describe the condition where the micronormal (2") resistivity curve reads a higher value than the microinverse (1" × 1") curve. This condition usually denotes the presence of mud-cake on the face of a porous, permeable formation. 3.4.3.

Poteclinometer:

a device for making a continuous measurement of the angle and direction of borehole deviation during a survey. 18.3.

ppm:

parts per million

Probe:

a downhole logging instrument or a sonde.

Proportional counter:

similar to an ionization chamber. It is designed for the detection of neutrons. A metal chamber is filled with gas, generally ^{3}He which produces a (n, α) reaction or BF_3 which produces the (n,p) reaction. A central electrode is maintained at a positive voltage with respect to the shell (1300 V for ^{3}He, 2500 V for BF_3). When a neutron enters the chamber the gas produces ionizing particles.

Proton:

the atomic nucleus of the element hydrogen. A positively charged hydrogen atom.

Proximity log:

symbol PL: 3.5.2. PL is a Schlumberger registered trademark.

Pseudo-geometrical factor:

symbol J: a coefficient used for estimating uninvaded formation resistivity from the response recorded by a laterolog device.

Pseudo-static SP:

symbol PSP: the static SP of a dirty rock.

p.u.:

porosity unit: one percent pore volume.

Pulsed neutron:

fast neutron (14 MeV) emitted by a neutron generator.

P-wave:

compressional wave. 13.2.1.

Radiation:

the emission and propagation of energy through space or matter. The speed of propagation is that of light. It requires no intervening medium for its transmission.

Radioactivity:

6.1.

Radioisotope:

a nuclide. A radioactive isotope which spontaneously emits particles (α or β) or electromagnetic radiation (gamma rays) as it decays to a stable state.

Rayleigh wave:

surface acoustic wave in which the particle motion is elliptical and retrograde with respect to the direction of propagation.

Reading:

a value taken from a curve at a specific depth.

Receiver:

a transducer used to receive a form of energy which has been propagated through the formation or induced in the formation.

Reciprocal sonde:

a sonde in which the current and measure electrodes are interchanged according to a specific rule.

Reciprocator:

an electronic module designed to convert conductivity measurements into resistivity (in induction logging).

Recorder:

a device which records well-log data on film, chart or tape.

Redox potential:

symbol Eh: a quantitative measure of the energy of oxidation. Oxidation is equivalent to a net loss of electrons by the substance being oxidized; reduction is equivalent to a net gain of electrons by the substance being reduced.

Red pattern:

a convention used in dipmeter interpretation to denote decreasing formation dip with decreasing depth with a near constant azimuth.

Reflection peak:

an increase in the value of resistivity recorded by a lateral device as the A electrode passes a thin highly resistive bed. 3.2.6.2.2.

Relative bearing:

symbol β: in dipmeter interpretation, looking down the hole, it is the clockwise angle from the upper side of the sonde to the reference electrode no. 1.

Relative permeability:

symbol: k_r. 1.4.5.1.3.

Repeat Formation Tester:

symbol RFT: 20.2.3.

Repeat section:

a short section of a log, that is recorded in addition to the main survey section in order to provide an inter-run comparison of log similarity, and, therefore, instrument stability and repeatability.

Residual:

that which remains after a removal or displacement process.

Residual oil:

oil remaining in the reservoir rock after the flushing or invasion process, or at the end of a specific recovery process.

Resistivity:

symbol R: specific resistance. 1.3.3.2.

Resistivity index:

symbol $I = R_t / R_0$. The ratio of the resistivity of a formation bearing hydrocarbon to the resistivity it would have if 100% saturated with formation water. 1.4.2.

Reversal:

an interval of characteristic distortion on a normal curve across a resistive bed which has a thickness less than the AM spacing. 3.2.6.1.2.

RFT:

repeat formation tester: 20.2.3.

Rugosity:

the quality of roughness or irregularity of the borehole wall.

Salinity:

refers to the concentration of ions in solution.

Saturated:

1. containing as much as it can contain under given conditions of temperature and pressure.

2. reached the limit of its measuring capacity.

Saturation:

symbol S: the percentage of the pore volume occupied by a specific fluid. 1.4.2.

Saturation exponent:

symbol n: the exponent of the saturation term in Archie's saturation equation. 1.4.2.

Sawtooth SP:

4.4.7.

SBR:

shoulder bed resistivity 3.3.1.5.

Scale:

1. depth scale 2.4.
2. grid scale sensitivity. 2.4.

Scattered gamma-ray log:

see density log. 11.

Schmidt diagram:

a polar plot where the azimuth indicates dip or drift direction and the distance from the origin indicates dip or drift magnitude.

Scintillation:

a flash of light produced in a phosphor by an ionization event.

Scintillation counter:

a type of gamma ray detector. 6.5.

Screen:

1. a view screen on the camera or recorder.
2. a video screen in a computerized logging truck.

Search angle:

in dipmeter interpretation, the angle which will define the depth interval along which a correlation will be searched. 19.5.4.3.

Search interval:

the depth interval defined by the search angle.

Secondary porosity:

porosity resulting from alteration of the formation caused by fractures, vugs, solution channels, dolomitization. 1.3.3.1.

Secondary porosity index:

symbol SPI: an estimate of the secondary porosity, calculated from sonic log values in conjunction with either density or neutron log values, or porosity resulting from a cross-plot of density and neutron porosities.

Section gauge:

a caliper. 17.

Sensitivity:

the magnitude of the deflection of a curve in response to a standard signal.

Shadow zone:

3.2.6.2.2.

Shale:

a fine grained, thinly laminated or fissile, detrital sedimentary rock formed by compaction and consolidation of clay, silt or mud.

Shale base line:

a line drawn through the deflections characteristic of shale on a SP curve, which is used as a reference in making measurements to determine the formation water resistivity and the shale percentage.

Shear modulus:

13.3.

Shear wave:

symbol S-wave. 13.2.2.

Short normal curve:

a resistivity curve recorded with a normal electrode configuration in which AM spacing is 16 inches.

Shoulder bed effect:

effect of adjacent beds on a well-logging measurement.

Sidewall core:

a formation sample obtained with a wireline tool. 20.1.

Sidewall neutron log:

an epithermal neutron log recorded with the neutron source and detector mounted in a skid which is pressed against the borehole wall and may cut into the mud-cake to minimize borehole effects on the measurement. 8.2.6.

Sidewall pad:

a measuring device mounted on the end of an arm which projects from the sonde body during the survey.

Sigma unit:

symbol s.u.: capture unit. 10.6.

Signal:

1. a meaningful response, to a well-logging instrument, which can be detected or measured.
2. any type of pulse sent into the formation.

Single receiver Δt curve:

a continuous record of the travel time for acoustic energy to pass from a transmitter to a single receiver separated by a specific distance called spacing. 14.1.

Skid:

refers to the projecting portion of the body of a sonde, containing emission and measuring devices, which is pressed firmly against the borehole wall.

Skin depth:

3.3.1.7.

Skin effect:

3.3.1.7.

Skip:

see cycle skip. 14.9.2.

Slowing down:

8.2.1.2.

Soda plot:

19.6.2.2.

Sonar:

a technique involving the measurement of the time interval between the emission of a focused acoustic signal and detection of the signal reflected from a distance surface.

Sonde:

a downhole instrument connected to a wireline. 2.3.3.

Sonde error:

an unwanted portion of the total conductivity signal sent to the surface by the downhole induction logging instrument. It is generated by imperfections in the coils in the sonde. It is isolated and measured during the calibration operation when the sonde is placed in a zero-signal medium (air). Once evaluated it is cancelled during the survey operation.

Sonic log:

symbol SL: an acoustic velocity log. 14.

Source:

refers to the source of radiation used in the nuclear logging (density, neutron, thermal decay time...).

SP:

spontaneous potential: 4.

Spacing:

the distance between certain electrodes or sensors, or between source and detector, or between transmitter and receiver.

Span:

the distance separating certain sensors, for instance two receivers on the acoustic sonde.

Specific activity:

the amount of radioactive isotope present per unit of element, generally expressed as curie per gram.

Spectral gamma-ray log:

7.

Spherically focused log:

symbol SFL: 3.3.3. SFL is a Schlumberger registered trademark.

Spike:

noise: a spurious, unwanted event which has been recorded on a curve.

Spin:

used to describe the angular momentum of elementary particles or of nuclei: 21.3.

Spine-and-ribs plots:

is used in the computation of the compensation to be added to the measured value of bulk density from the dual-spacing formation density logs. 11.7.

Spontaneous potential:
symbol SP: 4.

SSP:
static spontaneous potential. 4.4.

Stand-off:
the distance separating a sonde from the wall of the borehole.

Static:
at rest, immobile.

Static spontaneous potential:
symbol SSP. 4.4.

Statistical variations:
5.3.

Step:
step distance in dipmeter computation process: 19.5.4.

Step profile:
an idealized invasion profile which assumes an abrupt transition from the flushed zone to the uninvaded zone. 2.2.1.2.

Stick plot:
19.6.2.6.

Stoneley waves:
are boundary acoustic waves at a liquid-solid interface resulting from the interaction of the compressional wave in the liquid and the shear wave in the solid. By definition, the Stoneley wave must have a wavelength smaller than the borehole diameter.

Streaming potential:
4.1.

Strike:
the direction or trend that a structural surface (bedding plane or fault plane) takes as it intersects a horizontal plane.

Structure:
1.4.

Surface conductance:
electrical conductance occurring at the surfaces of some solid crystalline materials (such as clays) when they are exposed to aqueous solutions.

Survey:
1. to take and record borehole geophysical measurements.
2. the result of a well-logging operation, a well log.

S-wave:
shear wave. 13.2.2.

Symbols:
short forms or abbreviations used to identify well logging parameters. Appendix 7.

Tadpole plot:
arrow plot. 19.6.2.1.

TC:
time constant. 6.3.

TD:
total depth.

Televiewer:
see borehole televiewer 21.1.

Telluric currents:
earth currents originating as a result of variations in the earth's magnetic field, or resulting from artifical electric or magnetic fields.

Temperature log:
18.

Test loop:
a device used in the calibration of induction logging tools.

Test pill:
an encapsulated radioactive material which serves as a portable source of gamma radiation for the calibration of some radioactivity logging tools.

Texture:
1.4.

Thermal conductivity:
a measure of the ability of a material to conduct heat. 18.

Thermal decay time:
symbol τ: the time for the thermal neutron population to fall to $1/e$ (37%) of its original value. 10.2.1.1.

Thermal (neutron) decay time log:
symbol TDT. 10.2.4.2. TDT is a Schlumberger registered trademark.

Thermal neutron:
a neutron which has kinetic energy in the same order of magnitude as that due to the thermal motion of nuclei in the medium ($= 0.025$ eV).

Tight:
compact, having very low permeability.

Time constant:
symbol TC. 5.3.

Tool:
downhole instrument (sonde and cartridge).

Torpedo:
a quick-connecting and quick-disconnecting device, mounted near the head end of the survey cable, which provides strength and the means to manually connect electrical survey conductors to the bridle and head.

Tortuosity:
the crookedness of the pore pattern. 1.4.1.

Total depth:
symbol TD: total depth reached by a specific logging tool.

Total porosity:
symbol ϕ_t: the total pore volume in a rock. 1.3.3.1.

Track:
well log track on the API grid.

Transducer:
any device or element which converts an input signal into an output signal of a different form.

Transition profile:
a realistic profile in which the distribution of fluids in the invaded section beyond the flushed zone varies with increasing distance from the borehole.

Transit time:
see interval transit time.

Transmitter:
a device which emits energy into the environment of the logging instrument.

Transverse-wave:
symbol S-wave: shear wave. 13.2.2.

Travel time:
acoustic travel time over a specific distance.

True bed thickness:
thickness of the stratigraphic unit measured along a line normal to the direction of extension of the unit.

True resistivity:
symbol R_t: the resistivity of the undisturbed, uninvaded rock.

True vertical depth log:
symbol TVD: a log computed from well logs recorded in deviated holes, in which measured depths have been rectified to true vertical depth.

ULSEL:
ultra long spaced electric log. A Chevron trademark.

Ultra Long Spaced Electric log:
symbol ULSEL: a well log recorded with the use of a modified long normal electrode configuration on a 5,000-foot bridle. The AM spacing can be made 75, 150, 600 or 1,000 feet. Difference between measured resistivities and anticipated resistivities calculated from conventional resistivity logs indicate nearby resistivity anomalies (previous nearby cased hole, salt domes).

Unconsolidated:
:   pertains to a rock framework which lacks rigidity.

Variable density:
:   variable intensity.

Variable density log:
:   symbol VDL. 15.5.

Vector plot:
:   arrow plot

Vertical resolution:
:   the minimum thickness of formation that can be distinguished by a tool under operating conditions.

Viscosity:
:   symbol μ or ν:
:   resistance of a fluid to flow.

Water loss:
:   a mud property: the measure of filtrate loss in a water base drilling mud under typical pressure conditions.

Water saturation:
:   symbol S_w: the percentage of the pore volume of a rock occupied by water.

Wavelength:
:   symbol λ. 13.1.3.

Wavelet:
:   see wave train.

Wave train:
:   corresponds to a wavelet of several cycles resulting from the response of an elastic system to an acoustic energy impulse.

Wave train display:
:   the acoustic wave train can be displayed in different modes:
    1. intensity modulated-time mode in which the wave train is shown in the VDL form.
    2. amplitude-time mode in which the wave train is shown as a wiggle trace. 15.5.

Weak point:
:   a machined connector or calibrated cable designed to break under specific tensile stress.

Well bore:
:   a borehole

Well log:
:   a wireline log or borehole log.

Wiggle trace:
:   a representation of the acoustic wave train in the amplitude-time mode.

Wireline log:
:   a well log.

Young's modulus:
:   symbol E. 13.3.

Z/A:
:   ratio of atomic number to atomic weight.

Z/A effect:
:   11.3

Z axis:
:   a third dimension added to a cross-plot of two parameters in an X-Y plane. The Z axis is perpendicular to both X and Y axes.

REFERENCES

American Geological Institute, 1977. Glossary of Geology. Edited by R. Gary, R. McFee, Jr. and C. Wolf, Washington, D.C.

CRC Handbook of Chemistry and Physics, 1981-1982. 62nd Edition, CRC Press, Inc., Boca Raton, Florida.

SPWLA, 1975. Glossary of terms and expressions used in well logging.

Sheriff, V.E., 1970. Glossary of terms used in well logging. Geophysics, 35: 1116–1139.